DIFFUSION PHENOMENA IN THIN FILMS AND MICROELECTRONIC MATERIALS

MATERIALS SCIENCE AND PROCESS TECHNOLOGY SERIES

Editors

Rointan F. Bunshah, University of California, Los Angeles *(Materials Science and Technology)*

Gary E. McGuire, Microelectronics Center of North Carolina *(Electronic Materials and Processing)*

DEPOSITION TECHNOLOGIES FOR FILMS AND COATINGS; **Developments and Applications**: by *Rointan F. Bunshah et al*

CHEMICAL VAPOR DEPOSITION FOR MICROELECTRONICS; **Principles, Technology, and Applications**: by *Arthur Sherman*

SEMICONDUCTOR MATERIALS AND PROCESS TECHNOLOGY HANDBOOK; **For Very Large Scale Integration (VLSI) and Ultra Large Scale Integration (ULSI)**: edited by *Gary E. McGuire*

SOL-GEL TECHNOLOGY FOR THIN FILMS, FIBERS, PREFORMS, ELECTRONICS, AND SPECIALTY SHAPES: edited by *Lisa C. Klein*

HYBRID MICROCIRCUIT TECHNOLOGY HANDBOOK; **Materials, Processes, Design, Testing and Production**: by *James J. Licari* and *Leonard R. Enlow*

HANDBOOK OF THIN FILM DEPOSITION PROCESSES AND TECHNIQUES; **Principles, Methods, Equipment and Applications**: edited by *Klaus K. Schuegraf*

DIFFUSION PHENOMENA IN THIN FILMS AND MICROELECTRONIC MATERIALS: edited by *Devendra Gupta* and *Paul S. Ho*

IONIZED-CLUSTER BEAM DEPOSITION AND EPITAXY: by *Toshinori Takagi*

SHOCK WAVES FOR INDUSTRIAL APPLICATIONS: edited by *Lawrence E. Murr*

Related Titles

ADHESIVES TECHNOLOGY HANDBOOK: by *Arthur H. Landrock*

HANDBOOK OF THERMOSET PLASTICS: edited by *Sidney H. Goodman*

HANDBOOK OF CONTAMINATION CONTROL IN MICROELECTRONICS; **Principles, Applications and Technology**: edited by *Donald L. Tolliver*

DIFFUSION PHENOMENA IN THIN FILMS AND MICROELECTRONIC MATERIALS

Edited by

Devendra Gupta and Paul S. Ho

IBM Thomas J. Watson Research Center
Yorktown Heights, New York

np

NOYES PUBLICATIONS
Park Ridge, New Jersey, U.S.A.

Library of Congress Catalog Card Number: 88-22634
ISBN: 0-8155-1167-1
Printed in the United States

Published in the United States of America by
Noyes Publications
Mill Road, Park Ridge, New Jersey 07656

10 9 8 7 6 5 4 3 2 1

Library of Congress Cataloging-in-Publication Data

Diffusion phenomena in thin films and microelectronic materials / edited by Devendra Gupta and Paul S. Ho.
p. cm.
Bibliography: p.
Includes index.
ISBN 0-8155-1167-1 :
1. Diffusion. 2. Thin films. 3. Microstructure. I. Gupta, Devendra. II. Ho, P.S.
QC176.84.D54D54 1988
621.381'7--dc19 88-22634
CIP

Preface

During the last decade or so, rapid progress has been made in the understanding of diffusion processes in thin films in general. At the same time, the applications of thin films have become extensive and diverse with respect to the base material, structure, chemistry and environment. While the earlier thin films commonly consisted of single or bi-layer polycrystalline metallization on inert substrates, currently alloy, oxide (dielectric) and semiconductor thin films are being used in multilevel schemes as well as having variable structure such as polycrystalline, amorphous and even epitaxial. In these applications, one is often also confronted with the synergetic effects arising from diffusion, structure, chemistry within and the ambient.

After having watched the field grow and sometimes making significant contributions of our own, we felt that the time was ripe to attempt a book to catalogue these events in a scientific manner. We have focussed our attention not only on the ubiquitous diffusion in thin films but also on its synergism with microstructure and chemical and physical driving forces.

In Chapters 1 through 3, formal aspects of diffusion, short-range diffusion on atomic scale and the role of steep chemical driving forces resulting in nonlinear chemical diffusion are discussed mostly in metallic thin films. In Chapter 4, the various diffusion processes in the growth of oxide films are discussed at length. In Chapter 5, diffusion-induced grain boundary motion (DIGM) in thin films which is often accompanied by alloying and dealloying is discussed. The observation of DIGM is rather recent but its exposé is very thorough. Half way through the book, Chapters 6 through 10 refer directly to the problems and solutions of the technology. For instance, Chapter 6 deals in the ambient effects on the diffusion mediated thin film reactions. The electromigration in thin films discussed in Chapter 7, remains timely despite its beginnings some 20 years ago because the higher current densities and thin lines in VLSI make new and stringent demands. In Chapter 8, an exhaustive treatment of the diffusion barriers in VLSI is given with tables of materials. Finally, Chapters 9 and 10 deal in soft lead based metallurgies used for Josephson junctions and the Si chip to substrate solder interconnections. In these applications, the soft metallurgies are also subjected to mechanical stresses and thermal instability which are diffusion controlled in one manner or the other. We hope that this book will be useful to the technologist and academic workers alike.

We are indebted to our fellow contributors who had to put up cheerfully with our usually gentle prods and criticism of the manuscripts; to George Narita of Noyes Publications and Professor R.F. Bunshah of UCLA for encouraging us to edit this book. It is with great delight and deep appreciation that we thank Janis Tobias Riznychok for her unenviable task of secretarial work and typesetting the entire book.

Yorktown Heights, NY
July, 1988

Devendra Gupta
Paul S. Ho

Contributors

Alan Atkinson
Atomic Energy Research Establishment Harwell
Oxfordshire, United Kingdom

Brian S. Berry
IBM Thomas J. Watson Research Center
Yorktown Heights, New York

Chin-An Chang
IBM Thomas J. Watson Research Center
Yorktown Heights, New York

A. Lindsay Greer
University of Cambridge
Cambridge, United Kingdom

Devendra Gupta
IBM Thomas J. Watson Research Center
Yorktown Heights, New York

Carol A. Handwerker
National Bureau of Standards
Gaithersburg, Maryland

Paul S. Ho
IBM Thomas J. Watson Research Center
Yorktown Heights, New York

Chao-Kun Hu
IBM Thomas J. Watson Research Center
Yorktown Heights, New York

Hillard B. Huntington
Physics Department
Rensselaer Polytechnic Institute
Troy, New York

Hannu P. Kattelus
Technical Research Centre of Finland
Espoo, Finland

Thomas Kwok
IBM Thomas J. Watson Research Center
Yorktown Heights, New York

Masanori Murakami
IBM Thomas J. Watson Research Center
Yorktown Heights, New York

Marc A. Nicolet
California Institute of Technology
Pasadena, California

NOTICE

Contents

1

Some Formal Aspects of Diffusion: Bulk Solids and Thin Films

Devendra Gupta

1. INTRODUCTION

It is almost a decade since the chapter on grain boundary diffusion was published (1) which covered various aspects of diffusion phenomena in grain boundaries and thin films. During the intervening period the progress made in the understanding of the grain boundary structure and kinetics has been by leaps and bounds as evidenced by symposia held, on average, only about a year apart (2-5). At the same time the very large scale integration in the computer industry has been taking place in geometrical progression with densities doubling every year. Random access memory chips have already reached four megabit densities and the logic chips are approaching almost a mega gates (6). These developments have imposed stringent reliability demands on the metallic thin interconnects and joining materials on the silicon chip and the package alike. The solutions of the yesteryear are, obviously, not good enough for the technology of tomorrow. In fact, one wonders if the gap between the understanding of the various kinetic and diffusion processes in thin films on one hand and their applications on the other hand is narrowing or widening. Furthermore, materials other than polycrystalline thin films are being introduced in the VLSI. Here, mention may be made of amorphous thin films for diffusion barriers, the man made superlattices and even organic-polyimide films. Perhaps, the use of the recently discovered high temperature superconducting films (7,8) is not far off.

The scope of this chapter, therefore, has been broadened significantly. While the earlier chapter (1) will still be found useful particularly for polycrystalline metallic thin films, this chapter is expected to provide ready analytical, thermodynamical and atomic frame work for most materials new and old. A comprehensive discussion on the driving forces for diffusion both in linear and non-linear regimes has been included which should be relevant in the context of the current technological trends of using thinner and multilevel metallization schemes. Typical data on diffusion in bulk metals, semiconductors and thin films are given in tabulated form with discussions on the role of variable structure and chemistry which are expected to be useful to the materials community in general. Since the field of diffusion is vast, it was impossible to address all its aspects. The interested reader is referred to the recent literature (9-11) where discussion of the topics which could not be covered here will be found. Of course, this book contains chapters on specific subjects with detailed discussions by leading investigators in the individual fields.

2. MATHEMATICAL BASIS OF DIFFUSION

2.1 Single Crystals

The mathematical basis of diffusion was first established by A. Fick in 1865. In this section we discuss the general aspects of diffusion without considering the role of the microstructure of the samples, which will be taken up later. The specimens only contain lattice point defects such as vacancies, divacancies and interstitials in thermal equilibrium. For the detailed mathematical solutions of specific diffusion experiments, the interested reader is referred to the text books by Carslaw and Jaeger (12), Crank (13), and Shewmon (14).

Fick's First Law

The diffusion coefficient (D_1) of the component 1 is an inhomogeneous single phase binary alloy due to concentration gradient in steady state is defined by Fick's First Law as

$$J_1 = -D_1\left(\frac{\partial C_1}{\partial x}\right)_T + C_1 v, \qquad (1)$$

where J_1 is the flux at time t of atoms of the component 1, C_1 its concentration, v the velocity of mass which moves due to application of forces such as electromigration or a thermal or chemical potential gradient, and x is the distance into the sample taken parallel to the concentration gradient. If the concentration of the diffusant can be defined at the entrance and exit surfaces of the sample separated by distance x in steady state condition, the diffusivity can be easily measured by monitoring the flux in the absence of a driving force. A number of such simple experiments can be found in the literature for example diffusion of Ni in Si has recently been studied using such a scheme by Thompson et al. (15). Since the measurement of flux contains no information on the distribution of the diffusant inside the sample, an independent study of the microstructure of the specimen and its chemistry is made to describe the diffusion conditions and path in the physical sense.

It is possible to measure diffusion coefficient in the absence of external forces including the chemical gradient by employing isotope techniques. The diffusant is the isotope of the host species; it may be a radioactive tracer or a stable enriched isotope. The latter is detected by mass spectroscopy techniques such as the one employed in the secondary ion mass spectroscopy (SIMS). Such measurements are termed as self-diffusion and generally denoted by asterisk or by the isotope used. The Fick's first law then reduces to

$$J^* = -D^*\left(\frac{\partial C^*}{\partial x}\right)_T . \tag{2}$$

For diffusion in a chemical gradient, the measured diffusion coefficient is affected by the motion of all the atomic species and is consequently termed as interdiffusion coefficient. As will be seen later on, it contains thermodynamical contributions due to the changing chemistry of the alloy. The interdiffusion or chemical diffusion coefficient is generally denoted by $\tilde{D}$ and consequently the Fick's First Law is written as

$$J = -\tilde{D}\left(\frac{\partial C}{\partial x}\right)_T . \tag{3}$$

In the three dimensional vector notation, the general statement of the Fick's First Law is

$$J = -\tilde{D}\nabla C . \tag{4}$$

Fick's Second Law

The Fick's second law describes diffusion in a non-steady state condition when the concentration changes with time. In essence, it states that the concentration change over a period of time is equal to the gradient of the flux and hence Fick's Second Law can be written as

$$\frac{\partial C}{\partial t} = -\nabla J. = D\nabla^2 C, \tag{5}$$

if D is not a function of position. Fick's second law is the basis of most of the diffusion measurements in solids in general. It has been widely used in samples in rod, plates and thin geometries. It has been used for measurements of diffusion in single crystal specimens, along grain boundaries and dislocations. We will discuss these individual applications in the following sections.

Instantaneous Source Geometry

Imagine an infinitesimally thin layer of diffusant of strength M is deposited on the face of a cylindrical specimen so that its thickness, the Dirac δ, is lot less than the diffusion distance $(Dt)^{1/2}$. Assuming that there is no flux on the surface or in the imperfections such as grain-boundaries and dislocations and the boundary conditions

$$C(x,0) = M\delta(x) \text{ and} \tag{6}$$

$$\frac{\partial C}{\partial x}(0,t) = 0, \tag{7}$$

are maintained, the solution to the Fick's second law is given by

$$C(x,t) = (M/2\sqrt{\pi Dt}) \exp(-x^2/4Dt) , \tag{8}$$

where M is the total amount of the diffusant per unit area, t time of diffusion and C its concentration at x and t.

This geometry has been widely used in the radioactive tracer work where a Fermi-Dirac δ type of diffusant distribution can be initially obtained and the mass and thickness involved are infinitesimally small. Much of the basic diffusion work in solids has been carried out in this manner. We refer to the excellent article by S. Rothman in Ref. 10 in which elegant experiments have been described to characterize roles of vacancies, divacancies, interstitials, vacancy-tracer correlation effects, etc.

Thick Layer Geometry

Generally, it is not possible to obtain Dirac delta function in non-tracer diffusion experiments. Consequently, a thick geometry constitutes a situation where the initial thickness of the source of the diffusant (h) is of the order of the diffusion distance ($\sqrt{Dt}$). The initial source condition for the thick layer geometry for t = 0 is given by

$$C(x,0) = C_0 \text{ for } h \geq x \geq 0 \tag{9}$$

and

$$C(x,0) = 0 \text{ for } x > h \tag{10}$$

For t>0 the boundary condition is

$$\frac{\partial C}{\partial x}(0,t) = 0 \tag{11}$$

The solution is then given by

$$C(x,t) = \frac{C_0}{2}\left[\operatorname{erf}\left(\frac{x+h}{2\sqrt{Dt}}\right) - \operatorname{erf}\left(\frac{x-h}{2\sqrt{Dt}}\right)\right] . \tag{12}$$

Thick layer geometries commonly arise during diffusion in thin film couples as well which will be described later on.

Infinite Couple with Surface Composition Constant

When a diffusion couple is formed from two samples having uniform concentrations of C_0 and C_1, the initial condition near the interface at $x = 0$ and $t = 0$ is described as

$$C(x,0) = C_1 \qquad \text{for } x<0 \tag{13}$$

$$C(x,0) = C_0 \qquad \text{for } x>0 \tag{14}$$

and the solution is given by

$$\frac{C(x,t) - C_0}{C_1 - C_0} = \frac{1}{2}\left[\operatorname{erfc}\left(\frac{x}{2\sqrt{Dt}}\right)\right] . \tag{15}$$

Boltzmann-Matano Analysis

The solution (Eq. 15) is actually based on constant D. When the difference between C_1 and C_0 is large and $x>>2\sqrt{Dt}$, as is the case in real experiments, the diffusion coefficient will vary with composition and along the distance x. To determine D at a known composition, Boltzmann-Matano analysis is used (see p. 28, ref. 14). Figure 1 shows an infinite diffusion couple subjected to a diffusion anneal. The initial conditions are:

$$C = C_0 \qquad \text{for } x<0, \text{ at } t = 0 \tag{16}$$

$$C = 0 \qquad \text{for } x>0, \text{ at } t = 0 \tag{17}$$

Following the determination of the concentration profile, the Matano interface is determined either graphically or by numerical techniques such that the two hatched areas are equal. The Matano interface is the plane $x = 0$. The value of chemical diffusivity $\tilde{D}$ then computed at the composition C' according to the equation:

$$\tilde{D}(C') = -\frac{1}{2t}\left(\frac{dx}{dc}\right)\int_0^{C'} x\,dc . \tag{18}$$

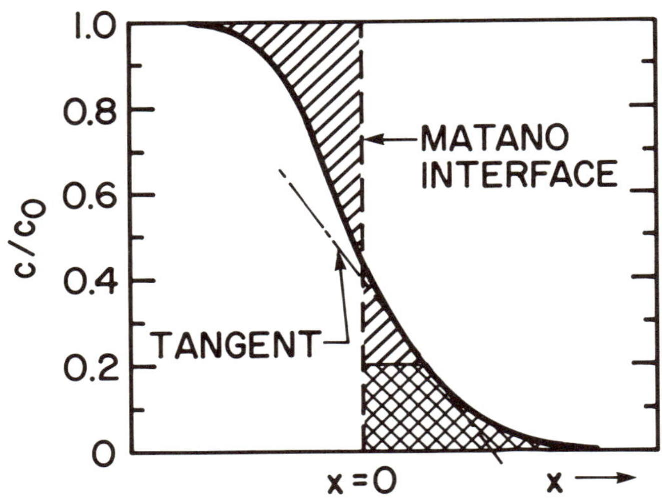

Fig. 1. The Matano Plot of the concentration ratio $C(x)/C_0$ versus the penetration distance x. The Matano interface is obtained by equalizing the area under the curve. A value for $\tilde{D}(C = 0.2\ C_0)$ is obtained by using the tangent there and the area shown cross-hatched in Eq. (18) (after Shewmon Ref. 14).

The term (dx/dc) is actually the slope of the concentration curve at C′. The integrated area under curve between $C/C_0 = 0$ and $C/C_0 = C'$ is then computed. Reasonable values for $\tilde{D}(C')$ can be computed in the range $0.1<C'<0.9$; outside this range the errors in computing the slope and the area under the curve become rather large. Diffusion coefficients for Zn in GaAs have been recently analyzed using Boltzmann-Matano method (16) owing to the strong composition and position sensitivity of the Zn diffusivity in GaAs.

It is also possible to determine the diffusivities for the individual components in an alloy using Matano type of analysis. For this purpose inert markers are placed at pure metal (C = 0)/alloy ($C = C_0$) interface. Concentration profiles for both the components of the alloy are measured employing analytical techniques such as electron-microprobe x-ray analysis. The shift of the markers with respect to the Matano interface is caused due to the unequal diffusivities of the two components. Campbell et al. (17) have measured interdiffusion in a bulk couple of Pb and Pb-50 wt% In alloy in this manner using Al wire inert markers. The marker velocity (v_m) is related to the diffusivities and the correlation factor f = 0.78 (see Sec. 4.2 for the definition of f) by

$$\frac{v_m}{f} = (D_{In} - D_{Pb}) \frac{\partial C(In)}{\partial x} , \tag{19}$$

Furthermore, for an alloy showing ideal solution behavior the interdiffusivities are related to the chemical diffusivity by Darken's equation (see Sec. 4.3)

$$\tilde{D} = [D_{In}C(Pb) + D_{Pb}C(In)] . \tag{20}$$

Since Eqs. 19 and 20 involve two unknowns, D_{In} and D_{Pb} can be determined for the specific composition. Campbell et al. (17) solved the above problem for $C(In) = 0.41$ atomic fraction. The interdiffusion coefficients thus determined should not be confused with the radiotracer diffusivities which have been recently determined in the PbIn alloys (18). Both the D_{In} and D_{Pb} diffusivities are 2 to 3 orders of magnitude larger compared to the radiotracer diffusivities $D_{In}{}^*$ and $D_{Pb}{}^*$ in Pb 43 at% In alloy. This difference actually stems from the thermodynamic factor and will be the subject of discussion in a subsequent section.

Interdiffusion studies using Matano analysis and employing inert markers have also been extended to thin films by Tu and co-workers (19). Hentzell et al. (20) have used isolated W thin film islands of 150Å diameter between Cu and Al thin film bilayer samples. Interdiffusion and reaction in the thin film couple are studied by Rutherford backscattering spectroscopy (RBS). The position of the W marker and its inert character are both observed in the same RBS spectra.

2.2 Structurally Inhomogeneous Samples

The discussion in the previous section assumed that the diffusion sample was structurally homogeneous i.e., free from such defects as dislocations and grain-boundaries (GB). Consequently, diffusion occurred in the lattice only due to equilibrium point defects such as vacancies, interstitial atoms, divacancies, etc. The engineering materials, however, are polycrystalline in nature and contain non-equilibrium defects such as dislocations, stacking faults and GBs. Diffusion along these defects is commonly six or more orders of magnitude more rapid than in monocrystalline samples. In thin film packages used in VLSI applications, the density of the structural defects, notably the GBs, is very high. Since the temperatures of device fabrication and performance are low ~100°C, the diffusive process in thin films may be largely controlled by the grain boundaries.

In polycrystalline materials, in general, the diffusion within the grains and along GBs occur simultaneously. The coupling of these processes takes place at the GBs which have an effective thickness δ of the order of atomic width. The extent to which the lattice atoms

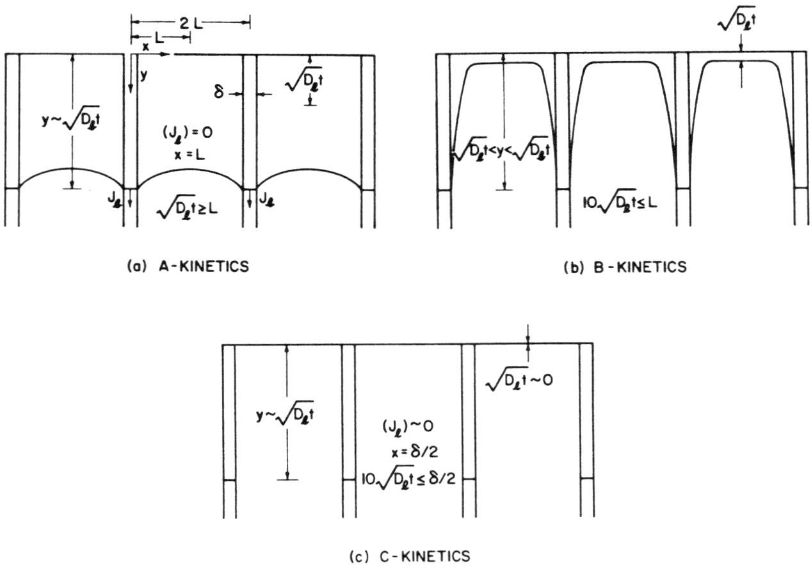

Fig. 2. Schematic representation of A, B and C-type of grain-boundary diffusion kinetics due to Harrison (21). The curved lines are isoconcentration contours. The diffusion source coincides with the top horizontal lines.

make excursions into the GBs, or vice-versa, determines the diffusion kinetic regime which prevails in the sample as a whole. Harrison (21) designated three types of kinetic regimes known as A, B, and C which are shown in Fig. 2.

The distinguishing feature of A-kinetics is the extensive lattice diffusion which causes the diffusion fields from the adjoining grains to overlap. The boundaries are shown here as parallel slabs with spacing 2L. In essence, the atoms in the lattice have made some excursions into the GBs so that their transport has been accelerated. The measurements in type-A kinetics would show enhanced apparent diffusivities (D_{app}) with lower activation energies which can be described by Hart's equation (22)

$$D_{app} \simeq D_\ell + gD_b \; , \tag{21}$$

where the subscript ℓ and b refer to the lattice and GB respectively and g is the site occupancy factor at the GBs so that $g = 4\delta/L$. Its value in bulk materials is $\leq 10^{-5}$ but in thin films values of 10^{-2} are not uncommon. Consequently, the second term becomes dominant in thin films.

In type-B kinetics, the grain boundaries are assumed to be isolated and the flux at large distance in the x-direction approaches zero. Effectively, the GB and lattice fields are confined individually to themselves. The leakage of mass from the grain-boundaries is only in the x-direction and the contribution of the GBs to the diffusion within the grains (at the top of Fig. 2b) is negligible. Type-B kinetics are commonly exploited in systematic experimental work which will be discussed later on in this section.

In type-C kinetics, lattice diffusion is considered negligible and significant atomic transport occurs only within the boundaries. Type-C diffusion experiments have been carried out by Atkinson and coworkers (23) and by Hwang and Balluffi (24). Type-C kinetics should, obviously, be of great significance for the mass transport in thin films at low temperatures.

From the foregoing discussion, it is clear that the mathematical analyses for GB diffusion would differ from situation to situation depending on the thickness of the sample, its grain-size and the relative contribution of the lattice diffusion which is dictated by the temperature of annealing. The various mathematical analyses have been discussed in the review paper by Gupta, et al. (1). We will only discuss here some limiting mathematical solutions which are commonly used in experiments and have yielded reliable diffusion data.

The first analysis of combined lattice and GB diffusion was carried out by Fisher (26) for a simple slab configuration in type-B kinetic regime shown in Fig. 3. The most serious approximation in Fisher solution was that the concentration C_b within the boundary changed so slowly with time t that the term $\partial C_b/\partial t$ could be set equal to zero. Subsequently, Whipple (27) and Suzuoka (28) have solved the problem exactly removing the above mentioned limitation.

For discussing the solutions, it is more convenient to use reduced variables which are defined as follows:

$$\eta = \frac{y}{\sqrt{D_\ell t}}, \ \zeta = \frac{x-(\delta/2)}{\sqrt{D_\ell t}}, \ \alpha = \frac{\delta}{2\sqrt{D_\ell t}},$$

$$\Delta = \frac{D_b}{D_\ell}, \ \beta = (\Delta - 1)\alpha \tag{22}$$

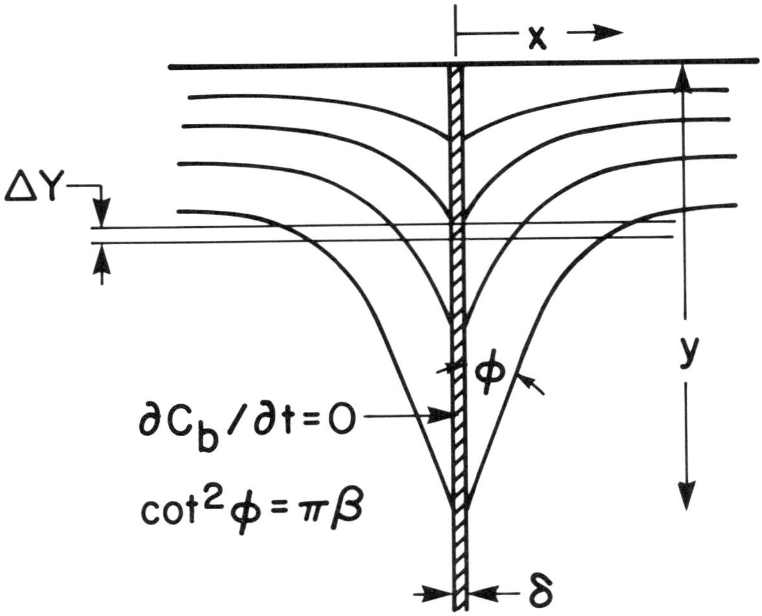

Fig. 3. The sectioning methods of grain-boundary diffusion average the isoconcentration contours at various depths along the grain-boundary slab of thickness δ. See Equations (24) and (25).

The Fisher solution for the slab geometry may be written as:

$$\delta D_b = 0.56\left(\frac{\partial \ell n\overline{C}}{\partial y}\right)^{-2}\left(\frac{4D_\ell}{t}\right)^{1/2}, \tag{23}$$

with the condition that $\partial C_b/\partial t = 0$ in the slab. Where $\overline{C}$ is the concentration of the diffusant in the sectioning method of determining the diffusivities. We have written the Fisher solution in the above form for easy comparison with Whipple and Suzuoka solutions.

The solution for the slab geometry of the GB according to Whipple (27) and Suzuoka (28) is given by

$$\delta D_b = \left(\frac{\partial \ell n\overline{C}}{\partial y^{6/5}}\right)^{-5/3}\left(\frac{4D_\ell}{t}\right)^{1/2}\left\{\frac{\partial \ell n\overline{C}}{[\partial(\eta\beta)^{-1/2}]^{6/5}}\right\}^{5/3}. \tag{24}$$

For large values of $\beta>30$, the third term converges to $(0.78)^{5/3}$ and $(0.72\beta^{.008})^{5/3}$ for infinite and instantaneous source conditions respectively. Consequently the asymptotic Whipple and Suzuoka solution can be approximately expressed as

$$\delta D_b = 0.661\left(\frac{\partial \ell n\bar{C}}{\partial y^{6/5}}\right)^{-5/3}\left(\frac{4D_\ell}{t}\right)^{1/2}, \tag{25}$$

for β values >30. The larger finite values of β imply better resolution between the GB and lattice diffusion processes. Physically, it ensures type-B kinetic regime as may be seen in Fig. 2.

LeClaire (29) and Canon and Stark (30) have pointed out the discrepancies between the Fisher and Whipple-Suzuoka analyses. Even for large β, the Fisher solution underestimates the GB diffusivity by a factor of 1.5-2.7. For $\beta<30$, the ratio D_b (Fisher)/D_b (true) depends on the value of β which results in erroneous measurement of the activation energy as well. Fisher solution has consequently been rarely used in the recent years. The Fisher solution, however, has the advantage that it involves linear power of the penetration distance and hence the origin of the penetration is not needed. It is consequently very useful in thin film couples where the interface may not be well defined. Furthermore, this author has never experienced any problem to keep β large in thin films since enough diffusant or radioactive tracer is trapped in the GBs due to the fine grain structure even at low temperatures. The penetration distance in thin film diffusion is also small <1 μm and measurements are conducted at low temperatures over a limited range. Under these conditions, the quality of linear or 6/5 power fit to the experimental data is usually comparable and the difference between the use of Fisher or Whipple-Suzuoka solutions is only about 10-50% being higher in the former case. For the GB diffusion measurements in polycrystalline bulk specimens, however, a large discrepancy is found between the use of the two solutions and use of only Whipple-Suzuoka is recommended with penetration distance not exceeding a few percent of the average grain diameter.

In thin film diffusion work, it is not always possible to maintain a semi-infinite specimen thickness condition compared to the diffusant penetration distance. Gilmer and Farrell (31,32) have shown that a correction must be made to account for the finite thickness (ℓ) of the thin film specimen in order to obtain true diffusivities. In Fig. 4 ratios of the effective D_b^{eff} to D_b are shown for various values of $\beta\eta_0^{-2}$

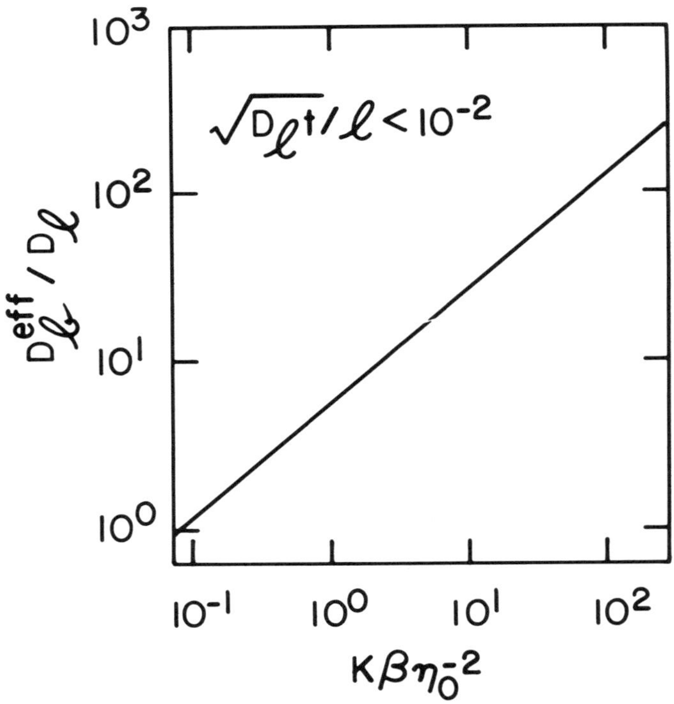

Fig. 4. Ratio of D_b^{eff} to D_b for various values of $K\beta\eta_0^{-2}$ (after Gilmer and Farrell, Ref. 32).

where $\eta_0 = \ell/\sqrt{D_\ell t}$. It is seen in Fig. 4 that true value of D_b occurs when $\beta\eta_0^{-2}\sim 0.1$. Physically, it implies that very large values of β result in flat diffusion profiles so that diffusant has reached the thin film-substrate interface and may have even been reflected. Consequently, β and η_0 need to be optimized so that the profiles have sufficiently large slope in the B-type kinetic regime.

Thus far, we have discussed GB diffusion in pure thin film or bulk specimens. If the specimens contain solute, the GB chemistry also affects the measured GB diffusion coefficient. Generally, if the solute content within the grain is C_0, its concentration at the GB is C_b so that $C_b/C_0 = K$. The factor K is known as a GB-solute segregation factor having values $K \geq 1$. It is also temperature and concentration dependent. Its full significance will be discussed in a later section. In this section, it should suffice to say that the factor K modifies Eqs. 23-25 so that the measured quantity is the combined diffusion coefficient $K\delta D_b$ (cm^3/sec) in the alloy samples instead of δD_b (cm^3/sec) for self GB diffusion in pure polycrystalline samples.

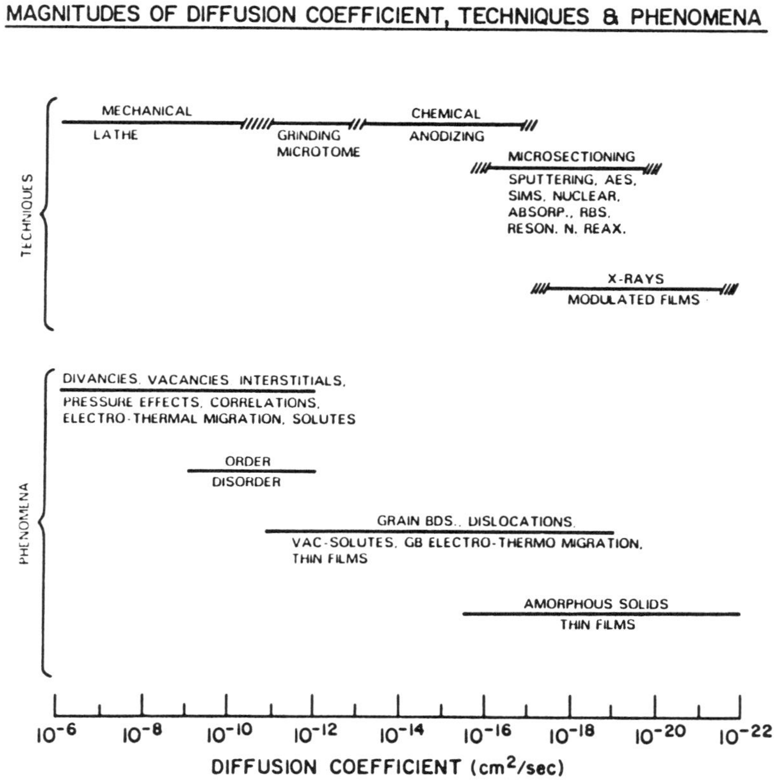

Fig. 5. Various diffusion techniques, their applicability in terms of the range of diffusion coefficient measured, and the phenomenon investigated.

3. EXPERIMENTAL TECHNIQUES

In Fig. 5, some important techniques for measuring diffusion are shown with their capabilities in terms of the magnitude of the diffusion coefficient measured. The diffusion scale is based on the bulk process such as due to vacancy and interstitials. It is readily seen that the developments of the various techniques, particularly those involving microsectioning have not only led to the measurements of smaller and smaller diffusion coefficients but rather new phenomena have been discovered. The individual experimental techniques have been reviewed by various authors (10,33-36). The techniques applicable to thin films involve detection and profiling of the diffusant essentially at micrometer or submicrometer depths. They are typified by sputtering techniques in conjunction with the radiotracer or Auger Electron Spectroscopy (AES), Secondary Ion Mass Spectroscopy (SIMS), Rutherford Backscattering Spectroscopy (RBS), x-ray diffraction of compositionally modulated thin films, etc. Description of these techniques will also be found in different chapters of this book. The application of these techniques can be broadly classified into two broad categories namely (1) which involve development of concentration profiles in the depth of the sample, and (2) in which surface accumulation of the diffusant is monitored during or after the diffusion anneal. The various techniques listed above, perhaps with the exception of the x-ray diffraction (37), may be used in one or the other mode. For example the radioactive tracer techniques can be used (i) to measure concentration contours of the diffusion adjacent to the grain-boundaries, (ii) to measure the angle ϕ (see Fig. 3) between the tangent of concentration contour and the grain boundary, (iii) to measure the total amount of indiffused material residing in thin sections cut parallel to the surface and finally (iv) in some situations radiotracers can also be used to measure the change in the surface concentration of the diffusant (38). The first two applications involve the use of autoradiographic techniques. Similarly, AES, SIMS, and RBS can also be used either to monitor the depletion or accumulation of the diffusant at the entrance and exit surfaces or for depth profiling of the total amount of the diffusant in sections at various depths. The mathematical treatment to extract the GB diffusivity differ for each situation and have been individually discussed elsewhere (1). The sectioning techniques are most widely used and have also generated much of diffusion information in thin films and along GBs. The surface accumulation techniques though less prevalent have the attractive feature of being non-destructive. We will discuss these two techniques briefly in this section.

3.1 Microsectioning Diffusion Measurements

The microsectioning diffusion measurements in thin films are carried out most fruitfully in the type-B regime where the lattice diffusion distance ($2\sqrt{D_\ell t}$) is small but finite and typically in the range 100-1000Å. As mentioned before, the total amount of indiffused material residing in thin sections cut parallel to the surface is measured in the sectioning techniques. The analytical procedures employed in microsectioning may be any one of the radiotracer, AES or SIMS techniques. The common features of the depth profiling techniques are the sputtering of the specimen surface by Ar^+ ions of ~1kV energy, and analysis of either the sputtered off material or the residual sample. In the AES, the specimen surface is analyzed while in the SIMS the sputtered-off material is sampled for mass spectroscopy. The same principles are inherent in the radioactive tracer microsectioning techniques where a known fraction of the sputtered-off material is collected onto planchets or plastic film for the subsequent radioanalysis. There are two versions of the radiotracer microsectioning techniques: In the one version (33) radiofrequency plasma of Ar^+ ions is used for sputtering in a chamber having a hexagonal rotary anode so that up to six sections can be taken in one pump down. After that, the planchets are replaced, specimen counted and weighed and the chamber evacuated for the subsequent cycle. In the Mehrer and Maier (39) version, a DC Ar^+ ion gun is employed and the sputtered-off material is continually collected on a myler film wound on a cassette-like arrangement. Following sputtering time schedule, frames are cut from

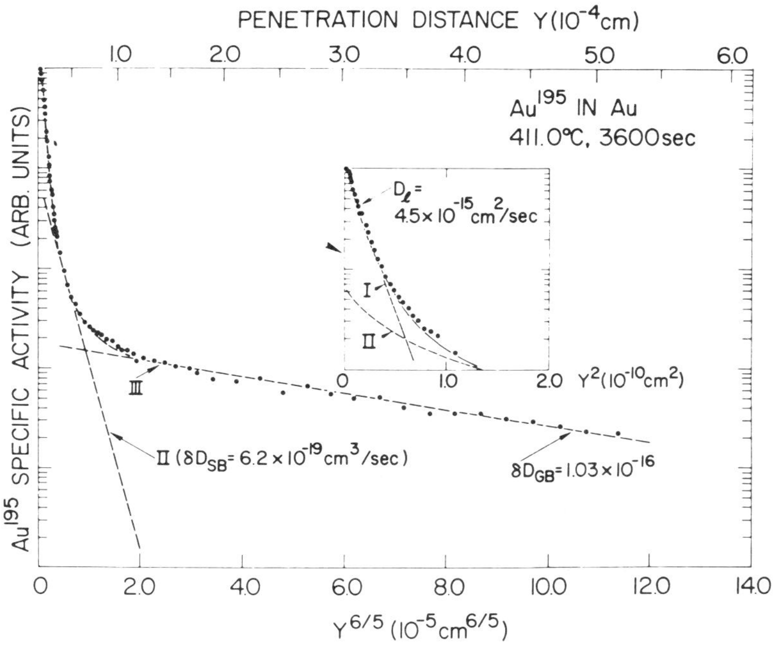

Fig. 6. Penetration plot of the diffusion of ^{195}Au in a Au polycrystalline specimen. A three-term exponential fitting procedure marked I, II and III yielded diffusion coefficients in the lattice, along subgrains and along large-angle grain boundaries respectively.

the exposed myler films and counted by the appropriate nuclear counter. The first method has a built-in calibration while the second one has to rely wholly on a prior calibration by developing a step which is measured by two beam interferometer or by simply measuring the weight loss as function of cumulative sputtering time. The second process is, however, much more expedient and convenient. In both the systems, layers as thin as 3nm may be removed reliably from a variety of materials including pure metals, dilute alloys and oxides.

In Fig. 6, a typical tracer profile in a polycrystalline Au specimen is shown. Since bulk Au specimens in general contain large angle grain boundaries and small subgrains in addition to the lattice defects, the penetration profiles is seen to have three distinct segments relating to diffusion along these paths. As the profile is traversed from right to left, the diffusivities along these paths are progressively smaller owing to the larger activation energies involved. The delination between the large angle GBs and the subgrains therein is possible due to a network of dissociated dislocations present in the latter, which are commonly observed in Au (1). In thin films, however, one or more diffusion paths may be absent depending on their structure. In Figs. 7 and 8, tracer penetration profiles are shown for epitaxial Au films grown on (001) MgO (40) and polycrystalline Au films grown on fused quartz substrates (41). Despite very long annealing times, the diffusion process in epitaxial films is very slow as these profiles maintain steep slopes. Furthermore, the epitaxial films display two regions of diffusion: at small penetration distance the diffusion is related to the lattice diffusion and at deeper depths to isolated (dissociated) dislocations. In polycrystalline Au films, however, there is only one diffusion path namely the GBs and diffusion there is extremely fast as indicated by short times even at lower temperatures. These penetration profiles were analyzed according to the Whipple-Suzuoka asymptotic solution (Eq. 25) to extract the combined diffusivity δD_b for either large angle GBs or for the low angle GBs having dissociated dislocations. The activation energy for diffusion along these two paths were determined to be 0.88 eV and 1.16 eV respectively; the latter is noticeably larger which has ramifications on core structures of GBs and dissociated dislocations.

3.2 Surface Accumulation Techniques

In the recent years, developments of several electron spectroscopy techniques such as AES, ESCA, SIMS, and ultraviolet photoemission spectroscopy (UPS) have made available surface-sensitive methods for diffusion studies in thin films. All these spectroscopy techniques employ energetic particle or photon beams to excite characteristic electrons from surface atoms for chemical analysis. The sampling

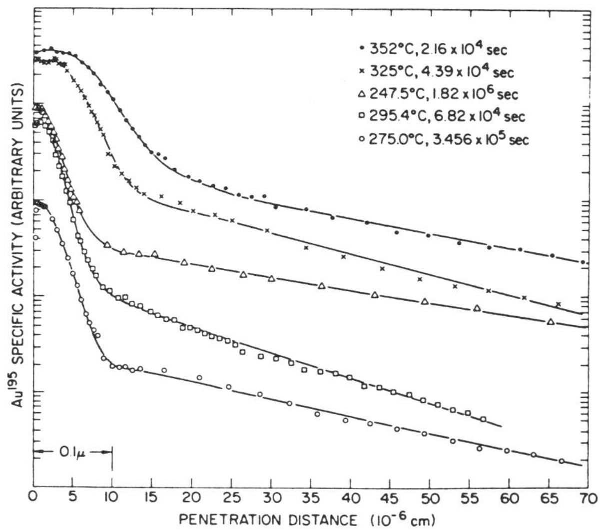

Fig. 7. Penetration plots of the diffusion of ^{195}Au in (001) epitaxial Au films on MgO substrates at several temperatures and times. Note the sharp transition from lattice to dislocation diffusion and absence of the grain-boundary diffusion component.

depth of these techniques is determined by the escape distance of the characteristic electrons which is usually limited to about 10 atomic layers for energies ranging up to about 2 keV. As alluded to earlier, these spectroscopy techniques have been employed for diffusion studies in thin films by measuring either the in-depth composition or the surface accumulation on thin films. The former is usually coupled with ion sputter sectioning of the material and consequently may be considered as a variation of the microsectioning techniques. Unlike in the radiotracer techniques, in the electron spectroscopy techniques impurity diffusion is measured which is accompanied by alloying. However, when AES is used in the surface accumulation method, owing to the analysis at the exit surface only, the amount of diffusant can be small within the specimen or in the GBs so that dilute alloys can be studied. Further, the accumulation AES technique with in-situ annealing is continuous and non-destructive in nature. Since the accumulated diffusant can be easily removed by sputtering, diffusion can be measured in a new time-temperature kinetic regime without having to withdraw the sample from the chamber. Using multicomplexing electronics, it is also possible to study several elements simultaneously.

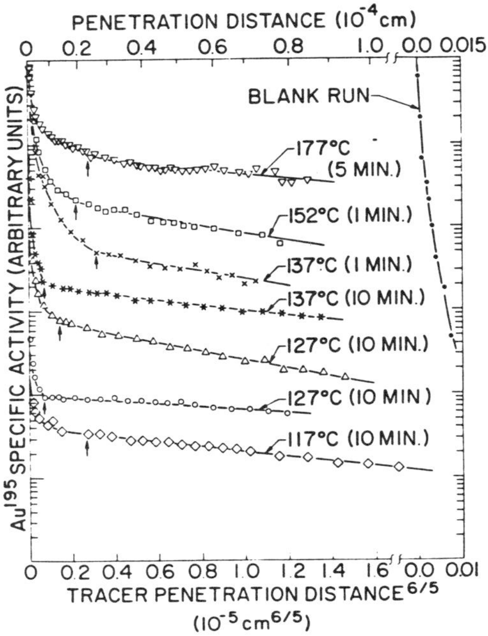

Fig. 8. Penetration plots of the diffusion of ^{195}Au in polycrystalline Au films grown on fused quartz substrates with a thin Mo adhesive layer. Note the change of scale on the right for the "blank" run. The regions between arrows are for grain-boundary diffusion. Other diffusion paths are absent.

The principal problem in these techniques is the beam-induced heating during analysis and possibly the knock-on of the diffusant ions. Both are likely to yield higher diffusivities and lower activation energies. The effect can be quite significant, particularly in low-temperature studies involving C-type kinetics and in very thin films.

The mathematical analyses of the surface accumulation experiments have been provided by Hwang and Balluffi (24,25) and Holloway et al. (42). The geometry of the accumulation method and the combined surface spreading is schematically shown in Fig. 9. In the analysis of Hwang and Balluffi, the GB to surface capacity ratio H and the weighted relative inverse diffusion time G are two important parameters which are related to the GB and surface properties as:

(a) $H = \delta\ell/2LW_s\sigma$, and (26)

(b) $G = H\tau_b/\tau_s$ (27)

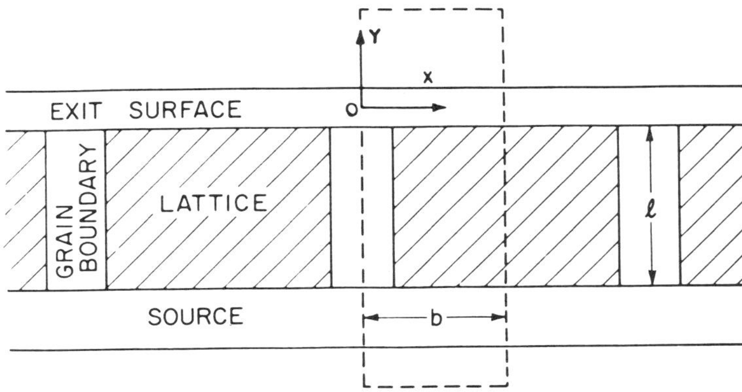

Fig. 9. Geometry of the permeation and combined surface spreading analysis used in surface accumulation technique. ℓ is the film thickness and b the grain radius (after Hwang et al., Ref. 24).

where σ = the equilibrium segregation ratio between GB and surface $\tau_b = D_b t/\ell^2$ and $\tau_s = D_s t/L^2$. δ and W_s are the effective width and depth of the grain boundary and surface respectively. If the GB capacity is small compared to that of surface accumulation (H<1), a quasi-steady state diffusion is maintained. Furthermore, since surface diffusion is rapid compared to GB diffusion (G<1), then the average surface concentration at the accumulation surface $\overline{C}_s$ as a function of time assumes the simple form

$$\overline{C}_s = 1 - \exp(-H\tau_b) \tag{28}$$

for the constant-source condition and

$$\overline{C}_s = [1 - \exp(-2H\tau_b)]/2 \tag{29}$$

for the instantaneous-source condition. Average Ag surface accumulation on thin Au films measured by Hwang et al. are shown in Fig. 10 as function of normalized time from which diffusion coefficients for GBs were extracted in type-C kinetic regime.

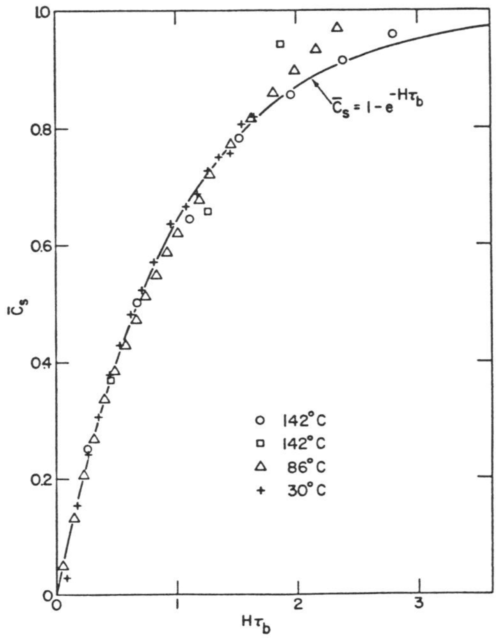

Fig. 10. Surface accumulation of Ag in Au films at various temperatures plotted as function of normalized time (after Hwang et al., Ref. 25).

4. THERMODYNAMICAL CONSIDERATIONS

It is well accepted now that diffusion in solids involves atomic exchange with some kind of lattice imperfections and that it is a thermally activated process. In this section, we will examine the interplay of the various thermodynamic factors in determining the diffusion coefficients in chemically inhomogeneous specimens be they bulk or thin films.

4.1 Driving Force F on Individual Atoms Is Zero

Figure 11 shows energy barrier as a function of atomic position for a jumping atom (43). Due to the absence of any driving force, the probability of the forward and backward jumps is the same. Hence, the velocity of mass term v in Eq. (1) is zero. A successful jump is executed when an atom gathers the free energy G_m, goes over the barrier and occupies an equivalent equilibrium position by an exchange

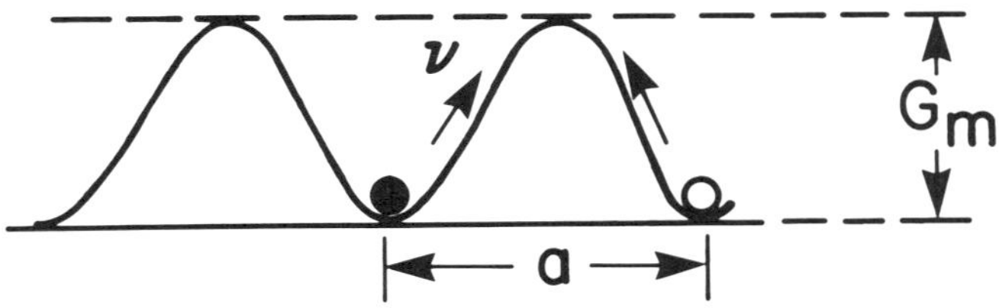

Fig. 11. Energy-distance diagram for diffusion in lattice when driving force F on individual atom equals zero. a is the atomic spacing (after Manning, Ref. 43).

with a defect like a vacancy. The probability W of acquiring the energy G_m is given by the Boltzman factor as

$$W = \nu_O \exp(-G_m/kT), \tag{30}$$

where ν_O is the atomic vibration frequency. The success of the jump, however, depends on the availability of a defect in the adjoining position which is also given by Boltzman factor. Hence, the average probability N_d of finding the proper defect may be written as

$$N_d = Z \exp(-G_f/kT), \tag{31}$$

where G_f is free energy necessary to form the defect and Z is a coordination factor which depends on the crystal type. The vacancy concentration in most metals and alloys is generally ~0.01% at the melting temperature and even smaller at lower temperatures. The frequency of jumping Γ then is obtained by multiplying Eqs. (30) and (31) as

$$\Gamma = Z\nu_O \exp -(G_m + G_f/kT) . \tag{32}$$

For the three-dimensional case, the diffusion coefficient is expressed as

$$D = \Gamma \frac{a^2}{6} f, \tag{33}$$

where a is the nearest neighbor atomic distance. For tracer diffusion measurements, a correction term $f \leq 1$ should be introduced for the non-random character of the jumps. The factor f is known as a *correlation factor*. Its full implications will be found in Manning's book (44), it will only be briefly discussed later on in Section 4.2.

Finally, the diffusion coefficient can be written as

$$D = \left(\frac{1}{6}\right) a^2 Z f \nu_O \exp\,[(S_m + S_f)/k] \times \exp -[(H_f + H_m)/kT]\ , \qquad (34)$$

where ΔH_f, ΔH_m are enthalpies for the formation and motion of the defect and S_f and S_m are the corresponding entropy terms. The temperature dependence of the diffusion coefficient is generally of the Arrhenius type and is written as

$$D = D_O e^{-(Q/kT)}\ , \qquad (35)$$

with $D_O = (1/6)a^2\, zf\nu_O \exp \dfrac{(S_f + S_m)}{k}$ and $Q = H_f + H_m$.

Thus, from the knowledge of the crystal structure and independent measurement of the correlation factor f, it is possible to evaluate the total entropy factor as well from the diffusion measurements. For the separation of formation and motion components of the entropy and enthalpy, separate experiments are needed such as the resistivity studies of the quenched in defects, positron annihilation and the simultaneous length and lattice parameters (45) etc. We wish to note that diffusion coefficient and the various factors involved therein are basic properties of the solids. In Table 1, diffusion data for several important pure elements are listed. It is seen that formation and motion entropies and enthalpies are about equal in close-packed FCC and HCP metals but the formation part is much larger in the BCC metals. For covalently bonded elements, like Si, the entropy and enthalpy for formation may constitute almost entirely the corresponding quantities observed in diffusion measurements which have implications with the diffusion mechanism operative.

Another common mode of mass transport in solids is by the movement of interstitial atoms directly in the lattice. Interstitial atoms may be of several types. When a lattice atom is displaced into an interstitial position and it is termed as self-interstitial; the vacancy left behind is annihilated at the interfaces, surfaces or by another interstitial atom. The formation energy associated with the self-interstitial atoms is usually much larger than that for vacancy formation. Self-interstitial atoms, indeed, appear to account for diffusion in Si at least at high temperatures (53). When an interstitial atom is formed and a vacancy is left behind, it is known as Frenkel defect. A notable example for the production of Frenkel defects is radiation damage in solids. The energy involved in Frenkel defect production equals the sum of the vacancy and interstitial formation energies less the binding energy.

TABLE 1

Lattice Self Diffusion in Some Important Elements[a]

Element	D_0(cm²/sec.)	Q_ℓ(eV)	H_f(eV)	H_m(eV)	$\Delta V/V_0$(at. Vol)	$\Delta S/k$
FCC						
Al	0.047	1.28	0.67[b]	0.62[b]	1.3	2.2
Ag	0.04	1.76	1.13[b]	0.66[b]	0.9	–
Au	0.04	1.76	0.95[b]	0.83[b]	0.72	1.0
Cu	0.16	2.07	1.28[b]	0.71[b]	0.9	1.5
Ni	0.92	2.88	1.58[c]	1.27[c]	–	–
Pb	1.37	1.13	0.54[d]	0.54[d]	0.73[d]	1.6[d]
Pd	0.21	2.76	–	–	–	–
Pt	0.33	2.96	–	1.45[b]	–	–
BCC						
Cr	970	4.51	–	–	–	–
α-Fe	0.49	2.95	–	0.68[e]	1.4	–
Na	0.004	0.365	0.39/0.42[f]	–	0.32	–
β-Ti	0.0036	1.35	–	–	0.33	–
V	0.014	2.93	–	–	–	–
W	1.88	6.08	3.6[b]	1.8[b]	–	–
β-Zr	0.000085	1.2	–	–	–	–
HCP						
Co	0.83	2.94	–	–	–	–
α-Hf	0.86/0.28(*)	3.84/3.62	–	–	–	–
Mg	1.0/1.5(*)	1.4/1.41	0.79/0.89	–	–	–
α-Ti	0.000066[g]	1.75[g]	–	–	–	–
Diamond	Lattice					
Ge	32[h]	3.1[h]	2.4[i]	0.2[i]	0.24/0.41[h]	10[h]
Si	1460[h]	5.02[h]	~3.9[j]	~0.4[j]		

* First number is parallel to c axis and second perpendicular to c axis.
a. Unless otherwise specified data are obtained from the compilation of Peterson (Ref. 46)
b. R. Balluffi (47)
c. Wycisk and Feller-Kneipmeir (48)
d. Knodle and Koehler (49)
e. Johnson (50)
f. Nieminen (51)
g. Dyment (52)
h. Frank et al. (53)
i. A. Seeger (54)
j. Kelly et al. (55)

The Frenkel defect is usually very mobile and is accompanied by mass transport. The interstitial atoms may also be of the impurity type which fit into the interstices in lattice easily by virtue of their small size. Consequently, their formation energy is negligible and in Eq. (34) the terms S_f and H_f may be omitted. Hence, diffusion of the foreign or impurity interstitials may be very fast. Commonly, gas atoms such as H, C, O, and N occur as interstitials in BCC metals such as Fe, W, Nb, and Ta and HCP metals such as Ti, Zr, and Hf. In the Si lattice as well, most transition metal impurities enter as interstitial atoms although their solubilities rarely exceed one part per million (15,53).

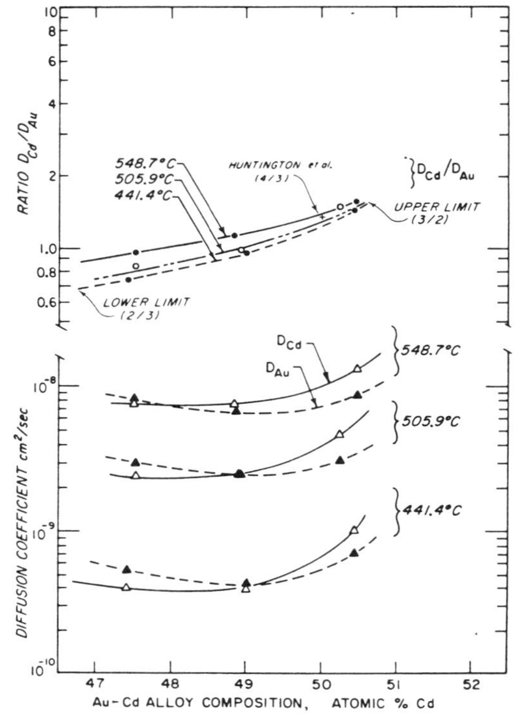

Fig. 12. Isothermal variation of the diffusion coefficients and relative rates of diffusion in ^{195}Au and ^{109}Cd in β-AuCd alloys. The alloys contain vacancies on the Cd rich side and anti-site atoms on Au rich side which are responsible for reversal of diffusivity ratios (after Gupta et al., Ref. 57).

Diffusion kinetics are drastically accelerated by the presence of non-equilibrium vacancies and interstitial atoms. The case of Frenkel defects during radiation damage in crystals was alluded to above. Their concentration can exceed the equilibrium vacancy concentration by orders of magnitude. Non-equilibrium vacancies also arise due to deviations from stoichiometry in intermetallic compounds such as NiAl, CoGa, AuCd, AgMg, etc. (56). To preserve the phase composition and the order, the excess concentration is usually accommodated by generation of additional lattice sites so that corresponding number of sites on the minority atomic sublattice are left vacant. This situation usually arises when the atomic sizes in the alloy are significantly different and excess concentration is of the larger species. In this manner the system reduces strain energy. Conversely, the excess concentration of the atomic species having smaller atomic diameter does not strain the lattice and need not generate additional lattice sites. The excess atoms, consequently, are readily accommodated as anti-structure atoms. In the former case, the off-stoichiometric vacancy concentration overwhelms the equilibrium vacancy concentration so that formation enthalpy and entropy, H_f and S_f in Eq. (34), become negligible. Consequently, diffusion rates in off-stoichiometric alloys richer in the species having larger atomic size may be enhanced by many orders of magnitude. The off-stoichiometric alloys having anti-structure atoms do not show such enhancement of diffusion rates. Accordingly, large enhancement of diffusion rates of Al, Ga, Cd, and Mg tracers have been observed in the NiAl, CoGa, AuCd, AgMg systems where the second atomic species in these alloys exceeded the stoichiometric ratio. In addition to the stoichiometry, diffusion in these systems must occur in a manner that the long range degree of order is maintained. This requirement results in a highly correlated sequence of atomic-vacancy interchanges detailed discussion of which is outside the scope of this publication. Briefly, the interplay of the correlated atom-vacancy interchanges between the two sublattice and the presence off-stoichiometric vacancies or anti-structure atoms leads to inversion of the ratios of diffusivities of the constituent species as shown in Fig. 12 for the AuCd system (57). The activation energy for diffusion in the off-stoichiometric vacancy side also show a sharp decline as the formation term H_f gets reduced in Eq. (34).

4.2 Pressure and Mass Dependence of Diffusion

Pressure Dependence

The diffusion coefficient may be expected to be pressure dependent in view of the thermodynamic relation

$$\left(\frac{\partial G}{\partial P}\right)_T = V \tag{36}$$

where the volume V is to be associated with the diffusion process. From Eqs. 34 and 35, the dependence of D on the hydrostatic pressure may be written as

$$\left(\frac{\partial \ell nD}{\partial P}\right)_T = -\frac{\Delta V}{kT} + \left(\frac{\partial \ell n a^2 \gamma f}{\partial P}\right)_T \tag{37}$$

Similar to the Eq. (34), one may separate the activation volume further into a sum of the terms $\Delta V_f + \Delta V_m$ for a process involving the defects in solids. Furthermore, the activation volume, the entropy and enthalpy are interrelated by the volume thermal expansion coefficient α and the isothermal compressibility β given respectively by Lawson (58) and Keyes (59) as

$$\Delta S = \alpha \Delta V/\beta, \tag{38}$$

and

$$\Delta V = 4\beta Q \tag{39}$$

The values of ΔV typically are of the order of an atomic volume (V_O) even in closed packed lattices which imply only a small relaxation around a vacancy. In Table 1 and 2, activation volumes for some pure metals and elemental semiconductors are listed. Rice and Nachtrieb (66) have shown that ΔV and Q can also be empirically related to the heat of melting L_{melt} and the volume of melting V_{melt} as

$$Q = L_{melt}\Delta V/V_{melt} \tag{40}$$

Experimentally, measurable changes in the diffusivity ($\geq 10\%$) occur at hydrostatic pressures of the order of a few kilobars in metals and at temperature $\geq 0.5T_{melt}$.

As mentioned earlier, the total activation volume is actually composed of formation and motion components for the mobile defect and analogous to Eq. 34. The individual components can be measured by monitoring resistance with pressure on and off. This technique is similar to quenching experiments (45). The measurements of the activation volumes have been very useful in determining the nature of the diffusing defect and, in fact, supplement the measurement of the isotope effect and the underlying correlation factor which will be discussed later on in this section. It may be mentioned in passing that thin metallic films held on substrate usually have a bi-axial tensile stress of the order 10^9 dynes/cm^2 which is of the above mentioned magnitude. Consequently, we may expect ~10% enhancement of the diffusivity which is, indeed, very minuscule and insignificant.

Mass Dependence

In Eqs. 33 and 34, a factor f known as the *correlation factor* was introduced to account for the non-random character of atom-vacancy jumps in tracer diffusion only. It measures the fraction of jumps that is effective in causing diffusion of the tracer. For self-diffusion in cubic lattices, f is a geometrical factor and depends on the crystal type and the mobile defect responsible for diffusion. It is also generally temperature independent. Its values for various diffusion mechanisms can be readily computed from the direction of jumps. In Table 2 the values of correlation factor for various diffusion mechanisms possible in FCC lattices are listed. Since correlation factors differ markedly for

TABLE 2

Isotope Effect for Various Diffusion Mechanisms in Some Elements

Mechanism(*)	0<f<1	ΔK	E	$\Delta V/V_0$
A Vacancy		0.87		
FCC Metals				
Ag(110,105)	0.782	0.92	0.718(1000K)[a] 0.639(1150K)[a]	0.9
Grain Boundary				
Ag(110,105)	?	?	0.46[b]	1.1[c]
Au(199,195)	0.782	0.9	0.704(1127K)[d] 0.653(1327K)[d]	0.9
Cu(67,64)	0.782	0.87	0.669(~1200K)[a]	
BCC Metals				
Na(24,22)	0.727	0.5[e]	0.348(300-350K)[f]	0.33
Diamond Lattice				
Ge(77,71)	0.5	0.86	0.26(1173K)[g] 0.30(1198K)	0.24/0.41
B Interstitial	1.0	0.68	–	–
C Divacancy	0.475	1.45	–	–
D Interstitialcy	0.667	1.99	–	–
E Atom Ring	0.969	2.66	–	–

a. Peterson (46)
b. Robinson et al. (60)
c. Martin et al. (61)
d. Herzig et al (62)
e. Feit (63)
f. Mundy (64)
g. Campbell (65)
* These numbers refer to the isotopes

different mechanisms, their measurement provides clues to the mobile defect responsible for diffusion. The correlation factor is measured by diffusing two radioisotopes simultaneously in a single crystalline sample. The resulting isotope effect ΔE is related to their diffusivities (D_α and D_β), masses (m_α and m_β) and a kinetic factor ΔK as

$$E \equiv [D_\alpha/D_\beta - 1]/[(m_\beta/m_\alpha)^{1/2} - 1] = f\Delta K. \tag{41}$$

The kinetic factor ΔK is defined as the fraction of kinetic energy at the saddle point associated with atomic motion in the jump direction. To obtain reasonable precision in ΔE, a mass difference of at least 5% is sought between the two isotopes. The isotopes are generally co-diffused so that the errors in temperature, time, diffusion distance

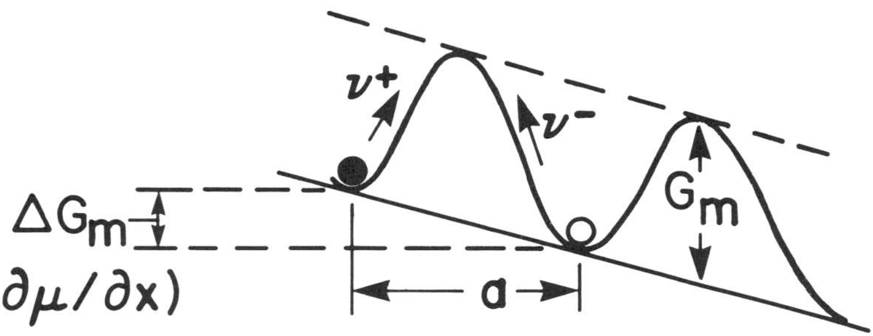

Fig. 13. Energy-distance diagram for diffusion in lattice when driving force F on individual atom is finite (after Manning, Ref. 43).

measurements, etc. are the same and do not figure in the relative diffusivities. Under ideal conditions, the relative difference between the two diffusivities can be measured with a precision of ~1%. This has been possible because a surprisingly large number of isotope combinations are found to fulfill the seemingly stringent mass criterion. In addition, imaginative schemes for nuclear counting have been employed based on discrimination of decay time, and energy of radioactive emissions which are discussed in detail by Rothman (67).

The kinetic factor ΔK has the lower and upper bounds of zero and unity respectively. For FCC and HCP lattices, the measured values of ΔK are of the order 0.9 for metals such as Cu, Ag and Zn. For BCC lattices, smaller values of ΔK = 0.68 are observed for metals like Na and Cr and are attributed to relatively open saddle-point structure rather than a change of diffusion mechanism. Finally, it is to be noted that the factor ΔK from the isotope effect measurements and the fractional activation volume (Eq. 36) show numerically similar values for the same crystal lattice and the diffusing defect. Thus, the values of ΔK and $\Delta V/V_O$ are considered complementary in deciding the diffusion mechanism. For vacancy mechanism in closed packed metals both ΔK and $\Delta V/V_O \simeq 0.9$. In open lattices like BCC, the corresponding values are of the order of 0.68 and 0.5 respectively for vacancy mechanism. For interstitial diffusion mechanism ΔK~1 and $\Delta V/V_O \leq 0$ may be expected.

4.3 Driving Force F on Individual Atoms Is Finite – Linear Chemical Diffusion Regime

In Fig. 13 the non-zero driving force F on individual atoms results in a net flux in the right hand direction. Consequently, the mass velocity term v in Eq. (1) cannot be neglected. The driving force F makes the energy at successive lattice sites differ by

$$\Delta G_m = aF \tag{42}$$

Assuming that the energy barrier undergoes an equal change on either side, the net number of jumps can be written as

$$\Gamma^{+} - \Gamma^{-} = \Gamma(e^{-\varepsilon} - e^{+\varepsilon}) \tag{43}$$

where Γ has been defined in Eq. (32) and the mobility is given by

$$\varepsilon = \frac{1}{2}\Delta G_m/kT = aF/2kT \tag{44}$$

Combining Eq. (43) and (44), the external force term may be written as

$$\frac{\Gamma^{+} - \Gamma^{-}}{2\Gamma} = \sinh aF/kT \tag{45}$$

Some atomic driving forces are listed in Table 3. These forces, with the exception of perhaps centrifugal force, are commonly encountered in the thin film metallization used in the interconnection technology of the modern computers. Each driving force deserves an exhaustive discussion in its own right which may be found in earlier review articles by Huntington (68), Shewmon (14), and Tu (69). Here, we discuss the case of the chemical composition variation which gives rise to chemical potential gradients in non-ideal solid solutions. The phenomenon, though old, has assumed greater significance in the context of the recent development of the artificially layered thin films known as man-made super-lattices and multilayered thin films in general.

The atomic force over N lattice planes due to the chemical potential gradient $(\partial\mu/\partial x)$ may be written as

$$F = \frac{kT}{aN}[\ln C_2\gamma_2 - \ln C_1\gamma_1] \tag{46}$$

where $\gamma_{1,2}$ are the chemical activity coefficients for the non-ideal solution and $C_{1,2}$ the concentration terms; the subscripts 1 and 2 refer to the specific component of the binary solution. Despite the large variation of concentration and activity factor e.g., $10 < C_2\gamma_2/C_1\gamma_1 < 1000$, over say 10 atomic planes, the force factor remains small in view of its logarithmic nature. Under these conditions, the Eq. (45) can be simplified as

$$\frac{aF}{kT} = \frac{\Gamma^{+} - \Gamma^{-}}{\Gamma} << 1 \tag{47}$$

TABLE 3

Driving Forces in Diffusion (after Manning, Ref. 43)

Driving Force	Atomic Force (F)	Governing Parameter	Examples
Electromigration	ZeE	Ze (effective charge)	Thin film failures with current
Thermomigration	$\frac{Q^*}{T}\frac{\partial T}{\partial x}$	Q* (heat of transport)	Soret Effect VLSI Solder failures
Chemical inhomogeniety ($\partial\mu'/\partial x$) (non-ideal chem. potential gradient)	$-kT\frac{\partial \ell n\gamma}{\partial x}$	γ (activity coefficient)	Kirdendall void Interface failures
Centrifugal Force	$m\omega^2 r$	m (effective molecular) mass	Creep in aero engines and centrifuges
Stress field	$\frac{\partial U}{\partial x}$	U (interaction energy)	Hydrostatic Pressure, Thin film bi-axial stresses, diffusional creep

which is known as the linear region. Hence, the drift term in Eq. (1) can be identified from its basic definition and random walk theory (43) as

$$C_1 v = C_1 \Gamma a^2 F/kT \tag{48}$$

$$\text{or } v = \frac{D_1 F}{kT} \tag{49}$$

The quantity D_1/kT is known as mobility B_1. The expression (49) is commonly known as Nernst-Einstein relation. For diffusion in concentrated alloys displaying non-ideal behavior, the Fick's first law for the first component may be then expressed in a generalized form as

$$J_1 = B_1 F\, C_1 \tag{50}$$

From the above generalized Fick's first law and the non-ideal behavior of the alloy, Darken derived the following equation for chemical diffusion (70):

$$\tilde{D} = (D^*_1 C_2 + D^*_2 C_1) \left[1 + \frac{d \ln \gamma_1}{d \ln C_1} \right], \qquad (51)$$

where D^*_1 and D^*_2 are the tracer diffusivities of the two components of the alloy and γ_1 and γ_2 the chemical activity coefficients. The terms in the square bracket are collectively known as *thermodynamic factor*. It should now be obvious that without the knowledge of the thermodynamical factor, it is not possible to account for the observed chemical diffusion fully. Consequently, Boltzman-Matano type of diffusion experiments coupled with marker velocity measurement alluded to in Section 2, see Eqs. 19 and 20, are likely to over-estimate the diffusivities of the individual components. In the regular solution-type alloys, the factor γ is independent of the composition but the thermodynamic factor still remains finite. Only in ideal solutions γ = 1 and the chemical diffusivities become identical with those obtained employing the tracer techniques. It may be mentioned in passing that Darken's equation (51) accounted for uphill diffusion against the composition gradients first observed in a Fe-0.4%C/Fe-0.4%C-4%Si couple. The explanation of this effect is that addition of Si increases the chemical potential of the Fe-C system (71). This observation underscores the fact that in chemical diffusion, the basic driving force is the chemical potential gradient and the concentration gradient becomes a limiting case of it which may be apparently violated. The uphill diffusion should be a common occurrence in thin film metallic alloy couples.

4.4 Non Linear Chemical Diffusion Regime

When sharp chemical potential gradients exist over small distances, ≤ 10 atomic spacings, the Eq. (45) cannot be linearly simplified since the boundary condition in the inequality (47) is not satisfied. Such a situation arises in the spinodal decomposition of single phase where the curvature of the free-energy composition plot $\partial^2 F/\partial C^2$, becomes negative. Decomposition by uphill diffusion occurs as it lowers the free energy. Other examples of the existence of sharp chemical potential are the artificially layered polycrystalline films investigation by Cook et al. (72) and compositionally modulated amorphous thin films studied by Spaepan and co-workers (73,74). The spinodal solid state transformation is known as continuous i.e. it does not involve any nucleation and is assisted by diffusion occurring under sharp chemical potential gradient. A brief discussion of this subject is provided here, for details the interested reader is referred to publications of Cahn and Hilliard (75) and the recent review article by Doherty (76) for background and the in depth understanding.

For spinodal decomposition, Martin and Doherty (76) have shown that the Darken's term is the controlling factor in determining the curvature of the free-energy and composition given by

$$\frac{d^2F}{dC^2} = RT\left(\frac{1}{C_1} + \frac{1}{C_2}\right)\left(1 + \frac{d \ln \gamma_1}{d \ln C_1}\right), \tag{52}$$

where the notations have the same meaning as previously defined. We only need to change the various chemical concentrations to keep them consistent with those used in this article so that $C_1 = 1-C_2 = 1-C$. Combining Eqs. (51) and (52), the chemical diffusivity $\tilde{D}$ in the spinodal decomposition may be written as

$$\tilde{D} = [D^*{}_1C_2 + D^*{}_2C_1]\frac{C_1C_2}{RT}\frac{d^2F}{dC^2} \tag{53}$$

and the diffusion mobility M_D as

$$M_D = [D^*{}_1C_2 + D^*{}_2C_1]\frac{C_1C_2}{RT} \tag{54}$$

To describe the thermodynamics of the inhomogeneous solid solution, a correction term is introduced as

$$\left(\frac{dF}{dC}\right)_{inh} = \left(\frac{dF}{dC}\right)_{homo} - 2\kappa\frac{\partial^2C}{\partial x^2}. \tag{55}$$

Herein, κ *is the gradient-energy coefficient* $= NkT_c\psi^2$ with T_c the critical temperature below which the homogeneous alloy tends to unmix and ψ is the 'interaction distance' of atoms (77).

From Eqs. (54) and (55), the Fick's two laws for the inhomogeneous solid solution may be written as

$$\tilde{J} = -M_D\frac{d^2f_0}{dC^2}C + 2M_D\kappa\frac{\partial^3C}{\partial x^3}, \tag{56}$$

and

$$\frac{dC}{dT} = M_Df''_O\frac{\partial^2C}{\partial x^2} - 2M_D\kappa\frac{\partial^4C}{\partial x^4} \tag{57}$$

where $f_O = (dF/dC)_{homo}$ and $f_0'' = \partial^2f_O/\partial C^2$.

The above non-linear differential equation has been solved by Cahn and Hilliard (75) and Cahn (77,78) for spinodal decomposition. An important spin off from the theory has been the understanding of diffusion at low temperatures in compositionally modulated thin films as discussed by Cook and Hilliard (72).

Accordingly, a harmonic composition modulation C(t) will grow or decay with time t as:

$$C-C_0 = A_0 \exp[-\tilde{D}_\lambda h^2 t]\cos hx, \qquad (58)$$

where C_0 is the average composition, A_0 is the initial amplitude of the modulation, h $(= 2\pi/\lambda)$ the wavenumber and $\tilde{D}_\lambda$ is the effective interdiffusivity. The macroscopic interdiffusivity and $\tilde{D}_\lambda$ are related as

$$\tilde{D}_\lambda = \tilde{D}[1 + 2\kappa h^2/f_0''], \qquad (59)$$

For small deviations from homogeneity $\tilde{D}$, f_0 and κ can be assumed to be composition independent and the measurement of $\tilde{D}_\lambda$ as a function of λ gives $\tilde{D}$ and κ/f_0, thereby providing information on the thermodynamic nature of the solid solution as well. Cook, deFontaine and Hilliard (79) have extended the model to cubic lattices and have measured diffusion in Ag/Au and Ni/Cu modulated films (79,80). Furthermore, see Chapter III of this book for measurement of diffusion in artificially modulated thin films.

5. STRUCTURE OF GRAIN BOUNDARIES, MOBILE DEFECTS AND DIFFUSION HIERARCHY

Our discussion on diffusion mechanisms in Section 4 was confined to idealized single crystalline specimens having point defects such as vacancies and interstitials. Motion of the point defects was, in fact, the principal mechanism of diffusion. Real materials, however, have structural defects such as grain boundaries and dislocations in addition to the aforementioned point defects within the grains or matrix. Thin films, in particular, have very small grain-size so that much of the diffusion may occur therein. Indeed, grain boundaries have recently been shown to be a world in themselves having a *potpourri* of defects such as vacancy, interstitials and dislocations. In this section we will examine briefly the structure of grain boundaries and the associated defects. The subject is exhaustive and complex and only a qualitative description of the gross features of grain-boundaries is possible in this article to develop general awareness. For in-depth understanding of the subject reference is made to the recent reviews by Balluffi (81) and several conference proceedings (3-5).

5.1 Grain Boundary Structure and State of Mobile Defects

A grain boundary is essentially a two dimensional transition region between two crystals having different orientations but of the same periodicity. This leads to a very complex crystallographic description of the grain boundary since 9 degrees of freedom or independent variables are involved which may be specified as 3 for the misorientations, 3 for the choice of the GB plane and 3 for the rigid body translation. An obvious consequence of bringing two periodic crystals in contact at an arbitrary plane is that a new two dimensional periodicity known as coincident site lattice would develop, structure and density of which will be dictated by the extent of the degrees of the freedoms used. Furthermore, perfect two dimensional periodic arrays would only develop at some discrete choices of the GB variables. To account for any arbitrary choice of orientations and translations at the chosen GB plane, incorporation of line defects will be required to make the grain-boundary. Extensive grain-boundary structure studies have been conducted over the last two decades or so, employing techniques such as high resolution transmission microscopy and computer stimulation from which the following broad features of the grain-boundaries have emerged.

Firstly, the grain-boundary is found to be relatively very narrow of the order of a few atomic distances only. Secondly, the atomic packing in the GB core is only slightly less dense than in the perfect crystal. Although, the core structure has finite excess volume but much of it may be considered to be relaxed by development of a variety of atomic groupings and localized configurations. Some boundaries having short wavelength periodicity run parallel to the coincident site lattice (CSL) and generally have low energy compared to the boundaries having long periodicity; the former may be termed as special boundaries. In the case of arbitrary misorientation between two crystals, the grain boundary minimizes its energy by preserving patches of the special low energy ordered boundary and introducing therein arrays of intrinsic grain boundary dislocations (GBDs) which are perfect having Burger vectors which preserve the CSL structure. It is also possible to have partial GBDs and extrinsic GBDs in the grain boundaries. The latter are punched out by the lattice dislocations and occur in more or less disorganized fashion. Intrinsic grain boundary dislocations have not been observed in grain boundaries having large misorientations owing to the fact that their spacings become very small and resolution is not possible.

Krakow and Smith (82) have recently demonstrated that image-enhanced high-resolution microstructure (HRM) of certain grain-boundaries in Au can be matched closely with the computer simulated relaxed model. The image-enhancement is carried out by Fourier filtering. The HRM of several grain-boundaries in Au is shown in Fig. 14 (a-d). In Fig. 15, superposition of image-enhanced HRM of $\Sigma = 19/[110]$ boundaries with their corresponding computer generated structure is shown (83). A close agreement is found between the computer predicted atom positions and the actual experimental observation. Furthermore, isosceles capped trigonal prisms separated by tetrahedra are observed at the interface. The former have been corre-

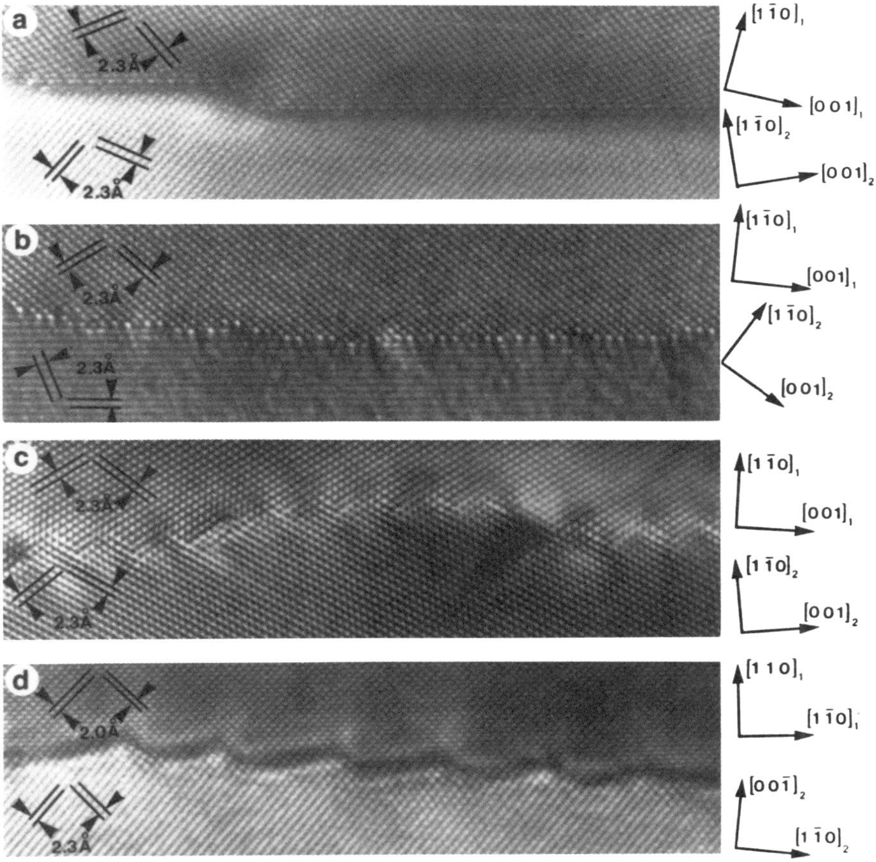

Fig. 14. High-resolution electron micrograph of assorted grain boundaries in Au.

(a) symmetrical [110]~27.5° tilt boundary,

(b) ~30.5° [110] asymmetrical tilt boundary,

(c) ~10° [110] tilt boundary,

(d) generalized boundary. All micrographs were taken at direct magnification of 800000×

(from Krakow and Smith, Ref. 82).

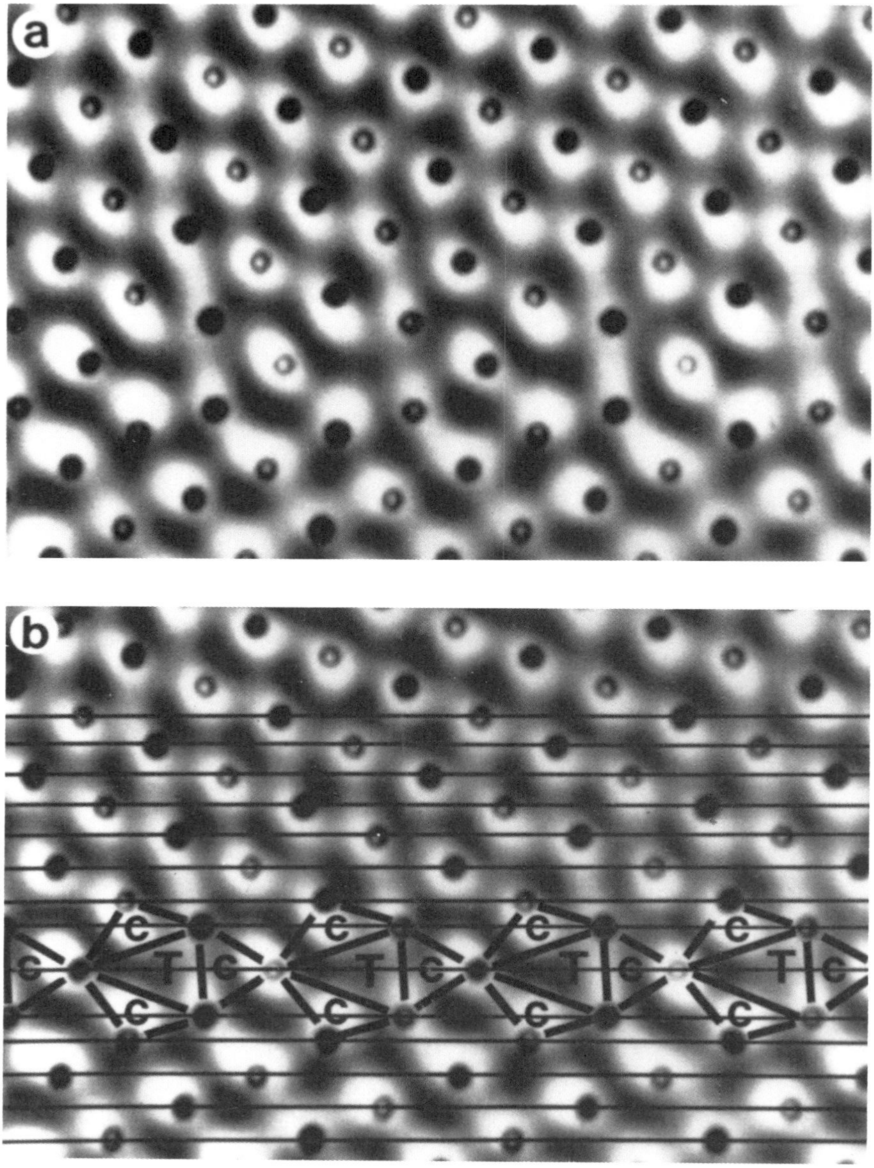

Fig. 15. (a) Superposition of the Fourier filtered experimental $\Sigma = 19$ boundary and relaxed model structure. (b) Image and model displayed in (a) with $(3\bar{3}1)$ planes parallel to interface designated. Note the arrangements of capped trigonal prisms (T) and caps (C) at the interface (from Krakow and Smith, Ref. 83).

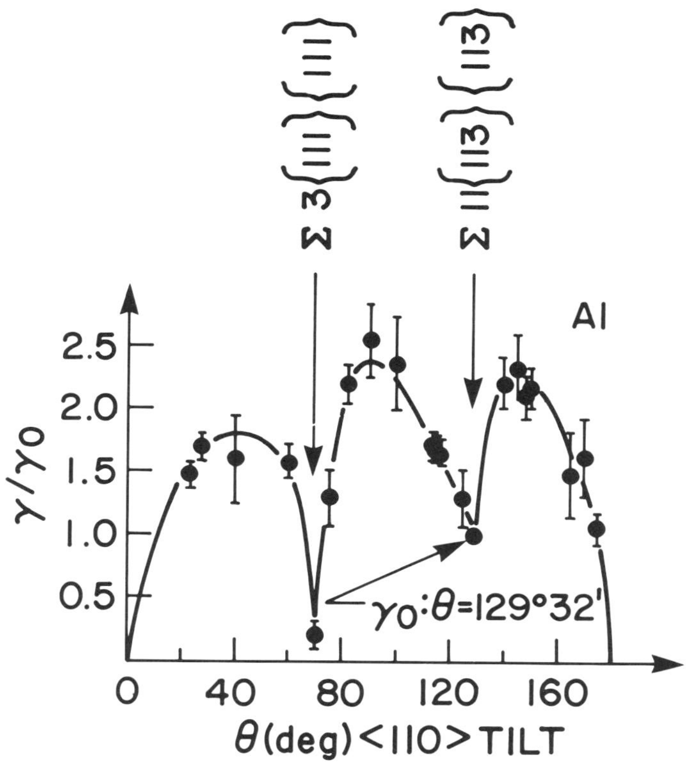

Fig. 16. Experimental variation in the energy of tilt boundaries about [001] in Al (from Hasson, et al., Ref. 85).

lated with the cores of primary ½ ⟨110⟩ grain boundary dislocations. These features at the interface also imply a dilation of about 45% which is localized to a region of three {331} planes thick centered on the boundary. These observations further lend support to the grain-boundary properties alluded to in the previous paragraph.

In view of the mixed structure of grain-boundaries dictated by CSL model, the GB energy is not expected to be a continuous function of the misorientation and, in fact, would show "cusps". These cusps were observed experimentally much earlier but their explanation remained elusive until recently (84). In Fig. 16 calculated and measured energies for symmetric tilt boundaries in Al are shown (85). The agreement between them is seen to be excellent.

There are several indirect experimental evidences from which a localized structure of vacancies may be inferred with little relaxation of the neighboring atoms. Firstly, mention may be made of the measurement of pressure dependence of the GB self diffusivities in Ag by Martin et al. (61) which resulted in a value of $\Delta V_b/V_O = 1.1$ where ΔV_b is the activation volume for GB self-diffusion defined analogous to Eq. 37 and V_O is the atomic volume. The value ΔV_b of the order of atomic volume in grain-boundaries compares very favorably to its counterpart in the lattice (Table I). Secondly, the isotope effect measurement for self-diffusion in Ag grain-boundaries by Robinson et al. (60) has also resulted in a relatively large value of $\Delta E_b \sim 0.46$ which is temperature insensitive. The corresponding value for self-diffusion in lattice is $0.65 < \Delta E_\ell < 0.71$, the variation is due to some temperature dependence (Table 2). If the mobile defects responsible for GB diffusion were partially dissociated or highly relaxed vacancy, it would require correlated movement of a number of atoms and hence only a small isotope effect value would have been observed. We mention in passing that diffusion in amorphous metallic phases proceeds by such cooperative movement of a number of atoms in the absence of discrete vacancies. Obviously, GB diffusion is more akin to lattice diffusion despite their grain-boundary dislocation structure.

It has recently been shown by computer modelling that point defects, notably the vacancies and interstitial atoms, are stable in grain boundaries (86). The binding energy for a vacancy to the GB core, although varying from site to site, was found to be generally attractive and of the order of -58 kJ/mole in Σ 5(001) tilt boundaries in BCC iron (87). The binding energies for interstitials in GBs are similarly attractive but considerably larger than those for the vacancies. Both these observations tie in well with the earlier notion that GBs are sources and sinks for point defects. Both kinds of point defects are mobile and account for diffusion under various conditions notably the low temperature creep (Coble creep), thin film electromigration, diffusion-induced GB migration (DIGM) etc. where net flux of atoms along grain boundary results in development of macroscopic or microscopic features. Typical vacancy jumping sequence in a $\Sigma 5$ [001] tilt boundary in BCC iron calculated by molecular statics or dynamics by Brokman (87) is shown in Fig. 17.

5.2 Mechanism of Grain Boundary Diffusion

In view of the above discussion, it is clear that mobile point defects such as vacancies and interstitials are responsible for diffusion in grain-boundaries as well. The diffusion coefficient for the GB can hence be written analogous to the lattice (Eq. 34) as

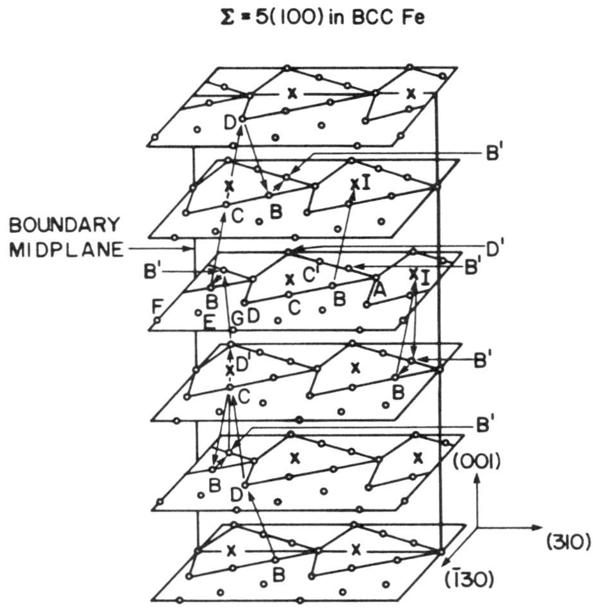

Fig. 17. Structure of $\Sigma = 5[001]$ tilt boundary ($\theta = 36.9$) in BCC iron calculated by molecular statics and dynamics. The preferred path of the typical vacancy excursions is shown by arrows. Note the sites E, F and G are not preferred (from Brokman et al., Ref. 87).

$$D_b = \left(\frac{1}{4}\right)a^2 Z\nu_0 f_b \exp\,[S_{f\ell} + S_{mb} - S_v^B]$$

$$\times \exp -[H_{f\ell} + H_{mb} - H_V^B]/kT, \qquad (60)$$

where the subscript b and ℓ are used to denote the various quantities related to the grain boundaries and lattice respectively, H_V^B is the binding enthalpy of a vacancy to a grain-boundary and S_V^B the corresponding entropy term. The factor 1/4 replaces 1/6 in view of the two dimensional diffusion in the GB plane. The pre-exponential terms for the GB are not known as precisely as in a perfect crystal. On the basis of the dense atomic packing in the grain-boundaries alluded to above, it is reasonable to assume that the nearest neighbor jump distance, the jump frequency and the atomic coordination remain unaltered in the GBs. Hence, no subscripts are used for a, Z and ν_O and these terms are assumed to remain unaltered in Eqs. (34) and

(60). A measurement of the correlation factor in the GB, $f_b = 0.54$ has been made in Ag by Robinson and Peterson (60) which is unambiguously smaller than the commonly accepted value of $f = 0.781$ for vacancy diffusion in the lattice of FCC metals; the correlation factor in both the cases are assumed to be temperature independent in the pure metals. It is therefore obvious that the product of the non-exponential terms, in essence, remains unperturbed and the diminished values of entropy and enthalpy in Eq. (60) due to vacancy-GB binding will determine the GB diffusivity. The reduced entropy will lower GB diffusivity while the reduced enthalpy will increase it. As will be seen later, the effect of reduced entropy is rather small on D_b as it may alter it at best by an order of magnitude, but the contribution of reduced enthalpy may be very drastic as it may increase D_b by typically 6 to 7 orders of magnitude compared to D_l in the lattice (Eq. 34).

The situation is exemplified by Au where precision diffusion measurements in both the lattice and grain-boundaries have been made. The self-diffusion in Au in the lattice (see Table 1) is expressed by $D_l = 0.04 \exp(-1.76\ \text{eV/kT})$. The self-diffusion along the grain-boundaries in Au (33) is described by $\delta D_b = 3.1 \times 10^{-10} \exp[-0.88\ \text{eV/kT}]\ \text{cm}^3/\text{sec}$. By assuming $\delta = a = 2.88\text{Å}$ and $\nu_0 = 4.0 \times 10^{12}\ \text{sec}^{-1}$ in the pre-exponent in Eq. (60) in the manner discussed above, the entropy for GB diffusion is calculated as $S_f + S_m - S_f^B = 0.7k$ which is about the same as in the lattice (see Table 1). On the other hand, the difference in activation energy between the lattice and GB in Au is ~50% and it enhances GB diffusivity by many many orders of magnitude. A detailed discussion on the relative magnitude of diffusion coefficients along the various paths such as vacancies in the crystal lattice, along the GBs, dislocations and surface will be found in Section 5.4.

5.3 Anisotropy and Orientation of Grain-Boundary Diffusion

Hoffman (88) and Couling and Smoluchowski (89) have studied the anisotropy of GB diffusion and its dependence on the orientation angle θ. In Ag [100] tilt boundaries diffusion of ^{110}Ag showed anisotropy of ~15 for parallel and perpendicular directions at 450°C. The anisotropy progressively declined for misorientations near $\Sigma = 13$ ($\theta = 23°$); $\Sigma = 17$ ($\theta = 28°$) and $\sigma = 5$ ($\theta = 36°$) etc. The anisotropy of diffusion within the GB plane can be understood on the basis of the distribution and nature of GB dislocations which predict channels of free volume parallel to the tilt axis in large-angle boundaries. The diffusion coefficients and penetration distance along GBs are also expected to show discontinuities and cusps as function of misorientation angle similar to the GB energies (see Fig. 16) since the contribut-

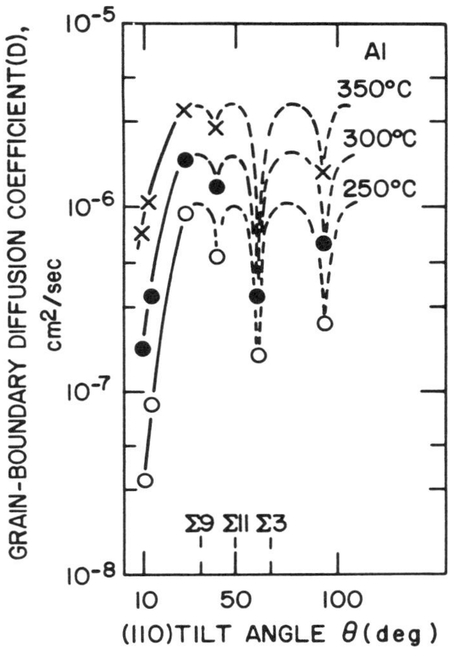

Fig. 18. Orientation dependence of ^{65}Zn diffusion coefficient in [110] tilt boundaries in Al for $\Sigma = 9$, 11 and 3 misorientation (Lange and Jurisch, Ref. 90).

ing factor in both the cases should be the coincident lattice sites and the GB dislocations. Lange and Jurisch (90) and Herbeuval et al. (91) have actually demonstrated the presence of such cusps shown in Figs. 18 and 19 which qualitatively agree with the GB energy cusps shown in Fig. 16. The activation energy for GB diffusion, however, appears to remain independent of the orientations and so that $Q_b/Q_\ell \approx 0.4$ ($\pm 20\%$). Thus, the anisotropy and orientation dependence of GB diffusion may be related primarily to changes in $D_b{}^\circ$.

5.4 Diffusion Data in Grain Boundaries and Thin Films: Diffusion Hierarchy

In view of the high density of grain-boundaries in polycrystalline thin films and generally the low temperatures of their applications $T<0.5T_m$, the diffusion in thin films may be expected to be largely controlled by grain-boundaries. Furthermore, due to the variable structures of the grain-boundaries, diffusion coefficients would vary from grain-boundary to grain-boundary according to their orientation differences as discussed above. Thus, diffusion in polycrystalline thin films would be some average for the ensemble of the orientations

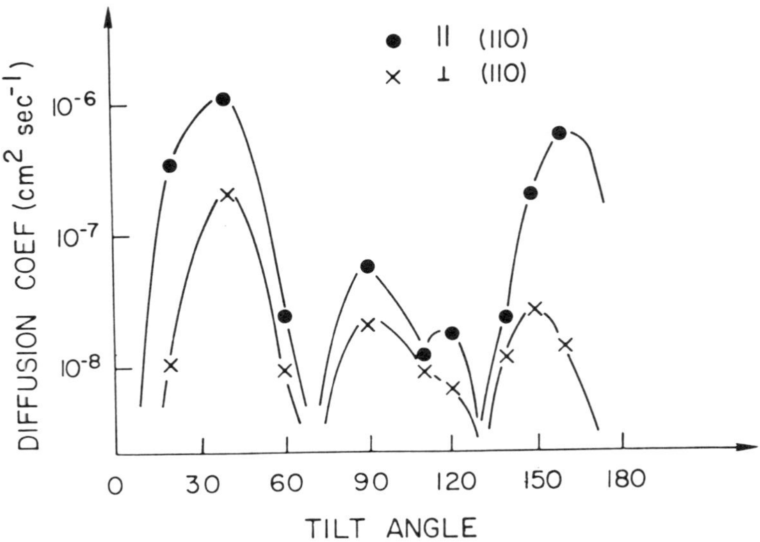

Fig. 19. Orientation dependence and anisotropy of ^{65}Zn diffusion coefficient along [110] tilt boundaries in Al at 250°C. Note the similarity with the energy for boundaries shown in Fig. 16 (Herbeuval et al., Ref. 91).

which is likely to change with the temperature assuming that no grain growth or grain boundary motion takes place. In pure metal polycrystalline thin films or bulk samples, there are three kinds of defects which may contribute to diffusion depending on the thermal conditions namely, the vacancies, dislocations and grain-boundaries. As discussed in Section 5.2, the dislocations and grain-boundaries themselves contain point-defects i.e. vacancies and interstitials. Thus the fundamental diffusive jumps either involve direct movement of the interstitial atoms or atom-vacancy exchanges. The formation and motion of point defects in the crystal lattice or along dislocations and grain-boundaries are, however, vastly different (see Eqs. 34-60) and actually result in a hierarchy of diffusion rates which was first proposed by Gjostein (92). In this section, we will examine critical diffusion data in pure thin films and grain-boundaries in the context of the proposed hierarchy of diffusion coefficients which in fact reflects the structural contributions in the samples.

In Fig. 20, the diffusivity spectra for various defects in metallic solids and liquids compiled by Gjostein (92) are shown. The most significant feature of the various Arrhenius plots is the activation energy for diffusion which is seen to be the highest for vacancy diffusion in lattice and it progressively diminishes as successively open

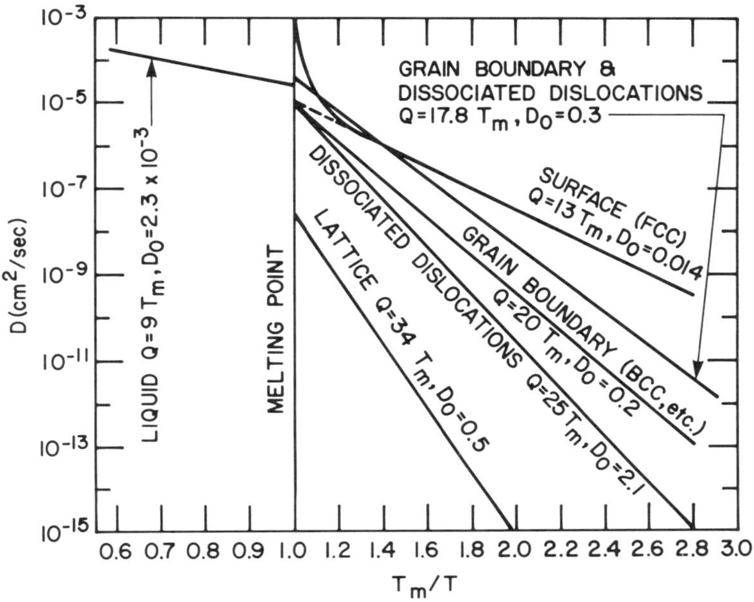

Fig. 20. The diffusion hierarchy for metallic solids and liquids shown as function of homologous temperature (after Gjostein, Ref. 92).

paths such as dissociated dislocations, single dislocations, arbitrary high angle boundaries and finally the open surface are available. In fact surface diffusion activation energy is not too large compared to that for the liquids. The pre-exponential factors for the "open paths" or short circuit diffusion are such that they appear to result in a diffusion coefficient of $\sim 10^{-5}$ cm^2/sec at the melting temperature which is also the diffusivity in the liquid phase. The observation may reflect on the core structure of the open paths such as grain-boundaries and dislocations alluded to in Section 5.1. In Fig. 21, the experimental self diffusion data in Au along various paths of diffusion in polycrystalline samples is shown (93). The various diffusion coefficients and the parameters for diffusion, Q and D^O, are in agreement with the spectra of diffusion shown in Fig. 20.

In Fig. 22 and Table 4, we show the grain-boundary diffusion data in thin films and some bulk materials obtained by various techniques employed in many laboratories. We have normalized these data by the absolute melting temperature of the host metal (T_m). It is noteworthy that these data split into groups: one is described by an

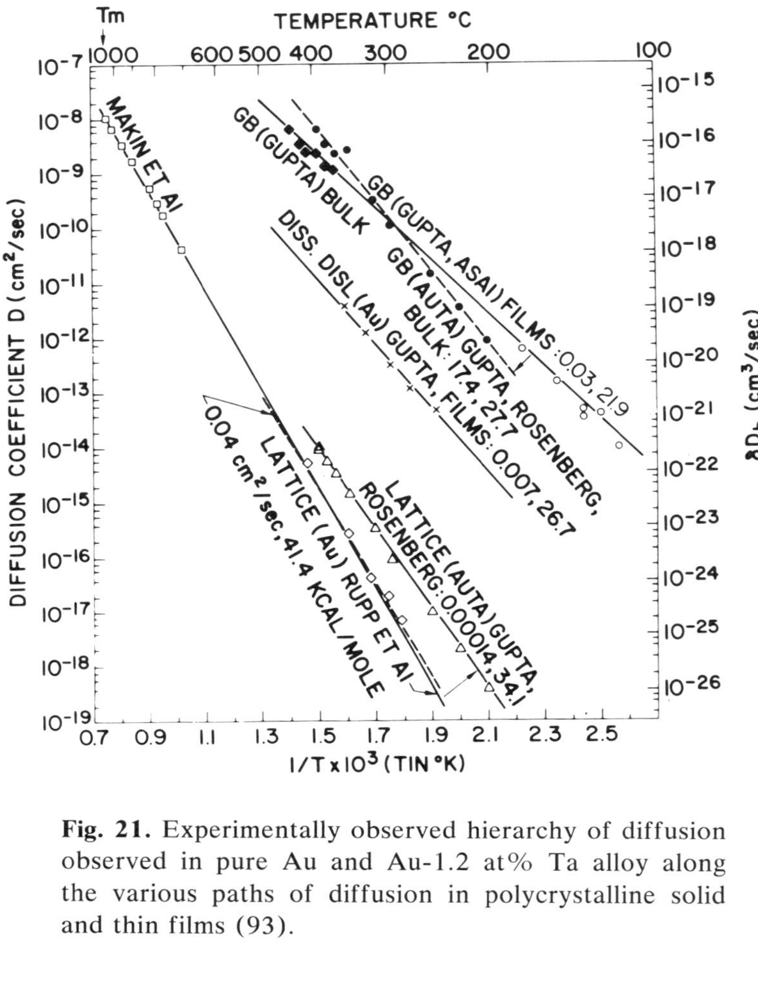

Fig. 21. Experimentally observed hierarchy of diffusion observed in pure Au and Au-1.2 at% Ta alloy along the various paths of diffusion in polycrystalline solid and thin films (93).

average activation energy of 17 T_m calories/mole and the other by 25 T_m calories/mole. These activation energies may be attributed respectively to high angle grain-boundaries and special or modified grain boundaries similar to those shown in Fig. 20 and once again reveals the role of the structure of the grain-boundaries.

Dopant GB diffusion data in Si shown in Fig. 22(c) and Table 4c show exception to the rule cited above where $Q_b \sim 40T_m$ appears to be the case. By the same token, the activation energies for self-diffusion in the diamond lattice (see Table 1 for Ge and Si) are almost double of those in metals, as $Q_\ell \geq 60\ T_m$ for these elements. In fact the high activation energies for the substitutional dopant GB diffusion in Si appear to be a reflection of the state of diffusion in the Si lattice. In essence, the rule of $Q_\ell \sim 34\ T_m$ is not obeyed in co-valently bonded

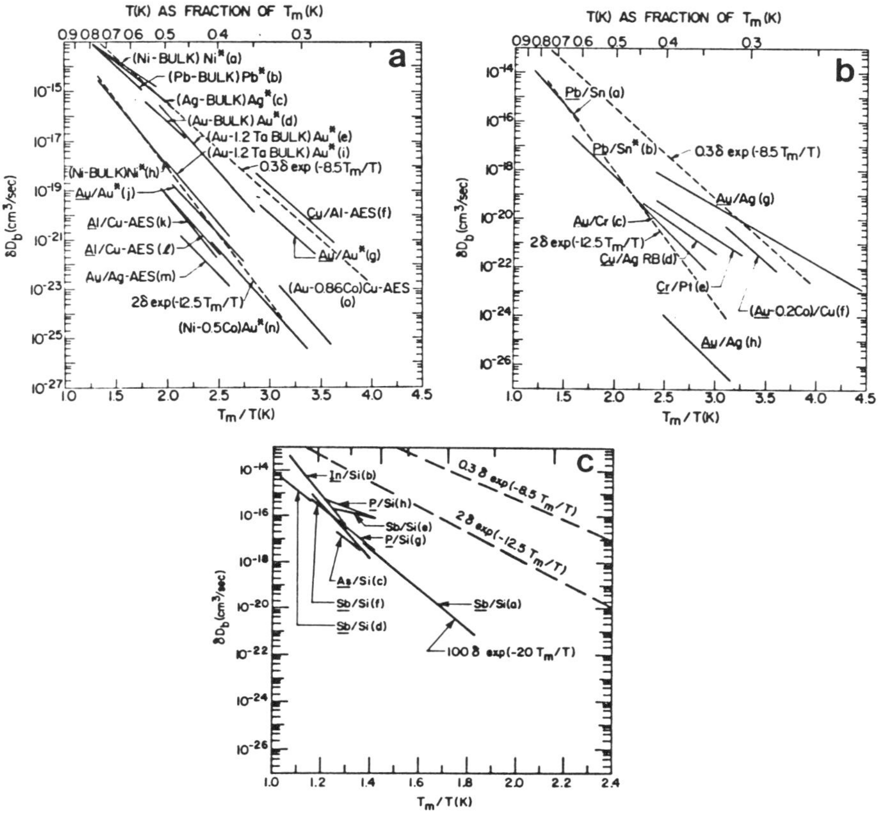

Fig. 22. Homologous Arrhenius plots of diffusion data in thin films and some bulk materials.
(a) grain boundary diffusion data obtained by sectioning techniques metallic samples,
(b) those obtained by permeation and surface accumulation techniques also in metallic systems, and
(c) dopant diffusion along dislocation and stacking faults in Si. Note the data in Si fall much lower compared to those in metals.
(See Table 4-A, B and C for source of data.)

TABLE 4A

Grain Boundary Diffusion Data Obtained by Sectioning Techniques*

Sample	Diffusant; matrix	δD_b° (cm^3s^{-1})	Q_b (eV)	Remarks
a	^{63}Ni; Ni	7×10^{-10}	1.08	45°⟨100⟩[a]
b	^{203}Pb; Pb	6.1×10^{-9}	0.46	Polycrystalline[b]
c	^{110}Ag; Ag	1.3×10^{-9}	0.80	Polycrystalline[c]
d	^{195}Au; Au	3.1×10^{-10}	0.88	Polycrystalline[d]
e	^{195}Au; Au-1.2%Ta	5×10^{-7}	1.26	Polycrystalline[d]
f	Al; Cu	5.1×10^{-7}	0.94	Polycrystalline films[e]
g	^{195}Au; Au	9.0×10^{-10}	1.0	Polycrystalline films[d]
h	^{63}Ni; Ni	2.2×10^{-8}	1.77	10°⟨112⟩[a]
i	^{195}Au; Au-1.2%Ta	1×10^{-9}	1.2	Low angle grain boundary[d]
j	^{195}Au; Au	1.9×10^{-10}	1.16	(100) epitaxial[d]
k	Cu; Al	4.5×10^{-8}	1.0	Polycrystalline films[e]
l	Cu; Al	1.8×10^{-10}	0.87	Polycrystalline films[e]
m	Ag; Au	9.3×10^{-5} (D_b°)	1.2	(111) epitaxial[e]
n	^{195}Au; Ni-0.5%Co	1.4×10^{-10}	1.6	Polycrystalline films[d]
o	Co; Au-0.186%Co	0.013 (D_b°)	1.20	Polycrystalline electro films[e]

[a] Autoradiography.
[b] Microtome.
[c] Lathe.
[d] Tracer ion sputtering.
[e] AES ion sputtering.
* From earlier compilation see Ref. (94)

TABLE 4B

Grain Boundary Diffusion Data Obtained by Permeation and Surface Accumulation Techniques*

Sample	Diffusant; matrix	D_b° (cm^2s^{-1})	Q_b (eV)	Remarks
a	^{119}Sn; Pb	1.1×10^{-3}	0.46	Foils[a]
b	Sn; Pb	$6.9 \times 10^{-8} (\delta D_b^\circ)$	0.62	Polycrystalline films[b]
c	Cr; Au	10^{-3}	1.09	Textured film[c]
d	Ag; Cu	3.1×10^{-6}	0.75	Polycrystalline films[d]
e	Pt; Cr	10^{-2}	1.69	Polycrystalline films[c]
f	Cu; Au-0.2	$Co(\delta D_b^\circ)$ 2×10^{-7}	0.56	Polycrystalline films[c]
g	Ag; Au	10^{-5}	0.63	Polycrystalline films[c]
h	Ag; Au	10^{-6}	1.1	(100) epitaxial films[c]
i	Ag; Au	4.07×10^{-17} 250°C ($K\delta D_b$)	--	Polycrystalline films[b]
j	Cu;Al	2.7×10^{-14} 250°C ($K\delta D_b$)	--	Polycrystalline films[b]
		1.3×10^{-13} 300°C ($K\delta D_b$)		Polycrystalline films[b]

[a] Nuclear adsorption and permeation.
[b] Cross-stripe and microprobe.
[c] AES surface accumulation.
[d] Surface accumulation and Rutherford backscattering.
* From earlier compilation see Ref. (94)

TABLE 4C

Discloation/Stacking Faults Diffusion in Silicon

Sample	Diffusant; matrix	δD_b° (cm^3/sec)	Q_b (eV)	Burger Vectors Reference
a	^{124}Sb; Si	2.25×10^{-5}	3.03	unspecified (95)
b	^{114}In; Si	5×10^{-4}	3.34	a/2[110] (96)
c	^{76}As; Si	29.4×10^{-8}	2.56	a/6[110] (97)
d	^{124}Sb; Si	2.25×10^{-5}	3.0	unspecified (98)
e	^{125}Sb; Si	2.25×10^{-13}	0.83	" (99)
f	^{125}Sb; Si	1.9×10^{-5}	2.9	" (97)
g	^{32}P; Si	6×10^{-6}	2.87	" (100)
h	^{32}P; Si	6×10^{-9}	1.4	" (100)

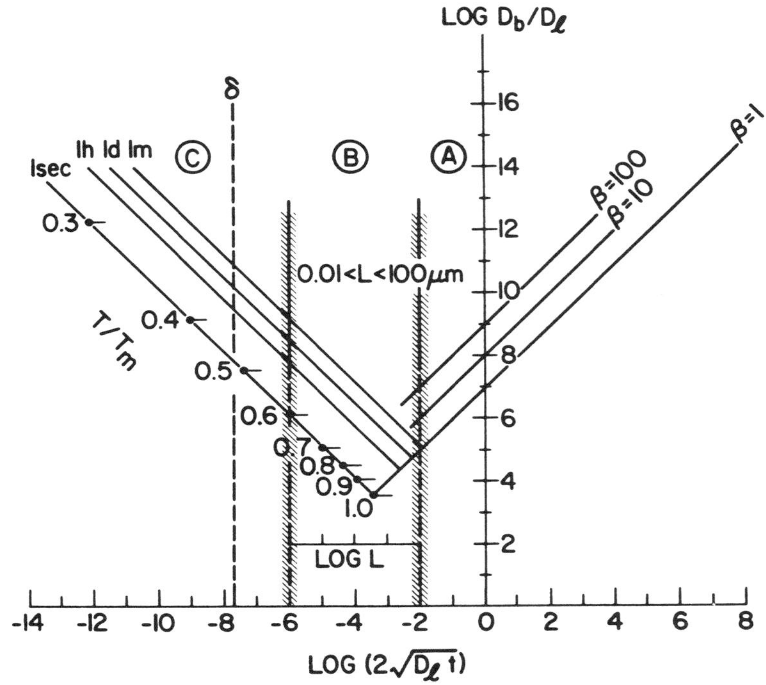

Fig. 23. Master curves for estimating diffusion in metallic polycrystalline thin films showing the role of temperature expressed as fraction of the absolute melting temperature T_m, time of annealing (1 sec, 1 hour, 1 day, 1 month) and grain-size (L). To estimate D_b/D_l, (a) project T/T_m on the appropriate time line and then draw a horizontal line. (b) Also draw a vertical line from the horizontal grain-size axis. (c) From their intersection point, draw a line parallel to the β-axis shown on the right hand side, its intersection with the Log D_b/D_l axis gives the desired value. The value of $2\sqrt{D_1 t}$ is read off from the perpendicular, drawn by the intersection point in (c).

elements while the relative magnitude of the GB activation energy still remains within the $0.5<Q_b/Q_\ell<0.7$ T_m bounds similar to the metallic systems. An atomistic explanation of observation of large activation energies must lie in the difficulty of forming vacancies in semiconductors in general, discussion of which is beyond the scope of this section.

Furthermore, in some instances the presence of impurities in dilute alloys appears to modify the high diffusivity path core structures so that diffusion data resemble those in the special boundaries. These effects will be discussed in the following sections.

We have used the above mentioned hierarchical diffusion concepts for estimating diffusion in thin films given their melting-temperatures (T_m), thickness ≃ the grain-size (L μm), the time of diffusion (sec) and the temperature. The diffusion parameters shown in Fig. 21 have been used. Figure 23 shows a set of master curves and procedures which can be used for estimating diffusion in situations where experimental data are not available. The films and foils with 0.01 $\mu m \leq L \leq 100$ μm shown in the shaded region diffuse in B regime at temperatures below 0.5 T_m with β values >100 even at long annealing times. At higher temperatures they are likely to diffuse in the A-regime if sufficient time is allowed and would, in fact, homogenize. The C-regime is shown on the left of the broken line indicating δ = 5Å, and can be realized at temperatures below approximately 0.3 T_m or at very short times below 0.5 T_m. Since $2(D_\ell t)^{1/2}<\delta$ = 5Å, the β values then tend to become infinite. Consequently, the thickness and grain-size of the thin film specimens in C-regime are of little consequence at temperatures below 0.3 T_m as the atomic transport is confined within the grain-boundaries. From Fig. 23, it is possible to estimate diffusion coefficient in the grain-boundaries of the film samples within ±20% using the criterion $D_\ell \approx 0.5 \exp(-34\ T_m/RT)$ cm^2/sec for computing D_ℓ itself. From the definition of β given in Eq. (22), it is also possible to estimate the broadening of the diffusion front adjacent to the grain-boundaries.

5.5 Relationship Among GB Diffusion, Lattice Diffusion and Grain-Boundary Energy

From the foregoing discussions in the previous sections, it is clear that the reason why GB diffusion is orders of magnitude larger than in the lattice, even though the atomic jump mechanism remains the same, is that it is mediated by the GB dislocations and the GB structure as a whole. So we posit that the information on GB structure is contained in GB diffusion coefficient itself when it is compared with that in the lattice! A semiempirical relationship due to Borisov, Galikov and

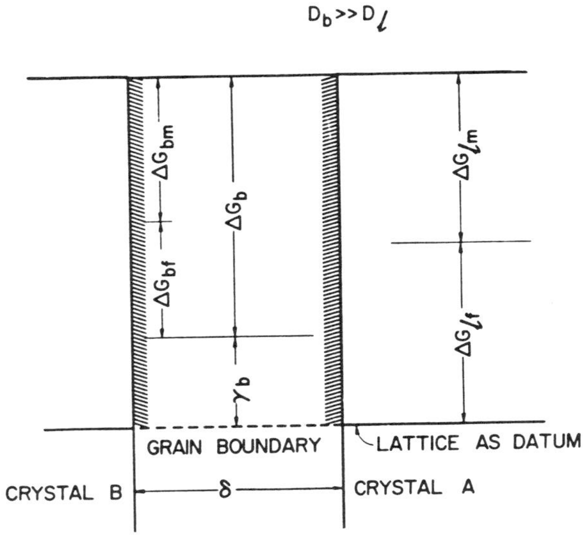

Fig. 24. Schematic representation of the relationship of the free energies of activation for atom vacancy jumps in the lattice and the grain boundary along with the grain boundary absolute energy (after Ref. 101).

Scherbedinskiy (101) has been available in the literature for some time which has attracted the attention of diffusion workers in recent years. It, indeed, seems to tie in the diffusion of the host species in the lattice and along the grain-boundaries with the absolute GB energy (γ_b). The basic postulate in this relationship is that the Gibbs free energy ΔG_b of an atom-vacancy jump in the GB, to the first approximation, is reduced by the absolute GB energy γ_b from its level ΔG_ℓ in the lattice as shown in Fig. 24. At this stage, we need not worry about the distribution of the total Gibbs free energy into the formation and motional components noted by the subscripts f and ℓ in Fig. 24. The Borisov *et al.* relationship therefore can be written at temperature T (°K) as

$$\gamma_b(\mathrm{mJ/m^2}) = \rho[\Delta H_\ell^b - T\Delta S_\ell^b], \tag{61}$$

where $\Delta H_\ell{}^b$ (in cal/mole) and $\Delta S_\ell{}^b$ (in cal/mole K) are respectively, the enthalpy and entropy difference associated with the diffusion processes in the lattice and along the grain-boundaries referred to by subscript ℓ and superscript b respectively. The conversion factor $\rho = 4.18 \times 10^7/2a^2N_O$ where a is the mean distance between the atoms at the GB is approximately equal to the lattice parameter and N_O is Avogadro's number. The factor 2 seems to arise from the multiplicity consideration of the GB between two crystals. From Eqs. 34, 60 and 61 therefore, the GB energy can be written as

$$\gamma_b(\mathrm{mJ/m^2}) = \rho[RT\ell n(D_b^o/D_\ell^o) + (\Delta H_\ell - \Delta H_b)], \quad (62)$$

or simply
$$\gamma_b = \rho RT[\ell n(D_b/D_\ell)], \quad (63)$$

In Table 5, GB energies obtained using Eq. 62 or 63 in some pure FCC metals are compared with those directly observed. Despite the semiempirical character of Eq. 62 or 63, a remarkable agreement, to within a few percent, is obvious between the computed GB energies and the ones directly observed.

The above relationships are also applicable to alloys when the individual quantities are taken for diffusion of the host species in the alloy. The extension of these relationships to alloys actually provides a handle on understanding the role of GB solute segregation effects on GB diffusion which are otherwise difficult or impossible to characterize and will be discussed in some details in the ensuing Section 6.

TABLE 5

GB Energies from Self-Diffusion in Some Pure FCC Metals and Directly Observed Values

Host (Diffusant)	$\gamma_b(\mathrm{mJ/m^2})^+$	$\gamma_b(\mathrm{mJ/m^2})$	Reference[e]
1. $Ag(Ag^{110})$	382 (1173K)	400 (1173K)[a]	Turnbull et al. (88)
2. $Ag(Ag^{110})$	392 (1173)	400 (1173)[a]	Robinson et al. (60)
3. $Cu(Ag^{110})$	638 (1123)	590 (1123)[b]	Barreau et al. (120)
4. $Au(Au^{195})$	378 (1123)	370 (1123)[b]	Gupta (118)
5. $Ni(Ni^{63})$	700 (1720)	690 (1523-1723)[c]	Lange et al. (94)
6. $Pb(Pb^{203})$	191 (588)	200(588)[d]	Kim et al. (112)

+ Extrapolated from the diffusion temperatures to the temperature of direct measurement

a. A. P. Greenough and R. King (102)
b. J. E. Hilliard, M. Cohen and B. L. Averbach (103)
c. E. R. Hayward and A. P. Greenough (104)
d. K. T. Aust and B. Chalmers (105)
e. from earlier compilation (see Ref. 112)

6. INFLUENCE OF SOLUTE SEGREGATION ON GB SELF-DIFFUSION AND ITS RELATIONSHIP WITH GB ENERGY, SOLUTE SEGREGATION AND GB DIFFUSION

The grain-boundaries in materials should in fact be viewed as single-phase solid-solid interfaces which are regions of high structural energy. The presence of solute in polycrystalline samples, bulk or thin films, will be prone to interfacial segregation as it would lower the GB energy through Gibb's adsorption. The GB solute segregation problem *per se* has reached a reasonable level of understanding and its discussions would be found in the literature. A notable source of reference is the ASM publication on Interfacial Segregation (2). In spite of the good understanding of the GB solute-segregation problem, its relationship with the GB diffusion of the host or the solute itself has been controversial. The addition of small amounts of solutes to materials, on the other hand, has resulted in remarkable improvements in their performance. In the VLSI Al thin film interconnects, a significant reduction in the electromigration damage has been accomplished through addition of 4 wt% Cu (106). The beneficial effect is attributed to reduction of GB diffusion of the Al host atoms through interaction of solute atoms with the GB defects (107). Similar examples can be found in fields of sintering of oxides, e.g. MgO/Al_2O_3, fatigue and fracture of super alloys like B doped Ni_3Al etc.

Not all solutes reduce GB diffusion. The older literature (108) in this regard showed all the possible trends — reduction of the GB diffusion by some solutes, increase by others and no observable effect in some cases. Gibbs in 1966 pointed out to the incorporation of the segregation factor $K = C_b/C_O$, where C_b is the solute content at the GB and C_O in the lattice, to the measured combined GB diffusion parameter δD_b (109) in alloys. For alloys, the measured parameter is $K\delta D_b$ for the solvent as well as solute tracers which takes into account the chemical environment in which GB diffusion is taking place. It was subsequently shown by Gupta (110) that it is possible to observe all the three effects of solute addition on GB diffusion in one and the same binary alloy system namely the Au-1.2 at% Ta. In Fig. 25, the combined GB diffusion coefficients for ^{195}Au tracer in polycrystalline Au and an Au-1.2 at% Ta alloy are shown; Ta enhances grain-boundary self-diffusion at high temperatures and inhibits it at lower temperatures, with no effect in between.

Since the publication of the Au and Au-1.2 at% Ta system, several binary alloys systems have been investigated over a range of composition and temperatures. Grain-boundary diffusion of the host and solute species has been measured in the Pb-In and Pb-Sn solders by Gupta and co-workers (18,111-113) covering composition ranges 0-44 at% In and 0-8.3 at% Sn respectively as listed in Table 6. Gas

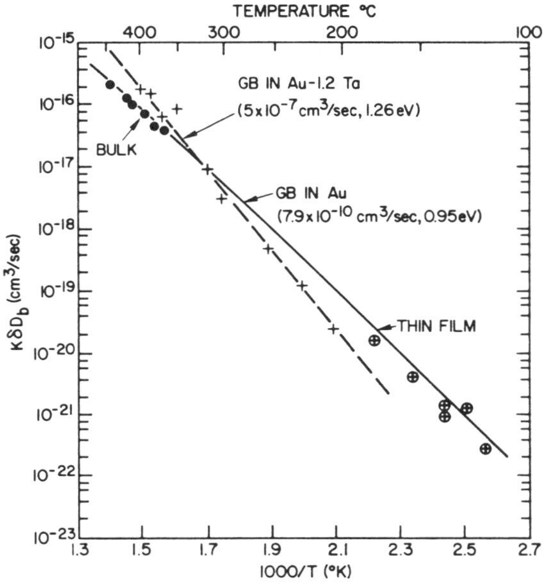

Fig. 25. Temperature dependence of $K\delta D_b$ in polycrystalline Au:bulk (•) and thin films (⊕) samples for ^{195}Au tracer. Also shown are the data in Au-1-2 at.% Ta polycrystalline bulk alloy samples (+).

and Bernardini (114) have studied ^{110}Ag and ^{113}Sn diffusion in Ag-Sn alloys in the 0-6 at% Sn composition range. The activation energy plots for lattice and GB diffusion in Pb-In alloys and Pb-Sn alloys are respectively shown in Figs. 26 and 27. Some isothermal values of the combined GB diffusion parameters for the Pb-In and Pb-Sn systems are shown in Figs. 28 and 29 respectively. From these data the following effects are clearly seen: (a) the activation energy for GB diffusion seems to show a minimum with composition which is unlike that for the lattice diffusion. (b) The isothermal values of $K\delta D_b$ go through maxima at low temperatures and show monotonic increases at higher temperatures. These effects are in fact caused by the interplay of the composition dependence of the segregation factor K, along with that of the GB diffusivity D_b and the solid solubility limit with the accompanying precipitation or pre-precipitation effects. These effects are discussed below in some details.

TABLE 6A

Tracer diffusion data of solute (^{114}In) and solvent (^{203}Pb) in Pb-In alloys (Ref. 18)

Host diffusant	D_ℓ (cm^2/sec)	Q_ℓ (eV) ±0.05	$K\delta D_b°$ (cm^3/sec)	Q_b (eV) ±0.05
Pb				
^{203}Pb or ^{210}Pb	0.89	1.11	6.1×10^{-9}	0.46
^{114}In	2.2	1.14	1.2×10^{-5}	0.69
Pb-8.7 at.% In				
^{203}Pb	0.015	0.89	1.7×10^{-10}	0.33
^{114}In	0.04	0.90	9.4×10^{-9}	0.44
Pb-19.8 at.% In				
^{203}Pb	0.0064	0.85	3.5×10^{-8}	0.45
^{114}In	0.00077	0.78	1.1×10^{-7}	0.50
Pb-4.36 at.% In				
^{203}Pb	0.00027	0.73	2.9×10^{-8}	0.46
^{114}In	0.0028	0.84	1.9×10^{-6}	0.58

TABLE 6B

Tracer diffusion data of solute (^{119m}Sn) and solvent (^{203}Pb) in Pb-Sn alloys (Ref. 111)

Host diffusant	D_ℓ (cm^2/sec)	Q_ℓ (eV) ±0.05	$K\delta D_b°$ (cm^3/sec)	Q_b (eV) ±0.05
Pb				
^{203}Pb or ^{210}Pb	0.89	1.11	6.1×10^{-9}	0.46
^{119m}Sn	0.41	1.03	7.3×10^{-9}	0.41
Pb-2 at.% Sn				
^{203}Pb	2.6×10^{-2}	0.98	4.0×10^{-10}	0.35
^{119m}Sn	–	–	–	–
Pb-3.5 at.% Sn				
^{203}Pb	2.2×10^{-3}	0.87	1.5×10^{-10}	0.31
^{119m}Sn	0.21	1.02	3.3×10^{-9}	0.39
Pb-8.3 at.% Sn				
^{203}Pb	2.7×10^{-3}	0.84	5.1×10^{-9}	0.43
^{119m}Sn	0.18	0.97	2.1×10^{-7}	0.54

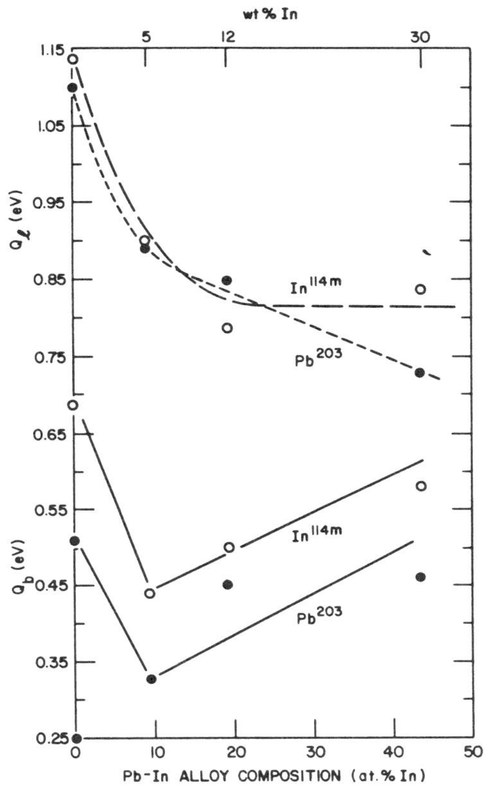

Fig. 26. Compositional dependence of the activation energy for lattice and grain-boundary diffusion of the solvent (^{203}Pb) and solute (^{114}In) radiotracers in Pb-In alloy samples (Ref. 18, 112).

6.1 Thermal Behavior of $K\delta D_b$

From Eq. 60, the activation energy Q_b for GB diffusion in pure host may be written as

$$Q_b \;=\; H_{f\ell} + H_{mb} - H_V^b. \tag{64}$$

Its value compared to the activation energy in the lattice (Q_ℓ) is generally found in the range $0.4 \leq Q_b/Q_\ell \leq 0.65$ in pure host depending on the distribution of the orientations and texture in the polycrystalline samples and films. In alloys, the presence of solute would affect the

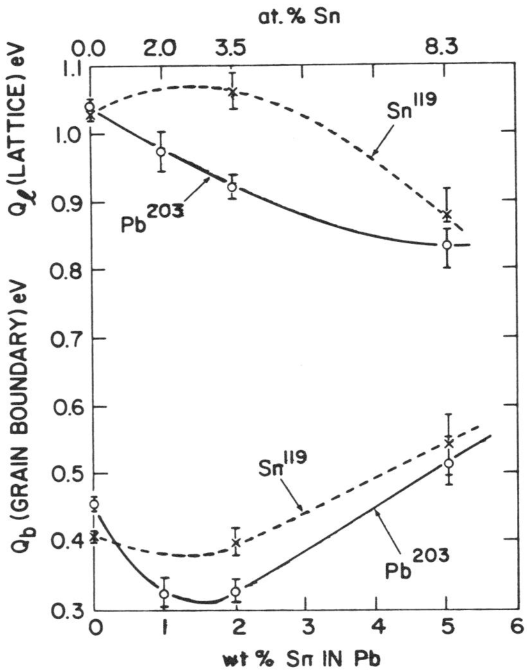

Fig. 27. Compositional dependence of the activation energy for lattice and grain-boundary diffusion of the solvent (^{203}Pb) and solute (^{119m}Sn) radiotracers in Pb-Sn alloy samples (Ref. 111).

GB diffusion of the solute as well as solvent. Equation (64), consequently is modified as

$$Q_b = H_{f\ell} + H_{mbi} - H_V^B - H_i^B + C \tag{65}$$

where H_i^B is the binding enthalpy between the impurity and the boundary, H_{mbi} is the migration enthalpy for vacancy-impurity atom exchanges within the boundary and $C = \partial \ell n f_i / \partial(1/T)$ is the temperature dependence of the correlation factor for impurity GB diffusion (see Section 4.2, Eq. 41 for its physical significance). The term H_i^B is in fact related to the Arrhenius-type temperature dependence of the GB segregation factor K so that:

$$K = K_O \exp(H_i^B/kT), \tag{66}$$

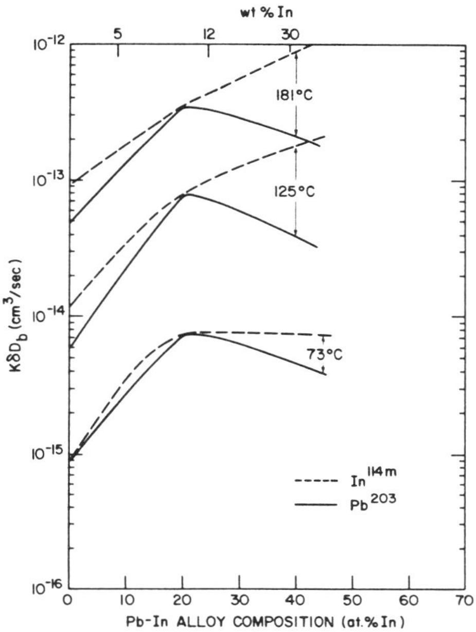

Fig. 28. Isothermal values of $K\delta D_b$ in polycrystalline Pb-In solder samples for the ^{203}Pb and ^{114m}In radiotracers. The $K\delta D_b$ values at 73,125 and 181°C were computed from the measured grain boundary diffusion parameters (see Table 6A).

Since, it is almost impossible to quantify the various terms in Eq. (65), the temperature dependence of $K\delta D_b$ is very complex and it is difficult to predict its behavior. *A priori,* the only valid conclusion which can be drawn is that solute addition will lower Q_b at least in dilute non-segregating type of alloys. The term C only partially accounts for the impurity-vacancy interaction. In fact, as H_i^B increases C will tend to increase for positive values of H_i^B (115). In concentrated alloys, clustering of the solute atoms in the grain boundary may in fact eliminate the vacancies. These effects, thus, appear to qualitatively explain the existence of minima in Figs. 26 and 27. In addition to thermal, all terms in Eq. (65) may be expected to have compositional dependence which is the topic of discussion in the following section.

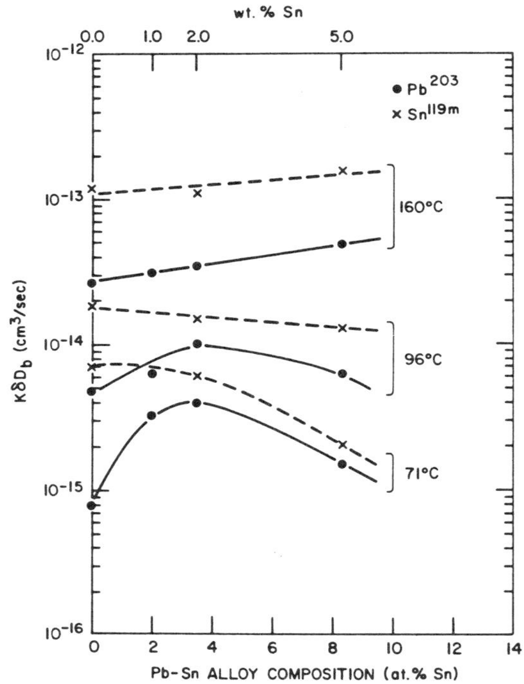

Fig. 29. Isothermal values of $K\delta D_b$ in polycrystalline Pb-Sn solder samples for the ^{203}Pb and ^{119m}Sn radiotracers. The $K\delta D_b$ values at 71, 96 and 160°C were measured at these temperatures (see Table 6B).

6.2 Composition Dependence of $K\delta D_b$

The composition dependence of $K\delta D_b$ results from the composition dependence of the individual factors K and D_b. The factor δ, the GB thickness, is assumed to remain constant and is taken of the order of atomic width which is borne out well in the recent literature.

Firstly, we discuss the composition dependence of the GB diffusivity D_b. In general, in an alloy it may be regarded as an increasing function of composition since lowering of Q_b on account of the melting temperature of the alloy is a strong effect which is discussed in Section 5.4. Some reduction in the pre-exponential factor, D_b^o, on account of the GB equilibrium segregation of the solute in an alloy which may allow mobile defects to be removed from the GB to the lattice would be trivial in comparison. Secondly, the segregation factor

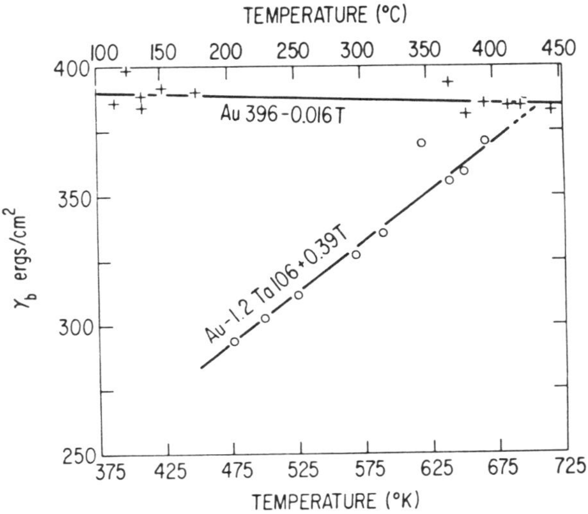

Fig. 30. Temperature dependence of the grain-boundary energies in Au (+) bulk and thin films and Au-1.2 at.% Ta alloy (o) determined from the self-diffusion (^{195}Au) measurements in the lattice and the grain boundaries. Note the change of slope in the latter.

K and the solute-GB binding energy H_i^B are also composition dependent. As solute content in an alloy increases, solute atoms do not remain isolated and binding energy decreases due to cluster formations. Indeed in very concentrated alloys, as the distinction between the solute and solvent atoms diminishes, the GB solute segregation may lose its meaning. Consequently, the segregation factor K is a decreasing function of the alloy composition. From the foregoing discussion, the composition dependence of the isothermal values of the combined $K\delta D_b$ shown in Figs. 27 and 28 can be easily understood.

The opposing behavior of D_b and K with respect to composition is seen more clearly in the Pb-Sn system (Fig. 28). At high temperature (160°C), the product $K\delta D_b$ increases for both the tracers primarily on account of D_b because the segregation effect is weak there due to extensive solubility and atomic mobility in the lattice as well as in the GB. The initial increase in the $K\delta D_b$ values at 71 and 96°C seen in Fig. 28 is the manifestation of the GB solute segregation in alloys

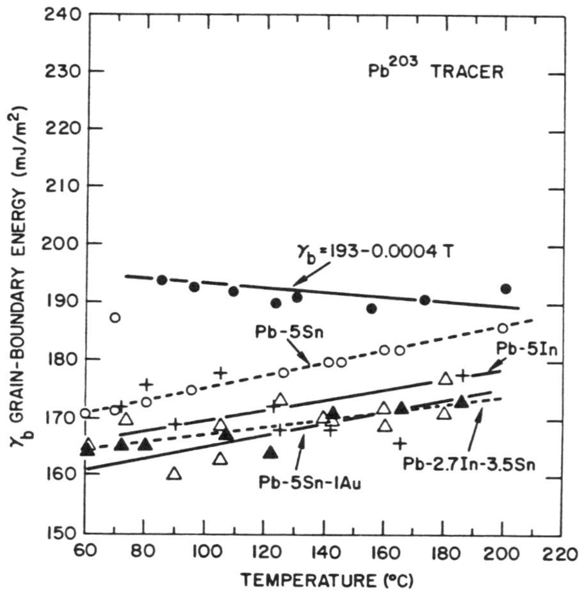

Fig. 31. Temperature dependence of the grain-boundary energies in the Pb, Pb-Sn, Pb-In and Pb-Sn-x alloys computer from ^{203}Pb diffusion data. The solute in all cases add to ~5 wt.%. See Table 6 for the source of the data.

having Sn content up to 4 at.%. However, as the phase boundary is approached in the case of the Pb-8.3 at% Sn alloy at 71 and 96°C, the value of the product $K\delta D_b$ declines. The reduction may occur in both terms due to the separation of the pro-eutectic Sn at the GBs; K would obviously decrease on this account. In addition D_b may also decrease due to the elimination of the mobile defects from the GBs alluded to earlier. Similar effects have been observed by Gas and Bernardini in the Ag-Sn alloys for ^{110}Ag and ^{113}Sn tracers (113,114) and by Hassner in Al-Zn alloys for the ^{65}Zn tracer (113).

6.3 Estimating Grain Boundary Segregation Parameters from Diffusion Measurements

In Section 5.5, a semi-empirical relationship of diffusion in the lattice and along the grain-boundaries with the absolute GB energy was discussed. The GB energies in pure metals obtained using Eq. 62 or 63 (see Table 5) were found to be in excellent agreement with those observed by other independent techniques. The effect of solute in lowering the GB energies in solids has been known for some time and the underlying thermodynamics bases have been discussed by Murr (116) and Hondros (117). Gupta and co-workers (112,118) have applied Eq. 62 to several alloy systems. The lowering of the GB energies in the Au-1.2 at% Ta and several Pb based alloys is shown in Figs. 30 and 31 respectively. It is to be noted that unlike the Arrhenius plot of the GB diffusivities shown in Fig. 25, the behavior of the GB energies in Fig. 30 is systematic and free from any cross-overs etc. The GB energies in the alloys show a smooth increase with temperature until they acquire values similar to that in pure metal which is attributed to elimination of the solute from GBs and dissolution into the adjoining lattice. This is, in fact, the Gibbs absorption problem. McLean (119) has derived the following expression for the equilibrium GB segregation in solids:

$$C_b \equiv C_O \exp(\Delta G_a/RT)/[1 + C_O \exp(\Delta G_a/RT], \qquad (67)$$

where C_b and C_O, as before, are the solute contents at the GB and in the lattice respectively, and ΔG_a is the associated free-energy associated with the solute-GB binding. Hondros (117) combined the concept of the Gibbsian energy lowering with Eq. 62 as follows:

$$\gamma_{bp} - \gamma_b = RT\phi \ell n[1 + C_o \exp(\Delta G_a/RT], \qquad (68)$$

where γ_{bp} is the GB energy of the pure solvent metal, and ϕ depends on the level of the GB saturation to be taken equal to unity. The free energy $\Delta G_a = \Delta E_a - T\Delta S_a$ where ΔE_a and ΔS_a are the corresponding enthalpy and entropy terms. It is to be noted that $\Delta E_a \sim H_i^B$ in Eq. (65) and $K = C_b/C_o$ as mentioned previously.

In Table 7, values of the GB solute binding energy, entropy and the relative size difference between the solute (γ_a) and the solvent (γ_o) atoms are listed. The classical Cu-Bi and Cu-Sb systems investigated by McLean (119) are also included for the sake of comparison. The binding energies, more or less, seem to scale with the relative atomic

TABLE 7

Values of Grain Boundary Solute Binding Energies, Entropies and Relative Atomic Size Difference for Pb, Cu and Au Based Alloys.

Alloy (wt%)	ΔH_a(KJ/mole)	ΔS_a(R units)	$(r_1-r_o)/r_1$
Pb-5In[a]	3.5	0.9	−0.05
Pb-5Sn[a]	16.0	3.4	−0.07
Pb-5Sn-1Au[a]	9.0	0.7	−0.07
Pb-2.7In-3.5Sn[a]	9.0	0.7	−0.05
Cu-Bi[b]	100	--	0.266
Cu-Sb[b]	65	--	0.194
Cu-1.0Cr[c]	84	4.0	0
Cu-0.13Ti[c]	63	1.3	0.13
Cu-0.08Zr[c]	65	1.6	0.20
Cu-S[b]	65	~0.5	0
Au-1.2Ta[d]	32	1.3	0

a. Gupta and J. Oberschmidt (112)
b. D. McLean, Grain Boundaries in Metals (119)
c. Barreau et al. (120)
d. Gupta (110)

size difference of the alloy constituent implying that elastic strain energy is a significant component of them. The entropies of binding of the order of R the gas constant and are reasonable. Furthermore, GB solute enrichment factors, C_b/C_o, obtained from Eq. (68) have been compared with those compiled by Hondros (121) in Fig. 32. It is seen that the enrichment factors obtained from diffusion measurements also follow the same general distribution as the data obtained from several direct techniques. Bernardini et al. (114) have shown the equivalence of the Sn GB segregation in the Fe-Sn system by comparing the results of direct technique of Auger Electron Spectroscopy with those obtained indirectly from GB and lattice diffusion measurements and the analyses discussed in this section.

6.4 Variable Effects of Solutes on Grain Boundary Diffusion

It is possible to explain and possibly predict the reported variable effects of the solute additions on the Gb diffusion from Eqs. (63) and (68) and taking $\phi = 1$. The GB solute content can be then written as

$$\left[\left(\frac{D_b^p}{D_b^a}\right)\left(\frac{D_\ell^a}{D_\ell^p}\right)\right]^{-1} = C_o \exp\left(\Delta G_a/RT\right), \qquad (69)$$

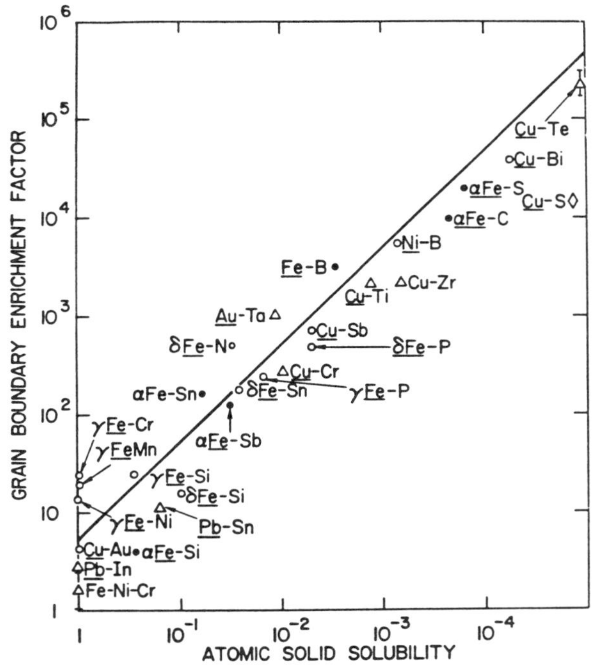

Fig. 32. Correlation of the grain-boundary solute enrichment factors with the atomic solid solubility at the temperatures of the experiments. Open circles are obtained by the interface energy approach, filled circles by Auger Electron Spectroscopy, open triangles by self-diffusion in the lattice and the grain-boundaries and open diamond by direct S diffusion (after Hondros: Ref. 121).

where the superscripts a and p refer to the alloy and the pure solvent respectively. It is obvious that the solute GB interaction free energy only partially explains the perturbation in the GB diffusivities. The changes in the lattice diffusion also need to be taken in account. Equation (69), in fact, contains all the three possibilities for solute effect on the GB diffusion seen in systems typical of the Au-1.2 at% Ta alloy shown in Fig. 25 namely the enhancement, suppression, or no effect.

For dilute alloys $C_0 \ll 1$, so that at high temperatures the right hand side (RHS) of Eq. (69) is sufficiently small and negligible.

Therefore

$$\left(\frac{D_b^p}{D_b^a}\right)\left(\frac{D_\ell^a}{D_\ell^p}\right) \sim 1 \; . \tag{70}$$

For alloys showing $D_\ell{}^a > D_\ell{}^p$, which is commonly observed, $D_b{}^p < D_b{}^a$ is required which is the case of the GB diffusion enhancement in the Au-1.2 at% Ta at high temperatures compared to pure Au.

At lower temperatures, on the other hand, despite $C_0 \ll 1$, the RHS of Eq. (69) $\gg 1$; and even though $D_\ell{}^a > D_\ell{}^p$, $D_b{}^p/D_b{}^a > 1$ would be required, thus resulting in *decreased* GB diffusion of the solvent due to solute addition. This effect is also seen in the Au-1.2 at% Ta system in Fig. 25. Between the two effects, of course, there would be a region where the solute will leave the GB diffusivity unaffected. It is possible to explain, at least qualitatively, the GB diffusion data in a large number of alloy systems, notably the Cu alloys (120) and Pb alloys (112) alluded to in this section, in this manner.

It will be very beneficial if it were possible to predict what solute in a given solvent would suppress GB diffusion for many technological applications such as diffusion barriers, electromigration in thin films, etc. A new concept based on the control of the structure and chemistry of the grain-boundaries is currently evolving in materials science which is being termed grain-boundary design (122). From the GB diffusion point of view, firstly one is required to guess the segregating systems. To this end, elastic strain energy based on the relative atomic size difference between the solute and solvent is a good starting point. Since elastic strain energy contributes only partially to the GB solute segregation effects, valency effects also need to be looked into. For instance, transition metal impurities in Al have greatly prolonged the electromigration life times as may be seen in Chapter 7 of this book. The shape of the liquidus in binary or ternary phase diagrams is another clue to the likelihood of GB solute segregation and possibility of reduction of GB diffusion. The phase diagrams which exhibit relatively sharp drop in the liquidus temperature with the increasing solute contents are considered the GB solute segregation-type. This criterion has often been used in zone-refining of metals. Gupta (123) has examined a large number of alloy systems which show variable degree of GB segregation and suppression of GB diffusion with attendant improvements in electromigration lifetimes. A correlation of over 80% is found with the above mentioned criterion of the steep nature of the liquidus.

7. CONCLUDING REMARKS

In this chapter some formal aspects of diffusion in bulk solids and thin films have been discussed. Considerable progress has been made in the techniques to measure smaller and smaller diffusion coefficients and also to understand complex materials such as thin films, amorphous alloys, man made superlattices etc. The technological need has been the prime mover in this effort. It is now possible to track only a few atomic jumps a day in diffusion measurements. This capability has led to good understanding of a myriad of diffusion phenomena. But much remains to be done. Understanding of diffusion processes in intermetallic systems currently is far from satisfactory. An important group of this class of materials is the silicides, where no systematic diffusion studies appear to have been made and yet technologically they are vital to the communication industry. The understanding of elemental and compound semiconductors also leaves much to be desired. Generally, the diffusion data along dislocations and grain boundaries in the intermetallic systems and semiconductors are sparse and usually imprecise.

REFERENCES

1. D. Gupta, D. R. Campbell and P. S. Ho, "Grain Boundary Diffusion," in *Thin Films-Interdiffusion and Reactions,* J. M. Poate, K. N. Tu and J. W. Mayer (Eds.), John Wiley (1978) p. 161.
2. W. C. Johnson and J. M. Blakely (Eds.) *Interfacial Segregation,* Amer. Soc. of Metals, Metals Park, Ohio (1979).
3. Proc. Symp. on *Grain-Boundary Structure and Kinetics,* Amer. Soc. of Metals, Metals Park, Ohio (1980).
4. *Proc. International Conf. on the Structure and Properties of Internal Interfaces,* J. de Phys. Colloq. C4, **46** (1985).
5. Proc. IV JIM International Symp. on *Grain Boundary Structure and Related Phenomena,* The Japan Institute of Metals, Sendai (1985).
6. P. B. Ghate, "Interconnections in VLSI," Phys. Today, **39**, 58 (1986).
7. M. K. Wu, J. R. Ashburn, C. J. Torng, P. H. Hor, R. L. Meng, L. Gao, Z. J. Huang, Y. Q. Wang and C. W. Chu, "Superconductivity at 93K in a new mixed-phase Y-Ba-Cu-O compounds system at ambient pressure," Phys. Rev. Lett., **58**, 908 (1987).

8. P. Chaudhari, R. H. Koch, R. B. Laibowitz, J. R. McGuire, R. J. Gambino, "Critical Current Measurements in Epitaxial Films of $YBa_2Cu_3O_{7-x}$," Phys. Rev. Lett. **58**, 2684 (1987).
9. A. S. Nowick and J. J. Burton (Eds.), *Diffusion in Solids – Recent Developments,* Academic Press (1975).
10. Graeme E. Murch and Arthur S. Nowick, *Diffusion in Crystalline Solids,* Academic Press (1984).
11. *Diffusion*, Amer. Soc., Metals, Metals Park, Ohio (1973).
12. H. S. Carslaw and J. C. Jaeger, *Conduction of Heat in Solids,* Oxford Univ. Press (1959).
13. J. Crank, *The Mathematics of Diffusion,* Oxford Univ. Press (1956).
14. P. G. Shewmon, *Diffusion in Solids,* McGraw-Hill Book Co. (1963).
15. R. D. Thompson, D. Gupta and K. N. Tu, Low-temperature diffusion and solubility of Ni in P-doped Czochralski-grown Si, *Phys. Rev. B* **33**, 2636 (1986).
16. S. E. Blum, M. B. Small and D. Gupta, Shallow Zn diffusion in liquid phase epitaxial GaAs and (GaAl)As at 600°C, *Appl. Phys. Lett.* **42**, 108 (1983).
17. D. R. Campbell, K. N. Tu and R. E. Robinson, Interdiffusion in bulk couple of Pb and Pb-50 wt.% In alloy, *Acta Metall.* **24**, 609 (1976).
18. D. Gupta, J. Oberschmidt, Solute and solvent diffusion in Concentrated Pb-In alloys, *Proc. International Symp. on Interface Migration and Control of Microstructure*, C. S. Pande, D. A. Smith, A. H. King and J. Walter (Eds.) Amer. Soc. Metals, Metals Park, Ohio (1986) p. 51.
19. W. K. Chu, S. S. Lau, J. W. Mayer, H. Müller and K. N. Tu, Implanted noble gas atoms as diffusion markers in silicide formation, *Thin Solid Films* **25**, 393 (1975).
20. H. T. G. Hentzell and K. N. Tu, Interdiffusion in Cu-Al thin film bilayers. II. Analysis of marker motion during sequential compound formation, *J. Appl. Phys.* **54**, 6929 (1983).
21. L. G. Harrison, Influence of dislocations on diffusion kinetics in solids with particular reference to the alkali halides, *Trans. Faraday Soc.* **57**, 1191 (1961).
22. E. W. Hart, On the role of dislocations in bulk diffusion, *Acta Metall.* **5**, 597 (1957).
23. A. Atkinson and R. I. Taylor, The diffusion of ^{63}Ni along grain boundaries in nickel oxide, *Philos. Mag. A* **43**, 979 (1981).
24. J. C. M. Hwang and R. W. Balluffi, Measurement of grain-boundary diffusion at low temperatures by the surface accumulation method I. Method and Analysis, *J. Appl. Phys.* **50**, 1339 (1979).

25. J. C. M. Hwang, J. D. Pan and R. W. Balluffi, Measurement of grain-boundary diffusion at low temperatures by the surface accumulation method II. Results for Au-Ag system, *J. Appl. Phys.* **50**, 1349 (1979).
26. J. C. Fisher, Calculation of diffusion penetration curve for surface and grain boundary diffusion, *J. Appl. Phys.* **22**, 74 (1951).
27. R. T. P. Whipple, Concentration contours in grain boundary diffusion, *Phil. Mag.* **45**, 1225 (1954).
28. T. Suzuoka, Lattice and grain boundary diffusion in polycrystals, *Trans. Jpn. Institute of Metall.* **2**, 25 (1961).
29. A. D. LeClaire, The analysis of grain boundary diffusion measurements, *Brit. J. Appl. Phys.* **14**, 351 (1963).
30. R. F. Cannon and J. P. Stark, Analysis of penetration data from grain boundary diffusion experiments, *J. Appl. Phys.* **40**, 4361 (1969).
31. G. H. Gilmer and H. H. Farrell, Grain-boundary diffusion in thin films I. The isolated grain boundary, *J. Appl. Phys.* **47**, 3792 (1976).
32. G. H. Gilmer and H. H. Farrell, Grain-boundary diffusion in thin films II. Multiple grain boundaries and surface diffusion, *J. Appl. Phys.* **47**, 4373 (1976).
33. D. Gupta, On the direct measurement of diffusion at temperature less than 0.5 T_m, *Thin Solid Films* **25**, 231 (1975).
34. P. F. Kane, and G. B. Larrabee (Eds.), *Auger Electron Spectroscopy*, Plenum Press, New York (1977).
35. A. Benning, J. Giber, J. László, M. Riedel and H. W. Werner (Eds.) *Secondary Ion Mass Spectroscopy SIMS III*, (Proc. III Internation Conf., Budapest, Hungary, 1981), Springer-Verlag (1982).
36. J. W. Mayer and J. M. Poate, Depth Profiling Techniques, Thin Films – Interdiffusion and Reactions, J. M. Poate, K. N. Tu, and J. W. Mayer (Eds.), John Wiley (1978) p. 119.
37. C. R. Houska, X-ray examination of diffused films, *Thin Solid Films,* **25**, 451 (1975).
38. D. Gupta and D. R. Campbell, Grain Boundary diffusion of ^{119m}Sn in Pb foils by nuclear absorption technique, *Philos. Mag. A* **42**, 513 (1980).
39. H. Mehrer, Measurement of Small diffusion Coefficients in Solids, *Solute-Defect Interaction: Theory and Experiment*, S. Saimoto, G. R. Purdy and G. V. Kidson (Eds.) Pergamon Press (1986) p. 162.
40. D. Gupta, Self-diffusion along dislocations ins single crystal Au films, *Phys. Rev.* **7**, 586 (1973).
41. D. Gupta and K. W. Asai, Grain-boundary self-diffusion in evaporated Au films at low temperatures, *Thin Solid Films* **22**, 121 (1974).

42. P. H. Holloway, D. E. Amos and G. C. Nelson, Analysis of grain-boundary diffusion in thin films: Cr in Au, *J. Appl. Phys.* **47**, 3769 (1976).
43. J. R. Manning, Theory of diffusion, in *Diffusion*, Amer. Soc. Metals, Metals Park, Ohio (1973) p. 1.
44. J. R. Manning, *Diffusion Kinetics for Atoms in Crystals,* Claredon Press, Oxford (1972).
45. R. W. Siegel, Vacancy concentration in metals, *J. Nuclear Maters.* **69&70**, 117 (1978).
46. N. L. Peterson, Self Diffusion in Pure Metals, *J. Nuclear Maters.* **69&70** (1978) 3-37.
47. R. W. Balluffi, Vacancy Defect Mobilities and Binding Energies Obtained from Annealing Studies, *J. Nuc. Mat.* **69&70** (1978) 240-263.
48. W. Wycisk and M. Feller-Kniepmeier, Quenching Experiments in High Purity Ni, *J. Nucl. Mat.* **69&70** (1978) 616-619.
49. W. C. Knodle and J. Koehler, The Quenching and Annealing of Pure Lead, *J. Nuc. Mat.* **69&70** (1978) 620-621.
50. R. A. Johnson, Interstitials and vacancies in α-iron, *Phys. Rev.* **134** (1964) 1329.
51. R. M. Nieminen, Energetics of Vacancies in Non-Transition Metals, *J. Nuc. Mat.* **69&70**, (1978) 633-635.
52. F. Dyment, Self and Solute Diffusion in Ti and Ti Alloys, in *Titanium 80* Proc IV Int. Conf. on Ti (Eds. H. Kimura and O. Izumi) The mat. Soc. of AIME (1980) p. 519.
53. W. Frank, U. Gösele, H. Mehrer and A. Seeger, Diffusion in Silicon and Germanium, in *Diffusion in Crystalline Solids,* Academic Press (1984) p. 63.
54. A. Seeger, Investigation of Point Defects in Si and Ge by Non-Irradiation Techniques, in *Radiation Effects in Semiconductor* (Eds., J. W. Corbett and G. D. Watkins), Gordon and Breach Sci. Pub (1971) p. 29.
55. P. J. Kelly, R. Car and S. T. Pantelides, Theoretical Determination of the Vacancy Migration Energy in Silicon, in *Defects in Semiconductors* (Ed. H. J. von Bardeleben) Materials Sci. Forum **10-12** (1986) p. 115-120.
56. H. Bakker, Tracer diffusion in concentrated alloys, in *Diffusion in Crystalline Solids,* G. E. Murch and A. S. Nowick (Eds.), Academic Press (1984) p. 189.
57. D. Gupta, D. Lazarus, and D. S. Lieberman, Self-diffusion in near-equiatomic ordered AuCd alloys, *Phys. Rev.* **153**, 863 (1967).
58. A. W. Lawson, Correlation of ΔS and ΔH in simple activated processes in solids, *J. Phys. Chem. Solids* **3**, 250 (1957).

59. R. W. Keyes, Continuum Models of the effect of pressure on activated processes, in *Solids under Pressure*, W. Paul, D. M. Warschauer (Eds.), McGraw Hill (1963) p. 71.
60. J. T. Robinson and N. L. Peterson, Correlation effects in grain boundary diffusion, *Surf. Sci.* **31**, 586 (1972).
61. G. Martin, D. A. Blackburn, and Y. Ada, Autodiffusion au joint de grains de bicristaux d'argen soumis à une pression hydrostatique, *Phys. Stat. Sol.* **23**, 223 (1967).
62. C. Herzig, H. Eckseller, W. Bussmann and D. Cardis, The temperature dependence of the isotope effect for self-diffusion and Co impurity-diffusion in Au, *J. Nuc. Maters.* **69&70**, 61 (1978).
63. M. D. Feit, Dynamical theory of diffusion II Comparison with rate theory and impurity isotope effect, *Phys. Rev.* **B5**, 2145 (1972).
64. J. N. Mundy, Effect of pressure on the isotope effect in Na self-diffusion, *Phys. Rev.* **B3**, 2431 (1971).
65. D. R. Campbell, Isotope effect for self-diffusion in Ge, *Phys. Rev.* **B12**, 2318 (1975).
66. S. A. Rice and N. H. Nachtrieb, Dynamical theory of diffusion in crystals II, *J. Chem. Phys.* **31**, 139 (1959).
67. S. J. Rothman, The measurement of tracer diffusion-coefficients in solids, in *Diffusion in Crystalline Solids*, G. E. Murch and A. S. Nowick (Eds.) Academic Press (1984) p. 2.
68. H. B. Huntington, Electro and Thermomigration in Metals, in *Diffusion*, Amer. Soc. Metals, Metals Park, Ohio (1973) p. 155.
69. K. N. Tu, Interdiffusion in Thin Films, in *Annual Rev. Mater. Sci.* **15**, 147 (1985).
70. L. S. Darken, Diffusion, mobility and their interrelation through free in binary metallic systems, *Trans. AIME*, **175**, 184 (1948).
71. L. S. Darken, Formal basis of diffusion theory, in *Atom Movements*, Amer. Soc. Metals,Cleveland (1951) p. 21.
72. H. E. Cook and J. E. Hilliard, Effect of gradient energy on diffusion in Au-Ag alloys, *J. Appl. Phys.* **40**, 2191 (1969).
73. M. P. Rosenblum, F. Spaepen and D. Turnbull, Duffusion and structural relaxation in compositionally modulated amorphous metal films, *Appl. Phys. Lett.* **37**, 184 (1980).
74. A. L. Greer, F. Spaepen, in *Synthetic Modulated Structure Materials,* L. Chang and B. C. Giessen (Eds.) Academic Press (1985) p.419.
75. J. W. Cahn and J. E. Hilliard, Free energy of a nonuniform system, I Interfacial free energy, *J. Chem. Phys.* **28**, 258 (1958).
76. R. D. Doherty, Diffusive phase transformation in the solid state, in *Physical Metallurgy*, R. W. Cahn and P. Haasen (Eds.) North-Holland Physics Pub. Co (1983) 933.

77. J. W. Cahn, Spinodal decomposition, *Trans, Met. Soc. AIME*, **242**, 166 (1968).
78. J. W. Cahn, On Spinodal decomposition, *Acta Metall.* **9**, 795 (1961).
79. H. E. Cook, D. de Fontaine and J. E. Hilliard, A model for diffusion on cubic lattices and its application to the early stages of ordering, *Acta Metall.* **17**, 765 (1969).
80. J. E. Hilliard, in *Phase Transformations*, Amer. Soc. Metals, Metals Park, Ohio (1970) p. 497.
81. R. W. Balluffi, Grain boundary diffusion mechanisms in metal, in *Diffusion in Crystalline Solids,* Graeme E. Murch and Arthur S. Nowick (Eds.) Academic Press (1984) p. 319.
82. W. Krakow and D. A. Smith, The observation of various grain boundary atomic structure in Au by high-resolution electron microscopy, *J. Mater. Res.* **1**, 1 (1986).
83. W. Krakow, J. T. Wetzel and D. A. Smith, Observation and interpretation of the atomic structure of a $\Sigma = 19/[110][3\bar{3}1]$ tilt boundary in Au thin films, *Philos. Mag. A* **53**, 739 (1986).
84. W. T. Read and W. Shockley, Dislocation models of crystal grain boundaries, *Phys. Rev.* **78**, 275 (1950).
85. G. Hasson, J.-Y. Boos, I. Herbeuval, M. Biscondi and C. Goux, Theoretical and experimental determinations of grain boundary structures and energies: correlations with various experimental results, *Surf. Sci.* **31**, 115 (1972).
86. T. Kwok, P. S. Ho, S. Yip, R. W. Balluffi, P. D. Bristowe and A. Brokman, Evidence for vacancy mechanism in grain boundary diffusion in BCC iron: a molecular-dynamics study, *Phys. Rev. Lett.* **47**, 1148 (1981).
87. A. Brokman and R. W. Balluffi, Coincidence lattice model for the structure and energy of grain boundaries, *Acta Metall.* **29**, 1703 (1981).
88. D. Turnbull and R. E. Hoffman, The effect of relative crystal and boundary orientations on grain boundary diffusion rates, *Acta Metall.* **2**, 419 (1954).
89. S. R. L. Couling and R. Smoluchowski, Anisotropy of diffusion in grain boundaries, *J. Appl. Phys.* **25**, 1538 (1954).
90. W. Lange and M. Jurisch, unpublished.
91. I. Herbeuval, M. Biscondi and C. Goux, *Mém. Sci. Rev. Metall.*, **70**, 39 (1973).
92. N. A. Gjostein, Short circuit diffusion in *Diffusion*, Amer. Soc. Metals, Metals Park, Ohio (1973) p. 241.
93. D. Gupta and R. Rosenberg, Grain boundary self-diffusion in Au and Au alloys, in *Grain Boundaries in Engineering Materials,* J. L. Walter, J. H. Westbrook, and D. A. Woodford (Eds.) Claitor's Pub. Co., Baton Rouge (1975) p. 181.

94. D. Gupta and P. S. Ho, Diffusion processes in thin films, *Thin Solid Films* **72**, 399 (1980).
95. P. V. Pavlov, V. A. Panteleev and A. V. Maiorov, Diffusion of Antimony Along Dislocations in Silicon, *Sov. Solid State* **6**, 305 (1964).
96. P. V. Pavlov, L. V. Lainer, V. A. Sterkhov and V. A. Panteleev, Proof of Existence of Self-Sustaining Diffusion flow Along Isolated Dislocation, *Sov. Solid State* **8** 580 (1966).
97. D. R. Campbell, K. N. Tu, and R. O. Schwenker, Diffusion of Arsenic along dislocations in Epitaxial Silicon Films, *Thin Solid Films* **25**, 213 (1975).
98. P. Fahrentholtz, and J. Mimkes, Diffusion of Sb along Dislocations in Si, *Phys. Stat. Solidi* **B78**, K137 (1976).
99. J. L. Liotard, R. Biberian, and J. Cabane, La diffusion intergranulaire dans le silicium, *J. de Phys.* **43**, C1-213 (1982).
100. F. H. M. Spit and H. Bakker, Diffusion of Donor Elements (^{125}Sb, ^{32}P and $^{74/73}$As) in Polycrystalline Silicon, *Phys. Stat. Solidi* **A93**, 135 (1986).
101. V. T. Borisov, V. M. Golikov and G. V. Scherbedinskiy, Relationship between diffusion coefficients and grain-boundary energy, *Phys. Metals Metallogr.* **17**, 80 (1964).
102. A. P. Greenough and R. King, *J. Inst. Met.* **79**, 415 (1951).
103. J. E. Hilliard, M. Cohen and B. L. Averback, Grain boundary energies in Au-Cu alloys, *Acta Metall.* **8**, 26 (1960).
104. E. R. Hayward and A. P. Greenough, *J. Inst. Meta.* **88**, 217 (1959).
105. K. T. Aust and B. Chalmers, Surface energy and structure of crystal boundaries in metals, *Proc. Roy. Soc.* A **204**, 359 (1950).
106. I. Ames, F. d'Heurle and R. Horstmann, Reduction of electromigration in aluminum films by copper doping, *IBM J. Res. and Dev.* **14**, 461 (1970).
107. J. K. Howard and R. F. Ross, Hillocks as structural markers for electromigration rate measurements in thin films, *J. Appl. Phys.* **18**, 344 (1971).
108. H. Gleiter and B. Chalmers, Grain boundary diffusion, in *Prog. in Mater. Sci.* **16**, 113 (1972).
109. G. B. Gibbs, Grain boundary impurity diffusion, *Phys. Stat. Solidi* **16**, K27 (1966).
110. D. Gupta and R. Rosenberg, Effect of solute addition (Ta) on low temperature self-diffusion processes in Au, *Thin Solid Films* **25**, 171 (1975).
111. J. Oberschmidt, K. K. Kim and D. Gupta Grain boundary diffusion in some Pb-Sn alloys, *J. Appl. Phys.* **53**, 5672 (1982).

112. D. Gupta and J. Oberschmidt, Grain-boundary diffusion and solute segregation effects in several Pb-In and Pb-Sn-X concentrated alloys, in *Diffusion in Solids: recent developments,* M. A. Dayananda, G. E. Murch (Eds.) The Met. Soc. of AIME (1985) p. 121.
113. D. Gupta, Grain boundary diffusion and influence of solutes in several FCC alloys, in *Proc. IV JIM International Symp. on Grain Boundary Structure and Related Phenomena,* The Japan Institute of Metals, Sendai (1985) p. 449
114. J. Bernardini, P. Gas, E. D. Hondros and M. P. Seah, The role of solute segregation in grain boundary diffusion, *Proc. R. Soc. Lond.* A **379**, 159 (1982).
115. N. L. Peterson, Grain boundary diffusion in metals, *International Metals Review* **28**, 65 (1983).
116. L. E. Murr, in *Interfacial Phenomena in Metals and Alloys,* Addison-Wesley Pub. 6, Reading, MA (1975) p. 33.
117. E. D. Hondros and M. P. Seah, Interfacial and surface, in *Physical Metall.* R. W. Cahn and P. Hassen (Eds.) North-Holland Physics Pub. (1983) p. 855.
118. D. Gupta, Influence of solute segregation on grain-boundary energy and self-diffusion, *Metall. Trans.* **8A**, 1431 (1977).
119. D. McLean, *Grain Boundaries in Metals* Oxford Univ. Press., London (1957) p. 116.
120. G. Barreau, G. Brunel, G. Cizeron and P. Lacombe, *Mem. Sci. Rev. Metall.* **68**, 357 (1971).
121. E. D. Hondros, Grain boundary segregation: the current situation and future requirements, *J. de Phys.* **36**, Coll-C4, 117 (1975).
122. T. Watanabe, Grain boundary design for new materials, in *Proc. IV Int. Symp. on Grain Boundary Structure and Related Phenomena,* The Japan Inst. of Metals, Sendai (1985) p. 73.
123. D. Gupta, Solute Segregation to grain-boundaries and bulk phase diagrams, in Proc. of *Thin Film Phenomena* – Interfaces and Interactions, J. E. Baglin and J. M. Poate (Eds.), Electro Chem. Soc. **78-2**, 498 (1978).

2

Anelastic Relaxation and Diffusion in Thin-Layer Materials

Brian S. Berry

1. INTRODUCTION

One of the most fascinating aspects of the mechanical behavior of solids is that certain structural features and kinetic processes occurring on an atomic scale can be studied in a sensitive and selective manner by measurements which detect small deviations from ideally elastic behavior. This chapter summarizes progress in the application of this approach to evaporated thin films and other thin-layer materials. The primary emphasis is on the use of mechanical damping or internal friction measurements to detect anelastic relaxation behavior associated with thermally-activated atomic movements. Since many readers interested in thin-layer materials may be unfamiliar with this branch of mechanical behavior, Section 2 includes enough introductory material to make the chapter reasonably self-contained and readable by the non-specialist.

It will become apparent that anelastic measurements provide a window on the world of atomic movements in solids that overlaps but is distinct from that provided by profiling techniques. Most anelastic relaxations are of a very short-range nature, and often involve quite specific atomic jumps. To mention the first and most famous example, it is now almost 50 years since J. L. Snoek explained an internal friction peak in α-iron with a simple model involving jumps of carbon or nitrogen atoms between adjacent interstitial sites. With hindsight, the Snoek effect can be seen as the first indication that anelastic measurements could be of value in the study of atomic movements in thin films, since the short-range nature of the relaxation mechanism implies that there is no theoretical limitation on the thickness of the sample employed. As events turned out, however, several decades elapsed before the first preliminary attempt to observe a Snoek peak in a thin film was reported. By way of a brief historical perspective, we may note that the development of anelasticity as a distinct discipline with broad application to the study of defects and atomic movements did not occur until after the discovery of the Snoek relaxation. The subject entered a period of rapid development following the appearance of Zener's pioneering monograph (1) in 1948, and reached considerable maturity by the early seventies (2)-(5). Up to that time, however, virtually all experimental work was performed with conventional bulk samples, frequently in the form of single crystals. It is only in the last ten years or so that the feasibility and general value of the application of anelastic measurements to thin-layer materials has been convincingly demonstrated.

This chapter falls into three main parts. Section 2 contains an introductory formal description of anelastic behavior that serves as a foundation for later discussion of mechanisms and results. Section 3 deals with the vibrating-reed and bending-cantilever methods that have so far dominated anelastic studies of thin layers. Experimental results and some specific mechanisms of relaxation are covered in Sections 4-9. A preliminary glance at these later sections should be helpful to the reader who is interested in obtaining a quick overall impression of the subject.

2. ANELASTIC RELAXATION BEHAVIOR

Certain types of mechanical stress-strain behavior, notably elasticity and plasticity, are familiar to us through everyday experience. For example, we are aware that the humble paper clip is intended to function simply as an elastic spring, but can fall victim to its ability to undergo plastic deformation. Viscoelasticity is another category of mechanical behavior where we have no difficulty anticipating the general nature of the subject, namely stress-strain behavior that includes both elastic and time-dependent nonelastic contributions to the total strain. On the other hand, the meaning of anelasticity and the nature of anelastic behavior are not self-evident. The term *anelasticity* was introduced by C. Zener to denote the special and restricted category of viscoelasticity produced by mechanisms with the character of stress-induced *relaxation* processes. Such processes arise whenever stress and strain are coupled thermodynamically to other internal variables, whose adjustment to new stress-induced equilibrium levels is rate-limited by a kinetic process such as diffusion.

Under a constant stress, the external manifestation of an anelastic relaxation process is a time-dependent nonelastic component of strain (the anelastic strain), superimposed on the initial elastic strain. The anelastic strain has the specific characteristics of *(a)* a unique equilibrium value for each level of applied stress, and *(b)* linearity, which for present purposes can be simply taken to mean a proportional relationship between stress and the equilibrium anelastic strain. This proportionality, when combined with that between stress and the initial elastic strain (Hooke's law) means that anelastic behavior is also characterized by a proportional relationship between stress and the total strain under fully relaxed or equilibrium conditions. It also follows that, upon unloading, an anelastic solid eventually exhibits a total recoverability of its original form. From these characteristics, it is possible to deduce the general nature of the *response functions* which describe the behavior of an anelastic material to various prescribed histories of stress or strain. As we shall see below, these include limited and recoverable creep, limited stress relaxation, and (in vibration) frequency-dependent behavior of both the internal friction and the dynamic modulus. It should be realized that all of these phenomena are quantitatively interrelated because each one merely represents a different aspect of the same internal relaxation process. Consequently, a material which shows any one of these phenomena must necessarily show all of them.

As noted above, an anelastic strain by definition possesses the characteristics of *(a)* a unique equilibrium relationship with the applied stress and *(b)* linearity. There is however no formal restriction on the form of the relaxation kinetics, and it is precisely in this area that many complexities can arise. On the other hand, the simplest and most important example of anelastic behavior is obtained when, in addition to *(a)* and *(b)*, a third condition *(c)* is added to restrict the relaxation kinetics to the simplest possible type, namely homogeneous first-order kinetics involving only a single rate-constant or relaxation time. A model solid which obeys all three of these conditions is said to be a *standard anelastic solid*. The behavior of such a solid is of prime importance and will dominate the remainder of this section.

2.1 Quasi-Static Response Functions

For an anelastic solid to be in *complete* static equilibrium, both stress and strain must assume constant values. This means, of course, that all the time derivatives of both stress and strain must become zero. The simplest case of *quasi-static* relaxation behavior arises when an anelastic solid is thrown out of equilibrium by a sudden step-like change in one of these variables, which is then held constant while attention is focused on the transient response of the conjugate variable. Quasi-static anelastic behavior is thus exemplified by strain relaxation at constant stress, or stress relaxation at constant strain. The strain relaxation produced by the application and eventual removal of a constant stress σ_0 is illustrated schematically in Fig. 1. In accord with the defining postulates of anelasticity, the immediate or unrelaxed elastic strain ε_U is followed by the anelastic strain ε_a which develops with time to a proportional equilibrium value $\bar{\varepsilon}_a$. The total or relaxed strain ε_R is thus $\varepsilon_U+\bar{\varepsilon}_a$. Upon subsequent removal of the stress, the immediate elastic springback is followed by a complementary creep recovery or *elastic aftereffect*, which eventually brings about total recoverability. By virtue of linearity, the ratio ε/σ_0 is independent of stress and for $t > 0$ defines a response function, the *creep function J(t)*, which varies with time t between the limits of an initial or *unrelaxed compliance* $J_U \equiv \varepsilon_U/\sigma_0$ and a final *relaxed compliance* $J_R \equiv \varepsilon_R/\sigma_0$. From Fig. 1 it is evident that

$$\delta J = J_R - J_U = (\varepsilon_R - \varepsilon_U)/\sigma_0 = \bar{\varepsilon}_a/\sigma_0, \tag{1}$$

where δJ is known as the *relaxation of the compliance.* An alternative dimensionless measure of the size of the relaxation is given by the *relaxation strength* Δ_A, defined as the ratio of the equilibrium anelastic strain $\bar{\varepsilon}_a$ and the equilibrium elastic strain $\bar{\varepsilon}_e$:

$$\Delta_A \equiv \bar{\varepsilon}_a/\bar{\varepsilon}_e. \tag{2}$$

For the case of constant stress $\bar{\varepsilon}_e = \varepsilon_U$, and

$$\Delta_A = \bar{\varepsilon}_a/\varepsilon_U = \delta J/J_U = (J_R - J_U)/J_U. \tag{3}$$

For the general anelastic solid, the form of $J(t)$ cannot be specified explicitly, since there is nothing in the general definition of anelasticity that restricts the form of the relaxation kinetics. On the other hand, the special case of the standard anelastic solid is by definition obtained with the added restriction of simple first-order kinetics. Here, the rate

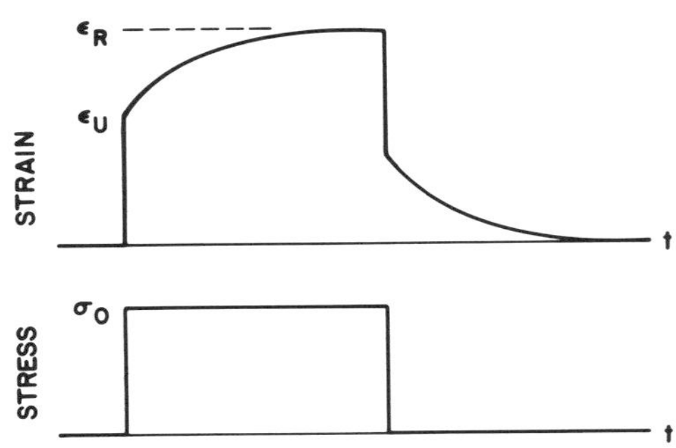

Fig. 1. Anelastic strain relaxation during the application and subsequent removal of the constant stress σ_0.

equation for the anelastic strain takes the form

$$\dot{\varepsilon}_a = -(\varepsilon_a - \bar{\varepsilon}_a)/\tau_\sigma, \tag{4}$$

where the dot notation denotes differentiation with respect to time and τ_σ represents the relaxation time for constant stress. Upon combining Eqs. (1) and (4) with the general relationship

$$\varepsilon = J_U\sigma + \varepsilon_a, \tag{5}$$

we obtain the differential stress-strain equation of the standard anelastic solid for constant stress:

$$\varepsilon + \tau_\sigma\dot{\varepsilon} = J_R\sigma_0, \tag{6}$$

whose solution is

$$J(t) \equiv \varepsilon(t)/\sigma_0 = J_U + \delta J(1 - \exp(-t/\tau_\sigma)). \tag{7}$$

The converse case of anelastic stress relaxation under constant strain is shown in Fig. 2. Here the imposed fixed strain ε_0 produces an initial unrelaxed stress σ_U which relaxes to an equilibrium value σ_R. For completeness, Fig. 2 also shows the behavior that occurs when, following relaxation under ε_0, the strain on the sample is suddenly returned to a constant zero value by superposition of the strain $-\varepsilon_0$. (A fuller discussion of superposition, including the important Boltzmann principle, can be found elsewhere (4)). As required by the unique equilibrium stress-strain relationship, the stress in Fig. 2 eventually returns to zero, after having undergone a reversal of sign.

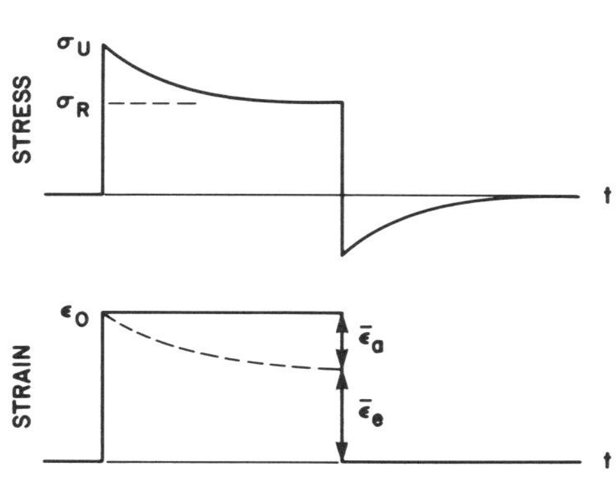

Fig. 2. Anelastic stress relaxation during the application and subsequent removal of the constant strain ε_0.

Returning to the interval of constant strain ε_0 in Fig. 2, the ratio σ/ε_0 defines a *stress relaxation function M(t)* which varies between the limits of an *unrelaxed modulus* $M_U \equiv \sigma_U/\varepsilon_0$ and a relaxed modulus $M_R \equiv \sigma_R/\varepsilon_0$. The *relaxation of the modulus*, δM, is given by

$$\delta M = M_U - M_R = (\sigma_U - \sigma_R)/\varepsilon_0, \tag{8}$$

and since $M_U = J_U^{-1}$ and $M_R = J_R^{-1}$, we also find that

$$\Delta_A = (J_R - J_U)/J_U = (M_U - M_R)/M_R. \tag{9}$$

It should be evident that stress relaxation reflects the conversion of the strain ε_0 from an initial wholly elastic character to a mixture of elastic and anelastic strains, as indicated by the broken curve in the lower section of Fig. 2. At equilibrium, we may write $\varepsilon_0 = \bar{\varepsilon}_e + \bar{\varepsilon}_a$, and think of the stress σ_R either as $M_R \varepsilon_0$ or as $M_U \bar{\varepsilon}_e$. Hence, we have

$$M_U/M_R = \varepsilon_0/\bar{\varepsilon}_e, \tag{10}$$

and thus find in a somewhat more insightful manner that

$$\Delta_A \equiv \bar{\varepsilon}_a/\bar{\varepsilon}_e = (\varepsilon_0/\bar{\varepsilon}_e) - 1 = (M_U - M_R)/M_R, \tag{11}$$

in agreement with Eq. (9).

As for the creep function, there is no general explicit form for the stress relaxation function $M(t)$. The special case of the standard anelastic solid is however obtained with the further restriction to first-order kinetics. The corresponding rate equation is

$$\dot{\sigma} = -(\sigma - \sigma_R)/\tau_\varepsilon, \tag{12}$$

where τ_ε is the relaxation time at constant strain. With the substitution $\sigma_R = M_R \varepsilon_0$, Eq. (12) becomes the differential stress-strain equation of the standard anelastic solid for constant strain:

$$\sigma + \tau_\varepsilon \dot{\sigma} = M_R \varepsilon_0, \tag{13}$$

whose solution is the simple decay function

$$M(t) = M_R + \delta M \exp(-t/\tau_\varepsilon). \tag{14}$$

To find the relation between τ_σ and τ_ε, we must realize that the conditions specified by a constant stress and a constant strain experiment only begin to diverge for t > 0. The initial step change can be thought of as producing an initial elastic strain ε_U and an initial relaxation rate $\dot{\varepsilon}_\sigma(t \to 0)$ or, equivalently, as an initial stress σ_U and an initial relaxation rate $-\dot{\sigma}_\varepsilon(t \to 0)$. Since both are alternative descriptions of

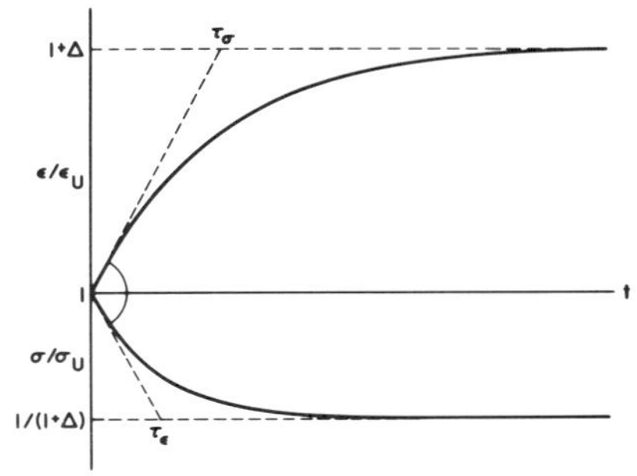

Fig. 3. Comparison of anelastic strain relaxation at constant stress with stress relaxation at constant strain. The initial slopes of both curves are equal and opposite, but the total change in σ/σ_U is less than in $\varepsilon/\varepsilon_U$ by the factor $1/(1+\Delta_A)$, and τ_ε is less than τ_σ by the same factor. The figure is drawn for the case $\Delta_A = 1$, $\tau_\sigma/\tau_\varepsilon = 2$.

the same initial response, it follows that, in general:

$$\dot{\varepsilon}_\sigma(t \to 0)/\varepsilon_U = -\dot{\sigma}_\varepsilon(t \to 0)/\sigma_U. \tag{15}$$

Applying Eq. (15) to the specific case of the standard anelastic solid, we find with the help of Eqs. (7) and (14) that

$$\tau_\sigma/\tau_\varepsilon = 1 + \Delta_A. \tag{16}$$

A useful graphical representation of these considerations is shown in Fig. 3. It can be seen that the result expressed by Eq. (16) is linked to the fact that, for the same starting conditions, the anelastic strain produced by creep is larger by the factor $1+\Delta_A$ than that produced by stress relaxation.

It is helpful to consider now some actual magnitudes for the relaxation strength Δ_A. In contrast to schematic diagrams such as Figs. 1 or 2, where for purposes of clarity the anelastic strain is shown as a substantial fraction of the elastic strain, most relaxations which provide information on the kinetics of atom movements have relaxation strengths below 0.1, and many are in the range 10^{-2} to 10^{-4}. In the latter cases the distinction between τ_σ and τ_ε is insignificant and the factor of greatest practical concern is that the anelastic strain is then so small that its direct measurement is either difficult or totally impractical. In such cases we may be solely dependent on the superior resolution afforded by internal friction measurements, discussed below.

2.2. Dynamic Anelastic Behavior: The Internal Friction Peak

Dynamic behavior in its most general form is obtained by lifting the restrictions of either constant stress or constant strain, so that both σ and ε exhibit non-vanishing time derivatives. In practice, the dynamic situation of prime importance is that in which stress and strain are simple harmonic functions of time, as obtained by setting a sample into vibration with some circular frequency ω. If we represent the stress by the real part of the rotating complex vector $\sigma_0 \exp(i\omega t)$, the presence of relaxation introduces a lag ϕ in the corresponding strain response $\varepsilon_0 \exp[i(\omega t - \phi)]$. Such a lag introduces a net expenditure of work over each stress-strain cycle, and an attendant damping of the motion which is said to arise from *internal friction*. From the viewpoint of formal theory, the most natural dimensionless measure of internal friction is provided by the tangent of the loss angle, tan ϕ. To see this, it is only necessary to write the strain in the equivalent forms

$$\varepsilon_0 \exp[i(\omega t - \phi)] = (\varepsilon_1 - i\varepsilon_2) \exp(i\omega t), \tag{17}$$

where ε_1 and ε_2 represent the components of strain that are in-phase and in quadrature with the stress. From Eq. (17), we have

$$\varepsilon_0 = (\varepsilon_1^2 + \varepsilon_2^2)^{1/2} \tag{18}$$

and

$$\tan\phi = \varepsilon_2/\varepsilon_1. \tag{19}$$

Two reasons why internal friction is pre-eminent as a sensitive detector of anelastic behavior can be inferred from Eq. (19). One is that, by virtue of linearity, tan ϕ is a dimensionless ratio that is independent of the absolute level of strain imposed on the sample. Unlike the case for quasi-static measurements, there is thus no loss of sensitivity associated with the use of small strain amplitudes. The other reason is that tan ϕ depends in a directly proportional manner on a component of strain that is entirely of anelastic origin. By contrast, relaxation produces only an incremental change in the creep and stress-relaxation functions $J(t)$ and $M(t)$, as is also the case for the dynamic compliance $J^*(\omega)$, which may be written

$$J^*(\omega) \equiv \varepsilon_0 e^{-i\phi}/\sigma_0 = (\varepsilon_1 - i\varepsilon_2)/\sigma_0 = J_1(\omega) - iJ_2(\omega). \tag{20}$$

The low and high-frequency limits of $J^*(\omega)$ are $J^*(0) = J_R$ and $J^*(\infty) = J_U$. Since these are both real quantities, it is evident that $J_2(\omega)$ vanishes at both these limits, as does $\tan\phi = J_2(\omega)/J_1(\omega)$. Although we know nothing in general about the frequency dependence of tan ϕ, it is evident from this result that tan ϕ must exhibit at least one intermediate maximum, or internal friction peak, as a function of ω. The form of this peak for the specific case of the standard anelastic solid can be obtained by inserting harmonic forms for σ and ε into the appropriate differential stress-strain equation. To obtain this equation when σ is not a constant, Eq. (6) must be generalized to

$$\dot{\varepsilon} = \dot{\varepsilon}_e + \dot{\varepsilon}_a = J_U\dot{\sigma} - (\varepsilon_a - \bar{\varepsilon}_a)/\tau_\sigma, \tag{21}$$

where ε_a and $\bar{\varepsilon}_a$ have the instantaneous values $\varepsilon_a = \varepsilon - J_U\sigma$ and $\bar{\varepsilon}_a = \delta J\sigma$. We thus find the standard anelastic solid to be represented by the differential stress-strain equation

$$\varepsilon + \tau_\sigma \dot{\varepsilon} = J_R\sigma + \tau_\sigma J_U \dot{\sigma}. \tag{22}$$

Upon inserting $\sigma = \sigma_0 \exp(i\omega t)$ and $\varepsilon = \varepsilon_0 \exp[i(\omega t - \phi)]$, we find

$$J_1(\omega) = J_U + \{\delta J/(1 + \omega^2\tau_\sigma^2)\} \tag{23}$$

and

$$J_2(\omega) = \delta J\omega\tau_\sigma/(1 + \omega^2\tau_\sigma^2). \tag{24}$$

Because of their similarity to the case of dielectric relaxation, Eqs. (23) and (24) are often referred to as the Debye equations of anelasticity. When plotted against log $\omega\tau_\sigma$, $J_1(\omega)$ is antisymmetric about the mean value $J_1(\omega) = J_U + (\delta J/2)$ for log $\omega\tau_\sigma = 0$, whereas J_2 is a symmetric bell-shaped function known as a Debye peak. At first sight, it may appear that if $J_2(\omega)$ is a Debye peak, the internal friction tan $\phi = J_2(\omega)/J_1(\omega)$ would be a Debye peak only for the limit of a small relaxation strength where $J_1 \doteq J_U$. Interestingly, however, it can be quickly shown by an appropriate change of variables that the general form of tan ϕ is a Debye peak, with a peak maximum shifted to the position $\omega\tau = 1$, where τ is a dynamic relaxation time given by the geometric mean of τ_σ and τ_ε:

$$\tau \equiv (\tau_\sigma\tau_\varepsilon)^{1/2}. \tag{25}$$

From Eqs. (16) and (25) we find $\tau = \tau_\sigma/(1+\Delta_A)^{1/2}$, and hence obtain from Eqs. (23) and (24) the following general result for the internal friction peak of the standard anelastic solid:

$$\tan\phi = \frac{\Delta_A}{(1+\Delta_A)^{1/2}} \frac{\omega\tau}{1+\omega^2\tau^2}. \tag{26}$$

When $\Delta_A \ll 1$, we may ignore the difference between τ_σ, τ_ε and τ. Moreover, the dynamic compliance J_1 becomes the reciprocal of the dynamic modulus M_1. Under these conditions the dynamic behavior of the standard anelastic solid is represented by the following simple expressions:

$$\tan\phi \doteq \Delta_A\omega\tau/(1 + \omega^2\tau^2) \tag{27}$$

$$J_1 \doteq J_U + \{\delta J/(1 + \omega^2\tau^2)\} \tag{28}$$

$$M_1 \doteq M_U - \{\delta M/(1 + \omega^2\tau^2)\}, \tag{29}$$

which are illustrated graphically in Fig. 4. It is to be noted that a variation of several orders of magnitude in the compound variable $\omega\tau$ is required to trace out the peak. For future reference, it is quickly found from Eq. (27) that the full width of the peak at half-maximum is given by:

$$\Delta(\log \omega\tau) = \log\{(2 + 3^{1/2})/(2 - 3^{1/2})\} = 1.1439. \tag{30}$$

In practice, internal friction measurements rarely involve a direct determination of the loss angle ϕ, although this can be done in subresonant forced vibration when the damping is large (4). Much more commonly, however, we are concerned with relatively low levels of internal friction (typically $\phi < 1°$), and with samples excited into resonance in some natural vibrational mode. Most frequently, the internal friction is then measured in terms of the logarithmic decrement of free decay, δ. Less commonly, use may be made of the bandwidth of the resonance curve in forced vibration to obtain the Q^{-1}-value, or of the fractional energy input per cycle $\Delta W/W$, required to maintain a constant amplitude of vibration. These different methods of internal friction measurement are illustrated schematically in Fig. 5. When the damping is small and amplitude-independent, the relations between the different measures of internal friction are (4):

$$\tan\phi = \delta/\pi = Q^{-1} = \Delta W/2\pi W. \tag{31}$$

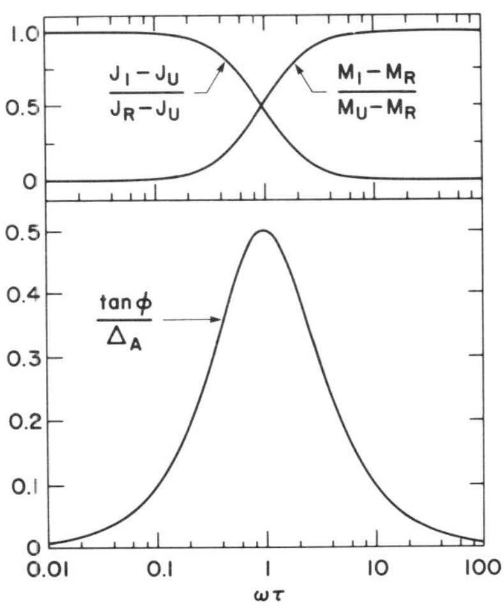

Fig. 4. Plots of Eqs. (27),(28) and (29), representing the Debye equations for a standard anelastic solid of small relaxation strength ($\Delta_A \ll 1$).

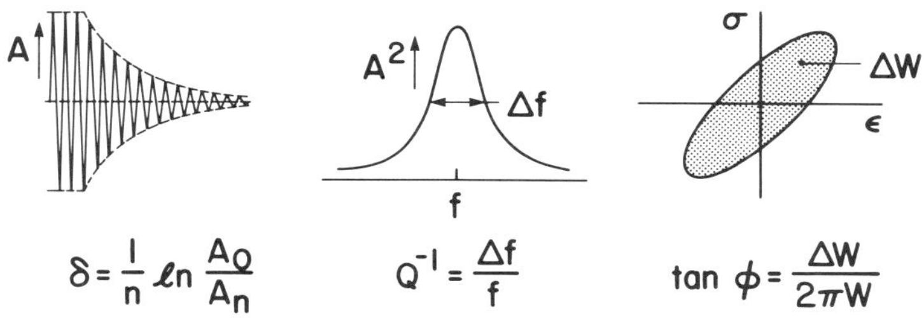

Fig. 5. Measures of internal friction. The logarithmic decrement δ is determined from the rate of decay of the amplitude A in free vibration; the Q^{-1} value is related to the width of the resonance curve in forced vibration, and the loss-angle ϕ of the stress-strain loop is related to the fractional loss of mechanical energy per cycle, $\Delta W/W$.

2.3 Anelastic Behavior with Multiple or Distributed Relaxation Times

We have previously emphasized that the definition of anelastic behavior places no restriction on the form of the relaxation kinetics. It should therefore be no surprise that Debye-type relaxation behavior, corresponding to a single or unique relaxation time, is a limiting ideal that is approached in practice only under particularly simple circumstances. Most internal friction peaks exhibit some degree of broadening, and the quasi-static response functions are correspondingly smeared out over wider intervals of time. It is often physically appealing to attribute such behavior to the existence of more than one relaxation time. In some cases, it may appear appropriate to consider a discrete spectrum, consisting of a set of values $\tau^{(i)}$ with corresponding magnitudes $\delta J^{(i)}$. In other cases, we may adopt a continuous distribution function $X(\ell n\tau)$, where $X(\ell n\tau)d\ell n\tau$ represents the relaxation magnitude associated with relaxation times in the interval $d\ell n\tau$. In place of Eq. (27) the internal friction is then represented either by the summation

$$\tan \phi \doteq \sum_i \delta J^{(i)} \omega\tau^{(i)} / \{1 + (\omega\tau^{(i)})^2\} \tag{32}$$

where

$$\delta J = \sum_i \delta J^{(i)}, \tag{33}$$

or by the integral

$$\tan \phi \doteq \int_{-\infty}^{+\infty} X(\ell n\tau)\{\omega\tau/[1 + (\omega\tau)^2]\} d\ell n\tau \tag{34}$$

where
$$\delta J = \int_{-\infty}^{+\infty} X(\ell n\tau) d\ell n\tau. \tag{35}$$

Analogous expressions are easily written down for the other response functions, such as the dynamic modulus or creep function. The choice of $\ell n\tau$ for the distributed variable in Eqs. (34) and (35) follows from the symmetry of the Debye peak when plotted against $\ell n\omega\tau$ (c.f. Fig. 4). It can be appreciated that a distribution function which is symmetrically peaked when plotted against $\ell n\tau$ will also produce a symmetrically-broadened internal friction peak. A particular example of such a distribution is provided by the lognormal distribution function

$$X(\ell nz) = (\delta J/\beta\pi^{1/2}) \exp(-(z/\beta)^2), \tag{36}$$

where β is the distribution parameter defining the width of the distribution, and $z \equiv \ell n(\tau/\bar{\tau})$ where $\bar{\tau}$ is the most probable relaxation time. Detailed computations of the form of the internal friction peak corresponding to Eqs. (34) and (36) have been tabulated by Nowick and Berry (4)(6). In practice, it can be very difficult to determine whether a given response function is more accurately represented by a discrete or a continuous distribution of relaxation times. In cases where the experimental data is of high precision and has been taken at close intervals, it can be of interest to employ Cost's Direct Spectrum Analysis method (7)(8) as an alternative to the adoption of a pre-conceived choice for the form of the distribution.

2.4 Thermally-Activated Relaxation Behavior

We have seen that a Debye internal friction peak spans more than two decades in the compound variable $\omega\tau$, and that an even greater range is needed when a distribution of relaxation times is involved. In principle, we may observe any relaxation peak by varying ω while keeping τ constant. However, the range of frequencies that can be realized experimentally is usually totally inadequate for the implementation of this approach. Fortunately, the relaxations of interest in this chapter are rate-controlled by thermally-activated atom movements, with the consequence that τ is strongly temperature dependent. This provides the opportunity to observe the peak at constant ω by variation of the temperature, an approach which is exploited whenever possible in practice. To start with the simplest case, we consider a relaxation that involves only a single type of jump, whose rate w is given by the standard rate expression

$$w = w_0 \exp -H/kT, \tag{37}$$

where w_0 is a composite prefactor containing the atomic vibration frequency ν_0 and other terms, H is the activation energy (enthalpy) of the jump, k is Boltzmann's constant (8.61735×10^{-5} eV/K) and T is the absolute temperature. The rate of relaxation can usually be written as

$$\tau^{-1} = bw, \tag{38}$$

where b is a parameter determined by specific details of the relaxation mechanism. Combining Eqs. (37) and (38), we have

$$\tau^{-1} = \tau_0^{-1} \exp -H/kT \quad \text{or} \quad \tau = \tau_0 \exp H/kT, \qquad (39a;39b)$$

where

$$\tau_0^{-1} \equiv bw_0 \quad \text{or} \quad \tau_0 = 1/bw_0. \qquad (40a;40b)$$

Debye-type relaxation behavior is obtained when both τ_0 and H are discrete single-valued quantities, and the function $\omega\tau/(1+\omega^2\tau^2)$ is then a symmetrical peak function when plotted against either $\ell n\omega\tau$ or $1/T$. Inserting Eq. (39b) into Eq. (27) we may write the Debye peak in the form

$$\tan \phi = (\Delta_A/2)\text{sech}\frac{H}{k}\left(\frac{1}{T} - \frac{1}{T_p}\right), \tag{41}$$

where T_p is the temperature of the peak maximum at which $\omega\tau = 1$. On the $1/T$ scale, the peak width of half-maximum $\Delta(T^{-1})$ is found from Eq. (30) to be

$$\Delta(T^{-1}) = 2.6339k/H. \tag{42}$$

An example of a Debye peak corresponding to Eq. (41) is shown in Fig. 6. A logarithmic ordinate scale has been included to bring out the

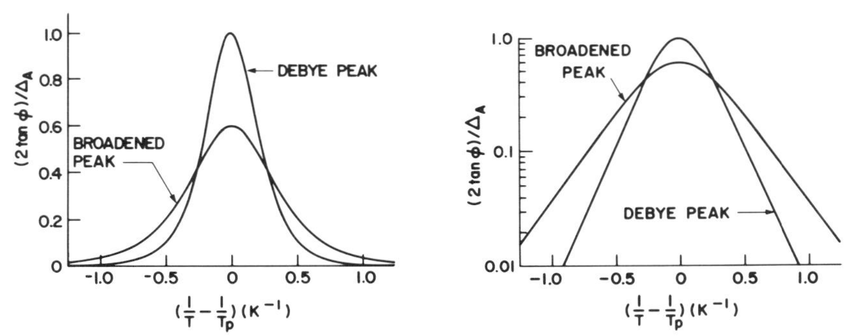

Fig. 6. An illustrative Debye peak calculated from Eq. (41) for the case H = 0.5 eV, compared with a broadened peak of the same strength and a relative width of 1.7. *Left:* Linear ordinate scale; *right:* Logarithmic ordinate scale.

point that the two sides of the peak are then linear, with equal and opposite slopes determined by the activation energy H, which also controls the peak width via Eq. (42). Figure 6 also shows for comparison a peak which involves a distribution of relaxation times. The peak height is reduced, the peak is broadened, and the sides become less steep. The erroneous application of Eq. (42) to this non-Debye peak produces a false or apparent activation energy which is less than the actual value. This comparison illustrates the fact that internal friction measurements made at only one frequency do not by themselves provide enough information for a rigorous analysis of the relaxation kinetics. However, if the relaxation strength Δ_A is large enough to be measured independently through the relaxation of the dynamic modulus, the peak can be tested for Debye behavior by conformity to the relationship $[\tan\phi]_{T_p} = \Delta_A/2(1+\Delta_A)^{1/2}$, which follows from Eq. (26).

In general, the most satisfactory and informative procedure for the determination of the activation energy is to observe the peak for a series of different frequencies, covering as wide a range as possible. As illustrated in Fig. 7, the basic effect of a change of frequency is to shift the peak position on the $1/T$ scale. Since the condition $\omega\tau = 1$ (or, more generally, $\omega\bar{\tau} = 1$) is satisfied at the temperature T_p of each peak maximum, such data enable the construction of a rate plot (log τ vs. $1/T_p$) whose slope and intercept provide the values of H and τ_0 (or, more generally, $\bar{H}$ and $\bar{\tau}_0$).

The dominant factor involved in the peak shift with frequency is that the Debye function $\omega\tau/(1+\omega^2\tau^2)$ depends only on the product $\omega\tau$. Consequently, equivalent points on two peaks measured at frequencies ω_1 and ω_2 must satisfy the condition $\omega_1\tau_1 = \omega_2\tau_2$, which leads, via Eq. (39b), to the result

$$T_1^{-1} - T_2^{-1} = (k/H)\ell n(\omega_2/\omega_1) \tag{43}$$

for the shift between the peaks. If H is a discrete single-valued quantity, this shift is the same for any pair of equivalent points, with the expected consequence that the peak should move bodily along the $1/T$

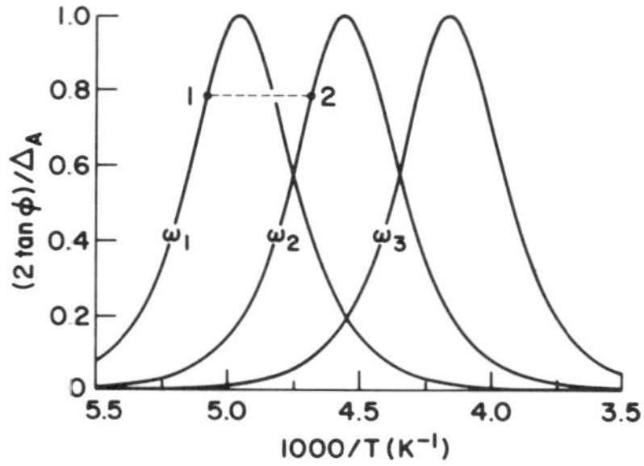

Fig. 7. The shift of a thermally-activated Debye peak with a change of measurement frequency, ω.

axis, without change of size or shape. (In reality, other factors, such as the temperature dependence of Δ_A, require consideration.) In most cases involving non-Debye behavior, however, the dominant source of peak broadening involves a distribution of activation energies. The width of the distribution of relaxation times is then temperature dependent, and the shape and height of the peak change gradually as the peak moves along the 1/T-axis.

So far, our discussion of thermally-activated behavior has only involved internal friction measurements. Usually, for reasons of sensitivity, this represents the preferred approach to relaxation studies of thin-layer materials. Nevertheless, it should be borne in mind that quasi-static measurements may provide a valuable alternative or supplement in some cases. In contrast to internal friction measurements, quasi-static measurements are performed isothermally. The relaxation time is determined from the time required to achieve some appropriate fraction of the total response, or from the initial slope of the normalized response function. The determination of the parameters H and τ_0 is accomplished in the usual manner from a rate-plot based on data obtained at a series of different temperatures.

In assessing the applicability of relaxation measurements as a tool for measurements of atomic mobility in thin-layer materials, it is useful to consider the range of activation energies that conveniently fall within the scope of presently available experimental methods. For this purpose, we shall focus on short-range relaxation processes, for which the prefactor τ_0 can be assigned a typical magnitude of 10^{-14}s. As discussed in Section 3, internal friction measurements on thin layers are presently performed by a vibrating-reed technique with a typical frequency range of 20 Hz to 5 kHz, corresponding to τ-values of approximately 10^{-2} to 3×10^{-5}s. Quasi-static measurements, on the other hand, provide an experimental window in the range of 10 to 5000 s. Figure 8 shows that, over a temperature range of 100-1000K, these techniques are potentially suitable for the study of relaxation processes with activation energies in the range 0.25 to 2 eV.

Finally, we should note that, in a few special cases, the jump w that produces a short-range relaxation is also the jump involved in long-range diffusion. If there are z_j different directions in which the jump w can occur from a given site (so that the total jump rate is $z_j w$), the diffusion coefficient D, obtained from the theory of random walk, can be written

$$D = z_j w a_j^2/6, \tag{44}$$

where a_j is the mean jump distance. Combining Eqs. (38) and (44), we obtain

$$D = z_j a_j^2/6b\tau \tag{45}$$

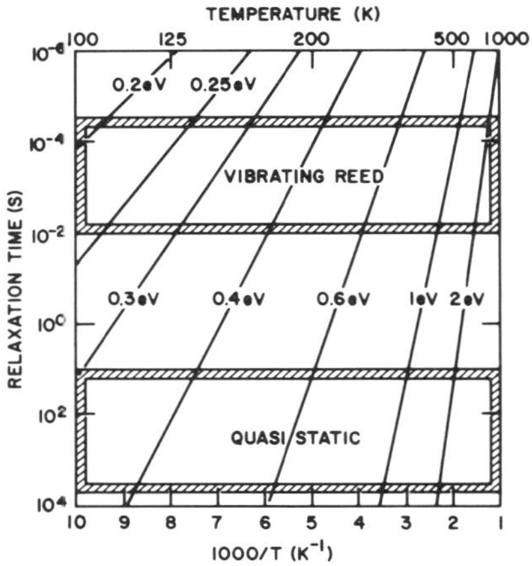

Fig. 8. The experimental windows (framed areas) provided by the vibrating-reed and quasi-static methods for observation of relaxations governed by various activation energies and a common prefactor of 10^{-14} s.

as the relation between the long-range diffusion coefficient and the short-range relaxation time τ. A particular case of Eq. (45) will be considered in Section 4.

3. EXPERIMENTAL METHODS

A specimen in the form of a thin cantilevered strip, exercized in bending either by internal or external forces, is a simple but remarkably versatile approach to the investigation of the mechanical behavior of thin-layer materials. To date, almost all internal friction studies of the anelastic behavior of thin layers have utilized the dynamic bending-cantilever or vibrating-reed method described in Section 3.1. In a few cases, this method has been applied to self-supporting single layers down to thicknesses of about 5μm. More commonly, however, the reed has the form of a multilayered composite in which the thin layers of interest are supported by a high-Q (low damping) substrate of about 50μm thickness. Provided the background loss in the substrate is not objectionably large, the use of a composite reed offers a number of advantages, and is the only approach which has so far been successfully applied to layers of submicron thickness.

Although not competitive with the vibrating reed in terms of sensitivity, isothermal quasi-static measurements of either relaxation or recovery can be a most useful experimental approach for some investigations. Bending-cantilever methods of this type are considered in Section 3.2. Finally, since complementary measurements of the internal accommodation stress in a supported layer can be a useful adjunct to relaxation studies performed under external loading, a simple and compatible form of the well-known bending-bilayer method for internal stress measurements is covered in Section 3.3.

Despite the present dominance of bending-cantilever methods, we may also take note here of other experimental approaches that may be useful for some investigations. For relatively thick polymer layers, there is the low-frequency torsional-braid technique developed by Gillham (9). High-frequency ultrasonic absorption measurements on detached silver and gold films have been reported by Uozumi and his co-workers (10)(11). Because of the widespread use of circular wafers as thin-film substrates, it is also relevant to mention the apparatus of Bordoni et al. (12) for dynamical measurements on disc-shaped samples excited into flexural vibration. Finally, high-Q torsional oscillators fabricated from silicon wafers (13) may also prove to be useful substrates for a variety of future studies related for example to oxidation, film deposition, or silicide reactions.

3.1 Vibrating-Reed Apparatus

The application of the vibrating-reed arrangement to the study of thin layers has attracted the attention of a number of workers (14)-(20). Apart from an obvious geometric compatibility with planar thin layers, a vibrating reed provides *(a)* relatively low (audio) frequencies that are frequently the most suitable for the investigation of thermally-activated processes, *(b)* the ability to excite a succession of overtones (Fig. 9) without remounting the sample and *(c)* an advantageous weighting of the intrinsic losses that occur in the near-surface regions of the reed. However, the successful realization of these advantages requires careful attention to apparatus design, in order to secure a low level of apparatus loss and a reliable method for safely mounting and handling the delicate samples. A refined vibrating-reed apparatus, designed expressly for internal friction measurements on thin films, was described by Berry and Pritchet in 1975 and has since proved useful in a continuing series of investigations. In this apparatus (16)(17), sample strips typically 2 cm to 4 cm in length and about 0.3 cm in width are attached to a supporting base pedestal so as to form a

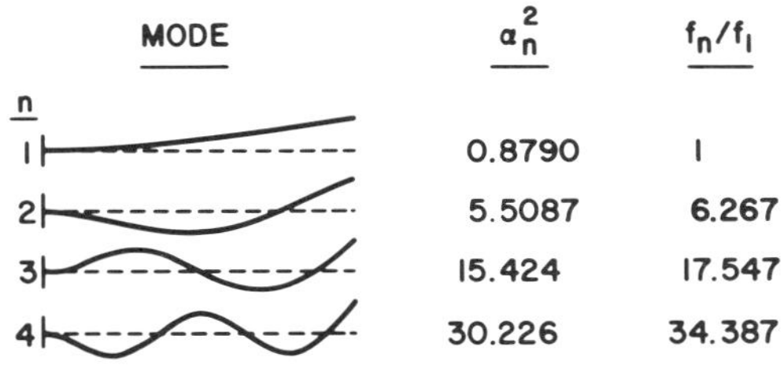

Fig. 9. Mode shapes and related parameters for a uniform reed in flexural vibration.

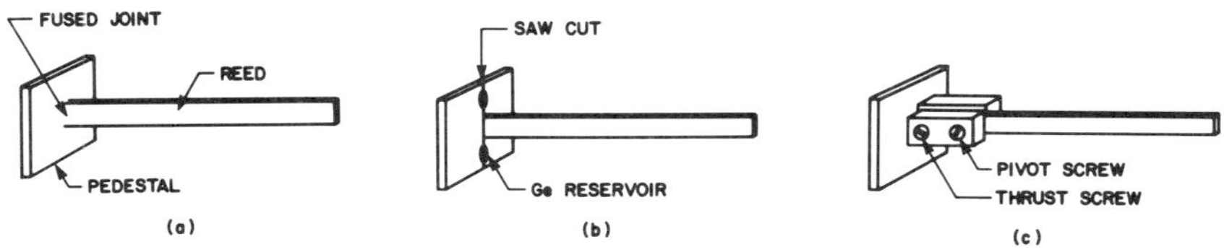

Fig. 10. Examples of pedestal-mounted reeds: (*a*) SiO_2 reed fused to a SiO_2 pedestal, (*b*) Si reed bonded to a Si pedestal with Ge used as a high-temperature solder, (*c*) preferred design of a mechanical clamp.

permanent vibrating-reed unit of the type shown in Fig. 10. Such units are easily handled without touching the reed, and can be repeatedly inserted and removed from the apparatus without affecting the measured frequencies or damping. Since both fused silica and single-crystal silicon have wide usage as substrate materials in microelectronic devices, it is a fortunate circumstance that both materials can be fabricated into high-Q ($>10^5$) vibrating-reed units by the methods shown in Figs. 10a and 10b. Over a wide range of temperature, the internal friction exhibited by such reeds is dominated by a calculable thermoelastic loss (4), which is particularly small in fused silica because of the low thermal expansion coefficient (Fig. 11). It should be noted, however, that below 100K the internal friction of fused silica increases sharply due to the presence of a strong intrinsic relaxation (21), with the consequence that silicon substrates become preferred for use at low temperatures.

For low-loss materials, such as Si or SiO_2, the use of an integral bond to the mounting pedestal is essential to avoid a high background due to rubbing friction. This degree of refinement may not be necessary for relatively high-loss materials such as metals, and in those cases a well-engineered mechanical clamp is often a suitable alternative (Fig. 10c). To reduce mechanical clamping losses, various more elaborate

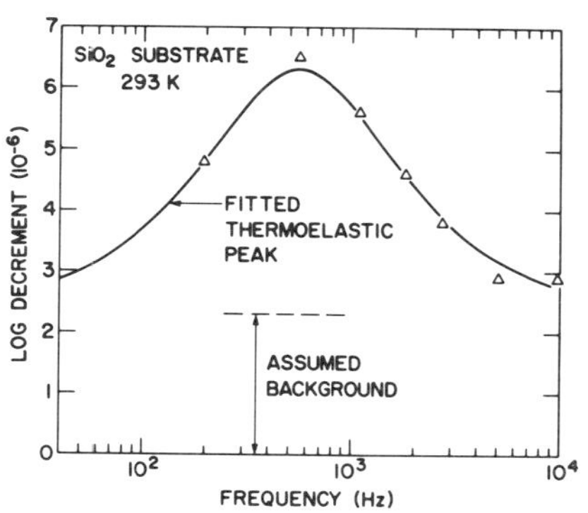

Fig. 11. Room temperature internal friction of a 50 μm fused silica reed, as measured at a series of overtones. From Ref. 17.

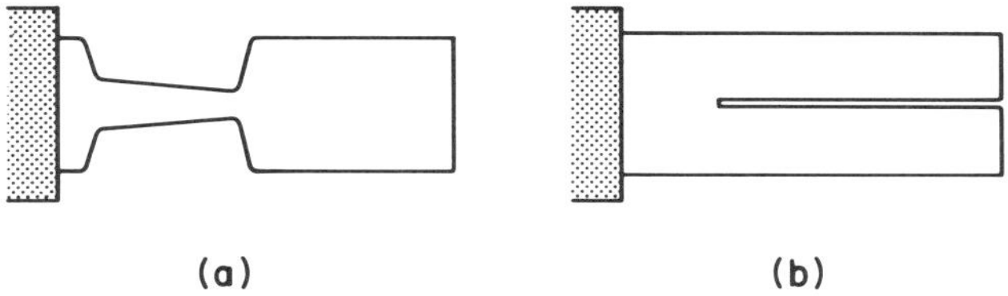

Fig. 12. Special vibrating-reed configurations. *Left:* Tapered gauge section with enlarged ends (Ref. 24); *right:* Modified tuning fork (Ref. 23).

reed configurations of the types shown in Fig. 12 have been employed by a number of different workers (22)-(24).

A sample-holder suitable for low-loss electrostatic excitation and detection of the natural modes of a reed is shown in Fig. 13. The pedestal base of the reed is secured in position by spring contacts that also serve to ground the reed. Two pairs of electrodes, mounted on tubular alumina insulators, are positioned around the reed with gap distances of typically 0.5 mm to 1 mm. The end electrodes are used to excite the sample via the push-pull electrostatic forces produced by application of a polarized ac voltage, as indicated in Fig. 14. The detection electrodes are usually located about one-third of the sample length from the fixed end, so as to avoid a nodal position for any of the tones of interest. These electrodes, which are dc biased from the intervening grounded reed by about 180 volts, provide detection in the manner of a differential condenser microphone. The arrangement is operated by tuning the drive oscillator to a selected mode frequency while observing the oscilloscope display of the filtered output voltage. Measurement of the logarithmic decrement is initiated by disconnect-

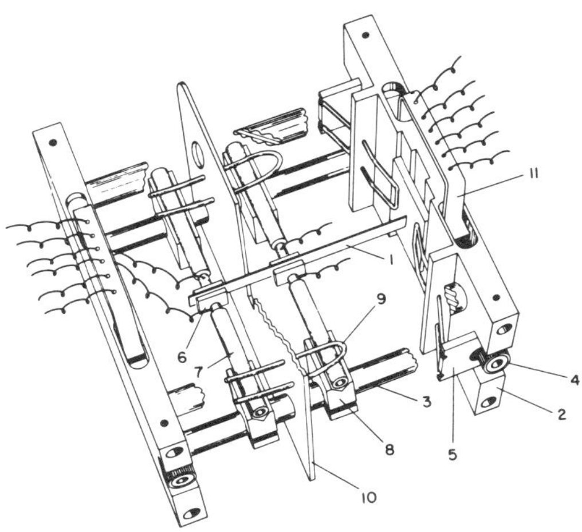

Fig. 13. Vibrating-reed sample holder (Ref. 16). 1 – reed, 2 – endplate, 3 – tierod, 4 – release screw, 5 – tension plate, 6 – electrode, 7 – electrode support, 8 – V-block support, 9 – spring clip, 10 – screen, 11 – terminal strip.

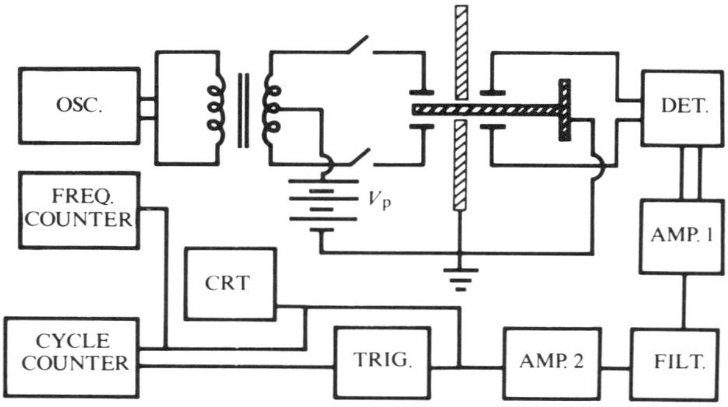

Fig. 14. Block diagram of the vibrating-reed drive and detection circuits. From Ref. 16.

ing the drive signal, thereby allowing the vibration to fall into free decay. Trigger circuits are used to control a counter which displays the number of cycles n in a half-amplitude decay, thereby permitting the calculation of the logarithmic decrement δ from the relationship $\delta = (\ell n2)/n = 0.693/n$.

Although designed initially for conducting (or semiconducting) samples, it has been found that the electrostatic drive and detection system described above can also be used with insulating samples, following a preparatory activation treatment which produces frozen-in charges on the reed (17). We also note in passing that a simple modification of the positions of the electrodes permits measurements to be made in the torsional, as well as the flexural modes (17).

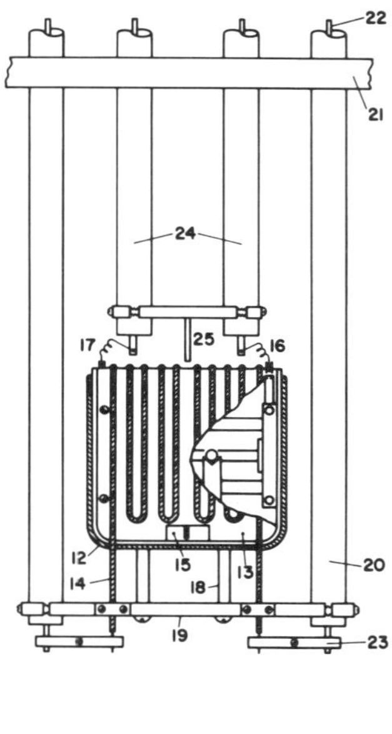

Fig. 15. Vibrating-reed heater and support assembly. 12,13 – interlocking heater sections, 14 – sheathed heater wire, 15 – pumpout access, 16,17,22 – electrical leads, 18 – heater stand offs, 19 – baseplate, 20 – support tubes (4), 21 – vacuum flange, 23 – heater clamps, 24,25 – screening tubes and plate. The cutaway in the heater reveals part of Fig. 13 in side-view.

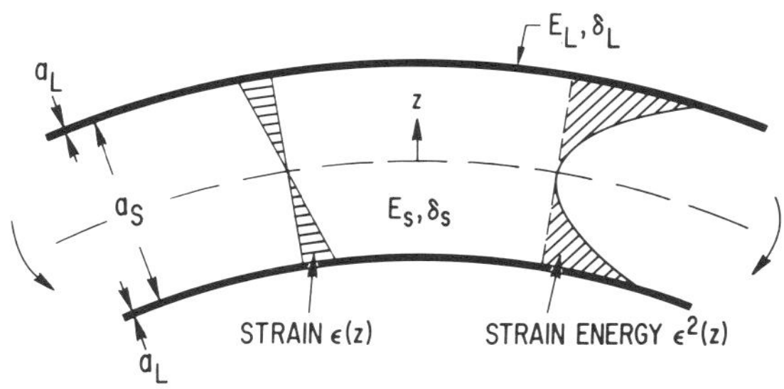

Fig. 16. The symmetrical trilayer with $a_L/a_S < 1$.

As shown in Fig. 15, the sample holder of Fig. 13 is assembled into a heater enclosure in the form of two interlocking U-sections. The heater is connected via tubular stand-offs to a stainless-steel platform, which is in turn connected to the main vacuum flange by stainless steel tubes terminating in standard electrical feed-throughs. Operation in vacuum is required not only to protect samples from possible contamination, but also to remove the air damping which can otherwise completely override the internal friction of a high-Q reed. The vacuum chamber may be cooled with liquid nitrogen to provide an operating range of 100K to 800K, and must be protected from external disturbances by an antivibration mounting (16).

As noted at the beginning of Section 3, the vibrating-reed method may be used for the investigation of single-layer samples and multilayer composites. The frequency and internal friction of a multilayer are discussed in some generality in the Appendix; however, we may remark immediately that in most cases the preferred multilayer configuration for internal friction studies is the symmetrical trilayer of Fig. 16, consisting of identical layers of thickness a_L on both sides of a substrate of thickness a_S. There are several reasons for this choice. First, it improves the sensitivity of the measurements relative to that for a single layer of the same thickness. Secondly, the symmetrical trilayer is not susceptible to the thermal curling behavior characteristic of the bilayer and other asymmetrical multilayer configurations. This has the important practical advantage that there is less tendency for accidental contact between the reed and the electrodes during a run. A final advantage, which is also a consequence of symmetry, is that it is possible to give simple general expressions for both the internal friction δ_C and frequencies f_C of the symmetrical trilayer without restriction to the thin-layer limit $a_L/a_S \ll 1$. The relations are

$$\delta_C = \left(\frac{1}{1+\gamma_{ST}}\right)\delta_S + \left(\frac{\gamma_{ST}}{1+\gamma_{ST}}\right)\delta_L \tag{46}$$

and

$$f_C^2 = \left(\frac{1+\gamma_{ST}}{1+\chi_{ST}}\right)f_S^2, \tag{47}$$

where

$$\gamma_{ST} = \{[1 + (2a_L/a_S)]^3 - 1\}(E_L/E_S), \tag{48}$$

$$\chi_{ST} = 2a_L\rho_L/a_S\rho_S, \tag{49}$$

and, from Eqs. (A11) and (A15):

$$f_S^2 = \frac{1}{3}\left(\frac{\alpha_n^2 a_S}{\pi \ell^2}\right)^2 \frac{E_S}{\rho_S}. \tag{50}$$

In these expressions, δ_S, E_S and ρ_S denote the internal friction, Young's modulus and density of the substrate, and similar meanings attach to δ_L, E_L and ρ_L for the surface layers. Values of the mode constant α_n are included in Fig. 5.

3.2 The Mandrel Method for Quasi-Static Measurements

Just as for the case of dynamical measurements, a clamped-free strip exercized as a bending cantilever is the most obvious choice of sample configuration for quasi-static measurements on thin-layer materials. We describe below a particular implementation of this approach, namely the *mandrel method* of Berry and Pritchet (25), which has proved useful in a number of applications on both single and multilayer samples. This method, illustrated in Fig. 17, is so-named because a cylindrical former or mandrel (Fig. 17a) is used to subject the sample to a fixed uniform curvature R_W^{-1}. The choice of the 'wind-up' radius R_W is usually based on a consideration of the maximum strain ε_{max} which can be employed without causing immediate plastic deformation of the sample. For a single layer of thickness a_L, we have $\varepsilon_{max} = a_L/2R_W$. Thus, for a material which can sustain a pre-yield strain of about 10^{-3}, a mandrel of about 1 cm radius is a convenient choice. After release from the mandrel, the residual curvature R^{-1} exhibited at some later instant may be found from the corresponding deflection ξ, measured along a line perpendicular to the initial straight portion of the strip and located at a distance ℓ from the clamped end. From the geometry shown in Fig. 17a, we have

$$R^{-1} = 2\xi/(\xi^2 + \ell^2). \tag{51}$$

The first phase of a mandrel test, during which the sample is held against the mandrel by a spring clip or push-rod, corresponds to stress relaxation at constant strain. In the terminology of the bending cantilever, this phase may also be described as the relaxation, at constant curvature, of the bending moment B or the bending stiffness $K \equiv B/R^{-1}$, as shown in Figs. 17c and 17d. Since the initial windup is presumed to take place rapidly under unrelaxed conditions, the slope of the windup line in Fig. 17c corresponds to the unrelaxed bending

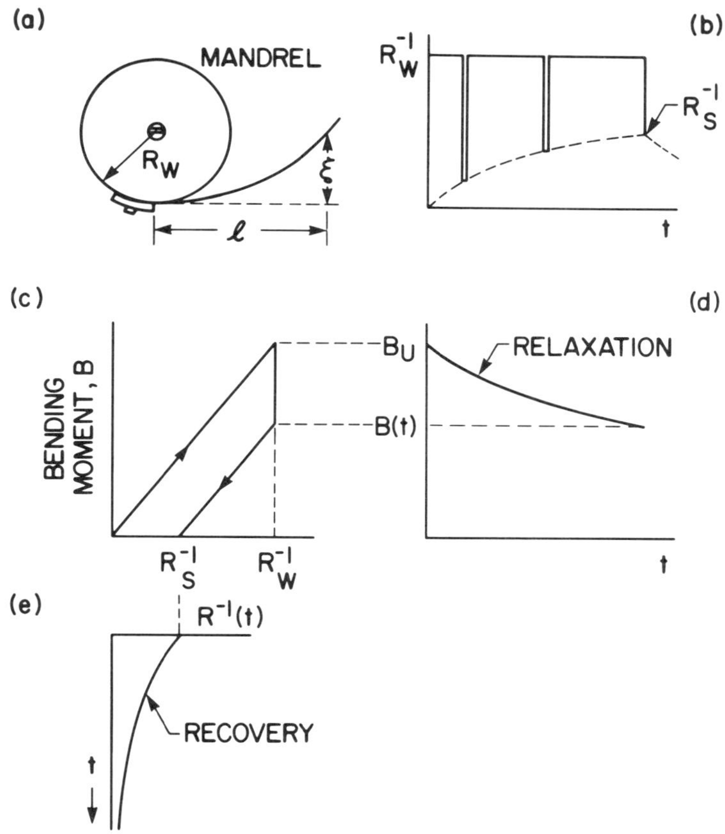

Fig. 17. Relaxation and recovery by the mandrel method. (*a*) The geometrical layout, (*b*) growth of the initial springback curvature R_S^{-1}, tracked by periodic momentary releases from the woundup position, (*c*) the moment-curvature diagram for windup and springback under unrelaxed conditions, (*d*) relaxation at constant curvature, (*e*) recovery at zero net moment.

stiffness K_U. To obtain the normalized relaxation function $K(t)/K_U$, we may, with only minor disturbance of the ideal constant curvature condition, periodically and momentarily release the sample from the wound-up configuration, for observation of the initial springback curvature R_S^{-1} (Fig. 17b). Since springback also occurs under unrelaxed conditions, it is evident that both the windup and springback lines of Fig. 17c are parallel. From similar triangles, we thus find

$$K(t)/K_U = B(t)/B_U = 1-(R_W/R_S). \qquad (52)$$

The second phase of the mandrel test consists of observation of the recovery, or flexural aftereffect, upon eventual complete release of the sample (Fig. 17e). Since recovery corresponds to a relaxation of curvature at constant (zero) moment, it should be noted that the two parts of the mandrel test are of a mixed or hybrid character. Thus, for a single layer of a standard anelastic solid, the relaxation $K(t)/K_U$ is governed by τ_ε, while the subsequent recovery $R^{-1}(t)/R_S^{-1}$ depends on τ_σ. However, as noted in Section 2, this distinction is unimportant if the relaxation strength Δ_A is small (c.f. Eq. 16).

A noteworthy feature of the mandrel test is that both relaxation and recovery data are obtained by measurements of a deflection. From the viewpoint of measurement sensitivity, we thereby capitalize on the fact that the curvature/strain ratio $R_W^{-1}/\varepsilon_{max}$ $(= 2/a_L)$ is large for thin layers. To give a specific example, relevant to the Gorsky relaxation of Section 5, if a layer with a relaxation strength of 1/20 is allowed to relax around a mandrel of 1 cm radius, the initial springback curvature is 1/21 cm^{-1} and the corresponding deflection ξ is about 1 mm at a distance ℓ of 2 cm. In such a case, a simple low-power microscope is entirely adequate for the deflection measurements. On the other hand, if normal experimental problems such as zero drift, vibration isolation etc. can be controlled satisfactorily, it is clear that measurements of much smaller relaxations, and/or measurements at smaller initial strains, can be realized with a more sophisticated displacement-monitoring system. For example, a capacitance detection system, such as that used by Verbruggen et al. (67), appears particularly attractive in view of the present availability of high-resolution self-balancing capacitance bridges with digital output.

So far, we have restricted attention to quasi-static measurements on single layers. For multilayers, we encounter a number of complicating factors even when the substrate behaves elastically, as will be assumed below. A substantial reduction in measurement sensitivity is brought about by two factors. As a consequence of the greater total thickness a_L+a_S, the curvature R_W^{-1} must be reduced to maintain ε_{max} at an acceptable value. This decrease is further compounded by the fact that the presence of the substrate also reduces the relaxation strength of the composite, Δ_C, to a value below that of the free layer alone, Δ_L. The strength Δ_C is given by $(K_U - K_R)/K_R$, where K_U and K_R denote the unrelaxed and relaxed values of the composite bending stiffness K. For selected specific geometries, Δ_C may be related to Δ_L using the formalism given in the Appendix. The choice of the symmetrical trilayer configuration again greatly simplifies the expressions involved, since the position of the neutral surface then remains fixed at the midplane of the sample. For quasi-static measurements the symmetrical trilayer has the added notable feature of eliminating thermal curling, which can complicate the investigation of non-symmetrical configurations such as the bilayer.

Another factor to be recognized in the testing of multilayers is that a distinction can be drawn between relaxation at constant *nominal* stress and relaxation at constant *actual* stress. Consider, for example, the tensile creep of a multilayer of total cross-sectional area A under a load L. For small strains, the nominal stress L/A remains essentially constant, whereas the actual stresses in the different layers can vary considerably, decreasing in those which relax the most and increasing in those which relax the least. In the mandrel method, this distinction enters in the recovery phase (Fig. 17e), where the nominal moment and stress are constant and zero, but where the actual stresses are finite and decrease hand in hand with the curvature. This complication, (which is again unimportant for small relaxation strengths) is absent during the first phase of the mandrel test. There, stress relaxation in each layer proceeds independently of the other layers by virtue of the condition of constant curvature.

3.3 Accommodation Stress and the Bending Bilayer

From the general viewpoint of stress-strain behavior, an important distinction exists between a detached single layer and one that forms part of a multilayer composite. The layers of a composite are, in general, subject to internal stresses as a consequence of their adherency and mutual accommodation. Compared to the low level and uniaxial character of the stress employed in the mandrel test, the self-induced accommodation stress shows two points of contrast. The accommodation stress can be large, and moreover forms, to a reasonable approximation, a biaxial stress system composed of equal orthogonal components in the plane of the layer.

The presence of an accommodation stress may have two consequences on the relaxation behavior of a supported layer. The primary effect is that relaxation can be induced directly by the accommodation stress. Another effect, linked to the first, is that the relaxation response to a small external stress can be modified by prior relaxation induced by the accommodation stress. (A specific example of both effects will be given in Section 4.) It is clear, therefore, that measurements of the accommodation stress can be a most useful adjunct to relaxation studies on supported layers.

As is well known, the use of a bending bilayer, consisting of the layer of interest on one side of an elastic substrate, is one of the standard methods for the measurement of the accommodation stress in thin films (26)(27). Fortunately, the same type of substrates that are employed for vibrating-reed measurements (Fig. 10) are also well-suited to the bending bilayer method. The addition of a stiff reference bar to the standard fused-silica pedestal, as shown in Fig. 18, produces a free-standing unit with a stable built-in reference for deflection measurements over a wide range of temperature.

While it is possible to perform vibrating-reed measurements on a bilayer sample produced primarily for internal stress measurements, we have already seen that the added sensitivity and absence of thermal curling provided by the symmetrical trilayer make this the configuration of choice for most internal friction studies. In many cases, it is often a simple matter to prepare samples of both types at the same time, thus eliminating the need for compromise. The use of separate substrates for bilayer and trilayer samples also introduces the possibility of adopting, if desired, a larger substrate thickness for the bilayer sample as a means of sensitivity control.

In contrast to the conditions that are usually striven for (26), it is to be noted that one of the characteristic features of stress measurements on the thin (50 μm) substrates of present interest is that the deflections ξ can be as large as a few millimeters, and are many times the substrate thickness a_S. Operation in this large-deflection regime has the experimental advantage that the deflections can be measured easily with a low-power microscope of long working distance, mounted outside the chamber containing the temperature enclosure in which the sample is mounted. On the other hand, the usual expression (26) derived for the film stress in terms of the curvature assumes by contrast the small-deflection regime $\xi < a_S$. It is only in this regime that a

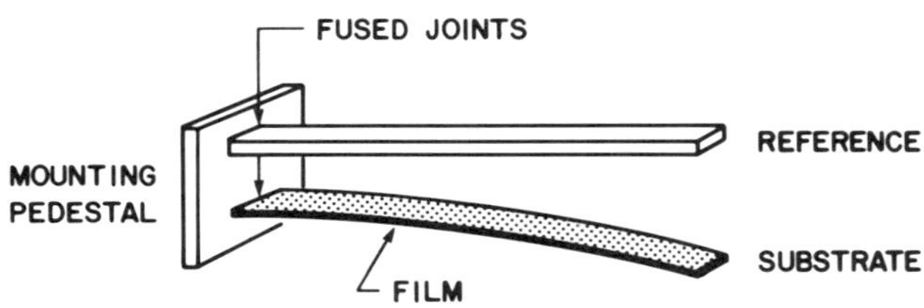

Fig. 18. Modification of Fig. 10(*a*) for measurements of the film accommodation stress by the bending bilayer method.

biaxial film stress produces a biaxial curvature, corresponding to the deformation of the substrate into part of a spherical cap. In the large-deflection regime a reed assumes the shape of a developable (cylindrical) surface, with a relative suppression of the transverse curvature. Although the details are complex, the effect of the transition from the small-deflection to the large-deflection regime appears essentially to involve a transition in the appropriate bending modulus from the biaxial modulus $E_S/(1-\nu_S)$ to the plate modulus $E_S/(1-\nu_S{}^2)$, where ν_S is Poisson's ratio of the substrate. For a bilayer reed of curvature R^{-1} satisfying the thin-film limit $a_L \ll a_S$, we may cover both situations by writing the film stress σ_F in the form

$$\sigma_F = E_S{}' a_S^2 / 6 a_L R, \tag{53}$$

where

$$E_S{}' \equiv E_S/(1-\nu_S{}^n), \tag{53a}$$

and the index n is taken as unity for the small deflection regime and as two for the large deflection regime of present concern. It is of interest to note that the Poisson's ratio of fused silica has the unusually small value of 0.167. When this material is used as a substrate, the difference in $E_S{}'$ between these two regimes is only 17%.

4. INTERSTITIAL SOLUTE DIFFUSION IN BCC METALS

This section is the first of several in which we consider the application of anelastic measurements to various categories of thin-layer materials. The order of presentation follows the sequence metals, semiconductors (silicon), and insulators. It is appropriate that our first example involves the use of the Snoek relaxation associated with interstitial atoms in bcc metals, since this was the first relaxation process to be understood in terms of a simple atomic mechanism, and was also the first to be applied to the study of thin films.

The Snoek relaxation serves as a prime example of a general category of relaxations that involve stress-induced directional (or orientational) ordering of a population of mobile, asymmetric point defects, or other structural groups of similar character (4). These anelastic relaxations represent the mechanical analogy of dielectric relaxation, which occurs under an applied electric field by the orientational ordering of centers possessing a permanent electric dipole moment. For this reason, point defects which produce asymmetrical displacements of their surroundings are often referred to as *elastic dipoles*.

The essential aspects of relaxation by stress-induced ordering can be illustrated in a simple formal way as follows. We denote the state of order of the defect population by a parameter q, whose value is chose as zero for the random configuration present in equilibrium under zero stress. Since the strain ε is now a function of both q and the applied stress σ we may write

$$\varepsilon = J_U\sigma + \kappa_1 q, \tag{54}$$

where κ_1 is the coupling coefficient between ε and q. If the value of q at equilibrium under the stress σ is $\bar{q}$, we may also write, for low stress-levels,

$$\bar{q} = \kappa_2\sigma, \tag{55}$$

where κ_2 is the coupling coefficient between $\bar{q}$ and σ. Finally, if we assume that at constant stress q relaxes towards $\bar{q}$ with a single relaxation time τ_σ, we have

$$\dot{q} = -(q-\bar{q})/\tau_\sigma. \tag{56}$$

By elimination of the order parameter from Eqs. (54)-(56) we obtain

$$\varepsilon + \tau_\sigma\dot{\varepsilon} = (J_U + \kappa_1\kappa_2)\sigma + J_U\tau_\sigma\dot{\sigma}, \tag{57}$$

which is equivalent to Eq. (22) for the standard anelastic solid with the identification $\delta J \equiv J_R - J_U = \kappa_1\kappa_2$. This relation for δJ can also be derived directly by noting from Eq. (54) that the equilibrium anelastic strain $\bar{\varepsilon}_a = \kappa_1\bar{q}$. Using Eqs. (1) and (55), we thus find directly that $\delta J = \bar{\varepsilon}_a/\sigma = \kappa_1\kappa_2$. We now turn our attention to the evaluation of δJ and τ_σ for the specific case of the Snoek relaxation.

4.1. The Snoek Relaxation

This relaxation, which was first observed in α-iron containing small amounts of carbon or nitrogen, is now known to be a general phenomenon that originates from the asymmetric or tetragonally-distorted character of interstitial solute-atom defects in the bcc structure. This asymmetric character is a direct consequence of the tetragonal symmetry of the available interstitial sites. We may remark immediately that in fcc and hcp metals the symmetry class of the interstitial sites is the same as the crystal symmetry, with the result that isolated solute interstitials in these metals do not constitute orientationally-distinguishable defects and do not produce a relaxation of the Snoek type. In the bcc structure, two distinct families of sites (octahedral and tetrahedral) are possible candidates for interstitial occupancy. However, both possess tetragonal symmetry and without loss of generality we may confine our attention to sites of just one type (4). Figure 19a illustrates the octahedral sites, which have been labelled with an orientation index (p = 1,2,3) to distinguish sites of different orientations. Figure 19b illustrates that the free-energy levels (or site-occupation energies) $g_p(0)$ of all orientations are equal in the stress-free crystal, and that all orientations are mutually accessible by nearest-neighbor jumps of equal activation energy. Examples of the energy-level splitting produced by the application of a uniaxial stress are shown schematically in Figs. 19c and 19d. Figure 19c illustrates the special case in which the stress is applied along one of the cube edges, and two of the three levels remain degenerate. The more general case of site-splitting under a uniaxial stress of arbitrary orientation is shown in Fig. 19d. For a dilute solution of non-interacting interstitials of mole-fraction C_0, the equilibrium concentrations $\overline{C}_p$ in each orientation p are given by Boltzmann statistics as

$$\frac{\overline{C}_p}{C_0} = \frac{\exp(-g_p/kT)}{\sum_p \exp(-g_p/kT)}. \qquad (58)$$

A particularly simple case of Eq. (58) corresponds to the case of the ⟨100⟩ directed stress (Fig. 19c). In this case we find Eq. (58) reduces to

$$\Delta\overline{C}_1 \equiv \overline{C}_1-(C_0/3) = (2/9)C_0\Delta g/kT, \qquad (59)$$

where $\Delta g = -(g_1-g_2)$ and we have used the limitation $\Delta g/kT \ll 1$

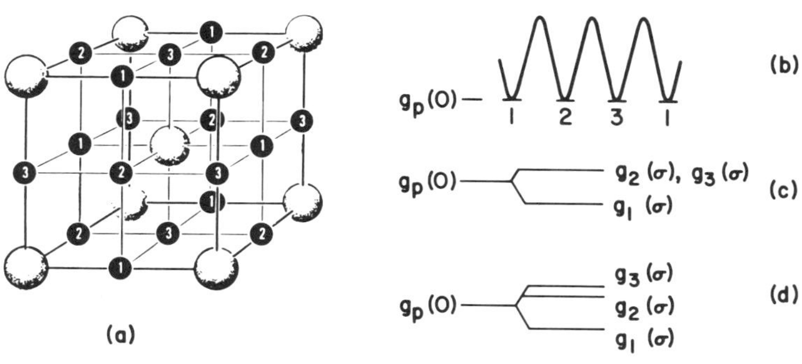

Fig. 19. (*a*) Octahedral sites in the bcc lattice, labeled with a site-orientation index p = 1,2,3. (*b*) Free-energy diagram for an interstitial in the unstressed crystal, (*c*) site-energy splittings for a ⟨100⟩ uniaxial stress and (*d*) site-splittings for an arbitrarily oriented uniaxial stress.

appropriate to the low-stress regime. We now require an expression for the equilibrium anelastic strain $\bar{\varepsilon}_a$ produced by the internal ordering $\Delta\bar{C}_1$. To this end, we make use of a quantity called the λ-tensor (4). This is derived from the tensor of macroscopic strains produced when a dilute mole fraction C of solute interstitials is introduced into the host crystal and homogeneously dispersed amongst sites of just one orientation. Such a distribution would cause a reference sphere to distort into an ellipsoid with principal strains ε_i (i = 1,2,3) proportional to C, and the derivatives $\lambda_i \equiv \partial\varepsilon_i/\partial C$ form the principal components of the λ-tensor. In this notation, the reorientation of dipoles by the stress $\sigma_{\langle 100\rangle}$ in Fig. 19c produces a conjugate equilibrium anelastic strain

$$\bar{\varepsilon}_{a\langle 100\rangle} = \Delta\bar{C}_1(\lambda_1-\lambda_2), \tag{60}$$

and the interaction energy Δg is given by

$$\Delta g = -\vartheta_0\sigma_{\langle 100\rangle}(\lambda_1-\lambda_2), \tag{61}$$

where ϑ_0 is the volume per atom of the host crystal. Combining Eqs. (59)-(61), we obtain

$$\delta E^{-1}_{\langle 100\rangle} = \frac{\bar{\varepsilon}_{a\langle 100\rangle}}{\sigma_{\langle 100\rangle}} = \frac{2}{9}\,\frac{C_0\vartheta_0}{kT}(\lambda_1-\lambda_2)^2, \tag{62}$$

where we have written the relaxation of the compliance δJ in the form $\delta E^{-1}_{\langle 100\rangle}$ in explicit recognition of the fact that $\sigma_{\langle 100\rangle}$ represents a uniaxial stress.

For the more general case of an arbitrarily oriented uniaxial stress, it can be shown (4) that δE^{-1} is given by

$$\delta E^{-1} = \delta E^{-1}_{\langle 100\rangle}(1-3\Gamma), \tag{63}$$

where Γ is the orientation function

$$\Gamma \equiv \gamma_1^2\gamma_2^2 + \gamma_2^2\gamma_3^2 + \gamma_3^2\gamma_1^2, \tag{64}$$

and γ_1, γ_2 and γ_3 are the direction cosines of the stress relative to the crystal axes. Since Γ has the extreme values of 0 and 1/3, corresponding to the ⟨100⟩ and ⟨111⟩ directions respectively, it is evident that the magnitude of the Snoek relaxation shows strong crystallographic anisotropy and in fact vanishes for a ⟨111⟩ stress. For a random polycrystalline aggregate, an averaging calculation based on the assumption of uniform strain in all crystals gives the result

$$\delta\bar{E}^{-1} = \frac{4}{45}\,\frac{C_0\vartheta_0}{kT}(\lambda_1-\lambda_2)^2, \tag{65}$$

Table 1: Parameters for Solute Interstitials in BCC Metals, taken mainly from Refs. 4 and 29.

System	$\lambda_1-\lambda_2$	$\lambda_1+2\lambda_2$	H(eV)	$\tau_0(10^{-15}$s)
Fe-C	0.87	0.70	0.82	10
Fe-N	0.80	0.69	0.80	5
Nb-O	0.55	0.40	1.15	3
Nb-N	0.67	–	1.58	1
Ta-O	0.55	–	1.11	9
Ta-N	0.78	–	1.67	4

from which it can be seen that the averaged tensile relaxation strength $\overline{\Delta}_E = \overline{E}_U\,\delta\overline{E}^{-1}$ can be expected to follow a relation of the type

$$\overline{\Delta}_E = AC_0/T, \qquad (66)$$

where A represents the lumped parameter $4\overline{E}_U\vartheta_0(\lambda_1-\lambda_2)^2/45k$. The term $(\lambda_1-\lambda_2)$ which appears repeatedly in Eqs. (60)-(65) represents the tetragonality or 'shape-factor' of an interstitial solute atom. For future reference, we may note that the sum of the principal components of the λ-tensor, $(\lambda_1+2\lambda_2)$, represents the volumetric size-factor $\Delta\vartheta/\vartheta_0$, where $\Delta\vartheta$ is the volume change upon insertion of a solute atom. As a consequence, $(\lambda_1+2\lambda_2)$ is also, to a first approximation, three times the ordinary or linear size-factor expressed by the derivative $a^{-1}\mathrm{d}a/\mathrm{d}C$, where a is the lattice parameter. Interestingly, λ_2 turns out to be negative in all three cases for which data is available (Table 1), with the consequence that the shape-factor $(\lambda_1-\lambda_2)$ is numerically larger than the size factor $(\lambda_1+2\lambda_2)$.

In contrast to the complexities associated with the relaxation strength, the kinetics of the Snoek relaxation are relatively straightforward. The proportionality factor b between τ^{-1} and w in Eq. (38) can be found quickly from standard kinetic considerations. If the three orientations in Fig. 19a are populated with the non-equilibrium concentrations C_1, C_2 and C_3, we first write the time-derivative of C_1 in the form

$$\dot{C}_1 = -4C_1w + 2w(C_2 + C_3). \qquad (67)$$

Upon using the conservation condition $C_0 = C_1+C_2+C_3$, and the equilibrium condition $\overline{C}_1 = \overline{C}_2 = \overline{C}_3 = C_0/3$, Eq. (67) becomes

$$\dot{C}_1 = -6w(C_1-\overline{C}_1), \qquad (68)$$

which is a simple relaxation equation of the same type as Eq. (56), with

$$\tau^{-1} = 6w. \tag{69}$$

This result, which shows that $b = 6$, may also be written in the form

$$\theta = \frac{3}{2}\tau, \tag{70}$$

where $\theta (= 1/4w)$ is the mean time-of-stay of an interstitial in a given site. For the case of octahedral site-occupancy, the jump distance a_j is $a/2$ (where a is the lattice parameter), and Eq. (45) becomes

$$D = a^2/36\tau. \tag{71}$$

The treatment of the Snoek relaxation for the alternative case of tetrahedral site occupancy is identical, except for a change in the jump distance to $a\sqrt{2}/4$, with the result that the factor 36 in Eq. (71) changes to 72. In either case, the activation energy of the Snoek relaxation is the activation energy for diffusion of the interstitial species, and the precise determination of both H and τ_0 using both dynamic and static measurements has represented one of the principal applications of the Snoek relaxation in bulk materials (4)(28)(29). Some illustrative results are included in Table 1. Because such data can be used to predict the location of peaks due to particular solutes, the Snoek relaxation has the potential for both the identification and semi-quantitative analysis of interstitial solutes in thin films of bcc metals. In particular, peaks may be expected in evaporated or sputtered films of the Group Vb metals V, Nb and Ta, which have a strong affinity for both oxygen and nitrogen.

4.2 Experimental Results

There appear to be only two reports by other workers of internal friction measurements on thin films of bcc metals. In an early exploratory investigation, Weiss and Smith (14) deposited a 0.5μm film of iron on a 1μm substrate of fused silica, and observed a damping maximum in the expected vicinity of the Snoek peak due to carbon. Vibrating-reed measurements on *detached* evaporated films of Ta, Nb, V and Re have been reported by Zolotukhin et al. (30). A number of peaks were observed, including Snoek peaks due to oxygen and nitrogen.

The excellent resolution attainable by refinement of the composite-reed approach was first demonstrated by Berry and Pritchet in an investigation of the effect of vacuum annealing treatments on the contamination of niobium films (16,31). In this work, films were prepared by electron-gun evaporation of high-purity niobium in a gettered, oil-free, ion-pumped system fitted with a substrate holder designed for the simultaneous preparation of both singe-sided and double-sided samples. The substrates were pedestal-mounted reeds of fused silica (Fig. 9a), maintained at approximately 400°C during film deposition. At the fundamental frequency of 36 Hz, the calculated locations of the oxygen and nitrogen peaks from the data of Table 1 are 202°C and 349°C respectively, the difference reflecting the distinctly higher activation energy for the diffusion of nitrogen. The results of Fig. 20 clearly reveal that subsequent annealing treatments in a vacuum of 10^{-5}-10^{-6} torr are predominantly associated with an uptake of oxygen, which is accommodated in interstitial solid solution. The small size of the peak indicated for nitrogen may in part reflect the relative slowness with which nitrogen can diffuse into the film. (At 300°C, for example, the diffusion coefficient for nitrogen is only 0.0043 of that for oxygen). The asymmetrical broadening on the high-temperature side of the oxygen peak is also seen in highly-doped bulk alloys, especially in the presence of small amounts of nitrogen (32)(33). While this broadening undoubtedly involves some type of solute interaction, recent work by Weller and his co-workers has shown that the details are not as well understood as previously supposed (29)(34).

The results described above were obtained with a symmetrical trilayer reed consisting of 0.45μm Nb films on a 50μm substrate of fused silica. The thickness ratio a_L/a_S is thus small enough to satisfy the thin-film limit in which $\gamma_{ST} = 6a_LE_L/a_SE_S$ (c.f. Eqs. (48) or (A22)). The excellent resolution of the measurements is indicated by the fact that the substrate damping ($\delta_S \simeq 10^{-5}$) is three orders of magnitude smaller than the largest peak in Fig. 20, even though the films form only 2% of the total thickness of the composite. Since we may neglect the term in δ_S in Eq. (46), the composite damping δ_C is

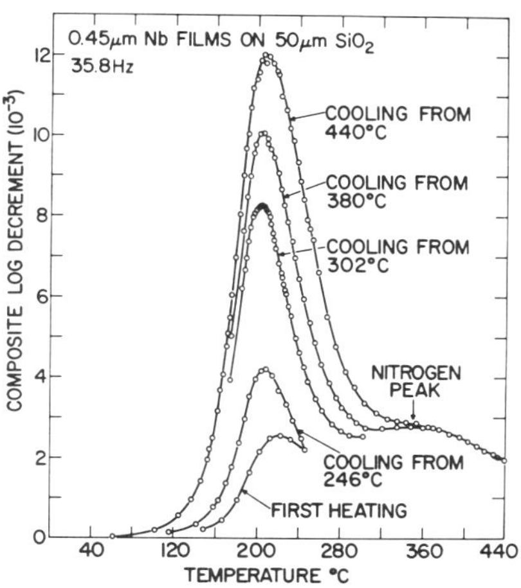

Fig. 20. Growth of the oxygen Snoek peak in thin-film niobium, caused by thermal cycling treatments in a vacuum of $10^{-5}-10^{-6}$ torr. From Ref. 31.

seen to be related to the film damping δ_L through the proportionality factor $\gamma_{ST}/(1+\gamma_{ST})$. From this result, together with the dipole shape-factor obtained from bulk studies (Table 1) and the assumption of a random texture in the film, it can be estimated that the as-deposited film contained about 0.3 a/o (atomic percent) of oxygen in solution, and that this increased to about 3 a/o after the 440°C exposure (Fig. 20). An independent check, based on the accommodation stress measurements described later, suggest that these estimates are probably too high by about a factor of 1.25.

To investigate the kinetics of relaxation under the simplest possible conditions, it is desirable to work at the lowest possible concentrations, to minimize complications from solute interactions. In the present study, the minimum concentration was not determined by sensitivity considerations, but by the level of contamination unavoidably produced by exposure to the maximum temperature of measurement. The results of Fig. 21 were obtained in one run on cooling from 280°C, using the first, second and fourth tones of the reed. Exposure to this temperature, which was just sufficient to permit observation of the peak maximum at the highest frequency, was estimated to result in an oxygen concentration of about 0.6 a/o. The data of Fig. 21 reveal a temperature dependence of the peak height which is roughly in

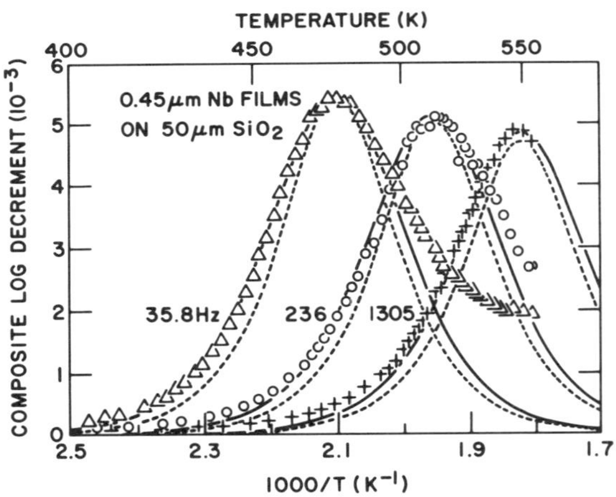

Fig. 21. Frequency dependence of the oxygen Snoek peak in thin-film niobium containing approximately 0.6 a/o oxygen. The peaks shown by the dashed lines are $1/T$-modified Debye peaks calculated for H = 1.12 eV and $\tau_0 = 7\times10^{-15}$ s. The peaks shown by the solid lines additionally include the effect of a Gaussian distribution in H with β_H = 0.035 eV. Both sets of calculated peaks are pinned to the data only at the maximum of the 35.8 Hz peak.

accord with the $1/T$ dependence predicted by Eq. (66). To remove this factor from the frequency-shift analysis, it is convenient to replot the peaks in the form of the temperature damping product $T\delta_C$ versus $1/T$, and then compute the activation energy from the average displacements between the normalized peaks. The result of this analysis, $H = 1.12$ eV, is about 4% smaller than the value previously reported by a less-complete analysis, but is in excellent agreement with the results of Table 1. To compare the observed peaks with those for ideal Snoek behavior, Fig. 21 includes the Debye peaks calculated for a single activation energy of 1.12 eV, and a single τ_0-value of 7×10^{-15} sec. In these calculations, the relaxation strength has been pinned to fit the largest peak, but contains the 1/T dependence required by Eq. (66). It is clear that the observed peaks exhibit a broadening which is not confined to the high-temperature side, and which appears to be an exaggerated version of that reported by Weller et al. on bulk samples of comparable concentration (32). The much closer fit obtained with modeling calculations which introduce a Gaussian distribution of activation energies is also shown in Fig. 21, where the peaks shown by the solid lines are computed for a mean activation energy $\bar{H}$ of 1.12 eV and a distribution parameter β_H (4) of 0.035 eV.

In contemplating the relaxation behavior of supported thin films, the most obvious new factor which may lead to differences from bulk behavior is the presence of a large accommodation stress σ_F in the film. In the present experiments, curvature measurements on a bilayer sample prepared concurrently with the symmetrical trilayer used for the internal friction measurements have shown that the initial value of the accommodation stress was about 0.5 GPa. (This was previously erroneously reported as 1.5 GPa (31)). One way the film stress may influence the kinetics of relaxation is through the addition of an interaction term of the type $-\sigma_F\Delta\vartheta^*$ to the activation energy H, where $\Delta\vartheta^*$ is the activation volume for diffusion. The absence of a clear shift in the expected location of the Snoek peak suggests that this term is not significant in the present experiments. A second possible kinetic effect may be the appearance of a dispersion δH in the activation energy, caused by site-splittings induced by the accommodation stress. For realistic parameters we have in this case $\delta H \simeq \vartheta_0\sigma_F(\lambda_1-\lambda_2) \simeq 0.03$ eV. Since the broadening shown by the peaks of Fig. 21 appear to involve a total (full-width) dispersion of about 0.07 eV, it appears that site-splitting may be a significant factor that augments an intrinsic broadening due to solute-solute interactions.

In contrast to the marginal effects of the accommodation stress on the kinetics of interstitial motion, complementary bending-bilayer experiments have demonstrated a direct coupling between the accommodation stress and both the concentration and the state of order of the interstitial population in the film. The first of these effects, the coupling between film stress and interstitial concentration, relates to the solute interstitial as a center of dilatation whose entry into the film tends to relieve a tensile accommodation stress. The second effect, a coupling between the film stress and the state of directional order, relates to the solute interstitial as an elastic dipole, and is manifest as a relaxation of the accommodation stress by Snoek ordering. To discuss both effects, we may first write the film stress in the form

$$\sigma_F = \sigma_F^{(I)} - E_F'\{\Delta\alpha\Delta T + \varepsilon(C_0,R)\}, \tag{72}$$

where $\sigma_F^{(I)}$ represents the initial tensile stress in the as-deposited film at some convenient reference temperature T_0 (e.g. room temperature), E_F' is the appropriate film modulus complementary to that of the substrate in Eq. (53), $\Delta\alpha$ is the difference in the expansion coefficients of the film and the substrate, $(\alpha_F - \alpha_S)$, and ε is the linear planar strain which would be exhibited by the unrestrained film on introduction of the solute concentration C_0 with a specified state of order represented by a parameter R. If we first neglect Snoek ordering and assume random occupancy, we have $C_1 = C_2 = C_3 = C_0/3$ and $\varepsilon = C_0(\lambda_1 + 2\lambda_2)/3$, with the result that Eq. (72) takes the simple form

$$\sigma_F(T_0) = \sigma_F^{(I)} - 3E_F'(\lambda_1 + 2\lambda_2)C_0, \tag{73}$$

which corresponds to a linear decrease of the film stress with added solute concentration. Since the size factor $(\lambda_1 + 2\lambda_2)$ is known (Table 1) and the value of E_F' for a particular test sample can be determined experimentally [c.f. Eq. (77)], Eq. (73) can be used to estimate the increment of solute concentration associated with an observed reduction of the film stress. (This procedure presumes a stable microstructure, which can be expected in the present case because the maximum annealing temperature only slightly exceeds the deposition temperature of the film). In the experiments of Fig. 20, the stress in the accompanying bilayer sample was reduced from 0.52 GPa to 0.26 GPa, and the corresponding increment in the oxygen concentration calculated from Eq. (73) is 1.7 a/o. This is smaller than the value of 2.7 a/o previously estimated from the growth in the strength of the peak, and suggests that the concentration derived from the peak may be too high as a result of a slight texture in the film.

As noted above, Eq. (73) is based on the assumption of random occupancy, and thereby ignores the effect of directional ordering induced by the accommodation stress. To examine the consequences of ordering in an extreme case amenable to simple analysis, we may consider a {100} single-crystal film, corresponding to Fig. 19a, subject to a horizontal biaxial tensile stress which favors occupancy of orientations 1 and 2 at the expense of orientation 3. In this case we have $C_1 = C_2$, and the order parameter R, chosen as the ratio C_1/C_3, has the equilibrium value

$$\overline{R} = \overline{C}_1/\overline{C}_3 = \exp\,(\vartheta_0\overline{\sigma}_F(\lambda_1-\lambda_2)/kT), \tag{74}$$

where $\overline{\sigma}_F$ now represents the relaxed or equilibrium stress given to the equation

$$\overline{\sigma}_F(T_0) = \sigma_F^{(I)} - E_F'\varepsilon(C_0,\overline{R}). \tag{75}$$

Illustrative calculations based on Eq. (75) are shown in Fig. 22 for the case $\sigma_F^{(I)} = 0.52$ GPa, corresponding to the measured value for the polycrystalline Nb film used in the present experiments. It will be seen that this level of stress is sufficient to produce a substantial degree of ordering, and that the associated relaxation of the accommodation stress introduces a characteristic curvature into the variation of the equilibrium stress with solute content. It should also be noticed that both stress curves in Fig. 22 intersect on the concentration axis. This reflects the fact that, in the absence of stress, the equilibrium configuration is the random configuration. As a consequence, the solute concentration to produce total stress relief is given by the simple general expression

$$C_0 = 3\sigma_F^{(I)}/E_F'(\lambda_1 + 2\lambda_2). \tag{76}$$

Although a non-linear variation of stress with solute content is one manifestation of Snoek ordering, it does not represent the most straightforward method of demonstrating the existence of Snoek order-

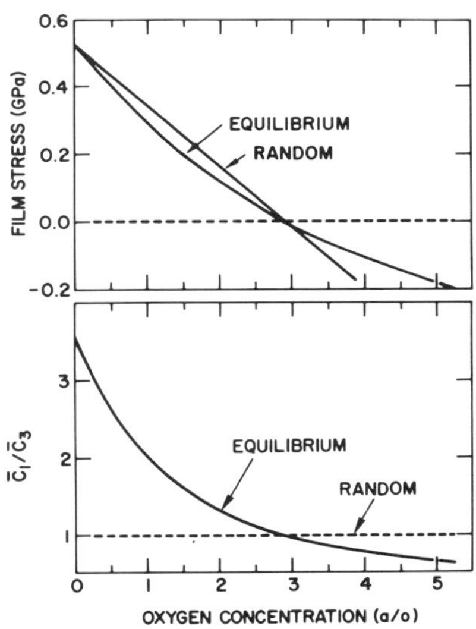

Fig. 22. Illustrative calculations of the stress relief and directional order predicted for the addition of oxygen to a {100} single-crystal niobium film. The initial stress corresponds to that observed in a polycrystalline film.

ing in a thin film. A simpler method, illustrated by the results of Fig. 23, involves the observation of a relaxation in the temperature dependence of the film stress, under conditions of fixed solute content. At low temperatures, where the solute atoms are immobile and the order parameter R is frozen, we find from Eq. (72) that the slope of the stress-temperature curve has the unrelaxed value

$$(\partial\sigma_F/\partial T)_{C_0,R} = -E_F{}'\Delta\alpha, \tag{77}$$

whereas at higher temperatures there is a changeover to the equilibrium behavior

$$(\partial\sigma_F/\partial T)_{C_0} = -E_F{}'\Delta\alpha-(\partial\varepsilon(\overline{R})/\partial T)_{C_0}. \tag{78}$$

The transition temperature between these two regimes defines a freezing-in temperature for directional ordering, which in the present experiments is seen to be in the vicinity of 330 K.

In addition to the detection of Snoek ordering induced by the accommodation stress, we may also note that Fig. 23 illustrates two other points of general interest. One is the use of temperature as a convenient means of generating a thermal stress in a composite in order to initiate a relaxation response. This provides an interesting alternative to the direct application of an external load, and is especially attractive because of the ease with which a relatively large stress can be induced in a supported thin film. The second point of interest relates to Eq. (77), which shows that thermal cycling measurements in

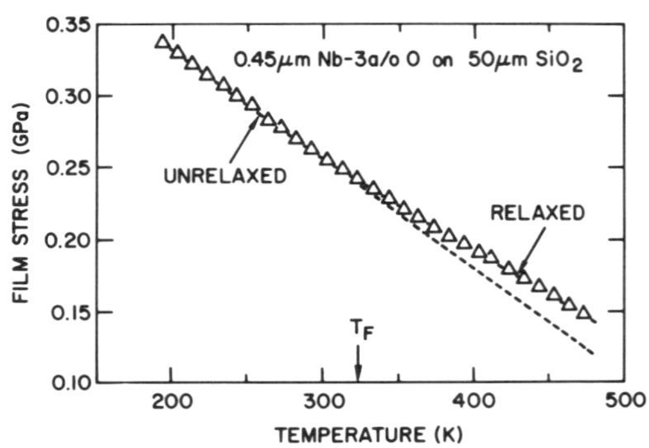

Fig. 23. Bending bilayer measurements of the stress in a niobium film containing approximately 3a/o oxygen. The change in slope at about 325K marks the transition to a relaxed regime in which the oxygen atoms have enough mobility to establish an equilibrium state of directional order.

the unrelaxed elastic regime provide a convenient method for the determination of E_F' for a film of known expansion behavior. This is especially useful for polycrystalline films of cubic metals, where α_F is isotropic but E_F' depends on the texture of the particular film investigated.

Having seen that a relatively large static accommodation stress in a thin film can induce a significant degree of directional ordering, we must now consider the effect this may have on the Snoek relaxation observed in a vibrating-reed experiment, where a much smaller uniaxial dynamic stress is superimposed on the film stress. A simple illustrative solution to this problem is again obtained by restricting attention to a {100} single-crystal film. In this case, it can be shown that Eq. (62) is replaced by the expression

$$\delta E^{-1}_{\langle 100\rangle} = \left\{\frac{(1+(1/\overline{R}))}{(2+(1/\overline{R}))^2}\right\} \frac{C_0\vartheta_0}{kT}(\lambda_1-\lambda_2)^2, \tag{79}$$

where $\overline{R}$ is the equilibrium order parameter $\overline{C}_1/\overline{C}_3$ given by Eq. (74). In contrast to the significant suppression of $\delta E^{-1}_{\langle 100\rangle}$ that can be produced by a strong compressive accommodation stress ($\delta E^{-1}_{\langle 100\rangle} \rightarrow 0$ as $\overline{R} \rightarrow 0$), it can be seen from Eq. (79) that a tensile accommodation stress has a relatively minor effect on the proportionality factor between $\delta E^{-1}_{\langle 100\rangle}$ and C_0. Compared with a stress-free value of 2/9, the bracketed term in Eq. (79) only increases to 2/8 even in the limit $\overline{R} \rightarrow \infty$, corresponding to complete depopulation of orientation 3 (Fig. 19a). Other noteworthy features of the Snoek relaxation in this limiting regime are a change in the direction of the null orientation $\delta E^{-1}(\Gamma) = 0$ from $\langle 111\rangle$ to $\langle 110\rangle$, and a change in the proportionality factor b between τ^{-1} and w from $b = 6$ to $b = 4$. This change reflects an interesting modification in the nature of diffusion in the film from a 3-dimensional to a 1-dimensional process, in which (by virtue of their exclusion from orientation 3) interstitial atoms are confined to linear chains of sites of the type 121212.... (Fig. 19a). Unfortunately, the approach to this interesting regime requires an accommodation stress approaching 3 GPa, which is several times larger than that available experimentally.

5. HYDROGEN DIFFUSION IN METALLIC GLASSES

Hydrogen occupies a special place as the smallest, lightest and most rapidly diffusing interstitial solute in metals (35)(36). Although not used as an intentional alloying element (except in special circumstances such as hydrogen storage), hydrogen can exert, particularly through the phenomenon of hydrogen embrittlement, important effects even when present at trace levels. Due to the ease with which it can be picked up unintentionally during film deposition, the role of hydrogen in thin films is likely to be even more pervasive than for bulk materials, and methods for its detection and characterization are therefore of considerable interest.

In this section, we survey the use of two relaxation processes that have been used to study hydrogen in a relatively new group of thin-layer materials, namely the metallic glasses (37)(38). These amorphous alloys are for the most part prepared by rapid solidification methods such as liquid-quenching or vapor condensation on a cold substrate, and are usually only available in thicknesses below $50\mu m$. Free-standing strips or ribbons of liquid-quenched glasses form excellent samples for vibrating-reed studies, and were under investigation as early as 1970 (39)(40). Significantly, however, the first hydrogen-related peak was not reported until 1978, following the first studies of amorphous films prepared by sputter deposition (41)(42). Although the characteristics of this new peak clearly pointed to a point-defect reorientation mechanism, the precise identity of the defect involved was initially by no means obvious, primarily because no similarity existed with the relaxation behavior associated with hydrogen in bcc metals. There, the available evidence was of two kinds. First, the Gorsky relaxation had revealed the hydrogen interstitial as an extremely fast non-classical diffuser ($H \simeq 0.1$ eV) which nonetheless possessed a normal size-factor or dilatation strength (43)(44). On the other hand, attempts to identify a hydrogen Snoek relaxation, either by the appearance of a complementary peak or a modulus defect, had ended in failure (45). This, supported by other evidence (46), had led to the conclusion that the shape-factor or dipole strength of the free hydrogen interstitial was unusually small, and that the hydrogen-related peaks that had been observed in bcc metals involved the reorientation of more complicated defect complexes consisting of hydrogen atoms bound to other solute atoms (45)(47). With this perspective, it will become apparent that the results obtained on hydrogenated metallic glasses represent a fascinating new turn of events. The hydrogen peak in these materials has proved to be of the Snoek type, and the nature of the hydrogen interstitial in metallic glasses has turned out to be quite different from that in bcc metals.

5.1 The Hydrogen Reorientation Relaxation

In view of the prominent role played by niobium in hydrogen studies of bulk bcc metals, it is an interesting coincidence that the first example of a hydrogen peak in a metallic glass was observed in an alloy of nominal composition Nb_3Ge, which is perhaps the closest practical approach to a stable form of amorphous niobium. The results shown in Fig. 24 were obtained on a detached sputtered film of about 8μm thickness. The first clue relating the peak to the accidental presence of hydrogen was obtained from the annealing behavior of the peak, which correlated with the desorption of hydrogen from the film. This connection was confirmed by the reintroduction of the peak using either hydrogen implantation or acid-immersion treatments (39). In contrast to the oxygen Snoek peaks of Figs. 20 and 21, the hydrogen-related peak of Fig. 24 is located below room temperature and provides, from the shift with frequency, a central or most probable activation energy $\bar{H}$ of 0.55 eV. The peak exhibits considerable dispersion, being about three times wider than a Debye peak, and also shows noticeable asymmetry with a greater width on the low-temperature side. This broadening, coupled with the calculated value $\bar{\tau}_0 = 5 \times 10^{-14}$ sec, immediately indicates that we are dealing with a short-range relaxation that is sensitive to the atomic-scale disorder of the host structure. With the assumption that each relaxation center contains one hydrogen atom, it was also possible to estimate a mean defect shape-factor from the strength of the peak. The result, $\delta\lambda \simeq 0.2$ (42), is entirely reasonable for a point-defect reorientation mechanism involving hydrogen, but again does not identify the specific nature of the defect involved. In particular, the very large difference in the jump rate of the hydrogen defect in glassy Nb_3Ge compared to that of the free interstitial in bcc Nb appeared to be a major factor pointing towards defects of a very different nature.

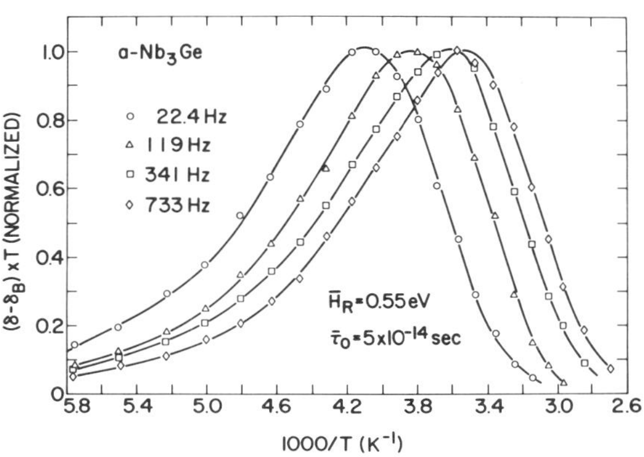

Fig. 24. The first example of the hydrogen reorientation peak in a metallic glass, discovered by vibrating-reed measurements on a detached 8 μm film of Nb_3Ge accidentally contaminated with hydrogen during sputter deposition. From Ref. 42.

Table 2: Parameters of the Hydrogen Peak in Some Binary Metallic Glasses

Metallic Glass	Peak Height $\delta(10^{-3})$	$\bar{H}$ eV	$\bar{\tau}_0$ $(10^{-15}s)$	Relative Width
$Nb_{75}Ge_{25}$	14	0.55	50	2.8
$Nb_{40}Ni_{60}$	21	0.47	30	2.8
$Ni_{64}Zr_{36}$	50	0.34	30	3
$Fe_{80}B_{20}$	10	~0.55	~1	~4
$Pd_{80}Si_{20}$	9	0.34	2	3.8

The discovery of the hydrogen peak in glassy Nb_3Ge was soon followed by observations of similar peaks in other metallic glasses, both of the metal-metalloid and metal-metal type (31)(48)(49). Peaks have now been observed in many different alloy compositions (Table 2), with activation energies in the range 0.25 eV to 0.55 eV, and the number of examples continues to increase (50)(51)(52). Notable examples include the well-known glasses $Pd_{80}Si_{20}$, $Fe_{80}B_{20}$ and several Ni-Zr compositions such as the $Ni_{64}Zr_{36}$ glass made prominent by the pioneering work of Spit et al. (53). During these exploratory studies, it was also soon learned that the kinetics of hydrogen absorption and desorption can depend greatly on the surface condition of the glassy metal, and can be strongly inhibited by surface poisons and oxide films.

Of the different alloys investigated, the Pd-Si glasses have proved to be very convenient for detailed studies because they can be doped to controlled levels simply by equilibration against various pressures of hydrogen gas. From a number of solubility studies (54)-(58), it is clear that hydrogen enters in the usual dissociated (atomic) form, and is accommodated in solid solution with no evidence of the formation of an intermediate (hydride) phase. The observed deviations from ideal solution behavior are well explained by Kirchheim's solution model based on a Gaussian distribution of site-occupation energies in the disordered host (59). Figures 25 and 26 show the growth of the internal friction peak in a $Pd_{80}Si_{20}$ glass after equilibration against various pressures of hydrogen or deuterium gas, which produce solute concentrations below 2 a/o. For equivalent pressures, deuterium produces smaller peaks than hydrogen, in a ratio that is in good agreement with the ratio of the isotopic solubilities determined both by resistometric and dilatational measurements (25). This equilibrium or solution isotope effect is very similar to that observed in pure crystalline Pd, and can be expected to have a similar explanation (60). Another significant feature of Figs. 25 and 26, which has turned out to be a common characteristic of hydrogen solution in metallic glasses, is the downward shift in peak temperature that accompanies an increase in peak height. This variation, shown in Fig. 27, is a manifestation of concentration-dependent relaxation kinetics, with a shorter mean relaxation time for higher solute contents. Significantly, it can be seen from Fig. 27 that, within experimental error, the results for hydrogen and deuterium fall on the same curve. In contrast to an earlier report (61) of an observable isotope effect based on measurements made after isobaric, rather than isoconcentration doping, Fig. 27 demonstrates that the kinetic isotope effect in this system is undetectably small (25).

Further insight into the concentration dependence of the relaxation kinetics is provided by Fig. 28, which shows the results obtained from a systematic series of frequency-shift measurements conducted on a hydrogen-doped sample of $Pd_{82}Si_{18}$. Although the dominant effect causing the change in peak temperature with concentration is found to be the concentration dependence of the mean activation energy $\bar{H}$, it is also seen from Fig. 28 that this effect is to some extent opposed or moderated by a contrary trend in the prefactor $\bar{\tau}_0$. The results shown in Fig. 28 also form the basis for three other important conclusions. First, in contrast to the startling difference in hydrogen mobility between hydrogenated crystalline Nb and glassy Nb_3Ge, the magnitude

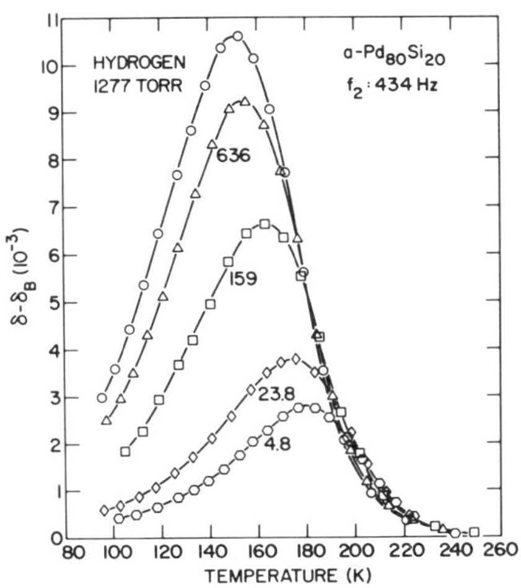

Fig. 25. The internal friction peak shown by glassy $Pd_{80}Si_{20}$ after equilibration against various pressures of hydrogen gas at 294K.

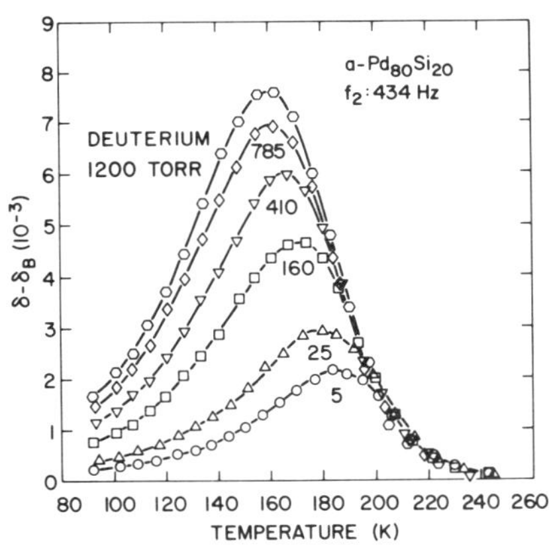

Fig. 26. Comparative data for the same sample as Fig. 25, after equilibration against deuterium gas. Both sets of peaks are corrected for the small background δ_B shown prior to exposure. From Ref. 25.

of $\bar{H}$ from Fig. 28 for glassy $Pd_{82}Si_{18}$ is not remarkably different from the value of 0.23 eV for hydrogen diffusion in fcc crystalline Pd. Secondly, as has been shown elsewhere (62) the observed concentration dependence of $\bar{H}$ is in satisfactory agreement with the predictions of Kirchheim's solution model, thereby implying that the reorientation relaxation arises from the general population of hydrogen atoms in solution. Third, by comparison with the diffusion parameters obtained from Gorsky relaxation measurements on the same sample, we can conclude that the hydrogen jumps which produce reorientation also produce long-range diffusion. To these kinetic results we may add the additional findings that, over the same range of hydrogen concentrations, both the volume dilatation and the reorientation relaxation strength increase linearly with the hydrogen content (62). The hydrogen size-factor $\Delta\vartheta_H/\vartheta_0 = \mathrm{tr}\lambda$ is found to have a normal value of 0.17, whereas the shape-factor $\delta\lambda$ estimated from Eq. (65) is almost equally large at 0.15. Taken together, all of the above results show quite convincingly that the reorientation relaxation is produced by hydrogen interstitials that are not only centers of dilation but are also elastic dipoles of comparable strength, which reorient by the same jumps which also produce long-range diffusion. Precisely similar conclusions have also been obtained from a $Ni_{64}Zr_{36}$ glass containing up to 20 a/o H (63). We may thus safely conclude that the hydrogen reorientation relaxation is an analog of the Snoek relaxation produced by the heavier interstitials C, N, and O in the bcc metals, and is a direct and general

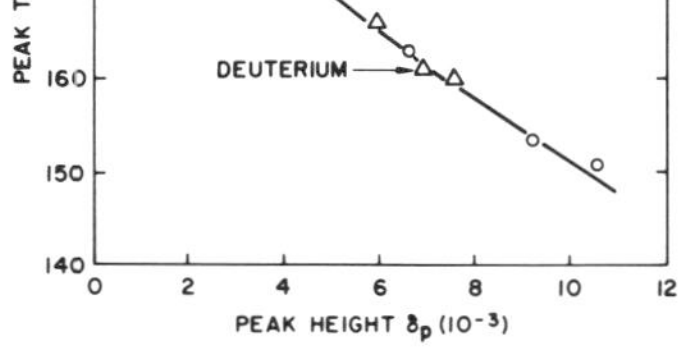

Fig. 27. Combined plot of results for hydrogen and deuterium, obtained from the peaks of Figs. 25 and 26. Within experimental error, all points fall on a common curve, demonstrating an undetectably small kinetic isotope effect.

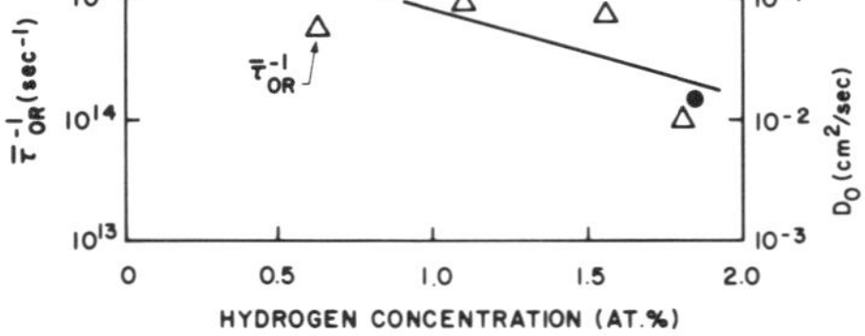

Fig. 28. Comparison of the rate parameters for the reorientation and Gorsky relaxations produced by hydrogen in glassy $Pd_{82}Si_{18}$. All measurements were made on the same sample. From Ref. 25.

indication that the sites occupied by hydrogen atoms in metallic glasses are characterized by a distinct asymmetry. The finding that the dipole strength is so large in relation to the size factor implies that most of the dilatation produced by a H-interstitial essentially occurs by a local atomic displacement along just one direction. If, as appears probable, the occupied sites are mainly of tetrahedral character, this further indicates that, as in the case of bcc metals, the site symmetry is basically tetragonal (62).

The discovery of the hydrogen reorientation peak in metallic glasses, and its identification as a Snoek-type relaxation, is significant from three points of view. First, it has shown that the unusually weak dipole strength exhibited in bcc metals is not a general character of the hydrogen interstitial, but rather is linked to a special nonclassical or delocalized configuration specific to bcc metals. Secondly, we now have the first example of hydrogen behaving as a quasi-classical defect analogous to the heavier solute interstitials in bcc metals, wherein simple Arrhenius behavior and a comparatively low mobility are coupled with a strong elastic dipole strength. (At present, the interesting question of whether quantum effects are associated with hydrogen in glassy metals at much lower temperatures than those used in the present work appears to be unexplored). The third area of significance relates to the use of the hydrogen reorientation relaxation as a tool for probing the structure of metallic glasses. Although the basic mechanism is of the Snoek type, the disordered nature of the glassy host can clearly be expected to introduce novel ramifications into the particular details of the relaxation process. One prominent example is provided by the wide and asymmetrical peak shape, which can be used to estimate the distribution of activation energies involved in the local motions of hydrogen atoms (51)(64). Apart from contributing to dispersion, it can also be expected that the distribution of site-energies in a metallic glass should also produce a distinct suppression of the strength of the reorientation relaxation. Recently, it has been argued that the absence of such an effect has important implications concerning the *spatial* distribution of disorder. Specifically, it has been suggested that the observed strength of the peak implies a pairwise near-degeneracy of neighboring interstitial sites, which is aided by the presence of common host-atom neighbors to both sites (62).

5.2 The Gorsky Relaxation

The Gorsky relaxation came into prominence through the work of Alefeld, Völkl and their co-workers, who first recognized its potentialities for studies of hydrogen in crystalline metals (65),(43),(44). In contrast to the reorientation relaxation, the Gorsky relaxation focuses on the hydrogen interstitial as a mobile center of dilatation, rather than as an elastic dipole, and involves long-range rather than short-range diffusion. The relaxation is typically observed in quasi-static bending experiments on strip samples, and may be explained with reference to the mandrel test of Section 3.2, starting with the sample in the wound-up position. The bending mode of deformation is now an essential ingredient of the arrangement, since it is the variation in the bending stress from tension to compression across the thickness of the strip that induces a complementary gradient in the chemical potential of the hydrogen in solution. This gradient is gradually eliminated and equilibrium restored by a net diffusion of hydrogen atoms from the compressed to the expanded side of the strip. Because hydrogen is a dilatant solute species (i.e. possesses a finite size-factor $\mathrm{tr}\lambda$), this redistribution leads to a complementary stress relaxation at constant curvature which may, if desired, be observed by the momentary spring-back technique described in Section 3.2. Alternatively, and more usually, the sample may be left undisturbed in the wound-up position until stress relaxation is complete. It is then released for observation of the complementary Gorsky aftereffect, during which the previously induced concentration gradient is removed by a reverse flow of hydrogen atoms. A detailed analysis of the relaxation shows that, to an excellent approximation, the Gorsky effect produces standard anelastic behavior with a single relaxation time given by

$$\tau = a_L^2/\pi^2 D_F, \tag{80}$$

where D_F denotes the chemical or Fickian diffusion coefficient. In metallic glasses, the existence of a single relaxation time for the Gorsky effect is a major contrast to the wide (typically 4-decade) distribution of local jump times indicated by the reorientation relaxation, and represents an important identifying feature. The relaxation was first observed in 1981 in an electrolytically-charged Pd-Si glass, where close agreement to single relaxation time behavior was confirmed (64). Since then, the relaxation has also been observed in hydrogenated Ni-Zr and Cu-Zr glasses, and has even been observed (with somewhat unusual characteristics) in a crystalline palladium silicide, which is possibly the first report of the Gorsky effect in a noncubic crystalline material (25)(66). Examples of Gorsky aftereffect curves obtained by the mandrel method from a ribbon of glassy $Ni_{64}Zr_{36}$ containing about 4 a/o of hydrogen are shown in Fig. 29. Diffusion data obtained by the use of Eq. (80) are shown for a range of temperatures and compositions in Fig. 30.

For an amorphous (isotropic) host, the relaxation strength of the Gorsky effect, Δ^G, is given by

$$\Delta^G = E_U \vartheta_0 (\mathrm{tr}\lambda)^2 / 9(\partial\mu/\partial C), \tag{81}$$

where the concentration derivative of the chemical potential μ is taken at constant stress and temperature. It can be seen that in a case where E_U, ϑ_0 and trλ have been obtained by independent measurements, the Gorsky relaxation strength provides a method for the determination of $\partial\mu/\partial C$. This quantity is of interest for two reasons. First, it may be used to test Kirchheim's solution model, as has been done recently for a Pd-Si glass (62). Secondly, it can also be used to obtain the thermodynamic factor F_T which represents a dimensionless measure of the non-ideality of the solution, and is given by

$$F_T = C(\partial\mu/\partial C)/R_G T, \tag{82}$$

where R_G is the molar gas constant. Once F_T is known experimentally, it may be used to convert the chemical diffusion coefficient D_F to the tracer diffusion coefficient D^* by means of the relation

$$D^* = D_F/F_T. \tag{83}$$

An excellent application of this procedure to crystalline Pd-H alloys has been described by Verbruggen et al. (67). More recently, the procedure has also been applied by Berry and Pritchet (62) to a glassy $Pd_{86}Si_{14}$ alloy containing up to 8 a/o H, with the results shown in Figs. 31 and 32. The derived tracer coefficients are smaller than the corresponding chemical coefficients, and show less concentration dependence. Although the differences between D_F and D^* are substantial, they are seen from Fig. 33 to translate into rather minor differences in terms of the activation energies, and no effect is apparent on the value of the prefactor.

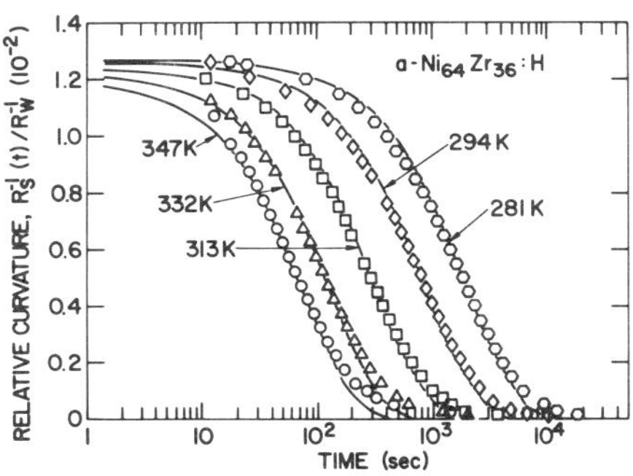

Fig. 29. The Gorsky aftereffect in a glassy ribbon of $Ni_{64}Zr_{36}$ containing about 4 a/o H. The results were obtained by the mandrel method; the solid lines are fitted simple exponential curves.

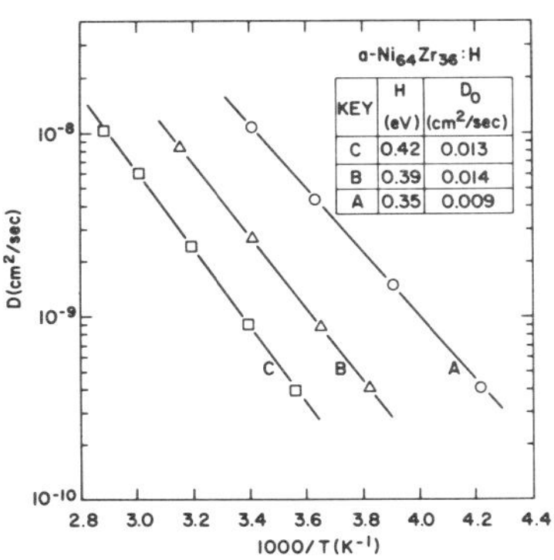

Fig. 30. Hydrogen diffusion in glassy $Ni_{64}Zr_{36}$, determined by the use of the Gorsky relaxation and Eq. (80). From Ref. 63.

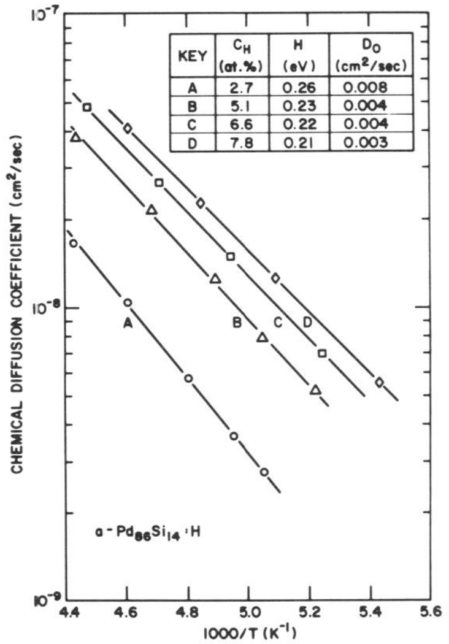

Fig. 31. Rate plots for chemical diffusion of hydrogen in glassy $Pd_{86}Si_{14}$, as derived from the kinetics of the Gorsky relaxation.

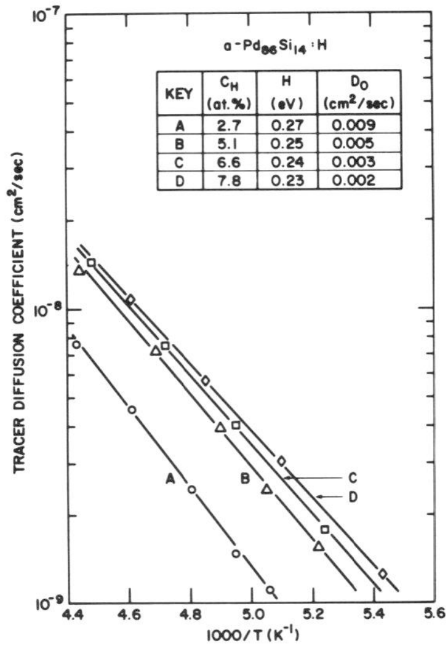

Fig. 32. Conversion of Fig. 31 to tracer diffusion, using the thermodynamic factor obtained from the strength of the Gorsky relaxation. From Ref. 62.

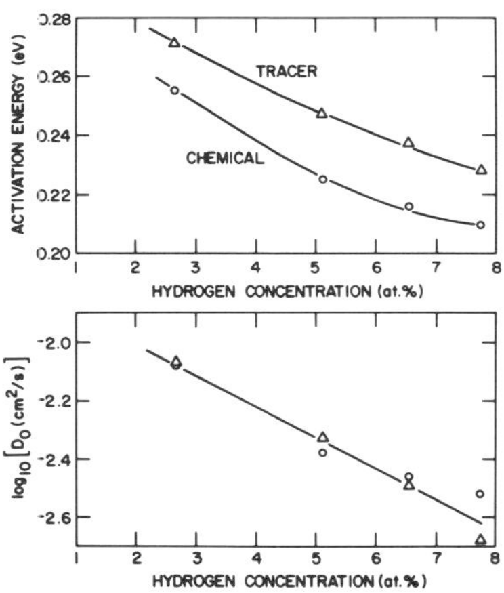

Fig. 33. Comparison of tracer and chemical diffusion parameters for hydrogen diffusion in glassy $Pd_{86}Si_{14}$.

6. OTHER THERMALLY-ACTIVATED PROCESSES IN METALLIC GLASSES

At the relatively low temperatures employed for the observation of relaxation behavior due to hydrogen, the host matrix of a metallic glass is usually configurationally frozen. This is not the case at higher temperatures, however, and internal friction associated with the motion of host atoms in non-hydrogenated glasses was already well known before the relaxations due to hydrogen were discovered (68)(69). In this section, we shall briefly summarize the status of this area, and describe one application of the vibrating-reed method to the kinetics of short-range ordering in a ferromagnetic glass.

The first indication of the onset of intrinsic atomic mobility in a metallic glass can be best obtained from processes which are sensitive to short-range atom movements. Thermally-activated anelastic behavior can be observed in metallic glasses well below the glass transition temperature T_g (e.g. $T/T_g<0.6$), and appears to originate from stress-induced directional ordering of a type roughly analogous to the Zener relaxation in crystalline solid solutions (4).

In contrast to the relatively narrow distribution found in crystalline alloys, it is not surprising to find that metallic glasses reveal a very wide distribution of relaxation times. Moreover, it appears that the distribution of activation energies associated with recoverable anelastic behavior merges with the distribution involved with non-recoverable viscous flow and long-range diffusion. As a result, anelasticity without overlap from viscous flow can only be expected in a relatively narrow temperature range below T_g, and the activation energies obtained should be less than or equal to those for long-range diffusion (70)(71).

In Pd-Si glasses which have been pre-annealed to produce a stabilized or structurally-relaxed configuration, intrinsic anelasticity is observed in the temperature range from about 125°C to 250°C (69). In quasi-static creep and recovery measurements, strong dispersion is evident from the highly smeared non-exponential form of the response. In internal friction measurements, the dispersion is evident from the fact that the apparent activation energy derived from the rate of rise of the internal friction with temperature is less than one-half of that obtained by the frequency-shift method (c.f. Fig. 6). From two independent investigations (68)(69), we may conclude that intrinsic host-atom anelasticity in glassy $Pd_{80}Si_{20}$ is controlled by activation energies in the range 1.2 eV to 1.4 eV. A general tendency for the activation energy to increase with the temperature of the prior stabilization anneal is evident from the behavior of the internal friction, which shows a general tendency to decrease with annealing. This behavior, which reflects a gradual shifting of the internal friction curves to higher temperatures, is a direct indication that structural relaxation is accompanied by a reduction of atomic mobility. A spectacular example of this behavior is encountered in the structural relaxation of silicon monoxide, described in Section 9.

Some of the most interesting metallic glasses, from both a scientific and technological point of view, are the iron-based soft ferromagnetic alloys (37)(72)(73). In addition to thermally-activated anelastic behavior, these alloys also undergo reversible directional ordering under the influence of the spontaneous magnetization, and develop an induced uniaxial magnetic anisotropy by annealing treatments in a magnetic field. (The purpose of the external field is to align the direction of the spontaneous magnetization along a selected direction, which is typically either the longitudinal or major transverse axis of the sample). The discovery of magnetic annealing in magnetic glasses was made during vibrating-reed studies of the ΔE-effect in a $Fe_{75}P_{15}C_{10}$ alloy (74). The ΔE-effect refers to the variation in Young's modulus E as a sample is taken from the demagnetized to the saturated condition. It is caused by stress-induced movements of magnetic domains, which add a magnetostrictive strain to the ordinary elastic strain, and thereby reduce the modulus by an amount ΔE below the saturation modulus E_S. Because the ΔE-effect is highly sensitive to the domain configuration of the material (69), it can be strongly influenced by prior magnetic annealing treatments which influence the domain configuration adopted in the demagnetized or remanent states (75). A novel application of the vibrating-reed method to obtain the kinetics of magnetic directional ordering is illustrated in Fig. 34. Prior to each isothermal run, the sample was first conditioned by a longitudinal magnetic anneal to induce a preferred axis of magnetization along the length of the sample. At the start of each run, the longitudinal field was replaced by a transverse field, which caused the anisotropy to decay in the longitudinal direction and then start to grow in the transverse direction. During this process, the ΔE-effect (which is measured during brief interruptions of the transverse field) is observed to pass through a maximum at the crossover point at which the induced anisotropy vanishes. As shown in Fig. 35, the results obtained by this technique are in excellent agreement with magnetic measurements on the same alloy by Luborsky (76).

At the time they were obtained, the activation energies of 1.0 eV and 1.35 eV for the two alloys of Fig. 35 were in striking disagreement with the results of internal friction measurements conducted at higher temperatures, which yielded the much larger values of 2.2 eV and 2.6 eV, respectively (74)(75). For several years, the probable cause of this discrepancy was not understood, although it was recognized that the prefactor associated with the internal friction was highly anomalous (69). It has since been pointed out (77) that the usually reliable frequency-shift method can give misleading results if the activation

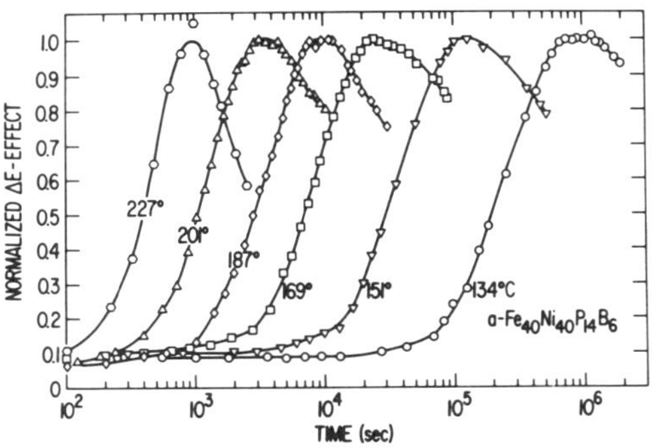

Fig. 34. ΔE-behavior produced by magnetically annealing glassy $Fe_{40}Ni_{40}P_{14}B_6$ along a previously-induced hard-axis direction.

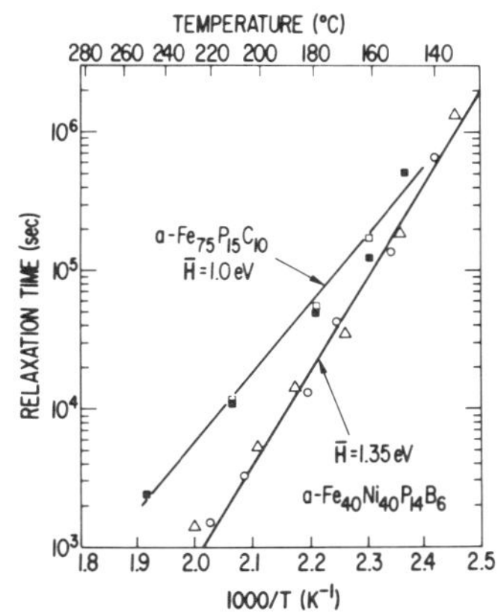

Fig. 35. Magnetic annealing kinetics for two amorphous alloys. Triangles and squares are ΔE-effect measurements from Ref. 75; the circles are Luborsky's magnetic measurements from Ref. 76.

energy becomes slightly temperature dependent over the range of the measurements, an effect which may be related to the onset of a reversible structural relaxation of the type discussed by Morito and Egami (78). A reanalysis of the internal friction results now suggests that they are compatible with the magnetic annealing behavior, and this in turn leads to the revised conclusion that the directional ordering that occurs in response to either stress or magnetization involves the same type of atom movements.

7. GRAIN-BOUNDARY DIFFUSION IN PURE AND ALLOYED ALUMINUM FILMS

So far in this chapter, we have dealt only with relaxations associated with point defects. In crystalline metals, it is known that prominent relaxations can also be produced by dislocations and by grain boundaries (4). Our last example from the area of metals deals with the internal friction of supported aluminum films, where the observed relaxation behavior can be associated with thermally-activated grain-boundary sliding (31). Here, we encounter the interesting situation that the results obtained from films show a striking difference from those previously obtained on bulk samples, and promise to be a major factor in the resolution of long-standing difficulties.

The possibility that anelastic behavior might be associated with grain boundaries (or other internal interfaces) in metals was first discussed by Zener, who recognized that time-dependent viscous-type sliding at a boundary was a means for local relaxation of a shear stress acting across the boundary (1)(79). If, in Fig. 36a, the slip offset u is elastically accommodated by the build-up of a back stress near the grain corners, this stress will lead to a reversal of the slip on unloading and anelastic behavior will result. In a rough analogy, the grain-boundary relaxation can be visualized in terms of a jigsaw puzzle whose interlocking pieces are coated with a viscous glue. The rigidity of such a structure relaxes when heated to a temperature at which the glue flows, allowing the pieces to slide until until they jam together at the corners. For a fine-grained polycrystalline aggregate, Zener deduced that the maximum relaxation strength should depend only on Poisson's ratio ν, and gave the following expression for the tensile relaxation strength Δ_E:

$$\Delta_E = (7-3\nu-5\nu^2)/(7 + 5\nu). \qquad (84)$$

For the typical value $\nu = 1/3$, it can be seen that Δ_E has the substantial value of 0.63. By assuming a formal continuum model for sliding at a boundary of width b, length g and viscosity η, the relaxation time τ_σ can be expressed as

$$\tau_\sigma = \left(\frac{\eta}{G_U}\right)\left(\frac{g}{b}\right), \qquad (85)$$

where G_U is the unrelaxed shear modulus of the aggregate. It should be noted that Eq. (85) predicts that the relaxation time should vary linearly with the grain size (boundary length), and exhibit a temperature dependence governed by the boundary viscosity η.

The first example of relaxation associated with grain boundaries was found by Kê in polycrystalline aluminum wires (80),(81). Figure 37 shows some internal friction measurements conducted at about 1 Hz using a torsion-pendulum apparatus. The prominent peak located near 300°C is of the predicted strength, and exhibits a dispersion which can be attributed to a distribution of boundary types and sizes.

Fig. 36. Schematic illustration of grain-boundary sliding (a) with elastic accommodation of the displacement u, and (b) with diffusional accommodation.

Table 3: Kinetic Parameters for Grain Boundary Relaxation in Bulk and Film Samples.

Sample	Nominal Grain Size(μm)	Peak I		Peak II		Ref.
		$\bar{H}$(eV)	$\bar{\tau}_0$(s)	$\bar{H}$(eV)	$\bar{\tau}_0$(s)	
Bulk Al	200	1.48	10^{-14}			80,81
Al film	0.1	0.51	4×10^{-13}	0.56	6×10^{-10}	Fig. 38
Al-Cu film (0.17a/o Cu)	0.1	0.97	10^{-15}	0.93	10^{-11}	Fig. 39

The peak shifts to higher temperatures as the grain size is increased by annealing, in the manner suggested by Eq. (85). Coarsely polycrystalline samples exhibit peaks of reduced strength, and no peak was observed in a sample consisting of only a few grains. While these results have generally been accepted as persuasive evidence in favor of the grain-boundary sliding model, the kinetic parameters of the relaxation (Table 3) exhibit some unexpected and puzzling features. First the value of τ_0 estimated from Eq. (85) is about six orders of magnitude larger than the observed value, which intriguingly turns out to be of a typical magnitude for a single atom jump. Secondly, the activation energy is close to that for lattice diffusion, whereas it might be expected that the boundary viscosity η should scale with a boundary diffusion coefficient controlled by a substantially lower activation energy. Subsequent work on bulk metals has done little to clarify this situation. In particular, although peaks have been observed in high-purity copper and silver with activation energies below that of lattice diffusion, no such reduction has been observed for aluminum (4).

Following the emergence of aluminum as the most important conductor in thin-film interconnection technology, the subject of

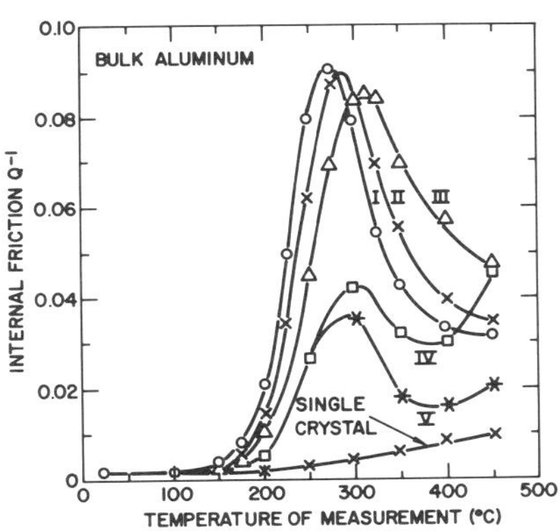

Fig. 37. Internal friction peaks associated with grain-boundary relaxation in polycrystalline aluminum wire of 0.084 cm diameter, tested in torsion at ~1 Hz. For curves I-III, the grain size was 0.02, 0.04 and 0.07 cm, respectively: For curves IV and V, the grain size exceeded the wire diameter. From Kê, Ref. 81.

grain-boundary diffusion in this material came into painful and prominent focus with the device failures associated with preferential electromigration at grain boundaries. These failures, and the investigations they provoked, provided clear evidence that the activation energy for grain-boundary diffusion in pure aluminum films was in the vicinity of 0.5 eV to 0.6 eV (82). It also became apparent that films possess a major advantage for studies of boundary phenomena by virtue of their much finer grain size, which is typically three orders of magnitude smaller than in bulk samples. The investigation of aluminum films therefore became an early target of internal friction studies, following the development of the composite vibrating-reed method (16)(31). The results of Fig. 38 were obtained with a symmetrical trilayer sample consisting of 0.1 μm films on both sides of a 50 μm substrate of fused silica. The films were obtained by electron-gun evaporation of zone-refined aluminum, deposited at a rate of 2 nm/s. Transmission electron microscopy of sister samples prepared on thin windows of silicon nitride showed the films to consist of a single layer of grains, with a grain size nominally equal to the film thickness, and with the boundaries mostly arranged as vertical walls perpendicular to the substrate. Despite the fact that the film to substrate thickness ratio is now only 0.004, it can be seen from Fig. 38 that the substrate damping level ($\sim 10^{-5}$) is again unimportant. The indicated frequencies correspond to the 1st, 2nd, 4th, and 8th tones, and provide a maximum frequency ratio of 154. The data indicate the presence of two broad overlapping internal friction peaks with the smaller peak (I) located below the larger peak (II). Peak I was more prominent in the as-deposited film, but becomes less well-resolved following the short 550°C anneal which preceded the measurements of Fig. 38. It should be noted that Peak I extends well below room temperature. Taken together, both peaks indicate thermally-activated behavior over a broad temperature range. As can be quickly appreciated from the near equality of the $1/T$ shifts at high and low temperatures, the mean activation energies of Peaks I and II are essentially identical (Table 3), and are in good agreement with the average value of 0.55 eV deduced for grain-boundary diffusion by electromigration experiments. Whereas Peak I has a τ_0-value which, within experimental uncertainty, can be taken as normal for a short-range relaxation, τ_0 for Peak II is substantially larger and indicates a mechanism involving grain-boundary diffusion over a larger distance.

On the basis of these and other experiments, particularly those concerning the effects of grain size, it has been concluded (31) that Peaks I and II correspond to the two stages of grain-boundary sliding shown in Fig. 36. Stage I, associated with Peak I, involves sliding with elastic accommodation in both the film and the underlying substrate, and corresponds to Zener's model for bulk materials (Fig. 36a). Stage

II, associated with Peak II, involves a second regime of sliding at higher temperatures where sliding is rate-controlled by diffusional accommodation in the film (83), as indicated schematically in Fig. 36b. In bulk samples, this type of sliding does not produce a grain-boundary peak because diffusional creep is not recoverable and contributes instead to a high-temperature background damping. In an elastically-supported thin film, however, the situation is greatly modified by the presence of the substrate, which enforces recoverability and a pseudo-anelastic response on the composite which the detached film would not by itself exhibit. Based on this interpretation, it should be noted that the near equality of the activation energies for Peaks I and II neatly demonstrates that sliding in the stage I (Zener) regime is indeed rate-limited by grain-boundary diffusion and points to the fact that in the larger-grained bulk samples sliding becomes rate-limited by some local type of pinning that is released only by a jump of much higher activation energy. There is abundant evidence that solute atoms segregated at or near a grain-boundary may play a key role in this pinning process. In studies of both bulk copper and bulk silver, small solute additions have been observed to increase the activation energy for grain-boundary relaxation up to values close to that for lattice diffusion (4). The role of impurities in retarding the migration of grain boundaries is another well-known effect in bulk metals. In films, the role of solute atoms in retarding electromigration damage was first demonstrated for the case of copper additions to aluminum films (84), and has since proved to be of considerable technological importance. The marked effect of small additions of copper on the grain-boundary behavior of

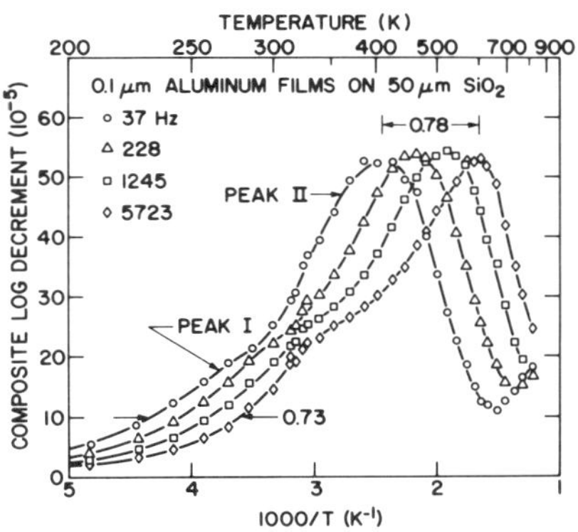

Fig. 38. Internal friction peaks associated with two stages of grain-boundary sliding in 0.1 μm supported aluminum films. From Ref. 31.

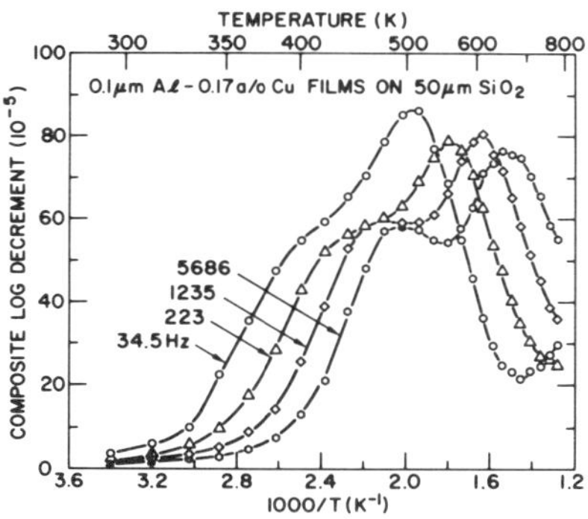

Fig. 39. Internal friction peaks associated with two stages of grain-boundary sliding in 0.1 μm supported aluminum films containing 0.17 a/o of copper.

aluminum films has also been strikingly demonstrated by vibrating-reed measurements. For comparison with Fig. 38, results obtained on a nominally identical sample containing 0.17 a/o of copper are shown in Fig. 39. The same general behavior of two overlapping peaks is again evident. Indeed, Peak I is now somewhat better resolved, perhaps in part because of the effect of solid-solution hardening in strengthening the grain corners. The important difference in the kinetics of relaxation is immediately indicated by the considerably higher temperatures marking the onset of relaxation behavior. From analysis of the frequency shifts, it is found (Table 3) that the addition of copper raises the mean activation energy of both Peaks I and II from about 0.55 eV to 0.95 eV. As yet, a detailed explanation of this remarkable difference is lacking. Indeed, in contrast to the readily understandable concept of solute-dragging associated with the retardation of boundary *migration*, it may even appear surprising that small solute additions should have such a marked effect on the *sliding* behavior of grain boundaries. The important conclusion suggested by the relaxation behavior is therefore that sliding and migration are coupled processes, and that a solute addition which retards one will necessarily retard the other. Atomistic models of grain-boundary dislocations in which sliding and migration are coupled processes were discussed some years ago by Ashby (85).

8. POINT-DEFECT RELAXATIONS IN ION-IMPLANTED SILICON

Point defects in semiconductors, and particularly silicon, have been a major area of investigation for about the last quarter-century. Nevertheless, basic issues concerning relatively simple defects still remain matters of debate. The situation is complicated by the fact that a given generic defect (a vacancy, for example) may exist in more than one charge state, which in turn may affect both the configuration and atomic mobility of the defect. In contrast to metals, many important techniques (e.g. electron paramagnetic resonance, infrared spectroscopy, photoconductivity, and deep-level transient spectroscopy) are available for the study of point defects in semiconductors. Nevertheless, the difficulties referred to above make it clear that there is still an opportunity for other techniques, such as anelastic relaxation measurements, to be of use. Particular interest centers on the discovery and interpretation of internal friction peaks associated with the reorientation of defects possessing an asymmetric or orientationally-distinguishable character. As we have seen earlier in this chapter, such peaks provide their own unique view of a defect through an interesting blend of kinetic, thermodynamic and structural information. Particularly for semiconductors, however, it must be recognized that the sensitivity of internal friction measurements may be much less than that of spectroscopic techniques which are sensitive to the electronic, rather than the atomic, features of the defect. For example, a defect concentration of 10^{-4} atom fraction is usually required to secure a good resolution of an internal friction peak above the background damping. On the other hand, there is another viewpoint from which the available detection sensitivity is more encouraging. Unlike techniques whose sensitivity is linked to the absolute number of defects in the sample, the height of an internal friction peak is not reduced by decreasing the sample size. The sensitivity therefore tends to become more competitive as the sample volume becomes small, as in the study of thin layers produced by ion implantation.

The sensitivity considerations involved in the application of the composite vibrating-reed method to ion-implanted layers are very similar to those involved for thin films, since the effective layer thicknesses produced by implantation are of typical thin-film dimensions (0.1 to 1 μm). Following implantation, a silicon reed of a convenient practical thickness (50 μm) effectively becomes a composite sample largely consisting of unimplanted material acting as a carrier substrate for the thin implanted near-surface layers. As for deposited thin films, the success of the approach is dependent on a low substrate damping, which fortunately is available for silicon if due regard is paid to the method of mounting the reed.

The potential of the vibrating-reed method for this type of application is well illustrated by the results of Fig. 40, which show the

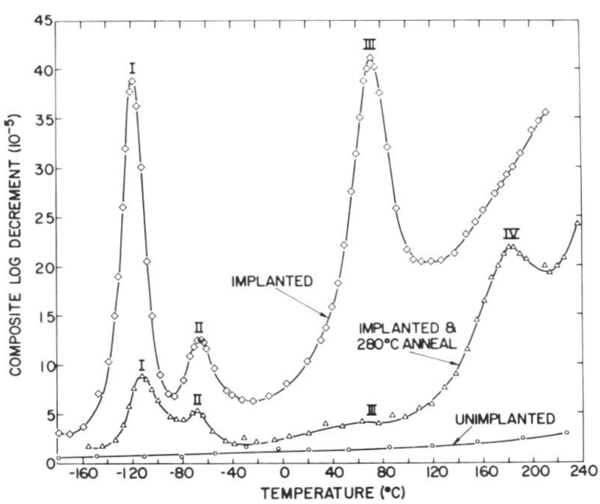

Fig. 40. Internal friction at 438 Hz of a ⟨100⟩ reed of Czochralski silicon, before and after boron implantation to 5×10^{15} B^{11}/cm^2 per face, and after a 1 h anneal at 280°C. From Ref. 86.

Table 4: Parameters for Internal Friction Peaks Observed in Boron-Implanted Czochralski Silicon (Fig. 40 and Ref. 86).

Peak No.	H (eV)	τ_0 (10^{-14}s)	Relative Width	Orientation Dependence
I	0.29	18	1.3	$\delta E^{-1}_{\langle 111 \rangle} \simeq 0$
II	0.40	7	1.2	$\delta E^{-1}_{\langle 111 \rangle} = 0$
III	0.70	4	1.2	$\delta E^{-1}_{\langle 111 \rangle} < \delta E^{-1}_{\langle 100 \rangle}$
IV	0.92	3	1.1	$\delta E^{-1}_{\langle 111 \rangle} = 0$

peaks produced by implantation of boron at 100 kV and 200 kV into a ⟨100⟩ single-crystal reed of Czochralski silicon (87). The implantation conditions were chosen to distribute the boron in a roughly uniform layer over a depth of about 0.6 μm from each surface. The overall configuration is therefore that of a symmetrical trilayer, satisfying the thin-film limit $a_L/a_S \ll 1$. Whereas the unimplanted reed showed no peaks, the uppermost curve of Fig. 40 shows three well-resolved peaks labeled I-III. While not evident initially, yet another peak (IV) makes a transient appearance during subsequent annealing, and is shown by the middle curve of Fig. 40 as it appears after a 1 hr anneal at 280°C. It can be seen that this treatment virtually eliminated peak III and caused a substantial reduction of peaks I and II. Complete elimination of all peaks was obtained by annealing at 400°C. Other important characteristics of the peaks are summarized in Table 4. The kinetic parameters τ_0 and H were obtained by the frequency-shift procedure, using the first four tones of vibration and covering a range from 77 Hz to 2300 Hz. The relative width r is the half-width of the peak relative to that of a Debye peak of the same activation energy. The tabulated peak heights refer to the values exhibited by the composite sample and should be multiplied by a factor of 11 for conversion to the values characteristic of the implanted layers themselves. The orientation dependence of the peak heights, shown in the last column of Table 4, summarizes the conclusions that can be drawn from measurements on identically treated ⟨100⟩ and ⟨110⟩ oriented samples. Three of the four relaxations (I, II and IV) reveal a strong anisotropy of similar type, corresponding to defect configurations of tetragonal or orthorhombic symmetry in which the principal axis lies along a ⟨100⟩ direction (4).

In an early interpretation of these results, it was suggested that all four peaks were associated with different types of dumbbell or split-interstitial defects (86). Peaks I and II were associated with different charge states of a boron-silicon dumbbell, and peaks III and IV were associated with different charge states of a silicon self-interstitial. The detailed rationale behind these assignments has been given by Frank (87). Since then, additional experiments have shown that the initial identification of Peak II is incorrect, and that the defect involved is not boron-related. It is now known that Peak II can be produced by either electron-irradiation or hydrogen implantation of Czochralski silicon. A strong link to oxygen was revealed by the virtual absence of the peak in parallel experiments on float-zoned (low-oxygen) silicon. Based on these results and other considerations involving the anisotropy of the defect, the reorientation kinetics and the annealing behavior, there is now a convincing case that peak II is due to the vacancy-oxygen (V-O) defect identified by Watkins and Corbett (88)(89) and shown in Fig. 41. This defect, which is formed by a mobile implantation-produced vacancy combining with an impurity interstitial oxygen atom, may alternatively be regarded as an off-center substitutional oxygen atom. The reorientation of the defect occurs by a local jump of the oxygen between six equivalent positions in the cage provided by the vacancy, and involves an activation energy of approximately 0.4 eV (Table 4). The ability to reorient with a jump of much lower energy than that which produces either dissociation or migration of the defect is clearly evident from the fact that the peak temperature is far below the annealing temperature of the peak.

To capitalize on the identification of the defect associated with Peak II, and the later calibration of the peak by the use of electron irradiation and EPR measurements, subsequent experiments have been performed in which oxygen ions were implanted to various fluences into a ⟨100⟩ reed of float-zoned silicon (90). In this work, Peak II provided a tool for determining how many of the implanted oxygen ions were detectable as V-O centers. Figure 42 shows the variation of the peak with oxygen fluence, and Fig. 43 gives the absolute and relative numbers of the V-O centers deduced therefrom. The maxi-

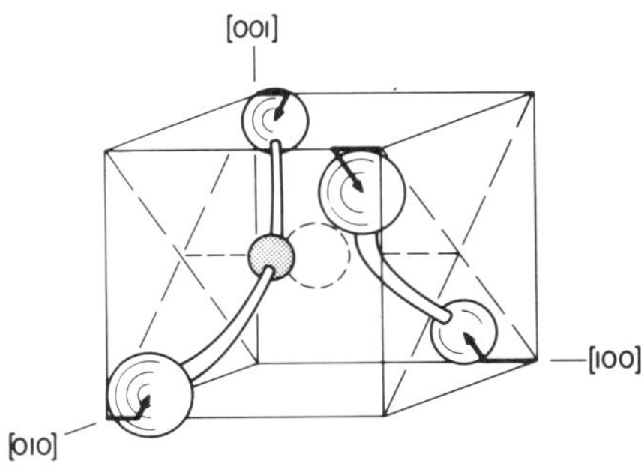

Fig. 41. The vacancy-oxygen defect in silicon. The vacancy is indicated by the dashed circle, the oxygen atom by the shaded circle. Displacements of the surrounding silicon atoms are indicated schematically by the arrows.

mum in the height of the peak occurs at a fluence of 3×10^{15} O^+/cm^2, which coincides with the coalescence of implantation-produced amorphous zones to form a continuous surface layer. This and the eventual leveling off in the peak height indicates a saturation of the defect damage in the transitional damaged region below the amorphous layer. The presence of a maximum in the V-O defect concentration indicates that the conversion of a V-O center from a crystalline to an amorphous environment inactivates the defect as far as a contribution to Peak II is concerned. While it is possible that this is caused by strong local distortions that lock the defect in a particular orientation, it may also imply a more extensive structural rearrangement in which the defect identity is destroyed. Quantitatively, it can be seen from Fig. 43 that at a fluence of 10^{14} O^+/cm^2, about one-third of the implanted oxygen ions are detected as V-O centers. As the oxygen fluence increases, this ratio falls dramatically. It is about 0.07 at the peak in the absolute number of V-O centers, and is only 0.002 for the highest fluence of 5×10^{16} O^+/cm^2.

Complementary kinetic data obtained by the frequency-shift method are in good agreement with Table 4. The results shown in Fig. 44 provide a mean activation energy of 0.38 eV and a τ_0-value of 2×10^{-13} sec. Interestingly, in contrast to the Debye shape obtained by electron irradiation, Peak II shows significant broadening when produced by implantation. The peaks of Fig. 44 have a relative width of 1.5, indicative of a substantial perturbation of the crystalline environment by implantation damage. As indicated by the absence of a peak shift with fluence (Fig. 42), this broadening is associated with a symmetrical dispersion of activation energies about a constant mean value.

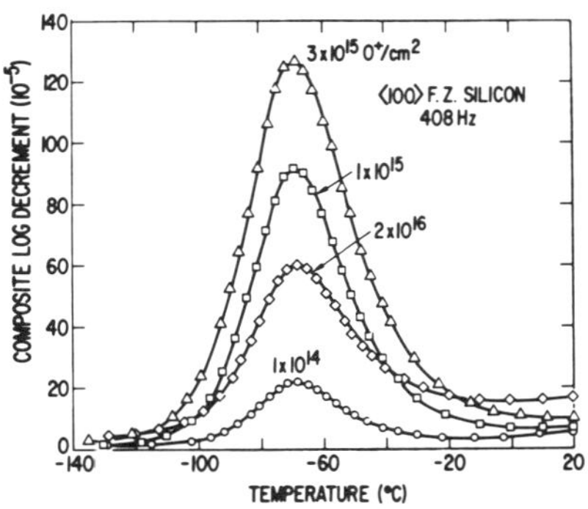

Fig. 42. Internal friction at 408 Hz of a ⟨100⟩ reed of float-zoned silicon, after oxygen implantation on each face to the fluences indicated.

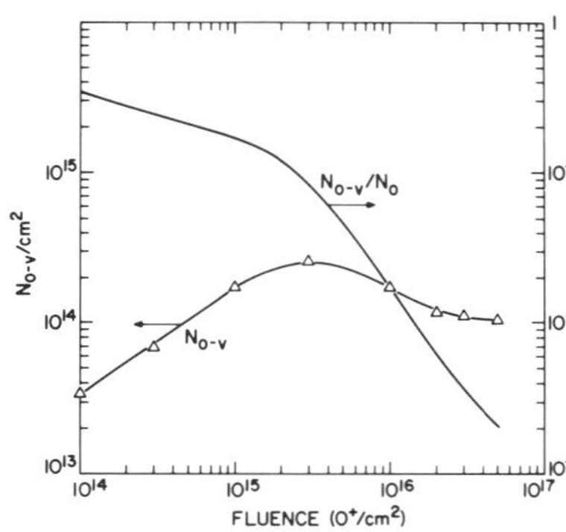

Fig. 43. The deduced absolute and relative numbers of V-O defects produced by oxygen implantation of float-zoned silicon. From Ref. 90.

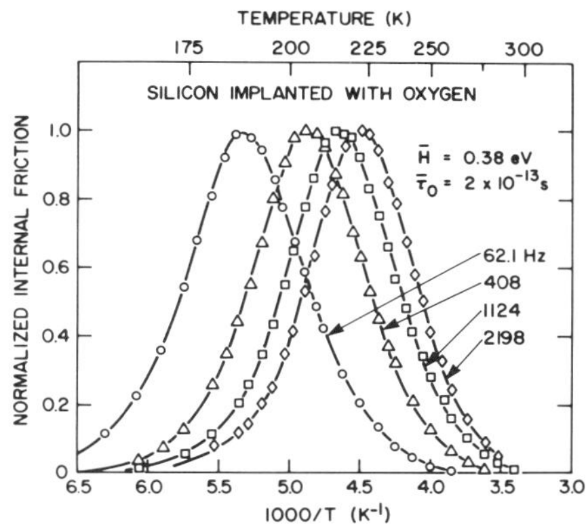

Fig. 44. Frequency dependence of the internal friction peak produced by the implantation of a float-zoned ⟨100⟩ silicon reed with 3×10^{15} oxygen ions/cm^2 per face. From Ref. 91.

To conclude this section, it is appropriate to consider the impact the new defect identity of Peak II may have on the defect identities suggested earlier for the other peaks listed in Table 4. Based on present evidence, further revisions appear likely. In particular, the defect responsible for Peak I now appears to be in doubt. We recall that there are strong similarities between Peaks I and II in terms of symmetry, reorientation kinetics, and annealing behavior (86). These similarities are so marked that they led originally to the suggestion that Peaks I and II are caused by basically similar defects. Since it still appears that Peak I is caused by a boron-related defect, the new defect identity of Peak II now suggests that Peak I is associated with a boron-vacancy defect rather than with a boron interstitial. A second reason for suspecting a misidentification of Peak I is provided by strong EPR work which assigns notably different characteristics to the boron interstitial (92).

9. STRUCTURAL RELAXATION IN SILICON MONOXIDE

For our last example of the application of anelastic and bending-cantilever methods to thin-layer materials, we consider an electrical insulator, and in doing so return to the realm of amorphous materials. Unlike its better known relative silicon dioxide, which is invariably deposited by sputtering, the lower monoxide centered on the stoichiometric composition SiO has a high enough vapor pressure to be deposited in vacuum from a sublimation source. This advantage, coupled with reasonably attractive dielectric and optical properties, has resulted in a continued interest in this material. Surprisingly, however, remarkably little work has appeared on the mechanical properties of SiO, and even its elastic modulus appears to be in doubt. One important unexplored aspect of this and other amorphous dielectrics produced by condensation techniques is that the as-deposited state at room temperature can be expected to be very far from configurational equilibrium. As a consequence, we can anticipate that such materials will undergo structural relaxation on subsequent annealing, with an accompanying densification and attendant changes in other properties of the film. While direct measurements of small density changes would be a difficult experimental problem on detached films, it is a natural application of the bending-bilayer method (Section 3.3), where elastic accommodation of even a small density increase leads to the generation of an appreciable tensile stress in the film. Moreover, as we shall see below, complementary vibrating-reed measurements that provide information on both the elastic stiffness and thermally-activated internal friction behavior of the film add an informative further dimension to such studies.

The internal friction results of Fig. 45 were obtained from a double-sided sample of the symmetrical trilayer configuration, consisting of 1.1 μm films of SiO on a 50 μm substrate of fused SiO_2. The films were deposited at 0.9 nm/s from a shielded source at 1140°C on to the substrate reed held nominally at room temperature. The data of Fig. 45 illustrate the marked movement in the internal friction behavior resulting from a sequence of 1-hour vacuum anneals extending up to 800°C. It will be noted that the largest changes are produced by the anneals at the lower temperatures, and particularly below 400°C. From measurements made in each run at up to five different frequencies between 36 Hz and 2054 Hz, the rise of internal friction has been shown to reflect a thermally-activated relaxation behavior governed by a wide distribution of relaxation times. The observed frequency-shifts reveal that the annealing of the internal friction is caused by a smooth monotonic rise in the effective activation energy, from about 1 eV to 2 eV, as shown in Fig. 46. As for metallic glasses, the relaxation behav-

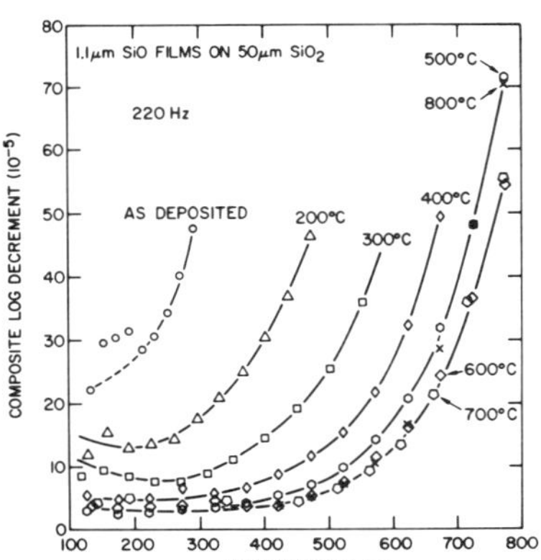

Fig. 45. Effect of a sequence of 1h anneals on the internal friction of supported 1.1 μm films of silicon monoxide.

ior can be attributed to stress-induced atomic motions occurring preferentially at sites of low coordination and high free-volume. The activation energy measurements of Fig. 46 not only provide a most striking illustration of the marked decrease in atomic mobility that occurs on structural relaxation, but more specifically appear to support a free-volume mechanism of diffusion, such as that discussed by Argon (93). Figure 46 also shows the complementary variation in the elastic modulus, as deduced from concurrent frequency measurements. Structural relaxation is accompanied by an increase of about 15% in the modulus, which follows the same trend as that shown by the activation energy. It may be noted that the absolute value of the modulus found in this work is twice that reported recently by Hoffman, based on tensile testing of thin detached layers (94). Apart from the magnitude of the modulus, its temperature coefficient $(1/E)\mathrm{d}E/\mathrm{d}T$ is also of considerable interest, in view of the unusual *positive* value $(+1.7\times10^{-4}/\mathrm{K})$ exhibited by fused SiO_2. The present measurements indicate that SiO is almost at the crossover point to more normal negative values, and has an unusually small but negative coefficient of $-2\times10^{-5}/\mathrm{K}$.

Having seen from Fig. 46 that structural relaxation occurs continuously over the whole temperature range, we now turn to the bending-cantilever results obtained from a sister sample of the one-sided or bilayer configuration. As shown in Fig. 47, annealing up to 400°C causes the stress to increase in the manner expected for densification associated with structural relaxation. The almost ten-fold increase in stress, which amounts to an increment of 30MPa, corresponds to a

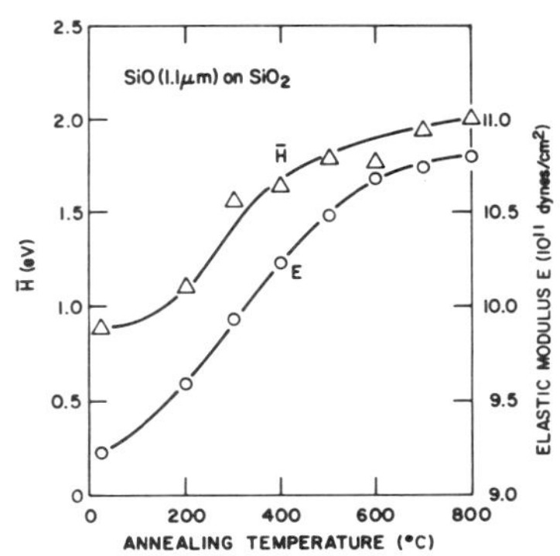

Fig. 46. Effect of annealing on Young's modulus E and the internal friction activation energy $\overline{H}$ of supported 1.1 μm films of silicon monoxide.

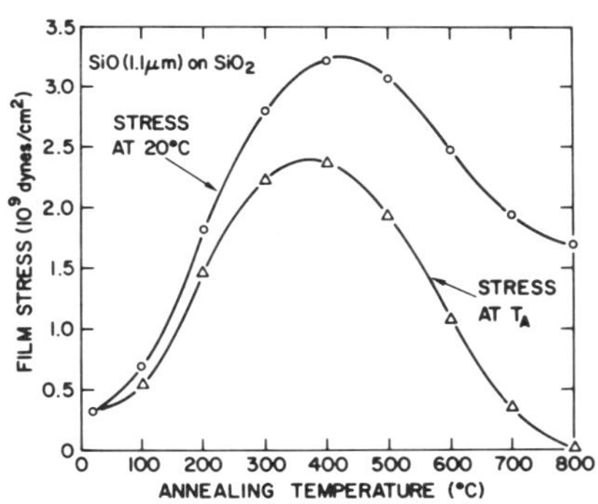

Fig. 47. Effect of a sequence of 1 h anneals on the at-temperature and room-temperature accommodation stress in a 1.1 μm film of silicon monoxide.

calculated density increase of 0.7%. Beyond 400°C, however, the stress variation reverses and the stress at temperature progressively falls to zero. From the knowledge provided by the vibrating-reed measurements concerning the monotonic course of structural relaxation, the interpretation of the maximum becomes straightforward. In contrast to the elastic accommodation of the density change which occurs at low temperatures with an attendant increase of stress, we can conclude that above 400°C viscous flow takes over as the accommodation mechanism. A final point of interest is provided by the difference between the two stress curves of Fig. 47. This difference represents the thermal stress generated on cooling from the annealing temperature. From the variation of this component of the stress with temperature, we may deduce the thermal expansion coefficient of silicon monoxide. Near room temperature, the value is 2.3×10^{-6}/K, which represents a surprisingly good match to that of crystalline silicon.

10. CONCLUDING REMARKS

The static and dynamic bending-cantilever methods described in this chapter have emerged as a useful and frequently complementary set of techniques with broad applications to the study of thin-layer materials. Although emphasis has been placed here on kinetic phenomena related to atomic diffusion, it should be apparent that the methods also embrace broader topics related to defects, microstructure, and the mechanical and other properties of thin layers. It should also be noted that the work described here represents a selective rather than a comprehensive survey, and that a number of other investigations covering such diverse fields as polymer films, compositionally-modulated metal films, and organic layered structures have also appeared in the literature (19)(20)(95)(96)(97). While experimental limitations cannot be ignored, the overall impression gained from a decade of experience is of the power and surprising versatility of bending-cantilever methods in general, and of the composite vibrating-reed method in particular.

As an investigative tool, anelastic relaxation measurements provide several important advantages. The approach is applicable to virtually any material, whether it be metallic, semiconducting, insulating, amorphous or crystalline. The equipment is relatively simple and of modest cost, yet provides for the selective and sensitive study of the structural and kinetic aspects of atomic-level defects. In the specific area of thin-layer materials, there is now an impressive discovery rate of significant scientific phenomena, as illustrated by the examples discussed in this chapter. With a seemingly endless list of future applications in sight, we can look forward with optimism to the continued success of these methods for the investigation of thin-layer materials.

ACKNOWLEDGMENT

It is a particular pleasure to acknowledge the participation of my colleague W. C. Pritchet in much of the work described in this chapter.

APPENDIX: FREQUENCIES AND INTERNAL FRICTION OF A MULTILAYERED REED

Expressions for the frequencies and internal friction of a uniform reed consisting of m adherent layers may be obtained by generalization of the treatment for a single layer. The logarithmic decrement δ_C of the composite may be written

$$\delta_C = \sum_m \Delta W_i / 2\pi \sum_m W_i = \sum_m (W_i/W_C)\delta_i, \tag{A1}$$

where

$$W_C = \sum_m W_i. \tag{A2}$$

In this notation, $\Delta W_i/W_i$ is the fraction of the mechanical energy in the i'th layer dissipated per cycle, and $\delta_i = \Delta W_i/2\pi W_i$ is the logarithmic decrement that would be exhibited by the i'th layer if tested alone under otherwise identical conditions. We consider here only multilayers composed of two materials, L and S. The treatment followed is applicable to the general configuration $L_1S_1L_2S_2 \ldots L_pS_q$ if all individual L-layers have the same internal friction δ_L, and all the S-layers have the same internal friction δ_S. Equation (A1) then reduces to

$$\delta_C = \left(\frac{W_S}{W_C}\right)\delta_S + \left(\frac{W_L}{W_C}\right)\delta_L, \tag{A3}$$

where $W_S = \Sigma_q W_{Sq}$ and $W_L = \Sigma_p W_{Lp}$. The cases we shall consider here are relatively simple, and involve layers of the same material L on one or both sides of a single substrate, S. However, in contrast to previous treatments (16)(31), we shall not immediately assume that the thickness of these layers is negligible compared with the substrate thickness.

The flexural strain energy of a multilayer reed of length ℓ is given by the integral

$$W_C = (K_C/2)\int_{-\ell/2}^{+\ell/2}(d^2y_n(x)/dx^2)^2 dx, \tag{A4}$$

where $y_n(x)$ is the vibration shape of the n'th mode, and K_C is the bending stiffness of the composite reed, defined by

$$K_C \equiv B/(1/R), \tag{A5}$$

where $(1/R)$ is the static uniform curvature produced by a bending moment B. Using the known mode shapes of the clamped-free reed (4), evaluation of Eq. (A4) gives the result

$$W_C = (2\alpha_n^4/\ell^3)K_C A_0^2, \tag{A6}$$

where α_n is the mode constant of Fig. 9, and A_0 is the maximum amplitude at the free end of the reed. Because all of the component layers have the same mode shape, similar expressions to Eqs. (A4) to (A6) can also be written for the partial energies W_S and W_L in terms of the partial bending stiffnesses K_S and K_L. We may thus write Eq. (A3) in the form

$$\delta_C = \left(\frac{K_S}{K_C}\right)\delta_S + \left(\frac{K_L}{K_C}\right)\delta_L, \tag{A7}$$

or

$$\delta_C = \left(\frac{1}{1+\gamma}\right)\delta_S + \left(\frac{\gamma}{1+\gamma}\right)\delta_L, \tag{A8}$$

where γ is a stiffness factor defined by

$$\gamma \equiv K_L/K_S = W_L/W_S. \tag{A9}$$

It is clear from Eq. (A8) that the experimental determination of δ_L from δ_C and δ_S requires a knowledge of γ. Before considering an explicit form of γ, it is useful to consider the frequencies of the reed. These may be found by Rayleigh's method, which equates the strain energy in the position of maximum amplitude (zero velocity) to the kinetic energy in the position of maximum velocity (zero displacement). The kinetic energy W_C' is given by

$$W_C' = (\mu_C\omega_C^2/2)\int_{-\ell/2}^{+\ell/2}(y_n(x))^2 dx = (\mu_C\ell/8)\omega_C^2 A_0^2, \tag{A10}$$

where μ_C is the mass per unit length and ω_C is the circular frequency. Combining Eqs. (A6) and (A10), we obtain the frequency equation

$$\omega_C^2 = (2\pi f_C)^2 = (2\alpha_n/\ell)^4 K_C/\mu_C. \tag{A11}$$

Since Eq. (A11) also covers as a special case the blank substrate of frequencies f_S, stiffness K_S and mass per unit length μ_S, we have

$$f_C^2/f_S^2 = (K_C/K_S)(\mu_S/\mu_C) \tag{A12}$$

or

$$f_C^2/f_S^2 = (1+\gamma)/(1+\chi), \tag{A13}$$

where χ is an inertia factor, complementary to the stiffness factor γ, and defined by

$$\chi \equiv \mu_L/\mu_S. \tag{A14}$$

In some situations, Eq. (A13) may provide a useful experimental approach to the determination of the stiffness factor γ appearing in Eq. (A8). As an alternative, we may seek an explicit form for the ratio K_C/K_S (= $1+\gamma$) in terms of the thicknesses and moduli of the component layers. For a single-layer reed, such as a blank substrate, the bending stiffness K_S has the well-known form

$$K_S = E_S I_S = E_S b a_S^3/12, \tag{A15}$$

where E_S is Young's modulus, I_S is the second moment of area of a cross-section about the neutral axis, b is the breadth and a_S is the thickness of the substrate. By contrast, the explicit form of K_C for an arbitrary multilayer is much more cumbersome. A case of sufficient generality for present purposes is provided by the general trilayer of Fig. 48. Various simplifications (specifically, to the cases of a bilayer or a symmetrical trilayer) are then easily made as special cases. On the other hand, the trilayer formalism clearly shows how the expression for K_C may be extended, if needed, to a larger number of layers. With the terminology of Fig. 48, the bending stiffness of the general trilayer is given by

$$K_C = E_1 b\{\tfrac{1}{3}z_1^3 - z_N z_1^2 + z_N^2 z_1\} + E_2 b\{\tfrac{1}{3}(z_2^3-z_1^3) - z_N(z_2^2-z_1^2) + z_N^2(z_2-z_1)\} + E_3 b\{\tfrac{1}{3}(z_3^3-z_2^3) - z_N(z_3^2-z_2^2) + z_N^2(z_3-z_2)\}, \tag{A16}$$

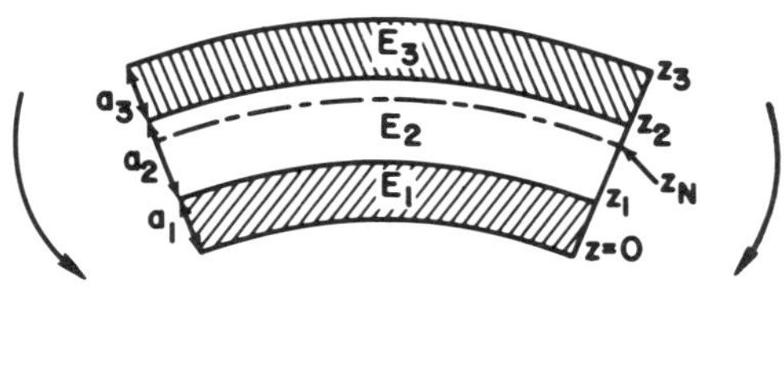

Fig. 48. General form of the trilayer reed.

where z_N, which defines the position of the neutral surface in Fig. 48, is

$$z_N = \frac{E_1 z_1^2 + E_2(z_2^2 - z_1^2) + E_3(z_3^2 - z_2^2)}{2(E_1 z_1 + E_2(z_2 - z_1) + E_3(z_3 - z_2))} . \tag{A17}$$

The substitutions $E_1 = E_3 = E_L$ and $E_2 = E_S$ produce the general case of a substrate carrying surface layers of unequal thickness but of the same material. The general case of the bilayer is obtained by setting $z_3 - z_2 = a_3 = 0$. Neither of these general situations is appealing from the viewpoint of simplicity. On the other hand, a substantial advantage is gained by the imposition of symmetry. The symmetrical trilayer is represented by the conditions $a_1 = a_3 = a_L$; $E_1 = E_3 = E_L$; $a_2 = a_S$ and $E_2 = E_S$. This has the immediate and obvious consequence that the neutral surface lies at the mid plane of the substrate, as may be confirmed by inserting the values $z_1 = a_L$; $z_2 = a_L + a_S$ and $z_3 = 2a_L + a_S$ into Eq. (A17) to obtain

$$z_N = a_L + (a_S/2). \tag{A18}$$

With this result Eq. (A17) simplifies to

$$K_C = K_S(1 + \gamma_{ST}), \tag{A19}$$

where the stiffness factor γ_{ST} of the symmetrical trilayer is given by

$$\gamma_{ST} = \{[1 + (2a_L/a_S)]^3 - 1\}(E_L/E_S). \tag{A20}$$

For completeness, we note that the complementary inertia factor χ_{ST} appearing in Eq. (A13) is given by

$$\chi_{ST} \equiv \mu_L/\mu_S = 2a_L\rho_L/a_S\rho_S, \tag{A21}$$

where ρ_L and ρ_S are the densities of the surface layers and the substrate, respectively. It should be emphasized that Eqs. (A20) and (A21) represent the case of the symmetrical trilayer without restriction on the thickness ratio a_L/a_S. However, an important special case is represented by the "thin-film limit" $a_L/a_S \ll 1$. Here, Eq. (A20) simplifies to

$$\gamma_{ST} = 6a_L E_L/a_S E_S \tag{A22}$$

and Eqs. (A13) and (A8) take the forms

$$(f_C^2 - f_S^2)/f_S^2 = n_L\left(\frac{3E_L}{E_S} - \frac{\rho_L}{\rho_S}\right)\frac{a_L}{a_S} \tag{A23}$$

and
$$\delta_C - \delta_S = 3n_L\left(\frac{a_L E_L}{a_S E_S}\right)\delta_L, \tag{A24}$$

where the coefficient $n_L = 2$ corresponds to the situation in which both sides of the substrate carry a layer a_L. In the thin-film limit, Eqs. (A23) and (A24) also apply, with $n_L = 1$, to the case of a layer a_L on only one side of the substrate.

REFERENCES

1. C. Zener, *Elasticity and Anelasticity of Metals,* University of Chicago Press, Chicago, 1948.
2. B. S. Berry and A. S. Nowick, in *Physical Acoustics,* W. P. Mason, ed., Academic Press, New York, 1966, Vol. 3A, p. 1.
3. C. Wert, in *Physical Acoustics,* W. P. Mason, ed., Academic Press, New York, 1966, Vol. 3A, p. 43.
4. A. S. Nowick and B. S. Berry, *Anelastic Relaxation in Crystalline Solids,* Academic Press, New York, 1972.
5. R. De Batist, *Internal Friction of Structural Defects in Crystalline Solids,* North-Holland Publishing Company, Amsterdam, 1972.
6. A. S. Nowick and B. S. Berry, IBM J. Res. Dev. **5**, 297 and 312 (1961).
7. J. R. Cost, in *Nontraditional Methods in Diffusion,* G. E. Murch, H. K. Birnbaum and J. R. Cost, eds., The Metallurgical Society of AIME, Warrendale, PA, 1984, p. 111.
8. J. R. Cost, J. de Physique, **46**, C10-31 (1985).
9. J. K. Gillham, in *Techniques and Methods of Polymer Evaluation,* P. E. Slade, Jr. and L. T. Jenkins, eds., Marcel Dekker, New York, 1970, Vol. 2, p. 225.
10. K. Uozumi and A. Kinbara, Jpn. J. Appl. Phys. **10**, 950 (1971).
11. K. Uozumi, T. Nakada and A. Kinbara, Thin Solid Films, **12**, 67 (1972).
12. P. G. Bordoni, M. Nuovo and L. Verdini, Nuovo Cim., **14**, 273 (1959).
13. R. N. Kleiman, G. K. Kaminsky, J. D. Reppy, R. Pindak and D. J. Bishop, Rev. Sci. Instrum. **56**, 2088 (1985).
14. G. P. Weiss and D. O. Smith, Rev. Sci. Instrum. **34**, 522 (1963).
15. V. S. Postnikov, V. K. Belonogov and I. V. Zolotukhin, Fiz. metal. metalloved., **23**, 946 (1967). Translated in Phys. Met. and Metallog. **23**, 173 (1967).
16. B. S. Berry and W. C. Pritchet, IBM J. Res. Dev. **19**, 334 (1975).
17. B. S. Berry and W. C. Pritchet, Rev. Sci. Instrum. **54**, 254 (1983).

18. K. Uozumi, H. Honda and A. Kinbara, Thin Solid Films **37**, L49 (1976).
19. R. H. Lacombe and B. M. J. Kellner, Nature, **289**, 661 (1981).
20. R. H. Lacombe and J. Greenblatt, in *Polyimides,* K. L. Mittal, ed., Plenum, New York, 1984, Vol. 2, p. 647.
21. S. Hunklinger and W. Arnold, in *Physical Acoustics,* W. P. Mason and R. N. Thurston, eds., Academic Press, New York, 1976, Vol. 12, p. 155.
22. W. W. Scott and R. W. MacCrone, Rev. Sci. Instrum. **39**, 821 (1968).
23. J. Baur and A. Kulik, J. Appl. Phys. **58**, 1489 (1985).
24. C. Borner and K.-H. Robrock, J. de Physique, **46**, C10-55 (1985).
25. B. S. Berry and W. C. Pritchet, in *Nontraditional Methods in Diffusion,* G. E. Murch, H. K. Birnbaum and J. R. Cost, eds., The Metallurgical Society of AIME, Warrendale, PA, 1984, p. 83.
26. D. S. Campbell, in *Handbook of Thin Film Technology,* L. I. Maissel and R. Glang, eds., McGraw-Hill, New York, 1970, p. 12-3.
27. R. W. Hoffman, in *Physics of Thin Films,* G. Hass and R. E. Thun, eds., Academic Press, New York, 1966, Vol. 3, p. 211.
28. R. W. Powers and M. V. Doyle, J. Appl. Phys. **30**, 514 (1959).
29. M. Weller, J. de Physique, **46**, C10-7 (1985).
30. I. V. Zolotukhin, V. V. Abramov and V. F. Turkin, Fiz. metal. metalloved. **39**, 591 (1975).
31. B. S. Berry and W. C. Pritchet, J. de Physique, **42**, C5-1111 (1981).
32. M. Weller, J. Diehl, G. Harz, R. Mann and K. Schulze, J. de Physique, **46**, C10-47 (1985).
33. B. Heulin, J. de Physique, **46**, C10-51 (1985).
34. M. Weller, J. X. Zhang, G. Y. Li, T. S. Kê and J. Diehl, Acta Met., **29**, 1055 (1981).
35. G. Alefeld and J. Völkl (eds.), *Hydrogen in Metals I and II*, (Vols. 28 and 29 in Topics in Applied Physics), Springer-Verlag, Berlin, 1978.
36. J. Volkl and G. Alefeld, in *Diffusion in Solids; Recent Developments,* A. S. Nowick and J. J. Burton, eds., Academic Press, New York, 1975, p. 231.
37. F. E. Luborsky (ed.) *Amorphous Metallic Alloys,* Butterworths, London, 1983.
38. S. Steeb and H. Warlimont (eds.,) *Rapidly Quenched Metals 5,* Vols. 1 and 2, North-Holland, Amsterdam, 1984.
39. H. S. Chen, H. J. Leamy and M. Barmatz, J. Non-Cryst. Solids, **5**, 444 (1971).
40. B. S. Berry and W. C. Pritchet, J. Appl. Phys. **44**, 3122 (1973).

41. B. S. Berry, W. C. Pritchet and C. C. Tsuei, Phys. Rev. Lett. **41**, 410 (1978).
42. B. S. Berry and W. C. Pritchet, in *Rapidly Quenched Metals 3,* B. Cantor, ed., The Metals Society, London, 1978, Vol. 2, p. 21.
43. G. Shaumann, J. Völkl and G. Alefeld, Phys. Stat. Sol. **42**, 401 (1970).
44. Zh Qi, J. Völkl, R. Lässer and H. Wenzl, J. Phys. F: Met. Phys. **13**, 2053 (1983).
45. J. Buchholz, J. Völkl and G. Alefeld, Phys. Rev. Lett. **30**, 318 (1973).
46. M. A. Pick and R. Bausch, J. Phys. F: Met. Phys. **6**, 1751 (1976).
47. H. K. Birnbaum, in *Internal Friction and Ultrasonic Attenuation in Solids 6*, R. R. Hasiguti and N. Mikoshiba, eds., University of Tokyo Press, Tokyo, 1977, p. 73.
48. B. S. Berry and W. C. Pritchet, Scripta Met., **15**, 637 (1981).
49. K. Agyeman, E. Armbruster, H. U. Künzi, A. Das Gupta and H.-J. Guntherodt, J. de Physique, **42**, C5-535 (1981).
50. O. Yoshinari, M. Koiwa, A. Inoue and T. Masumoto, Acta Metall. **31**, 2063 (1983).
51. L. E. Hazelton and W. L. Johnson, J. Non-Cryst. Solids **61/62**, 667 (1984).
52. U. Stolz, M. Weller and R. Kirchheim, Scripta Met. **20**, 1361 (1986).
53. F. H. M. Spit, J. W. Drijver and S. Radelaar, Z. Phys. Chem. (Weisbaden), **116**, 809 (1979).
54. B. S. Berry and W. C. Pritchet, in *Rapidly Solidified Amorphous and Crystalline Metals,* B. H. Kear, B. C. Giessen and M. Cohen, eds., Elsevier, New York, 1982, p. 249.
55. R. S. Finocchario, C. L. Tsai and B. C. Giessen, in *Rapidly Solidified Amorphous and Crystalline Metals,* B. H. Kear, B. C. Giessen and M. Cohen, eds., Elsevier, New York, 1982, p. 243.
56. R. S. Finocchiaro, C. L. Tsai and B. C. Giessen, J. Non-Cryst. Solids **61/62**, 661 (1984).
57. R. Kirchheim, F. Sommer and G. Schluckbier, Acta Metall. **30**, 1059 (1982).
58. R. Feenstra, B. M. Geerken, D. G. de Groot and R. Griessen, J. Phys. F: Met. Phys. **16**, 1965 (1986).
59. R. Kirchheim, Acta Metall. **30**, 1069 (1982).
60. G. Sicking, Ber. Bunsenges. Phys. Chem. **76**, 790 (1972).
61. H. U. Künzi, E. Armbruster and H.-J. Guntherodt in *Rapidly Quenched Metals 4*, T. Masumoto and K. Suzuki, eds., The Japan Institute of Metals, 1982, p. 1653.
62. B. S. Berry and W. C. Pritchet, in *Hydrogen in Disordered and Amorphous Solids,* G. Bambakidis and R. C. Bowman, Jr., eds., NATO ASI Series B: Vol. 136, Plenum, New York, 1986, p. 215.

63. B. S. Berry and W. C. Pritchet, J. de Physique, **46**, C10-457 (1985).
64. B. S. Berry and W. C. Pritchet, Phys. Rev. B **24**, 2299 (1981).
65. G. Schaumann, J. Völkl and G. Alefeld, Phys. Rev. Lett. **21**, 891 (1968).
66. B. S. Berry and W. C. Pritchet, in *Rapidly Quenched Metals 5*, S. Steeb and H. Warlimont, eds., North-Holland, Amsterdam, 1984, Vol. 2, p. 1529.
67. A. H. Verbruggen, C. W. Hagen and R. Griessen, J. Phys. F: Met. Phys. **14**, 1431 (1984).
68. T. Soshiroda, M. Koiwa and T. Masumoto, J. Non-Cryst. Solids **22**, 173 (1976).
69. B. S. Berry in *Metallic Glasses,* J. J. Gilman and H. J. Leamy, eds., American Society for Metals, Metals Park, Ohio, 1978, p. 161.
70. B. Cantor and R. W. Cahn, in *Amorphous Metallic Alloys,* F. E. Luborsky, ed., Butterworths, London, 1983, p. 487.
71. B. Cantor, in *Rapidly Quenched Metals 5*, S. Steeb and H. Warlimont, eds., North-Holland, Amsterdam, 1984, Vol. 1, p. 595.
72. C. D. Graham, Jr., and T. Egami, Ann. Rev. Mater. Sci. **8**, 423 (1978).
73. H. Warlimont, in *Rapidly Quenched Metals 5,* S. Steeb and H. Warlimont, eds., North-Holland, Amsterdam, 1984, Vol. 2, p. 1599.
74. B. S. Berry and W. C. Pritchet, Phys. Rev. Lett. **34**, 1022 (1975).
75. B. S. Berry and W. C. Pritchet, AIP Conf. Proc. **34**, 292 (1976).
76. F. E. Luborsky, AIP Conf. Proc. **29**, 209 (1975).
77. B. S. Berry, Scripta Metall. **16**, 1407 (1982).
78. N. Morito and T. Egami, Acta Metall. **32**, 603 (1984).
79. C. Zener, Phys. Rev. **60**, 906 (1941).
80. T. S. Kê, Phys. Rev. **71**, 533 (1947).
81. T. S. Kê, Phys. Rev. **72**, 41 (1947).
82. F. M. d'Heurle and P. S. Ho in *Thin Films – Interdiffusion and Reactions,* J. M. Poate, K. N. Tu and J. W. Mayer, eds., John Wiley and Sons, 1978, p. 243.
83. R. Raj and M. F. Ashby, Met. Trans. **2**, 1113 (1971).
84. I. Ames, F. M. d'Heurle and R. Horstmann, IBM J. Res. Dev. **14**, 461 (1970).
85. M. F. Ashby, Surface Science **31**, 498 (1972).
86. S. I. Tan, B. S. Berry and W. F. J. Frank, in *Ion Implantation in Semiconductors and Other Materials,* Billy L. Crowder, ed., Plenum, New York, 1973, p. 19.
87. W. F. J. Frank, Rad. Effects **21**, 119 (1974).
88. G. D. Watkins and J. W. Corbett, Phys. Rev. **121**, 1001 (1961).

89. J. W. Corbett, G. D. Watkins, R. M. Chrenko and R. S. McDonald, Phys. Rev. **121**, 1015 (1961).
90. B. S. Berry and W. C. Pritchet, in *Ion Implantation in Semiconductors, 1976*, F. Chernow, J. A. Borders and D. K. Brice, eds., Plenum, New York, 1977, p. 435.
91. B. S. Berry, Phys. Teach. **21**, 435 (1983).
92. G. D. Watkins, Phys. Rev. B **12**, 5824 (1975).
93. A. S. Argon, Acta Metall. **27**, 47 (1979).
94. R. W. Hoffman, Mat. Res. Soc. Symp. Proc. **69**, 95 (1986).
95. B. S. Berry and W. C. Pritchet, IBM J. Res. Dev. **28**, 662 (1984).
96. B. S. Berry and W. C. Pritchet, Thin Solid Films **33**, 19 (1976).
97. M. Barmatz, L. R. Testardi, A. F. Garito and A. J. Heeger, Solid State Comm. **15**, 1299 (1974).

3

Diffusion in Artificially Modulated Thin Films

A. Lindsay Greer

1. INTRODUCTION

Artificially layered thin films are made by sequential deposition of alternating layers of two compositions. The deposition can be by evaporation (1,2), sputtering (3,4), molecular beam epitaxy (5), or chemical vapor deposition (6). The materials become of interest, and distinct from macroscopic composites, when the layer thicknesses approach atomic, or molecular, dimensions. They can be classified according to their structural type. The first type to be produced (1) consisted of polycrystalline metal layers; these materials were not well characterized, nor were their properties easy to analyze. In contrast, the simplest materials to consider are those in which the layers have the same, single crystalline, structure, and are coherent. Such materials are usefully thought of as arising from a one-dimensional composition modulation imposed on a single lattice. Later work on fcc metals, e.g. (2), has used materials that approximate to this ideal. Such *compositionally modulated* materials have also been produced in semiconductor systems, either by layering two different semiconductors with the same structure (5), or by modulating the doping (7). When the lattice parameters of the materials are significantly different, the resulting strains may be relieved by the introduction of dislocations at the interfaces, making them partially coherent. Layered amorphous materials, which have been produced artificially in metallic (4) and semiconductor (6) systems, also belong in the compositionally modulated category. Because materials in this category have a uniform structure, their interdiffusion behavior is easier to analyze, and most interdiffusion measurements have used materials of this type. Materials can also be made in which the two materials have different crystal structures; when the layers are effectively single crystal and display an orientation relationship, such materials have been called *layered ultrathin coherent structures* (LUCS) (8,9). In some cases only the crystal axes perpendicular to the film plane are fixed, and there is little or no preferred orientation in the film plane; in other cases true coherency is observed in which the in-plane orientation is defined, and successive layers of the same composition have identical orientations.

The properties of artificially layered materials are of great interest. It would not be appropriate to attempt here a comprehensive survey, but an introduction to the reasons for interest will be given. Compositionally modulated metals have shown spectacular enhancements in elastic modulus (10,11), and novel magnetic behavior (12). Layered metals with two different crystal structures have demonstrated a cross-over from three-dimensional to two-dimensional superconductivity (13), and anomalous melting (14). Multilayers of materials with widely different atomic scattering factors (e.g. W and C) are now applied as grazing incidence and normal incidence mirrors for soft x-rays (15). These are used in spectrometers and as focussing elements, e.g. in x-ray microscopes. In the latter application, artificially layered materials as normal incidence mirrors have sufficiently high reflectivities to make them superior to grazing incidence mirrors and zone plates (16). The greatest commercial potential, however, lies with artificially layered semiconductors. The electronic properties of these materials can be tailored in a remarkably complete way (6,7,17).

In this chapter, diffusion in artificially layered thin films will be examined in detail. After a discussion of the theory, experimental studies are surveyed. These studies are almost exclusively on compositionally modulated materials. The objectives of diffusion studies in artificially layered materials are: to obtain measurements of very low diffusivities; to assess the stability of novel, and potentially useful, structures; to understand interdiffusion and reactions at single interfaces; to observe novel effects on diffusion; and to obtain information on the thermodynamics of mixing. Layered materials also occur naturally, and our understanding of these structures may be improved by studies on artificial materials.

2. THEORY

2.1 Interdiffusion in a Steep Concentration Gradient

In Section 2 it is the intention to introduce those elements of theory which are essential for the interpretation of experimental results on artificially modulated thin films. The theory applies also to naturally modulated materials, and in many cases was originally developed for such systems. Natural composition modulations are usually three-dimensional, but only one-dimensional modulations, as appropriate for artificial layers, will be considered here. The theory is not specific to a material with any particular type of bonding, but in some aspects does depend on the structure (i.e. coherent, incoherent, or amorphous layers).

The simplest case to consider is that of a binary alloy with a harmonic composition modulation in one dimension. The modulation, with wavelength (repeat distance) λ, is described by:

$$c - c_o = A \cos hx, \tag{1}$$

where c is the atom fraction of one of the components, c_o is the average atom fraction of that component, A is the amplitude of the composition wave, h ($= 2\pi/\lambda$) is the wavenumber, and x is a spatial coordinate normal to the plane of the thin film. The decay, or growth, of this modulation can be analyzed using the diffusion equation:

$$\frac{\partial c}{\partial t} = \tilde{D} \frac{\partial^2 c}{\partial x^2}, \tag{2}$$

where t is time and $\tilde{D}$ is the interdiffusion coefficient. Solving this equation gives:

$$\frac{d}{dt} \ell n A = -\tilde{D} h^2. \tag{3}$$

A general periodic composition modulation can be Fourier analyzed, and if $\tilde{D}$ is independent of c, Eq. (3) can be applied to each harmonic. In this linear treatment each harmonic acts independently.

The simple treatment outlined above breaks down for a number of reasons described in the following sections. In Section 2.2 it is pointed out that the simple diffusion equation, Eq. (2), is not valid in very steep concentration gradients, with the result that $\tilde{D}$ in Eq. (3) must be replaced by an effective interdiffusion coefficient $\tilde{D}_\lambda$ dependent on the modulation wavelength. Even with the modified diffusion equation, however, the analysis is linear and individual harmonic components of a general modulation can be considered independently. In Section 2.3 the effects of strain in the artificial layers are described. It is shown that the structure of the material is particularly important in the consideration of strain effects. A variety of non-linear diffusion effects (e.g. a composition-dependent $\tilde{D}$) are discussed in Section 2.4. When non-linear effects are taken into account, it is mostly necessary to substitute numerical for analytical methods. The basis of the experimental methods used for the detection of composition modulations and the determination of $\tilde{D}_\lambda$ is described in Section 2.5. The relationship between $\tilde{D}_\lambda$ determined from artificially modulated thin films and other diffusion coefficients is discussed in Section 2.6. Finally in the consideration of theory, Section 2.7 deals with some aspects of the stability of modulated materials.

2.2 Gradient Energy Effects

Diffusion in a steep concentration gradient was first analyzed by Hillert (18). He used a regular solution model, with nearest neighbor interactions only, to calculate the free energy of a single crystal solid solution with a one-dimensional composition modulation. Numerical solutions were obtained to a set of difference equations describing diffusion between discrete atomic planes. Hillert (18,19) used this treatment to derive equilibrium compositional variations and to demonstrate the kinetics of changes in composition modulation. In contrast, Cahn and Hilliard (20) considered a continuum. Later, Cahn (21,22) derived the continuum diffusion equation and obtained analytical solutions for small departures from homogeneity. However, a continuum diffusion equation cannot provide an accurate description of diffusion on a lattice when substantial composition variations occur on a scale comparable with the interatomic spacing. This problem was solved for a single crystalline lattice by Cook, de Fontaine and Hilliard (23), who developed a microscopic model for diffusion on a lattice. Unlike Hillert (19) they did not restrict their treatment to a particular solution model.

For composition modulations on a single crystalline lattice the most appropriate analysis is that of Cook et al. (23). Their analysis cannot be applied, however, to amorphous or polycrystalline samples. Since artificial composition modulations may be of a wavelength comparable to the atomic, or molecular, size, a continuum analysis also is not valid for these cases. At present a rigorous analysis for amorphous or polycrystalline samples with a short modulation wavelength does not exist. Until it does, it seems best to use the continuum analysis. For this reason, both the continuum analysis of Cahn and Hilliard (20) and the discrete analysis of Cook et al. (23) will be described.

Cahn and Hilliard considered a non-homogeneous binary solution and gave an expression for the Helmholtz free energy per unit volume f of a local volume of composition c. The local free energy consists of the free energy $f_o(c)$ of a homogeneous solution of composition c and an excess term which is the sum of Taylor series of local composition derivatives. For a one-dimensional modulation the local free energy is given by:

$$f\left(c, \frac{\partial c}{\partial x}, \frac{\partial^2 c}{\partial x^2} \ldots\right) = f_o(c) + \kappa_{11}\frac{\partial c}{\partial x} + \kappa_{12}\left(\frac{\partial c}{\partial x}\right)^2 \ldots\ldots$$
$$+ \kappa_{21}\frac{\partial^2 c}{\partial x^2} \ldots\ldots \quad \vdots \qquad (4)$$

Cahn and Hilliard (20) assumed that higher order terms would be negligible. Since the first coefficient κ_{11} must be zero if the free energy is to be invariant on reflection, the total free energy of the non-homogeneous material is given by:

$$F = \mathscr{A}\int \left[f_o(c) + \kappa_{12}\left(\frac{\partial c}{\partial x}\right)^2 + \kappa_{21}\frac{\partial^2 c}{\partial x^2} \right] dx, \qquad (5)$$

where $\mathscr{A}$ is the area of the system normal to the x-axis. Cahn and Hilliard showed that this may be reduced to:

$$F = \mathscr{A}\int \left[f_o(c) + \kappa\left(\frac{\partial c}{\partial x}\right)^2 \right] dx, \qquad (6)$$

where $\kappa = \kappa_{12} - \partial\kappa_{21}/\partial c$. Thus the local free energy has a term proportional to the square of the local concentration gradient, a result first stated by Hillert (18). The constant of proportionality κ is the *gradient energy coefficient.*

The gradient energy term modifies the analysis of interdiffusion. In a binary solution of A and B atoms it is possible to choose a frame of reference in which the flux of B atoms is equal and opposite to the flux of A atoms. In such a frame, the interdiffusion flux $\tilde{J}$ is equal to the flux of A atoms and is given by:

$$\tilde{J} = -M\frac{\partial}{\partial x}(\mu_A - \mu_B), \qquad (7)$$

where M is the mobility (always positive) and μ_A and μ_B are the partial free energies per unit volume of A and B. From Eq. (6),

$$\mu_A - \mu_B = f'_o - 2\kappa\frac{\partial^2 c}{\partial x^2}, \qquad (8)$$

where $f_o' = \partial f_o/\partial c$ and c is the concentration of A atoms. Substituting into Eq. (7), the interdiffusion flux is

$$\tilde{J} = -Mf''_o\frac{\partial c}{\partial x} + 2M\kappa\frac{\partial^3 c}{\partial x^3}, \qquad (9)$$

where $f_o'' = \partial^2 f_o/\partial c^2$. Taking M, f_o'' and κ to be independent of c, and defining the interdiffusion coefficient $\tilde{D}$ to be Mf_o'', the modified diffusion equation is obtained:

$$\frac{\partial c}{\partial t} = \tilde{D}\frac{\partial^2 c}{\partial x^2} - \frac{2\tilde{D}}{f_o''}\kappa\frac{\partial^4 c}{\partial x^4} . \tag{10}$$

In considering periodic composition modulations, the important solution of this equation is:

$$c - c_o = A_o \exp\left[-\tilde{D}h^2\left(1 + \frac{2\kappa h^2}{f_o''}\right)t\right] \cos hx. \tag{11}$$

This predicts the evolution of a harmonic modulation, Eq. (1), with initial amplitude A_o. The composition wave may decay or grow in amplitude A to the sign of the *amplification factor* R:

$$\begin{aligned} R = \frac{d}{dt}\ell nA &= -\tilde{D}\left(1 + \frac{2\kappa h^2}{f_o''}\right)h^2 \\ &= -\tilde{D}_\lambda h^2. \end{aligned} \tag{12}$$

In the presence of steep concentration gradients, then, there is an effective interdiffusion coefficient $\tilde{D}_\lambda$ which is dependent on the modulation wavelength. Since Eq. (10) is linearized, the evolution of the harmonic components of a general periodic modulation can be considered independently. Because it is necessary to assume that M, f_o'' and κ are independent of composition, this useful analytical treatment can be applied only to composition modulations of small amplitude. The analysis with the composition dependence of M, f_o'' and κ is taken into account is discussed in Section 2.4.

Clearly, the form of the variation of the effective interdiffusivity with wavelength (Eq. (12)) depends on the sign of κ and f_o''. These must be evaluated from a solution model. Cahn and Hilliard (20) chose a *regular solution* model in which the Helmholtz free energy per unit volume of a homogeneous solution is given by:

$$f_o(c) = N_v[\omega c(1 - c) + k_B T[c\ell nc + (1 - c)\ell n(1 - c)]], \tag{13}$$

where N_v is the number of atoms per unit volume, k_B is Boltzmann's constant and T is temperature. The coefficient of the energy term ω can be evaluated from nearest neighbor interactions only, or from pairwise interactions summed throughout the system. When ω is negative, the variation of $f_o(c)$ with c shows a single minimum at all temperatures, i.e. the favored state of the solution is always homogeneous. This is an *ordering* system in which A-B bonds are in general preferred to A-A and B-B. When ω is positive there is again a single minimum in $f_o(c)$ at high temperatures, but at low temperatures when the entropy term no longer dominates there are two minima and *phase separation* into two immiscible phases is favored. Within the miscibility gap there is a region in which f_o'' is negative. In all other cases f_o'' is positive. When f_o'' is negative the homogeneous solution is unstable to composition fluctuations of arbitrarily small amplitude and will undergo *spinodal decomposition.* Cahn and Hilliard (20) showed that the gradient energy coefficient in a regular solution is given by:

$$\kappa = \omega\Lambda^2/2, \qquad (14)$$

where Λ is an effective rms interaction distance. When only nearest neighbor interactions are considered $\Lambda = a_o/\sqrt{3}$, where a_o is the interatomic distance. A simple derivation of Eq. (14) by summing nearest neighbor bond energies is given by Greer and Spaepen (24). It can be seen from Eq. (14) that κ is positive for phase-separating systems and negative for ordering systems.

Using the continuum diffusion model, and the values of κ and f_o'' from the regular solution model, the evolution of a harmonic composition modulation can be calculated. In Figures 1 to 3 the amplification factor R and the effective interdiffusion coefficient $\tilde{D}_\lambda$ are shown as a function of modulation wavelength for the three cases: κ positive and f_o'' negative; κ positive and f_o'' positive; and κ negative and f_0'' positive. The fourth combination, κ negative and f_o'' negative, is unlikely, and is impossible in a regular solution with only nearest neighbor interactions.

The first case, κ positive and f_o'' negative, applies to spinodal decomposition. In Fig. 1 it can be seen that when λ is large $\tilde{D}_\lambda$ tends to the macroscopic interdiffusion coefficient $\tilde{D}$ which in this case is negative. As λ gets smaller, however, gradient energy effects become more important and eventually $\tilde{D}_\lambda$ becomes positive. Below a critical wavelength λ_c both R and $\tilde{D}_\lambda$ change sign and the system is stable to composition fluctuations of small amplitude, although it could lower its free energy by macroscopic phase separation. This stability arises because the total interfacial energy in the system would be too great to permit phase separation at wavelengths less than λ_c. The solution is unstable to fluctuation with $\lambda>\lambda_c$ $(= 8\kappa\,\pi^2/f_o'')^{1/2}$ and in an infinite system the absolute limit of stability is $f_o'' = 0$. The correction for non-infinite samples is negligible. Thus, although the gradient energy term restricts the wavelength of the composition modulations which may grow, it does not alter the limit of stability, i.e. the spinodal line. In a phase-separating system κ is positive and consequently the interfacial energy between the two phases is positive. The interface between the phases is diffuse and has an equilibrium width which is determined by minimizing the integral in Eq. (6). The interface width is approximately $\lambda_c/2$ (20).

Figure 2 shows the behavior of R and $\tilde{D}_\lambda$ for a phase-separating system (κ positive) outside the spinodal, i.e. with f_o'' positive, as would

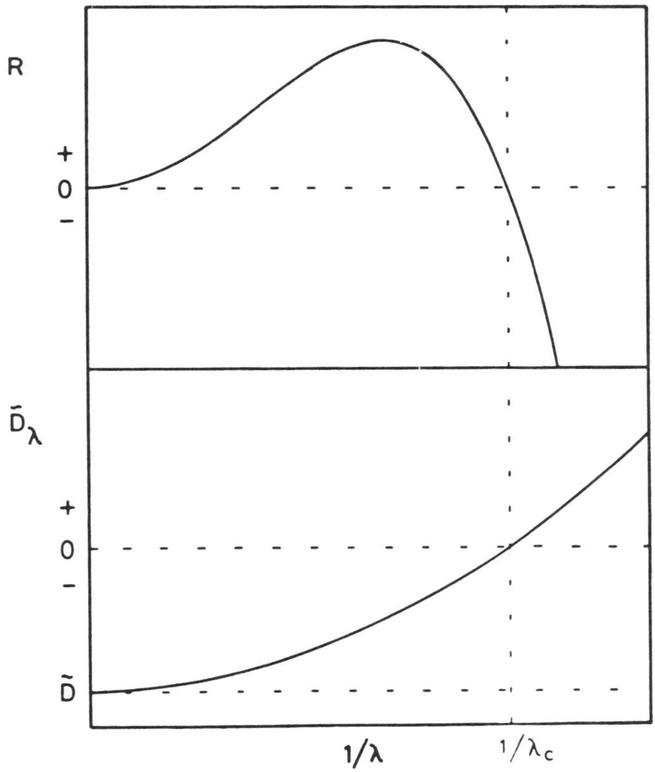

Fig. 1. The amplification factor R and the effective interdiffusivity $\tilde{D}_\lambda$, as a function of $1/\lambda$, for a system showing spinodal decomposition (κ positive and f_o'' negative).

be the case at higher temperatures or far from the equiatomic composition. Although the solution might still be within the miscibility gap and capable of phase separation, it is stable to fluctuations of any wavelength, i.e., any fluctuation would decay.

The third case to be considered is that of an ordering system with κ negative and f_o'' positive. As shown in Fig. 3, the macroscopic diffusion coefficient is positive. As λ is decreased, however, gradient energy effects cause $\tilde{D}_\lambda$ to become negative. Thus at small wavelengths a composition modulation would grow. This arises at wavelengths comparable with the interatomic distance and may be thought of as a one-dimensional ordering.

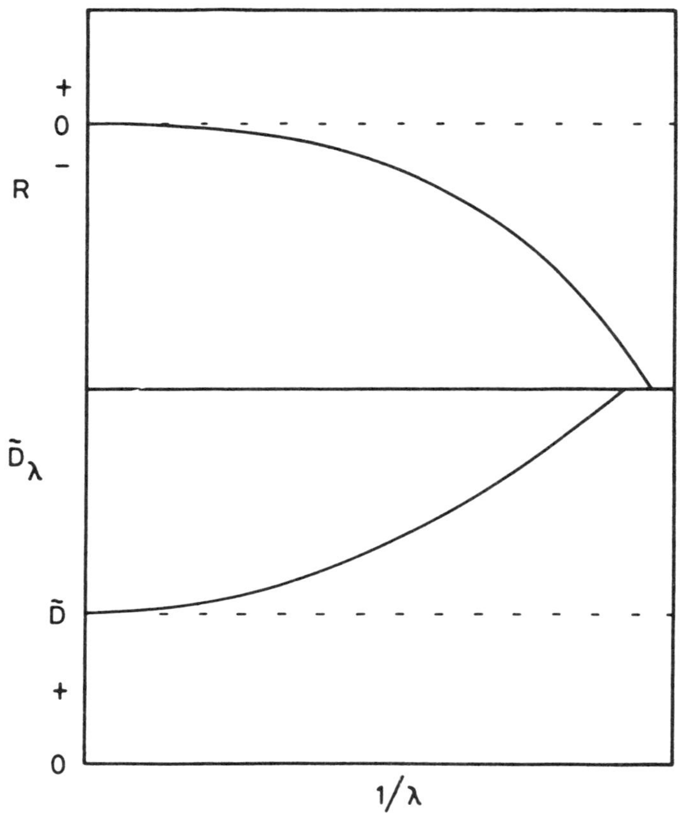

Fig. 2. The amplification factor R and the effective interdiffusivity $\tilde{D}_\lambda$, as a function of $1/\lambda$, for a phase-separating system outside the spinodal (κ positive and f_o'' positive).

The continuum analysis is not valid for very short wavelengths and the predictions when the wavelength is less than a few atomic spacings are at best qualitative. A particular problem is the divergence of R and $\tilde{D}_\lambda$ at low wavelength, seen in Figs. 1-3. Yamauchi and Hilliard (25) showed that even in the absence of gradient energy effects the use of a continuum equation to describe diffusion on a discrete lattice requires a correction term in the amplification factor for a harmonic composition wave. This correction term has the same λ-dependence as the gradient energy and so acts as a correction to κ in the general case. The correction term is anisotropic, and for cubic lattices it has a magnitude 1/6 to 1/2 that of the gradient energy term,

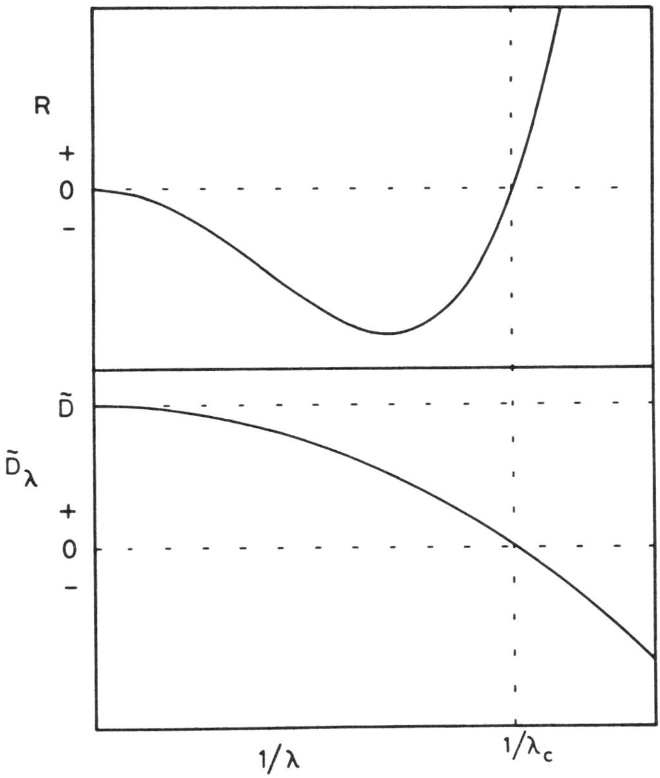

Fig. 3. The amplification factor R and the effective interdiffusivity $\tilde{D}_\lambda$, as a function of $1/\lambda$, for an ordering system (κ negative and f_0'' positive).

depending on the type of lattice and the direction of the modulation wave vector. This correction is significant, but it does not extend the validity of the continuum analysis to lower wavelengths.

Cook et al. (23) tackled the problem of diffusion on a lattice by taking a fully microscopic approach. Following Cahn and Hilliard (20), they took the local free energy to be the sum of the free energy of a homogeneous solution of the same composition and an excess term proportional to the square of the local concentration gradient. They derived a linear differential-difference equation which is the discrete counterpart of the continuum Eq. (10). Diffusion in a general direction in simple cubic, bcc and fcc lattices was considered, but the results are simplest for diffusion along ⟨100⟩ or ⟨110⟩ directions on bcc lattices or along ⟨100⟩ or ⟨111⟩ on fcc lattices. In these special cases the amplification factor for a harmonic composition wave is given by:

$$R = -\tilde{D}\left(1 + \frac{2\kappa H^2}{f_o''}\right)H^2 = -\tilde{D}_\lambda H^2, \qquad (15)$$

where

$$H^2 = (2/d^2)[1 - \cos(2\pi d/\lambda)], \qquad (16)$$

and d is the spacing between atomic planes perpendicular to the diffusion direction. When $\lambda \gg d$, the cosine can be expanded and $H^2 = 4\pi^2/\lambda^2 = h^2$, the wavenumber squared. Substitution of h^2 for H^2 in Eq. (15) gives the continuum Eq. (12). The continuum analysis is a reasonable approximation for $\lambda > 6d$.

The general characteristics of the evolution of a harmonic composition modulation as predicted by the discrete analysis are illustrated in Figures 4 to 7. The amplification factor is shown for diffusion normal to the {200} planes in bcc and fcc lattices plotted against a/λ where a is the lattice parameter (= 2d). In each figure the behavior is symmetric about $\lambda = a$. The two values of λ giving the same value of R correspond to composition waves in a discrete lattice which are physically equivalent.

Figure 4 illustrates the amplification factor for a phase-separating system inside the spinodal. In the range $a < \lambda < \infty$ the behavior is

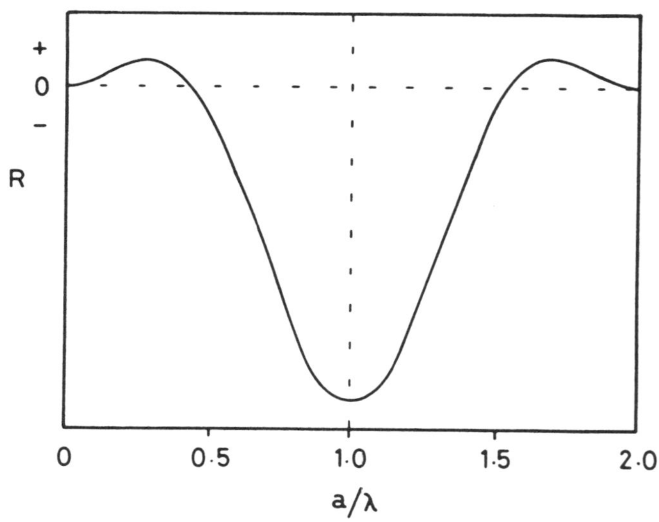

Fig. 4. The amplification factor, R, vs. a/λ for a phase-separating system inside the spinodal (after (23)).

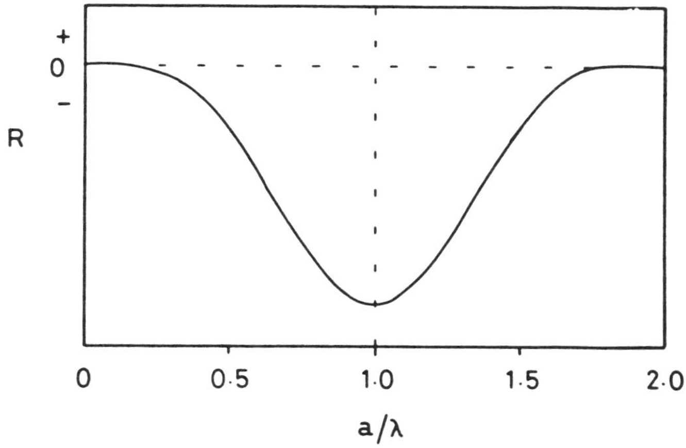

Fig. 5. The amplification factor, R, vs. a/λ for a phase-separating system outside the spinodal (after (23)).

similar to that shown for the continuum analysis in Fig. 1a, but in this case the divergence of R to $-\infty$ is avoided. Figure 5 shows the behavior for a phase-separating system outside the spinodal. As in the continuum analysis (Fig. 2a) the amplification factor is always negative, but again the divergence of R is avoided in the discrete model. Figure 6 applies in the case of an ordering system below its critical temperature. Composition waves of small wavelength will grow and the favored wavelength is $\lambda = a$, corresponding to successive {200} planes of alternating composition. In the continuum analysis (Fig. 3a) R is positive and tends to infinity as the modulation wavelength tends to zero. This divergence is avoided in the discrete analysis and as a result the maximum value of R (at $\lambda = a$) may be negative, as shown in Fig. 7. This represents an ordering system above its critical temperature which is stable to small composition fluctuations of any wavelength. Thus, unlike the continuum treatment, the discrete analysis is capable of describing ordering systems both above and below their ordering temperature.

In closing this section, it should be emphasized that both the continuum analysis of Cahn and Hilliard (20) and the discrete analysis of Cook et al. (23) apply only for composition modulations of small amplitude. If that condition is met, however, the analyses are very useful. In using artificially modulated thin films to determine macroscopic interdiffusion coefficients gradient energy effects may be viewed as a nuisance, but on the other hand the variation of the effective interdiffusion coefficient with modulation wavelength does provide valuable information on the thermodynamics of the binary solution.

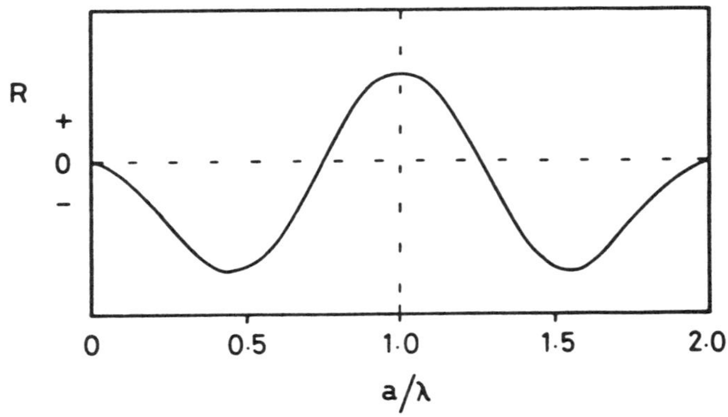

Fig. 6. The amplification factor, R, vs. a/λ for an ordering system below the critical temperature (after (23)).

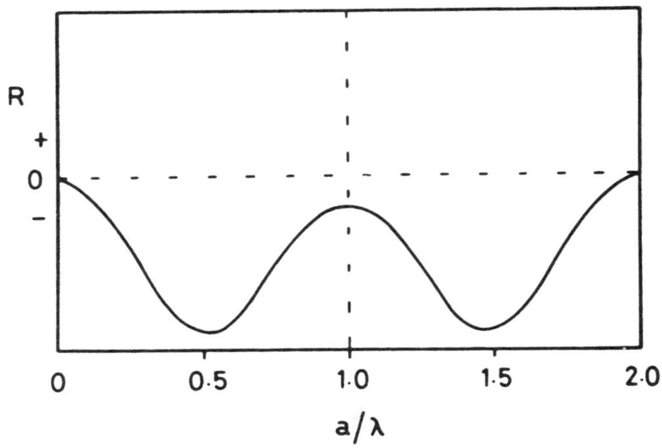

Fig. 7. The amplification factor, R, vs. a/λ for an ordering system above the critical temperature (after (23)).

2.3 Strain Effects

Differential stresses may arise in the layers of an artificially modulated material and if they do the interdiffusion kinetics will be modified. The most analyzed case is that of a single crystalline lattice with sites occupied by two atomic species. If the atoms are of different size, i.e., if the molar volume of the solid solution is a function of composition, stresses will arise when the composition is not uniform. Assuming that the lattice remains free of defects, the instantaneous stress is determined by the instantaneous composition distribution. In this case, independent of the chemical effects discussed in Section 2.2, the elastic strain energy always makes the uniform composition state more favored than it would be otherwise. It will thus impede the amplification of an imposed modulation in a spinodal system, or enhance the homogenization of layers in an ordering system or a phase separating system outside the spinodal. The same is true of an amorphous material if the topology of the amorphous network is unchanging as diffusion proceeds. The essential feature is that there is no mechanism for relaxation of stresses arising from a non-uniform composition, other than homogenization.

The next important case is that where a rapid stress relaxation mechanism does exist. When crystalline layers are partially coherent or incoherent the interface dislocations completely, or at least substantially, relieve the internal stresses. When the composition modulation changes on annealing the dislocation configuration may be able to change (e.g. by climb) to maintain a state of negligible stress. In a modulated amorphous system viscous flow may relieve the stresses, which is the assumption normally made for such systems. When a rapid stress relaxation mechanism does exist the interdiffusion kinetics are unaffected by the composition dependence of the molar volume and should be describable by the equations in Section 2.2.

In between these two extreme cases, it is important to consider the situation when stresses are only partially relaxed. This appears to be the case for spinodal decomposition of an initial uniform state in some oxide glasses (26). It may also be possible to have initially deposited layers which are stress-free (because of interface dislocations in a crystal, or a suitably relaxed amorphous network), but which develop stresses on interdiffusion because changes in the dislocation configuration, or viscous flow are not sufficiently rapid.

For composition modulations of small amplitude it is reasonable to assume that the stress-free molar volume varies linearly with composition. The strain coefficient η for a crystal is given by

$$\eta = \frac{1}{a}\frac{da}{dc}, \qquad (17)$$

where a is the lattice parameter. For an amorphous material this can be generalized to

$$\eta = \varepsilon^{SF}/(c-c_o), \qquad (18)$$

where ε^{SF} is the stress-free strain.

The case in which there is no stress relaxation was first considered by Cahn (21). He assumed an isotropic linear elastic continuum. For an artificially layered material the lateral distortion in the layers is zero unless the lateral extent of the layers is comparable with their thickness. Parallel to the modulation axis the distortion is unrestricted, but the stress component is zero. Cahn showed that for a modulation as in Eq. (1) the elastic strain components are:

$$\begin{aligned} \varepsilon_{xx} &= A\eta \frac{2\nu}{1-\nu} \cos hx \\ \varepsilon_{yy} &= \varepsilon_{zz} = -A\eta \cos hx \\ \varepsilon_{xy} &= \varepsilon_{yz} = \varepsilon_{zx} = 0 \end{aligned} \tag{19}$$

where ν is Poisson's ratio. The elastic strain is the difference between the total strain and the stress-free strain. The stress components are:

$$\begin{aligned} \sigma_{xx} &= 0 \\ \sigma_{yy} &= \sigma_{zz} = \frac{-AE\eta}{1-\nu} \cos hx \\ \sigma_{xy} &= \sigma_{yz} = \sigma_{zx} = 0 \end{aligned} \tag{20}$$

where E is the Young's modulus for the average composition c_o. From these components it can be derived that the average elastic energy per unit volume is independent of the wavelength of the composition modulation and has the value $A^2\eta^2E/2\,(1-\nu)$. In the presence of this elastic strain energy the amplification factor for a harmonic wave becomes:

$$R = -\tilde{D}\left[1 + \frac{2\kappa h^2}{f_o''} + \frac{2\eta^2 E}{f_o''(1-\nu)}\right]h^2 \tag{21}$$

Note that for a spinodal system the limit of stability is altered to:

$$f_o'' + \frac{2\eta^2 E}{1-\nu} = 0, \tag{22}$$

illustrating the point that strain effects tend to stabilize the homogeneous solution, which must be undercooled below the spinodal temperature (at which $f_o'' = 0$) to obtain decomposition. The required undercooling can be large, e.g. 200°C for Au-Ni, when the atomic sizes are quite different.

Next, Cahn (27) applied his continuum treatment to cubic crystals in which the elastic strain energy is anisotropic. The amplification factor in this case is given by:

$$R = -\tilde{D}\left[1 + \frac{2\kappa h^2}{f_o''} + \frac{2\eta^2}{f_o''}Y_{\langle UVW\rangle}\right]h^2, \tag{23}$$

where $Y_{\langle UVW\rangle}$ is a modulus dependent on the direction $\langle UVW\rangle$ of the wave vector of the composition modulation. For the high symmetry directions $\langle 100\rangle$, $\langle 110\rangle$ and $\langle 111\rangle$ in a cubic crystal, $Y_{\langle UVW\rangle}$ is given by:

$$Y_{\langle UVW\rangle} = \frac{1}{2}(c_{11} + 2c_{12})\times$$

$$\left[3 - \frac{c_{11} + c_{12}}{c_{11} + 2(2c_{44} - c_{11} + c_{12})(\ell^2 m^2 + m^2 n^2 + n^2 \ell^2)}\right] \tag{24}$$

where c_{11} etc. are the stiffness coefficients and ℓ, m and n are the direction cosines of the directions $\langle UVW\rangle$ with respect to the cubic axes. With stiffness coefficients $[2c_{44}-c_{11}+c_{12}]>0$ the elastic energy is minimum for composition modulations with wave vectors parallel to $\langle 100\rangle$ directions, and with $[2c_{44}-c_{11}+c_{12}]<0$ the energy is minimum for wave vectors parallel to $\langle 111\rangle$. In most metals the former case applies, and in natural spinodal decomposition the observed microstructure is often a superposition of {100} composition waves. This is a confirmation of the role of strain effects, since in cubic crystals the elastic energy is the only anisotropic factor. Because of this anisotropy, the diffusion results in artificially layered thin films should be dependent on the crystal orientation of the films.

As with gradient energy effects, the continuum approach is no longer valid for modulation wavelengths close to the interatomic spacing. To calculate the strain effects, particularly for small modulation wavelengths in order-disorder reactions, a discrete analysis must be used. This has been done by Cook and de Fontaine (28,29) who used the microscopic theory of elasticity, as formulated by Born and Huang (30) to derive the elastic free energy. For a binary solution on a lattice rather than a continuum, the excess free energy due to a harmonic composition modulation depends not only on the direction but also on the magnitude of the wave vector of the modulation. The modulus $Y_{\langle UVW\rangle}$ in Eq. (23) is replaced by an effective modulus $M_{\langle UVW\rangle}$:

$$R = -\tilde{D}\left[1 + \frac{2\kappa H^2}{f_o''} + \frac{2\eta^2}{f_o''}M_{\langle UVW\rangle}\right]H^2. \tag{25}$$

In the long wavelength limit the effective moduli tend to the values given by the continuum analysis, but for short wavelengths inconsistencies arise if the continuum analysis is used (29). Cook and de

Fontaine described the dependence of $M_{\langle UVW\rangle}$ on λ using a Taylor expansion about $1/\lambda = 0$. The first term in the expansion is $Y_{\langle UVW\rangle}$ and the second term is proportional to the second derivative (the first derivative is zero), which leads to a gradient energy coefficient κ_E for the elastic energy. This coefficient is analogous to the chemical gradient energy coefficient κ. Cook and de Fontaine (29) conclude that κ_E for metals will be negligible unless the strain coefficient $\eta > 0.1$, i.e. unless the atomic size difference exceeds 10%. Although κ_E is small the contribution of the elastic energy to the bulk free energy of a solid solution is significant, at least for $\eta \gtrsim 0.04$. As a result, the elastic contribution must first be subtracted from the total energy before using the method of Cahn and Hilliard (20) to calculate the chemical gradient energy coefficient. Otherwise an overestimate of κ will be obtained (29).

All the above treatments apply when there is no relaxation of the internal stresses. Recently, Stephenson (26) has developed an extension of the Cahn-Hilliard continuum analysis to take account of stress relaxation. He considered an isotropic amorphous material in which the stresses may relax by viscous flow. The local strain in the system was assumed to be the sum of stress-free strain due to the instantaneous deviation of the local composition, elastic strain due to the instantaneous local stress, and plastic strain from previous viscous flow. The plastic strain which develops to relieve the stress will be in phase with the stress-free strain. For a composition modulation as in Eq. (1) the plastic strain normal to the layers ε^p_{xx} will have the form

$$\varepsilon^p_{xx} = 2\eta B(t) \cos hx, \qquad (26)$$

where B(t) is a composition amplitude. Under the visco-elastic relations for a Maxwell solid B(t) evolves according to

$$\frac{\partial B(t)}{\partial t} = \frac{-E}{6\mu(1-\nu)}[B(t)-A(t)], \qquad (27)$$

where μ is the shear viscosity. Stephenson derived kinetic coefficients Q_A and Q_B for atomic diffusion and for stress relaxation by viscous flow respectively. He showed that:

$$Q_A = Mh^2 \qquad (28)$$

and

$$Q_B = 1/12\mu\eta^2. \qquad (29)$$

The important point is that the kinetic coefficient for atomic diffusion Q_A is proportional to the square of the wavenumber, whereas Q_B is independent of h. The derivation of the interdiffusion behavior consists of solving the coupled linear differential equations describing the evolution of the composition and the plastic strain. In special cases when B(t) is proportional to A(t) the solutions are simplified. The treatment predicts that when μ is infinite, the amplification factor for a harmonic composition wave is given by:

$$R = -Q_A(h)[f_o'' + 2\kappa h^2 + 2E\eta^2/(1-\nu)]. \tag{30}$$

This is the result of the zero relaxation model, seen earlier in Eq. (21), and in this case the plastic strain amplitude is zero. The other important limit is that when the chemical terms are small, specifically $Q_A | f_o'' + 2\kappa h^2 | \ll 2(Q_A+Q_B)E\eta^2/(1-\nu)$. In this case the stresses are relieved, and the plastic strain amplitude B(t) is equal to the composition amplitude A(t). The amplification factor is given by:

$$R = -\frac{Q_A(h)Q_B}{Q_A(h)+Q_B}[f_o'' + 2\kappa h^2]. \tag{31}$$

At small wavenumbers $Q_B \gg O_A$ and then $R = -Q_A(h)[f_o'' + 2\kappa h^2]$. This is the usual zero stress model as seen earlier in Eq. (12), and in this case atomic diffusion is the rate-controlling step in the overall interdiffusion. At large wavenumbers $Q_A \gg Q_B$ and viscous flow is the rate-controlling step. In this case the kinetics are significantly different from those of the zero relaxation and zero stress models. For example, Fig. 8 shows the maximum in the amplification factor for spinodal decomposition is shifted towards larger wavelengths. Stephenson (26) suggests that viscous flow control is likely to operate in an amorphous system when the two components have very different mobilities. The less mobile component moves mainly by viscous flow in response to stresses generated by the diffusion of the more mobile component.

Although Stephenson's treatment is for a continuum and applies only for composition modulations of small amplitude, it points the way to a general derivation of strain effects in interdiffusion. An adaptation of the treatment for a discrete lattice would be very useful. It would then be possible to investigate visco-elastic effects at modulation wavelengths close to the interatomic spacing.

2.4 Non-Linear Diffusion Effects

A possible origin of non-linear diffusion effects is the composition dependence of any of the parameters M, f_o'', κ, and, if strain effects are included, η, $Y_{\langle UVW \rangle}$ and μ. The non-linearity will be particularly important for composition modulations of large amplitude. A dramatic instance of the importance of non-linear effects is found in spinodal decomposition. The early stages of the reaction are adequately described by the linear analysis, assigning to each composition-dependent

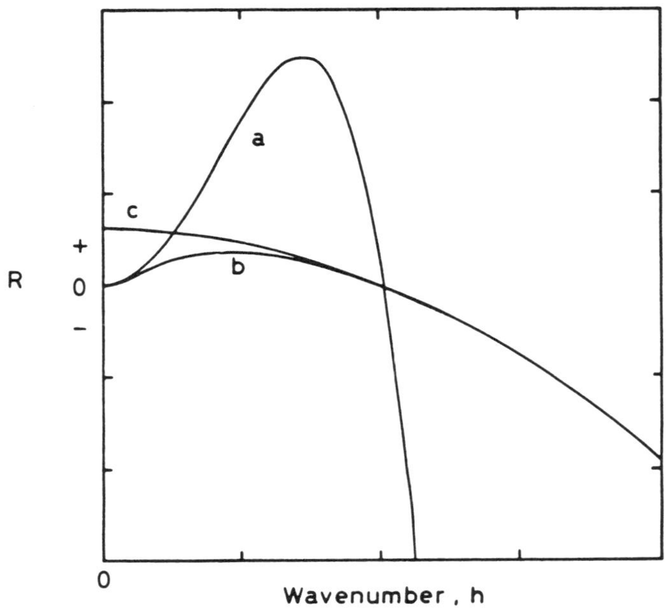

Fig. 8. The amplification factor, R, vs. wavenumber, h, for a system undergoing spinodal decomposition at a rate limited by: (a) atomic diffusion; and (c) viscous flow; with (b) as an intermediate case (after (26)).

parameter the value it would have at the average composition. According to the linear analysis the composition fluctuations will continue to grow exponentially, but in fact the growth must slow down in the later stages as the equilibrium compositions are reached. The breakdown of the linear analysis is largely due to the divergences which arise as the spinodal compositions are approached ($f_o'' \rightarrow o$).

Hillert's analysis of diffusion in steep concentration gradients (Section 2.2) included non-linear effects. Although the parameters $M/c(1-c)$ and κ were taken to be constant, Hillert allowed f_o'' to alter according to a regular solution model. Hillert (18,19) used a numerical technique to model the kinetics of spinodal decomposition, considering composition modulations in one dimension only. The modeling demonstrated that growth slowed down as the equilibrium compositions were approached. At long times the minimum energy configuration for a given modulation wavelength is achieved. These "stationary states" have also been described analytically by Tsakalakos (31) whose solutions are in good agreement with Hillert's numerical results. The amount of "squaring" of the composition profile of the stationary states is negligible near the critical wavelength λ_c, but

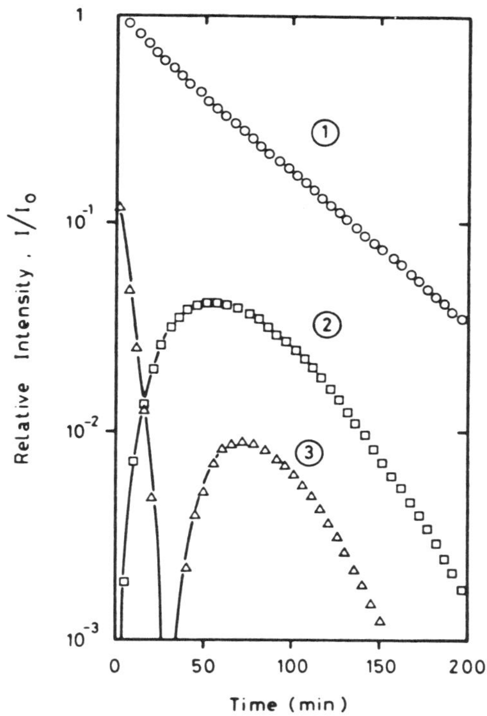

Fig. 9. Calculated relative intensities vs. annealing time of the first three satellites from an Ag-Au compositionally modulated material in which the diffusion is non-linear (after (32)).

becomes appreciable at larger wavelengths. Both Hillert and Tsakalakos showed that the free energy of the stationary states is maximum at λ_c. It is then favorable for the average wavelength to increase, and the composition modulation will coarsen, as observed in Hillert's numerical modeling (19).

De Fontaine (32) studied the evolution of a one-dimensional composition modulation in a homogenizing system. He described the variation of the free energy with composition by fourth-degree polynomial and the variation of the interdiffusivity $\tilde{D}$ by a quadratic function. (A fourth-degree polynomial is the simplest expression which will model non-linear behavior; a constant f_0'' implies a quadratic variation of free energy with composition). De Fontaine took parameters for the Ag-Au solid solution and assumed the gradient energy coefficient to be zero. He obtained numerical solutions to the non-linear diffusion equation by successive iterations. Figure 9 shows his results for the evolution of the first three Fourier components of a modulation in Ag-Au. The decay of the first component is nearly exponential, indicating that non-linear effects do not dominate. The original rectangular composition profile is taken to be symmetrical, and the second Fourier component is therefore zero. Non-linear effects are manifested by the appearance of the second component which indicates a developing asymmetry in the composition profile. This asymmetry arises because diffusion is occurring more rapidly in the Au-rich layers than in the Ag-rich layers. The third Fourier component also exhibits non-linear effects by not having an exponential decay. Tsakalakos (31) has given an approximate analytic solution to the non-linear diffusion equation which is in good agreement with de Fontaine's numerical results.

Tsakalakos (31,33) has considered the effect of including higher order terms in the expansion for the free energy of a non-uniform solution (Eq. (4)) and has derived the interdiffusion coefficient:

$$\tilde{D}_\lambda = \tilde{D}\left[1 + \frac{2\eta^2}{f_o''} Y_{\langle UVW \rangle} + \frac{2}{f_o''} \sum_{\zeta=1}^{\infty} K_\zeta H^{2\zeta}\right] \tag{32}$$

In the long wavelength limit the interdiffusion coefficient takes the value in Eq. (23) with $K_1 = \kappa$, the gradient energy coefficient derived by Cahn and Hilliard (20). The other coefficients K_ζ become important at short wavelengths and then lead to non-linearity in $\tilde{D}_\lambda$ vs. H^2. The coefficients can be expressed in terms of interaction potentials by including interactions beyond the nearest neighbors. For the first coefficient the expression is

$$K_1 = \kappa = \frac{1}{6}\sum_i z_i r_i^2 V_i, \tag{33}$$

where z_i, r_i and V_i are, respectively, the coordination number, radius and potential of the ith coordination shell. If only nearest neighbor interactions are taken into account (Eq. (14)), K_1 has the same sign as and is proportional to the heat of mixing. However, there are systems (such as Cu-Ni) in which K_1 is negative although the heat of mixing is positive. Tsakalakos (33) has suggested that such effects may be explained by the longer range interactions in Eq. (32), which are heavily weighted.

Non-linear effects can also arise when contributions to the free energy other than purely chemical are considered. For example, Purdes (34) has proposed the interaction of the Fermi surface with the Brillouin zone boundaries in compositionally modulated metals as an explanation for abrupt changes in $\tilde{D}_\lambda(H)$ at certain modulation wavelengths. Tsakalakos (31) has suggested that such Brillouin zone effects would account for the position of a sharp peak in $\tilde{D}_\lambda(H)$ found in Cu-Ni modulated films.

In closing this section, it is worth noting that apparent non-linear effects arise in multicomponent diffusion. The single interdiffusion coefficient $\tilde{D}$ for a binary system becomes an (n-1) × (n-1) matrix for an n-component system. A linear multicomponent diffusion theory has been developed in which the resulting equations are decoupled by matrix diagonalization. The particular case of ternary systems has been extensively studied. The four components of the diffusion matrix can be determined from two artificially modulated thin films of the same composition (35). Alternatively, one modulation will suffice if the thermodynamics of the ternary solution are sufficiently well known (36). If such ternary systems are viewed as pseudo-binaries, and if the behavior of the apparent first Fourier component of the binary modulation is monitored, marked non-linear effects are observed (35,36).

2.5 Measurement Techniques

The measurement of the composition profile in a multilayered material can be attempted in a number of ways. For example, Röll and Reill (37) used ion milling and Auger analysis to monitor composition as a function of depth in an artificially layered material. Also, Petroff (38) used cross-sectional transmission electron microscopy and saw image contrast from the layers. In each of these cases, the decrease in amplitude of the composition modulation on annealing was observed. However, by far the most sensitive technique to study the evolution of a composition modulation is X-ray diffraction. This derives its sensitivity from the coherent diffraction from many layers in a periodic structure, and indeed the layers may be thought of as artificial Bragg planes.

For a binary solution with a one-dimensional modulation only, as in Eq. (1), there is a modulation of the average atomic scattering factor:

$$f(x) = f_o + \Delta f \cos hx, \tag{34}$$

where $f_o = c_o f_1 + (1-c_o) f_2$ and $\Delta f = A(f_1 - f_2)$ with f_1 and f_2 being the atomic scattering factors of the two species. In a compositionally modulated crystal the lattice parameter is a function of composition, and there will be a modulation also in the spacing of the planes:

$$d(x) = d_o[1 + \psi \cos hx], \tag{35}$$

where d_o is the average spacing. The amplitude of this modulation is governed by the parameter ψ, defined by

$$\psi = AW\eta \tag{36}$$

where η is the strain coefficient (= da/adc) and W is a numerical factor whose value depends on the degree of coherence of the modulations. For incoherent modulations, with no stress in the layers, W = 1; the local plane spacing has the value characteristic of a uniform material of the same composition. For fully coherent modulations the Poisson effect ensures that the amplitude of the plane spacing modulation is greater than in the previous case. From Eq. (19) it is derived that $W = (1+\nu)/(1-\nu)$ for isotropic materials.

The diffraction effects arising from the combined modulations of scattering factor and plane spacing were first calculated by Preston (39). A more detailed examination of the effects has been given by de Fontaine (40). The most commonly used analysis is the simple one due to Guinier (41). An outline of Guinier's approach is given below and the main results for a harmonic modulation are quoted.

For a compositionally modulated single crystal in which the modulations in scattering factor (Eq. (34)) and plane spacing (Eq. (35)) are functions of only one position coordinate, diffraction effects arise only in that direction in reciprocal space. It is appropriate, therefore, to conduct the diffraction experiment such that the scattering vectors $\underline{k}$ are parallel to this direction. For an artificially layered thin film this is very conveniently achieved in a conventional diffractometer. With the thin film in the normal setting, a $\theta-2\theta$ scan gives scattering vectors normal to the plane of the film.

For an infinitely thick modulated crystal the amplitude of the scattered radiation as a function of h is proportional to the Fourier transform of the function

$$f(x) \cdot \sum_{n=-\infty}^{+\infty} \delta(x-x_n), \tag{37}$$

where x_n is the position of the nth atomic plane which can be derived from Eq. (35). The Fourier transform of this function can be obtained by the convolution of the Fourier transforms of $f(x)$ and of the position function. The transform of the position function is a series of reciprocal lattice points at $(2\pi n/d_o)$, each flanked by symmetrical satellites at $\pm h$, with an amplitude $k\psi/2$ h times that of the central peak. The transform of the scattering factor modulation consists of a peak at $k = 0$ flanked by antisymmetrical satellites at $\pm h$, with amplitudes $\pm\Delta f/2f_o$ times that of the central peak. The convolution of these two transforms, assuming that products of ψ and $\Delta f/f_o$ are negligible, is the series of reciprocal lattice points, each flanked by asymmetrical satellites at $\pm h$, the amplitudes of the positive and negative satellites being in the ratio $\mathscr{R}$ given by:

$$\mathcal{R} = (\psi k/h - \Delta f/f_o)/(\psi k/h + \Delta f/f_o). \tag{38}$$

Thus from the spacings of the satellite peaks in a diffractometer scan the modulation wavelength ($\lambda = 2\pi/h$) can be determined, and from the satellite intensities ψ and Δf can be determined. Knowing ψ, the degree of coherency in the layered film can be assessed (Eq. (36)). Note that the satellites about the forward scattering ($k = 0$) are particularly easy to interpret, being symmetrical with intensities proportional to Δf^2 and hence also to the square of the composition amplitude, A.

The analysis outlined here applies in modified form to amorphous materials. In the position function in Eq. (37) the sum is over the positions of individual atoms, not atomic planes. The Fourier transform of the function no longer gives reciprocal lattice points except for the forward scattering at $k = 0$. The satellites about this point are the same as before, depending only on the scattering factor modulation and not on the atomic structure.

The diffraction effects of a general, non-harmonic, composition modulation can be derived by Fourier analysis. The modulation is expressed as:

$$c(x) = \sum_{m=-\infty}^{+\infty} A_m \sin(mhx + \phi_m), \tag{39}$$

where ϕ_m is the phase of the mth component. The repeat distance λ is $2\pi/h$. Each component gives satellites at $\pm mh$ around each sharp reflection and so a series of satellites arises. The profile of the modulation can be obtained from the satellites about the forward scattering; their intensities, I_m, are proportional to A_m^2. If the diffusion process is linear, the Fourier components evolve independently. For each there is an effective interdiffusion coefficient $\tilde{D}_{\lambda/m}$ which can be determined by measuring the intensity of the appropriate satellite. In considering the evolution of the intensity of a particular satellite, a number of factors are held constant, and the intensity is proportional to A_m^2, even for satellites not about the forward scattering direction. For a continuum Eq. (12) then leads to:

$$\frac{d}{dt}\ell n I_m = -2\tilde{D}_{\lambda/m}(mh)^2 \tag{40}$$

For a diffusion on a discrete lattice Eqs. (15) and (16) lead to:

$$\frac{d}{dt}\ell n I_m = \frac{-4\tilde{D}_{\lambda/m}}{d^2}[1 - \cos(mhd)], \tag{41}$$

where d is the plane spacing perpendicular to the diffusion direction.

As pointed out in Section 2.4, the linear approximation used here may often not be valid. When the diffusion is non-linear, the time dependence of x-ray satellite intensities is usually analyzed by direct comparison with numerical simulations (5,31).

So far only modulated materials with a single atomic structure, crystalline or amorphous, have been considered. The diffraction effects which arise in materials with layers of different structures have been calculated, e.g. (8,9), but diffusion studies have not been made in such materials. In any case, the analysis would be complicated since the transition to a uniform material would not be possible without a first-order phase transformation.

For most modulated materials of interest the "infinite thickness" approximation is useful. However, for samples consisting of just a few ($\lesssim 15$) repeat thicknesses, the diffraction effects can be calculated, and it is found that there are subsidiary peaks between the main satellites around the forward scattering direction (42). Analysis of the entire diffraction pattern, including the subsidiary peaks, can yield useful one-dimensional structural information, and the method has been used to study Langmuir-Blodgett layers of manganese stearate, a (Ga,Al)As superlattice and an amorphous material with W, Re and C layers (43). Although the full potential of the method has not yet been realized it is clear that it could, if necessary, be applied to diffusion studies. In studying artificially layered materials, however, it would be preferable to deposit many layers and to use the simpler Fourier analysis.

In designing high reflectivity mirrors for x-rays, it is natural to extend to shorter wavelengths the principles applied to optical coatings (15). Although the fundamental theory (44) is the same as for Bragg reflection, the development of the two approaches has been to a large extent separate. At optical wavelengths there are materials with different refractive indices that are essentially absorption-free. With such materials the most widely used design for high reflectivity mirrors is a *quarter wave stack* in which the optical thickness of each layer is L/4, where L is the wavelength of the radiation. In a quarter-wave stack reflected waves from each boundary add in phase. The same principle can be applied for soft x-rays (L = 2-10 nm).

For x-rays, since the materials are not absorption-free, the reflectivity can be somewhat (~10%) improved by adopting a quasi-periodic design in which the layers of the optically denser component are thinner towards the top of the multilayer (15). According to a simplified theory (15) the reflectivity ξ of a multilayer, periodic or quasi-periodic, is reduced by the effective roughness σ of the interfaces:

$$\xi = \xi_o \exp -(2\pi\sigma/\lambda)^2, \qquad (42)$$

where ξ_o is the reflectivity with ideal smooth, sharp boundaries and λ

is the multilayer period. The effective roughness may be a true roughness, i.e. two times the rms height of irregularities in a sharp interface, or a measure of the diffusion of a smooth interface. Effective roughnesses as low as 0.2 nm have been reported for amorphous materials (15). The reflectivity for soft x-rays is a sensitive measure of the effective roughness of interfaces, and if the interfaces are smooth it may be useful for studying diffusion.

2.6 Relationship to Other Diffusion Measurements

It is important to know how interdiffusion coefficients determined using artificially modulated thin films are related to diffusion coefficients determined by other techniques. Usually the coefficients are measured in homogenizing systems. The first point to note is that the diffusion distances in an artificially modulated material are so short that the transport is dominated by *bulk* diffusion. Short circuit paths, such as dislocations and grain boundaries, have a negligible effect because the distances between them are much greater than the modulation wavelength. The only exception is that *moving* grain boundaries, by sweeping through the material, can have a significant effect because the volume swept through may be completely homogenized. Such effects are quite likely in polycrystalline films with imperfect texture, when grain growth may occur on annealing. Other more conventional diffusion measurements are dominated by short circuit diffusion at temperatures less than about two thirds of the absolute melting point.

Faults, such as partial layers, in the modulation sequence can arise. Because their spacing is again much larger than the modulation wavelength, they have a negligible effect on homogenization kinetics and on the effective interdiffusion coefficient determined from the kinetics. On the other hand, such faults may have a significant effect on the coarsening kinetics in phase-separating systems, as described in Section 2.7.

The interdiffusion coefficient $\tilde{D}$ in a binary system is related to the tracer diffusivities of the two components, D^*_1, D^*_2, by the expression due to Darken (45):

$$\tilde{D} = [cD^*_1 + (1 - c)D^*_2]\left(1 + \frac{\partial \ell n\gamma_2}{\partial \ell nc}\right), \tag{43}$$

where c and γ_2 are the concentration and activity coefficient of component 2. The temperature dependence of the thermodynamic factor in Eq. (43) must be taken into account in extrapolating data from bulk diffusion couples to lower temperatures for comparison with data on modulated films. The thermodynamic factor can be determined using the expression

$$\left(1 + \frac{\partial \ell n \gamma_2}{\partial \ell n c}\right) = \frac{c(1-c)f_o''(T)}{N_v k_B T} \tag{44}$$

if a solution model for $f_o''(T)$ is available. If sufficient thermodynamic data are not available, approximate methods must be used. The problem has been considered in detail by Hilliard (46) who points out that the temperature dependence of the thermodynamic factor in homogenizing systems is weak. Typically, it contributes only a few kJ/mole to the total activation energy of $\tilde{D}$. However, in spinodally decomposing systems the temperature dependence of the thermodynamic factor is very significant. In such systems $f_o''(T)$ is zero at the spinodal temperature T_s and has a different sign on either side; it can be expressed as an expression about T_s:

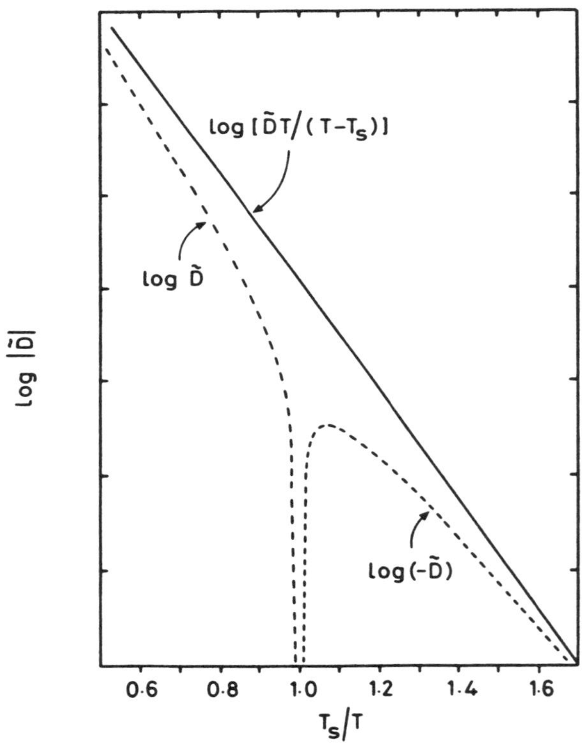

Fig. 10. A schematic Arrhenius plot showing that the quantity $\tilde{D}T/(T-T_s)$ and not $\tilde{D}$ should be used to extrapolate high temperature data into the spinodal (after (46)).

$$f_o''(T) = (T - T_s)\frac{\partial f_o''}{\partial T} = -(T - T_s)s'', \qquad (45)$$

where $s'' = \partial^2 s/\partial c^2$ and s is the entropy per unit volume. Since s'' has a negligible temperature dependence, Eq. (45) can be usefully rewritten as

$$\frac{\tilde{D}T}{T - T_s} = \frac{-c(1-c)s''}{N_v k_B}[cD^*_1 + (1-c)D^*_2] \qquad (46)$$

If the activation energies of D^*_1 and D^*_2 are similar, an Arrhenius plot of $\tilde{D}T/(T-T_s)$ is linear, and Hilliard (46) has suggested that such a plot should be used to compare data over a wide temperature range, especially inside and outside the spinodal. Figure 10 is a schematic comparison of the temperature dependence of $\tilde{D}$ and Eq. (46).

Using a modulated film and with simple x-ray diffraction measurements it is possible to determine the effective interdiffusion coefficient after only ~0.2 jumps per atom. Taking 10^6s (~10 days) as the maximum reasonable anneal time, this corresponds to a minimum measurable $\tilde{D}$ of $\sim 10^{-27} m^2 s^{-1}$. Table 1 gives an approximate comparison of this sensitivity with that of other diffusion measurement techniques. In each of the other techniques a single composition profile is measured and the sensitivity is determined by the spatial resolution of the profiling technique. In mechanical profiling slices of material are removed by microtome or by lapping with a depth resolution of ~3μm (47,48), and the composition of the removed material is determined by chemical analysis or by measuring the activity of radiotracers. In anodic oxidation successive layers of minimum thickness ~0.2μm (49) are oxidized and dissolved off. Material can also be removed by sputtering and this permits the determination of composition profiles as shallow as ~30Å. The analysis in sputter profiling has to be very sensitive and is normally accomplished by Auger electron spectroscopy (50), mass spectrometry of the sputtered atoms (SIMS) (51), or by radiotracer counting on the removed material (52). Each of these profiling techniques is destructive. Similar depth resolution to that of sputter profiling can also be achieved non-destructively using Rutherford back scattering (53) or nuclear reaction techniques (54).

It is clear from Table 1 that by using x-ray diffraction from modulated films, lower interdiffusivities can be measured than by any other technique. In addition, the technique is non-destructive, so that successive measurements can be made on the same sample. Both attributes are essential in some experiments, for example, in taking account of structural relaxation in metallic glasses (Section 4), and in detecting spinodal decomposition in Cu-Ni (Section 3.6).

TABLE 1: Comparison of Techniques for Measuring Low Diffusivities.

Technique	Minimum no. of jumps/atom	Minimum* D/m^2s^{-1}
mechanical profiling	10^9	5×10^{-18}
anodic oxidation	2×10^6	10^{-20}
sputter profiling	200	10^{-24}
RBS	200	10^{-24}
nuclear reaction	200	10^{-24}
modulated films	0.2	10^{-27}

* for an anneal time of 10^6s.

2.7 Stability of Modulated Structures

On annealing compositionally modulated materials they try to lower their excess free energy by reducing the high density of interfaces. The stability of materials which tend to homogenize on annealing is determined by the effective interdiffusion coefficient and is readily analyzed. Phase-separating systems tend to stationary states on annealing and can lower their free energy further only by coarsening. In this section the theory of coarsening will be briefly outlined.

Langer (55) has shown that even perfectly periodic stationary states are unstable to thermal fluctuations. The simplest continuous coarsening process is one in which there is thinning and thickening of alternate layers of the same component, leading eventually to a doubling of the wavelength. This process is likely to be very slow. In a spinodal system, phase separation could also occur in directions perpendicular to the modulation wave vector. For linear diffusion the growth kinetics of such orthogonal modulations would be independent of the original modulation. However, non-linear effects are likely to prevail and in that case Spaepen (56) has shown that the growth of the orthogonal modulations is hindered.

Coarsening can also occur by a non-uniform process of fault migration, analogous to that demonstrated by Graham and Kraft (57) in eutectic structures. A partial layer in the modulation sequence will shrink laterally and the surrounding layers will thicken. This process has a very large driving force arising from the interfacial curvature at the layer edge. If modulation faults are frequent, their migration is likely to be the dominant coarsening mode. The detailed kinetics of the process may have to be modified for very thin layers (56) and have yet to be worked out.

3. EXPERIMENTAL RESULTS ON CRYSTALLINE METALS

3.1 Early Work

The first diffusion measurements on artificial compositionally modulated materials were made by DuMond and Youtz in 1940 (1). They made by evaporation a polycrystalline copper film with a modulated gold content ($\lambda \approx$ 10nm). Presumably because of migrating grain boundaries this material exhibited an anomalously high interdiffusion coefficient of $\sim 5 \times 10^{-24}$ m^2s^{-1} at room temperature. Dinklage and Frerichs (58) attempted to develop modulated films for use in a soft x-ray spectrometer. They used polycrystalline layers of magnesium (2.66nm thick) and lead (0.25nm thick) and again found very rapid interdiffusion, precluding their intended use. The first systematic diffusion studies, using materials made specifically for the purpose were conducted by the Hilliard group at Northwestern University. The majority of the work cited in this section is from that group. The Hilliard group have worked on fcc metals evaporated from two sources on to heated mica substrates, giving excellent texture with $\langle 111 \rangle$ directions perpendicular to the substrate. The binary systems were chosen to exhibit complete solid solubility in the composition ranges used in order to avoid problems with compound formation.

3.2 Gold-Silver

Cook and Hilliard (2,59) deposited alternating layers of gold and silver with repeat distances in the range 0.7 - 3.5nm, and measured the intensities of the first order 000+, 111+, 111−, 222+ and 222− satellites (+ and − denote peaks at higher and lower angle than the Bragg reflection). They found that the satellites decayed on annealing as shown in Fig. 11. The initial non-linearity in these plots was attributed to grain growth in the films, but after some time the plots became linear, indicating a constant interdiffusion coefficient which was calculated using Eq. (15).

For a given average composition the effective interdiffusion coefficient $\tilde{D}_\lambda$ varied with λ according to Eq. (12) and was of the form shown in Fig. 3. From this variation κ/f_o'' was determined and using

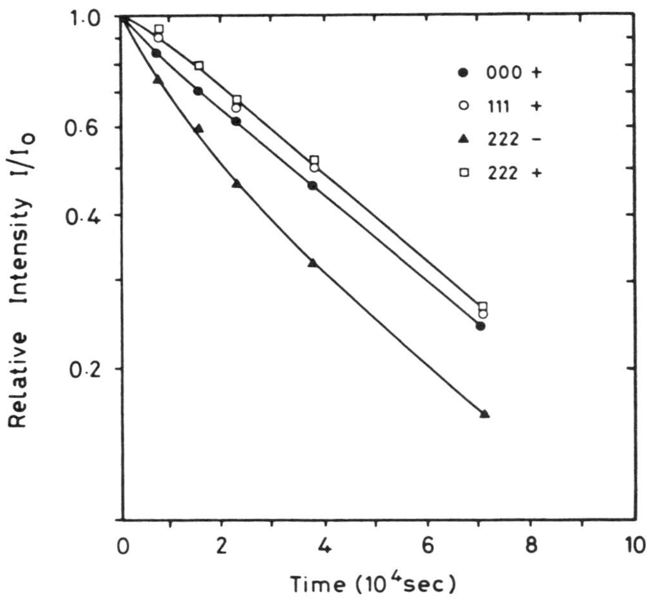

Fig. 11. Plots of log I/I_o vs. annealing time at 228.3°C for the 000+, 111+, 222− and 222+ satellite intensities in an Ag-Au modulated film. The film had an average composition of 32 at.% Au and a modulation wavelength of 3.35 nm (after (2)).

f_o'' from thermodynamic data κ was estimated. Gold-silver is an ordering system and as expected for such a system it was found that $\tilde{D}$ is positive and κ is negative. The gradient energy coefficient was fairly independent of temperature and composition and had an average value of $-(2.6\pm0.5)\times10^{-11}$ Jm^{-1} in excellent agreement with the value, -2.5×10^{-11} Jm^{-1}, calculated on the basis of nearest neighbor interactions. The values of $\tilde{D}$ at various temperatures were found to be in good agreement with the extrapolation of high temperature data from bulk diffusion couples (60) as shown in Fig. 12. This confirms that short circuit diffusion is not important in the modulated films. As the lattice parameters of gold and silver are so close ($\Delta a/a \approx 0.15\%$), strain effects are negligible in this system.

3.3 Copper-Palladium

Philofsky and Hilliard (61,62) studied the copper-palladium system, which again has a tendency to order. In this case $\Delta a/a$ is substantial (~7%) and strain effects were observed. Satellite decay curves similar to those in Fig. 11 were measured. A plot of $\tilde{D}_\lambda$ vs. H^2, which was expected to be linear, was found to have a step with re-

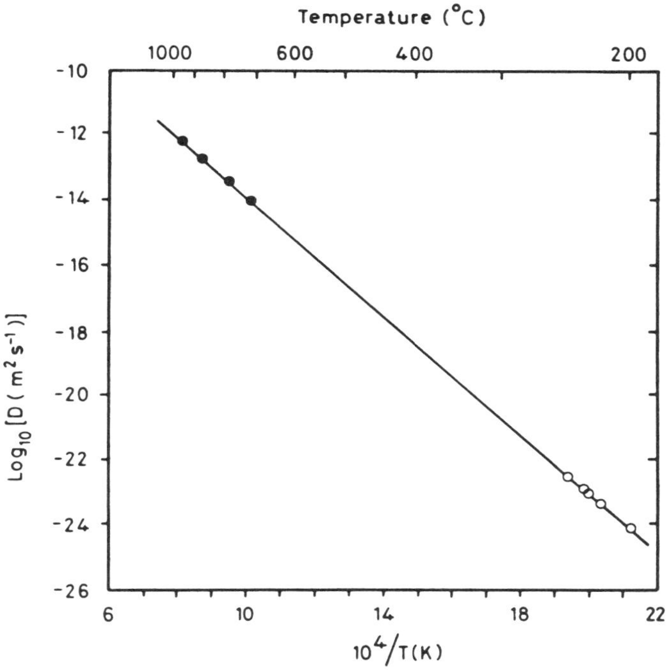

Fig. 12. A comparison on an Arrhenius plot of the interdiffusivity $\tilde{D}$ in $Ag_{50}Au_{50}$ alloys at high temperatures from conventional measurements (60) and at low temperatures from modulated films (2). (After (2)).

duced interdiffusion coefficients at long wavelength as shown in Fig. 13. An analysis of the satellite intensity ratio $\mathscr{R}$ (Eq. (38)) revealed that the layers were coherent for $\lambda \lesssim 2.8$nm and incoherent for $\lambda > 3.8$nm. On Fig. 13 the variation of $\tilde{D}_\lambda$ with H^2 is consistent with this, being governed by Eq. (25) for coherent modulations and Eq. (15) for incoherent modulations. The transition from coherent to incoherent layers is expected, and the wavelength at which it occurs is consistent with a simple energy balance (61) between the strain energy of coherent layers (independent of λ) and the energy of interfacial dislocations (inversely proportional to λ). Philofsky and Hilliard (62) determined κ from a comparison of $\tilde{D}_\lambda$ for two modulations of different wavelengths in the same specimen. By considering cases in which both modulations were coherent and in which one was coherent and one incoherent they were able to measure $\kappa f_o''$. Using f_o'' from thermodynamic data they calculated κ and found it was negative as expected. However, the value, -10.3×10^{-11} Jm^{-1}, was not in particularly

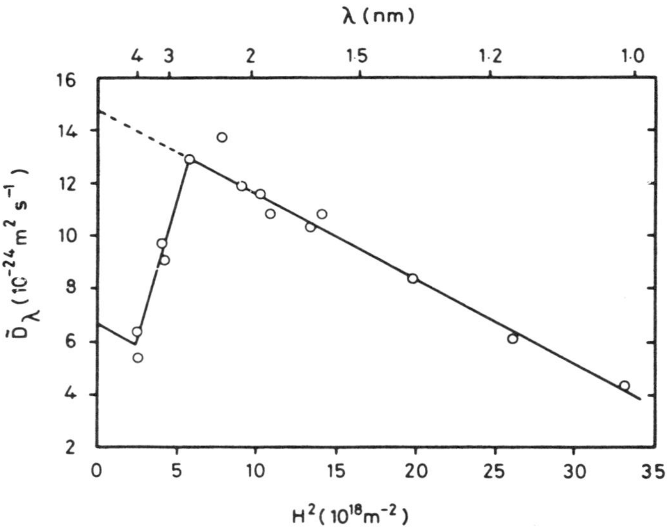

Fig. 13. The effective interdiffusivity $\tilde{D}_\lambda$ vs. H^2 for Cu-Pd modulated films (82 at.% Pd) annealed at 389°C. The layers were coherent for $\lambda<2.8$ nm and incoherent for $\lambda>3.8$ nm. (After (62)).

good agreement with the estimate, -6.3×10^{-11} Jm^{-1}, from nearest neighbor interactions. If the elastic contribution to the total heat of mixing is subtracted, as it should be, the theoretical estimate for κ is -8.2×10^{-11} Jm^{-1} in somewhat better agreement with experiment (29). In the copper-palladium system the bulk interdiffusivities were again found to be in good agreement with high temperature measurements on bulk diffusion couples.

3.4 Gold-Nickel

This system was studied by Yang and Hilliard (63). It is a phase-separating system with a large lattice parameter difference ($\Delta a/a$ = 15.7%). It was found, as for copper-palladium, that films of longer wavelength were incoherent. For this system some satellite intensities increased on annealing as shown in Fig. 14 and the variation of the amplification factor R with modulation wavelength is illustrated in Fig. 15. The form of this variation is exactly as expected for a system inside the spinodal, and matches Eq. (15) since the layers in the wavelength range of the figure were incoherent. The ratio of H_c for which

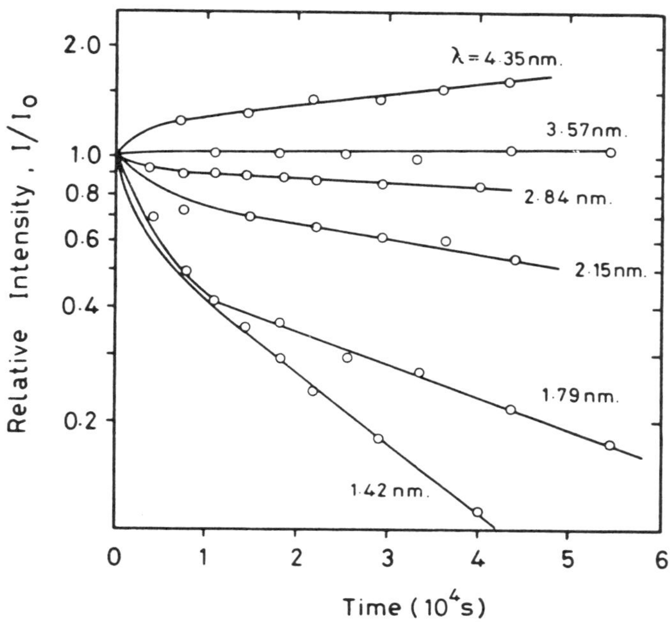

Fig. 14. Plots of log I/I_0 vs. annealing time at 275°C for 000+ or 111− satellite intensities in Au-Ni modulated films (44 at.% Ni). (After (63)).

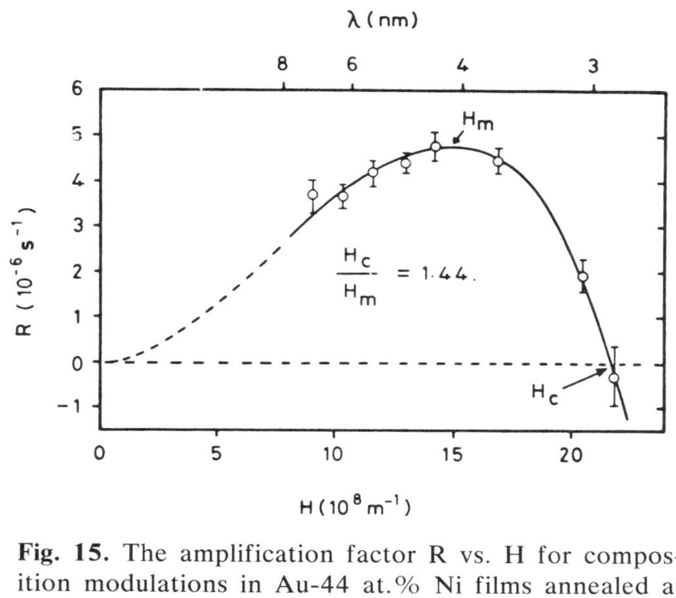

Fig. 15. The amplification factor R vs. H for composition modulations in Au-44 at.% Ni films annealed at 250°C (after (63)).

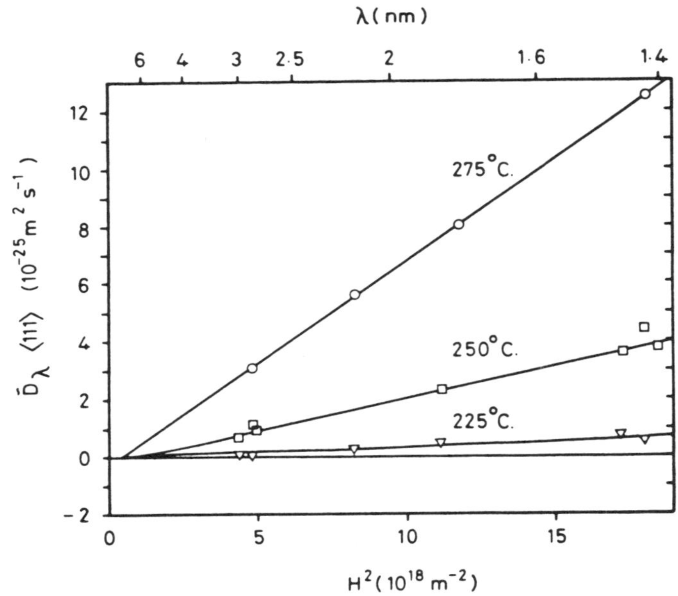

Fig. 16. Effective interdiffusivity $\tilde{D}_\lambda\langle 111\rangle$ vs. H^2 for coherent modulations in Au-44 at.% Ni films (after (63)).

R is zero to H_m for which R is maximum is close to the theoretical value of $\sqrt{2}$. Yang and Hilliard also studied films in which the layers were coherent. In that case Eq. (25) should apply and a plot of $\tilde{D}_\lambda$ vs. H^2 should be linear with a slope of $2\kappa\tilde{D}/f_o''$. The experimental plots are linear (Fig. 16) and permit an extrapolation to $\tilde{D}_\lambda = 0$ to find the critical wavelengths. The values so obtained, 5.9nm and 9.7nm, are significantly higher than the value for incoherent modulations in Fig. 15, reflecting the positive strain energy contribution to $\tilde{D}_\lambda$. Yang (63) derived that $H_c{}^2$ should vary with temperature according to:

$$(T_s - T)s'' + 2\kappa H_c^2 = 0, \qquad (47)$$

where T_s is the coherent spinodal temperature and s″ is as in Eq. (45). Figure 17 shows that $H_c{}^2$ is indeed linearly dependent on temperature. Extrapolating to $H_c = 0$ gives $T_s = 284 \pm 3°C$ for 44.0 at.% Ni and

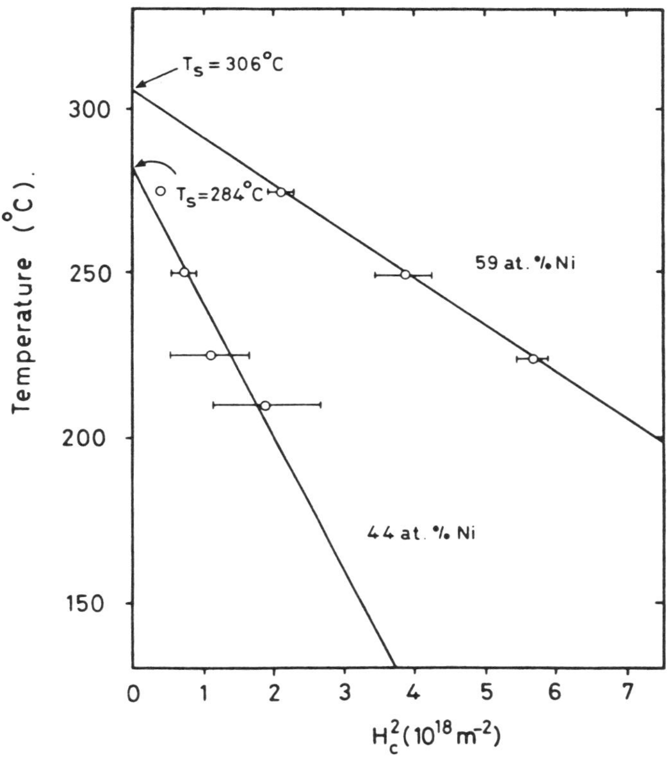

Fig. 17. Temperature vs. H^2 for Au-44 at.% Ni and Au-59 at.% Ni films. Extrapolation to $H_c = 0$ gives the coherent spinodal temperature (after (63)).

306 ± 1°C for 59.0 at.% Ni, values which are about 420°C and 500°C lower than the calculated corresponding chemical spinodal temperatures.

Yang (63) calculated s'' and derived values of the overall gradient energy coefficient from the slopes of the plots in Fig. 17. He estimated the elastic gradient energy coefficient from the expression of Cook and de Fontaine (28), and by subtracting this from the overall coefficient, obtained values for the chemical gradient energy coefficient. These were $\kappa = 3.46 \times 10^{-10}$ Jm^{-1} at 44 at.% Ni and $\kappa = 1.01 \times 10^{-10}$ Jm^{-1} at 59 at.% Ni. The dependence of κ on the average composition of the system was not expected. Furthermore, the values are not in agreement with the value, 0.19×10^{-10} Jm^{-1}, calculated on the basis of nearest neighbor interactions. It seems that for the gold-nickel system a more sophisticated solution model is needed which includes longer range interactions.

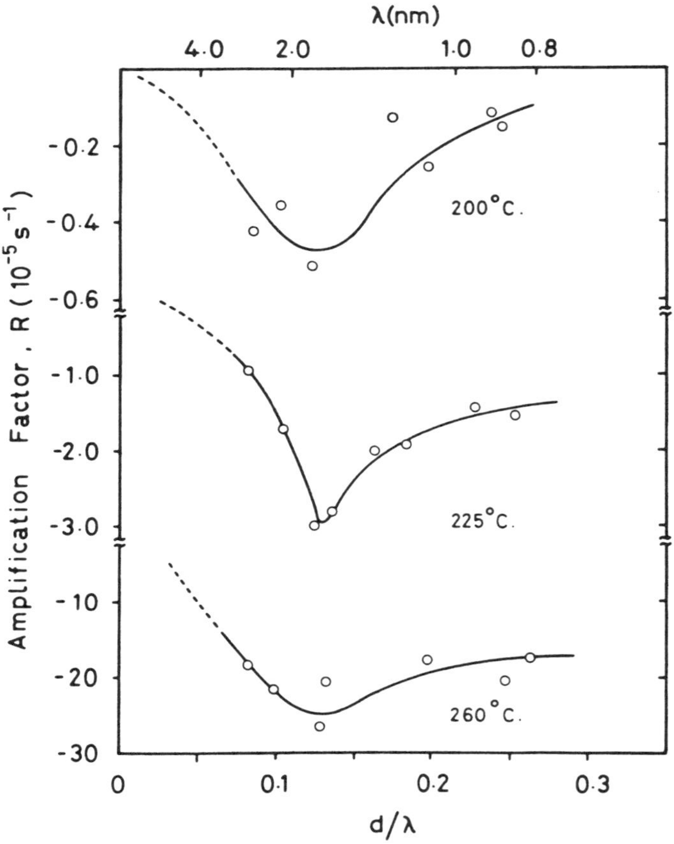

Fig. 18. The amplification factor R vs. d/λ (d is the interplanar spacing) for Cu-16 at.% Au modulated films annealed at 200, 225 and 260°C. (After (65)).

3.5 Copper-Gold

Paulson and Hilliard (64,65) studied the copper-gold system in order to test the discrete analysis of Cook et al. (23) for an ordering system with very short modulation wavelengths. The average composition was chosen to be 16 at.% Au, which is outside the ordered Cu_3Au phase field. The intensities of the 000+ satellites were measured as a function of annealing time at selected temperatures, and the amplification factors R are plotted in Fig. 18 as a function of d/λ where d is the interplanar spacing along ⟨111⟩. The amplification factor is always negative and its variation with d/λ is in qualitative agreement over the corresponding range ($0 \leq a/\lambda \leq 1.0$), with the prediction of the discrete analysis shown in Fig. 7.

It was found that, except at long wavelengths, plots of $\tilde{D}_\lambda$ vs. H^2 were not linear as expected from Eq. (25). Paulson and Hilliard (64) attributed this non-linearity to higher order terms in the diffusion equation or to the flux from random solute movements (i.e. Brownian motion (66)), but no experimental justification was offered for either explanation. Using a solution model with nearest neighbor interactions only the chemical gradient energy coefficient in copper-gold is calculated to be $\kappa = -3.84\times10^{-11}$ Jm^{-1} (64). By including interaction energies out to third nearest neighbors a value of -7.17×10^{-11} Jm^{-1} is obtained (65). From experiment, and by subtracting the elastic gradient energy coefficient, κ is estimated to be in the range -6.5×10^{-11} to -9.0×10^{-11} Jm^{-1} (65). Thus reasonable agreement between the measured and theoretical gradient energy coefficients is possible provided that elastic energy effects and longer range chemical interactions are taken into account. Values of the interdiffusion coefficient obtained from the modulated films were again in good agreement with extrapolated high temperature data.

3.6 Copper-Nickel

This system has been extensively studied by Tsakalakos (31,33,67). It is a phase-separating system, but direct observation of the phase separation is prevented due to low atomic mobility below the critical temperature which is estimated to be about 370°C (68). Figure 19 shows the wavelength dependence of $\tilde{D}_\lambda$ in a film of equiatomic composition annealed at 400°C. Tsakalakos attributes the non-linearity of the plot to higher order terms K_ζ. His values correctly predict the minimum in $\tilde{D}_\lambda$ (H) at $\lambda = 1.6$nm, but are about an order of magnitude lower than the values required to fit the whole curve.

At $H = 0$, only the term $K_1 = \kappa$ is significant. The slope of the plot at $H = 0$ is negative, indicating a negative κ, which is opposite to the sign expected for a phase-separating system (Eq. (14)). Tsakalakos suggests that this is due to longer range interactions in Eq. (33). From interatomic potentials he calculated $\kappa = -4.6\times10^{-12}$ Jm^{-1}, which though negative is again about an order of magnitude below the experimental value $\kappa = -4.3\times10^{-11}$ Jm^{-1}. He attributes the peak in $\tilde{D}_\lambda$ at $\lambda = 2.5$nm to Brillouin zone effects, and by estimating the position of the Fermi surface from positron annihilation results could account for the peak position quantitatively.

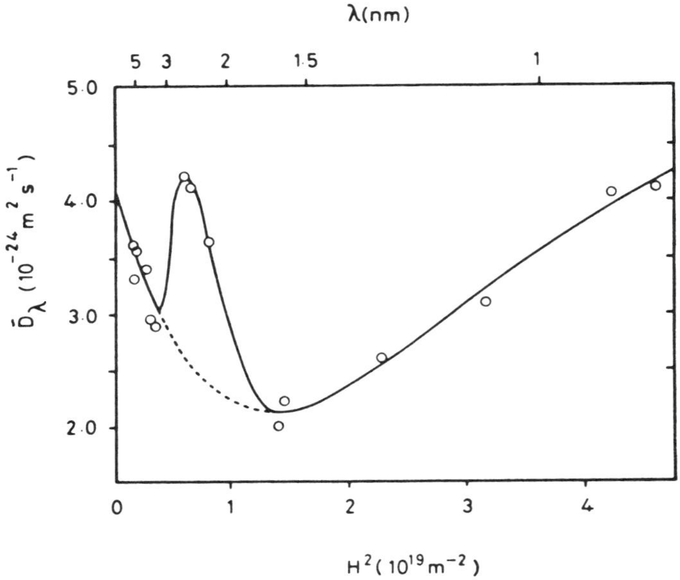

Fig. 19. Effective interdiffusivity $\tilde{D}_\lambda$ vs. H^2 for equiatomic Cu-Ni – modulated films annealed at 400°C (after (33)).

Appropriate extrapolation (Fig. 10) shows that the interdiffusivities at $\lambda = \infty$ estimated from the modulated films studies are in excellent agreement with bulk diffusion measurements at high temperature. In addition to the measurement of first order satellite intensities used to obtain $\tilde{D}_\lambda$, Tsakalakos measured the intensities of the second and third order satellites of the (111) peak. Their evolution on annealing was qualitatively very similar to that shown in Fig. 9. By applying his approximate analytic solution to the non-linear diffusion equation, Tsakalakos (31) was able to calculate κ from the satellite decay curves. He found it to be -2.3×10^{-11} Jm^{-1}, a factor of two less than the value obtained from the wavelength dependence of $\tilde{D}_\lambda$.

Tsakalakos (67) demonstrated phase separation in the copper-nickel system by annealing a sample successively at three temperatures. At 450°C the first order satellite intensity decayed; at 325°C it grew; and at 340°C it again decayed. The coherent spinodal temperature is therefore between 325°C and 340°C, in excellent agreement with the estimate of Vrijen and Radelaar based on clustering measurements (68).

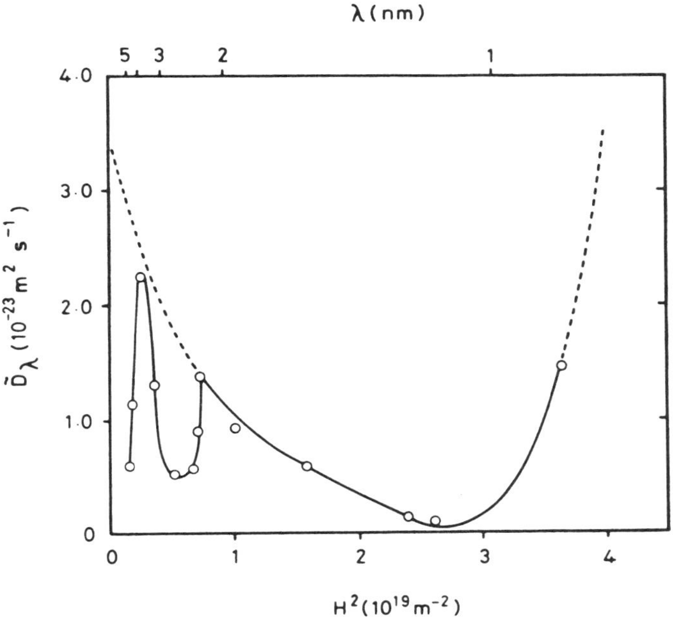

Fig. 20. Effective interdiffusivity $\widetilde{D}_\lambda$ vs. H^2 for equiatomic Ag-Pd films annealed at 375°C (after (69)).

3.7 Silver-Palladium

This is a homogenizing system. It was investigated by Henein (69), who found the variation of $\widetilde{D}_\lambda$ with wavelength shown in Fig. 20. The curvature was again analyzed according to the Tsakalakos expression, Eq. (32), and the dashed line in the figure is a calculated fit. The gradient energy coefficient was found to be $\kappa = -5.0 \times 10^{-10}$ Jm^{-1}. In contrast to some other cases, particularly copper-nickel, this value is in reasonable agreement with the calculation based on the heat of mixing and assuming nearest neighbor interactions. The minimum in $\widetilde{D}_\lambda$ at ~2.5nm was attributed to Fermi surface effects as it did not appear to correlate with the enhanced biaxial modulus of other films in the same wavelength range. Note, however, that the sign of these effects on $\widetilde{D}_\lambda$ is different from that found for copper-nickel (Fig. 19). The drop in $\widetilde{D}_\lambda$ at larger wavelengths (3.9 to 5.1 nm) appeared to be due to a loss of coherency, as seen in copper-palladium (Fig. 13).

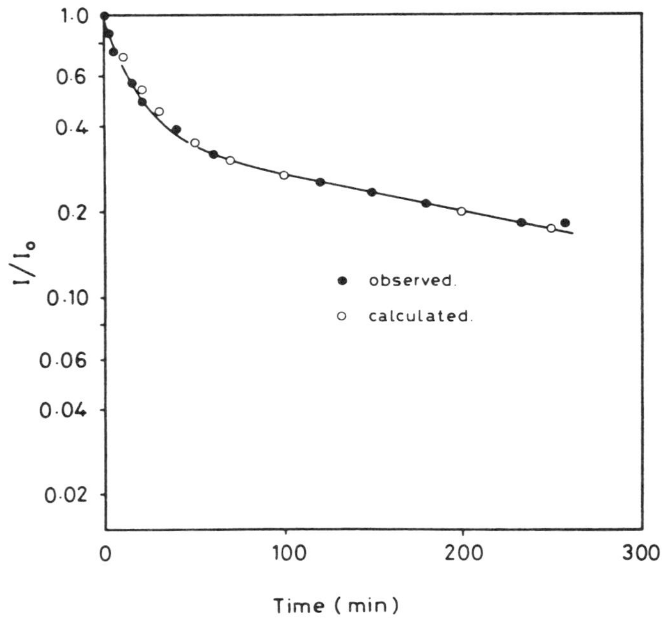

Fig. 21. Plots of log I/I_0 vs. annealing time at 300°C for 111-satellite intensity (measured relative to the Bragg peak) in a Au- (Ag,Pd) modulated film of average composition $Ag_{60}Au_{30}Pd_{10}$ (at.%) and modulation wavelength 1.13 nm. The calculated points were obtained using a linear diffusion theory (after (35)).

3.8 Ternary Systems

Murakami et al. (35) have studied the silver-gold-palladium system. They made modulated films of two types: alternating layers of gold and a silver-palladium alloy; and alternating layers of silver and a gold-palladium alloy. The films were all of the same average composition, 60 at.% Ag, 30 at.% Au, 10 at.% Pd, and were used to determine the four components of the diffusion matrix at that composition. Figure 21 shows observed and calculated satellite intensities for annealing one of the films. The calculation assumes linear diffusion and ignores gradient energy and strain effects; it is evidently a good fit to the data. The satellite decay curve is not linear because in a ternary system there are two characteristic time constants.

The silver-gold-copper system was similarly studied by Murakami et al. (36). In this case, a single modulated film sufficed to determine the diffusion matrix because the thermodynamics of this ternary solution are well known.

Recently the various types of composition modulation in a ternary system have been classified by Manikopoulos and Tsakalakos (70). In order to investigate spinodal decomposition in the alloy $Cu_{61}Ni_{33}Cr_6$ (composition in atomic %), Manikopoulos and Tsakalakos produced a layered film of that overall composition composed of alternating layers of compositions $Cu_{44}Cr_6$ and $Ni_{94}Cr_6$. Thus the structure is a superposition of a Cu-Ni compositionally modulated film and a uniform Cr deposition. No diffusion data have been reported as yet, but a numerical study by Tsakalakos and Dugan (71) may be useful in interpreting the kinetics of spinodal decomposition in a ternary system.

4. EXPERIMENTAL RESULTS ON AMORPHOUS METALS

4.1 Introduction to Amorphous Metals

The coverage given here to amorphous metals is somewhat greater than would be warranted on the basis of the volume of experimental work done. However, consideration in some depth is justified because of the novelty of these materials, because diffusion measurements using artificially modulated thin films are particularly valuable in this case, and because there are a number of worthwhile projects for the future.

Amorphous metals, or metallic glasses, can be produced by rapid quenching of the melt, evaporation, sputtering, electrodeposition or chemical deposition. Their structures are liquid-like in that they exhibit no long range order. This novel *solid glassy metallic state* has become increasingly well studied over the past twenty years and it is now clear that amorphous metals can have soft magnetic properties, strength, wear and corrosion resistance, that make them of commercial interest. A comprehensive recent review of their properties and applications may be found in Luborsky (72). Only alloys and not pure metals can be produced in the amorphous state. The alloys fall into two main categories: noble or transition metals with 15-25 atomic % of metalloids; and combinations of early and late transition metals, which typically form glasses over wider composition ranges. Examples of alloys in these categories are: $Fe_{80}B_{20}$, $Fe_{40}Ni_{40}B_{20}$, $Fe_{40}Ni_{40}P_{14}B_6$, $Pd_{82}Si_{18}$, $Pd_{77.5}Cu_6Si_{16.5}$; and Cu_xZr_{1-x} ($25<x<75$), Nb_xNi_{1-x} ($30<x<70$). (All compositions in atomic %.)

Amorphous metals are metastable with respect to the crystalline state and given sufficient atomic mobility will crystallize. The crystallization is rapid at temperatures above the glass transition temperature T_g which is the temperature of configurational freezing in the liquid. To avoid crystallization, diffusion measurements must be made below

T_g and the diffusion coefficients are normally less than $10^{-20}m^2s^{-1}$. While such low values can be measured by composition profiling techniques (Table 1), the use of compositionally modulated films is valuable.

As-produced amorphous metals can lower their free energy not only by crystallizing but also be relaxing their glass structure. This *structural relaxation* involves a slight densification ($\lesssim$0.4%) and leads to changes in all the physical properties. The effect is most marked in atomic transport properties; for example, Taub and Spaepen (73) found that the rate of viscous flow decreased by many orders of magnitude on relaxation. It is expected that the rate of atomic diffusion would be similarly affected, but conventional measurements by composition profiling have not shown a clear relaxation effect (74). However, such profiling techniques (Table 1) require at least 200 jumps per atom to get a measurement. In this time presumably most of the relaxation has taken place. On the other hand, a measurement of $\tilde{D}$ is possible after only 0.2 jumps per atom using a modulated film. As described in the next section this sensitivity permits time-resolved measurement of $\tilde{D}$ while structural relaxation is taking place. Thus the modulated film technique is essential for a study of the relationship between diffusive transport and structural relaxation. It is of interest to note that viscous flow rates can be measured after only 10^{-4} jumps per atom and that relaxation effects are then readily detected.

4.2 Interdiffusion Measurements

Rosenblum et al. (4) produced modulated amorphous metallic films by DC sputtering from two targets of different compositions. The films had compositions of $(Pd_{80}Au_7Si_{73})_{70}/Fe_{30}$ or $(Pd_{85}Si_{15})_{61}/(Fe_{85}B_{15})_{39}$ and modulation wavelengths in the range 0.2 - 7.0nm. The films showed only a first-order satellite about the 000 beam. The satellite intensity decayed on annealing and $\tilde{D}_\lambda$ was derived using Eq. (12). It was found that $\tilde{D}_\lambda$ decreased in the early stages of an anneal and then stabilized. Rosenblum et al. attributed this behavior to structural relaxation.

Greer et al. (75) and Cammarata and Greer (76) made films of composition $(Pd_{85}Si_{15})_{50}/(Fe_{85}B_{15})_{50}$ using the sputtering technique of Rosenblum et al. (4). The modulation wavelengths of the films were in the range 0.22 - 7.46nm. All the results quoted in this section are from these films. Since the metals dominate the x-ray scattering the measured $\tilde{D}_\lambda$ is an effective Pd-Fe interdiffusivity. The results on these films have been analyzed according to a binary solution model. Since there are four components present, this may be questioned. If the diffusivities of metalloids were quite different from those of the

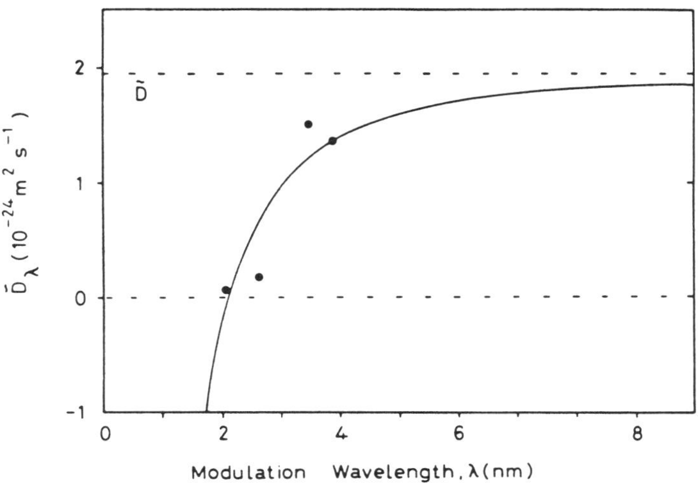

Fig. 22. The effective interdiffusivity $\tilde{D}_\lambda$ vs. modulation wavelength for amorphous $(Pd_{85}Si_{15})_{50}/(Fe_{85}B_{15})_{50}$ modulated films annealed at 250°C. (After (76)).

metals, there could be measurable effects on the Pd-Fe interdiffusivity. An example of such an effect, in a ternary system, has been shown in Fig. 21. Fortunately, in amorphous metals the diffusivities of the metalloid and metal atoms are remarkably similar (74). Since in addition the metalloids are minor components, treating the $(Pd_{85}Si_{15})_{50}/(Fe_{85}B_{15})_{50}$ films as pseudo-binary systems should be a good approximation.

Four films with modulation wavelengths between 2.06 and 3.94nm were annealed at 250°C and the intensities I(t) of the first order satellites were monitored. In each case (as found earlier by Rosenblum et al. (4)) there was an initial rapid non-linear decrease of ℓn I(t) with time. As discussed later, most of this non-linearity can be attributed to structural relaxation. When the decay of ℓn I(t) became linear with time $\tilde{D}_\lambda$ was determined from Eq. (12), and the results are plotted in Fig. 22. The diffusivities in the films can be compared in this way since all the films had been annealed for about the same time (~15 hours) and therefore should be similarly relaxed. The variation in $\tilde{D}_\lambda$ can be attributed to gradient energy effects, and it is evident from a comparison of Figs. 3 and 22 that amorphous PdSi/FeB is an ordering system. This is not unexpected since there are ordering

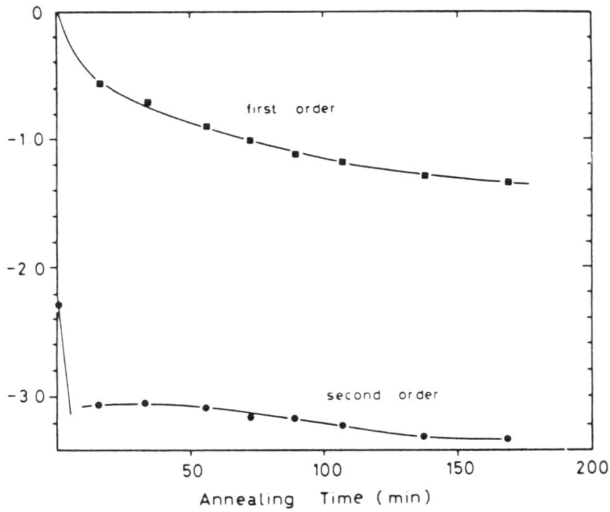

Fig. 23. Plots of I/I_o vs. annealing time at 300°C for the first and second order satellite intensities from an amorphous $(Pd_{85}Si_{15})_{50}/(Fe_{85}B_{15})_{50}$ modulated film with a wavelength of 3.22 nm. (After (75)).

reactions in the crystalline Pd-Fe system (77). For highly undercooled, metastable systems such as amorphous metals, basic information such as this on the thermodynamics of mixing may be obtainable only from modulated film studies. From a linear regression fit of $\tilde{D}_\lambda$ versus $1/\lambda^2$ to Eq. (12) it was found (76) that the macroscopic interdiffusion coefficient $\tilde{D} = 1.94 \times 10^{-24}$ m^2s^{-1} and that $\kappa/f_o'' = -5.2 \times 10^{-20}$ m^2. The critical wavelength for this amorphous system is 2.02 nm, which is considerably larger than in crystalline metallic systems.

The initial non-linear decay in ℓn I(t) versus t is unlikely to be due to the composition dependence of $\tilde{D}$ which appears to have little effect on first-order satellites (Fig. 9). Nor does it appear to correlate with structural changes in the films which cease after about 2 hours (75). The effects of moving grain boundaries, contributing to the non-linearity observed by Cook and Hilliard (2), cannot be present in an amorphous system. Accordingly, the decrease in $\tilde{D}_\lambda$ on annealing has been attributed to structural relaxation. The kinetics of the relaxation were studied by annealing a film with λ = 3.22nm at 300°C, and the decay curves of the first and second order satellites are shown in Fig. 23. The decay of the second order satellite shows non-linear effects attributable to the composition dependence of $\tilde{D}$ (see Fig. 9). However, the variation of $\tilde{D}$ with time could be determined from the instantaneous slope of the first order decay curve and the value of κ/f_o'' obtained above. It was found that $1/\tilde{D}$ increased linearly with annealing time, from 1×10^{22} sm^{-2} to 8×10^{22} sm^{-2} in 20 minutes.

In determining an activation energy for diffusion in amorphous metals it is obviously important to ensure that in measurements at different temperatures the degree of structural relaxation is the same. Such *isoconfigurational* measurements are greatly facilitated using the modulated film technique. Greer et al. (75) measured $\tilde{D}_\lambda$ in a film with λ = 2.06nm after a long anneal at 250°C. They then made measurements at other temperatures, but after each such measurement returned to 250°C and remeasured $\tilde{D}_\lambda$ to make sure that it had not changed. In this way they determined a true isoconfigurational temperature variation of $\tilde{D}_\lambda$. The variation was Arrhenius with an activation energy of 195 kJ mol^{-1} and a pre-exponential of 5.6×10^{-5}m^{2}s^{-1}. This pre-exponential is similar to those found for substitutional diffusion in fcc metals (74). Cammarata and Greer (76) showed that the temperature-dependent values of $\tilde{D}$ determined using the modulated film technique were in good agreement with values obtained on similar glasses by conventional profiling techniques at somewhat higher temperatures.

The linear rise on annealing found for $1/\tilde{D}$ parallels the linear rise in viscosity found by Taub and Spaepen (73) in $Pd_{82}Si_{18}$ glass, and suggests a link between the mechanisms of diffusive and viscous flow. Furthermore, the activation energy for viscous flow found by Taub and Spaepen was 192 kJ mol^{-1}, very similar to that found above. In liquid metals it is found that the viscosity η and the self-diffusivity D are related by the Stokes-Einstein equation:

$$D = k_B T/6\pi\eta r, \tag{48}$$

where r is approximately the ionic radius. In using Eq. (48) to compare $\tilde{D}$ for η for amorphous metals, the effect of the thermodynamic factor (Eq. (44)) on $\tilde{D}$ has to be taken into account. For the present case Spaepen (56) estimates that $\tilde{D} \simeq 3.8D$. By comparing the constant rates of increase of $1/\tilde{D}$ and η the effects of structural relaxation can be eliminated. It is found that the diffusive transport is about 150 times faster than would be predicted from η using Eq. (48). Greer (74) has suggested that, at least for metal-metalloid glasses, this disparity may arise because of strong chemical bonding leading to a network-like structure on which movement of vacancy-type defects can occur.

4.3 Discussion

There are obviously many possibilities for the application of artificially modulated thin films in studying amorphous metals. Much useful information may be gained on the stability and thermodynamics of these systems. In particular, it would be valuable to study interdiffusion in systems with only two components, as this would be easier to analyze. Such work, on Cu-Zr and Ni-Zr, is underway (78,79). In future analyses it may be useful to include non-linear effects, or to work only at low modulation amplitudes to ensure that the linear analysis is valid. Also, it may be necessary to include some consideration of strain effects analogous to that used by Stephenson (26) for oxide glasses (Section 2.3).

Amorphous metals have particular advantages in the study of composition modulation effects. The modulation wavelength can be adjusted continuously since there is no interference with a crystal periodicity. There are no coherency strains due to lattice matching, and such strains as do arise are likely to be small which should facilitate the interpretation of results. Finally, in amorphous alloys the composition can be adjusted continuously; in crystalline systems this is unlikely to be possible because of the formation of stoichiometric compounds.

The diffusivities measured in amorphous metals are slightly greater than those in comparable single crystals, but much less than in polycrystals. This is because of the absence of short-circuit diffusion paths such as dislocations and grain boundaries in an amorphous material. The absence of mobile grain boundaries in particular should enhance the stability of composition modulations in amorphous metals. The uniformity and low diffusivity possible with amorphous metallic layers has led to their being considered as diffusion barriers in semiconductor devices (80). This is a promising application, as discussed in Chapter VIII by Kattelus and Nicolet, and valuable preliminary information may be obtained from modulated films studies.

Recently it has been demonstrated that when multilayers of two polycrystalline metals are annealed an amorphous phase can be formed (81,82). This reaction is of fundamental interest, but may also be of technological importance in providing another production route for amorphous metals. Since it has recently been shown by cross-sectional transmission electron microscopy (83) that the reaction layers of amorphous phase are planar and uniform, the kinetics of the reaction might usefully be studied with the x-ray techniques described in Section 2.5.

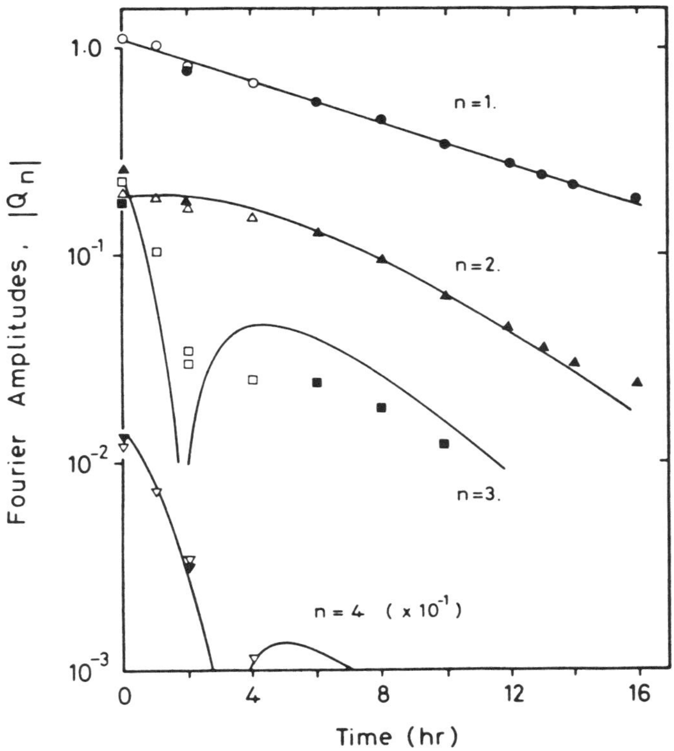

Fig. 24. Fourier amplitudes of the first four satellites of the (002) Bragg reflection from a $(GaAs)_{12}(AlAs)_9$ modulated structure vs. annealing time at 860°C. The open and closed symbols are for two different samples. The lines are calculated using a concentration-dependent interdiffusivity to obtain the best fit. (After (5)).

5. EXPERIMENTAL RESULTS ON SEMICONDUCTORS

5.1 GaAs-AlAs

Fleming et al. (5) have made multilayered films with very great control using molecular beam epitaxy. Annealing studies were made on a film with a repeat of 12 atomic layers of GaAs and 9 atomic layers of AlAs, having a wavelength of λ = 5.9nm. The decay at 860°C of the satellites, up to fourth order, about the (002) Bragg reflection is shown in Fig. 24. The non-linear behavior of the higher harmonics was attributed to a concentration-dependent interdiffusion coefficient. Fleming et al. (5) assumed that the dependence of the interdiffusion coefficient on the gallium concentration c was of the form:

$$D(c,T) = D(T) \exp [((c - 1)\Delta F_{Al} - c\Delta F_{Ga})/k_B T], \qquad (49)$$

where ΔF_{Al} and ΔF_{Ga} are the vacancy formation energies in AlAs and GaAs respectively. This interdiffusion coefficient represents Ga diffusion in AlAs when c = 0 and Al diffusion in GaAs when c = 1. Using Eq. (47) and treating ΔF_{Al} and ΔF_{Ga} as adjustable parameters the solid lines in Fig. 24 were fitted to the data. The fit is reasonably good, but it should be noted that gradient energy effects were not taken into account. It may be that such effects are negligible for the comparatively large wavelength of 5.9nm. However, Fleming et al. also studied a sample with a repeat of 1.1 atomic layers of GaAs and 1.1 atomic layers of AlAs (λ = 0.62nm). In this case only a first order satellite was observed and its decay at 800°C was monitored. If the interdiffusivity at 800°C is extrapolated to 860°C, assuming an activation energy of 4.3 eV as determined by Chang and Koma (84), a value of 6.3×10^{-24} m^2s^{-1} is obtained. This is significantly less than the value, 2.4×10^{-24} m^2s^{-1}, determined from the larger wavelength sample annealed at 860°C. Greer and Spaepen (24) have suggested that the discrepancy may be due to gradient energy effects. Assuming this to be so, they have calculated that the bulk interdiffusion coefficient is 2.44×10^{-23} m^2s^{-1}, and $\kappa/f_o'' = -1.5 \times 10^{-20}$ m^2. The value of κ/f_o'' is similar to those obtained for crystalline metallic systems. It seems likely that gradient energy effects exist in semiconductor systems and that they should be taken into account. A comparison of the interdiffusion results of Fleming et al. (5) with other results on the GaAs-AlAs system obtained by different means has been given by Greer and Spaepen (24).

5.2 GaSb-InSb

Eltoukhy and Greene (85) have made compositionally modulated films of GaSb-InSb by r.f. sputter deposition. The films had strong (110) texture. The decay of the first order satellites of the (220) Bragg peak was monitored in the temperature range 230°C - 340°C, and interdiffusion coefficients were obtained from the final linear portions of the decay curves. Because all the films had modulation wavelengths greater than 3.6 nm Eltoukhy and Greene ignored gradient energy effects in their analysis, although no wavelength-dependent data were presented to justify this. They found that the interdiffusion coefficient had an Arrhenius temperature dependence, with a pre-exponential of 4×10^{-18} m^2s^{-1} and an activation energy of 77 kJ mol^{-1}. Their data appear to be in broad agreement with other measurements (24), but extend to lower temperatures because of the sensitivity of the modulated film technique.

An important conclusion arising from the work of Eltoukhy and Greene (3,85,86) and Eltoukhy et al. (87) is that the interdiffusivity in the GaSb-InSb system during sputter deposition is enhanced by low energy ion bombardment. The enhancement is estimated to be a factor of 10^5 over the interdiffusivity measured on subsequent annealing. If a similar enhancement were to apply for other systems, this would set a lower limit on the wavelength of modulated films prepared by sputtering.

5.3 Amorphous Si-Ge

In recent years there has been much interest in amorphous semiconductors. There have been a number of studies of impurity diffusion, but so far only one study of diffusion of the network-forming components. This study, by Prokes and Spaepen (88), used a compositionally modulated thin film since it was expected (as for amorphous metals, Section 4.1) that a sensitive technique would be required to detect interdiffusion before the onset of crystallization.

Alternating thin films of amorphous silicon and amorphous germanium were deposited by ion beam sputtering. The multilayered films had an average composition of $Si_{50}Ge_{50}$ (atomic %) and a composition modulation wavelength of 5.83nm (88). With only one wavelength it was not possible to study gradient energy effects, but it was presumed that these would be small at the relatively large wavelength used. It was found that interdiffusion could be detected (from the decay of the first order satellite about the 000 beam) well before any crystallization. In the temperature range 277 - 357°C the effective interdiffusivity is given by:

$$\tilde{D}_\lambda = 1.07 \times 10^{-10} \exp(-1.6\text{eV}/k_BT)\,m^2s^{-1} \quad (50)$$

Initial and final measurements of $\tilde{D}_\lambda$ were made at 357°C; the final value was only 5% lower than the initial value, demonstrating that the measurements were essentially isoconfigurational.

The interdiffusivity values found by Prokes and Spaepen are surprisingly high. They point out that at 357°C the interdiffusivity in the amorphous phase is 3.86×10^{-23} m^2s^{-1}, while extrapolated values for the diffusivity of Ge in crystalline Si and crystalline Ge at the same temperature are 4.35×10^{-37} m^2s^{-1} and 5.3×10^{-30} m^2s^{-1} respectively. The mechanism of rapid diffusion in amorphous Si-Ge has not yet been elucidated. In particular it will be important to test the effects of deposition conditions on the interdiffusivity and on changes in the interdiffusivity due to structural relaxation.

5.4 Discussion

Artificially layered thin films for the study of interdiffusion have not been as widely applied to semiconductors as they have to metals, and indeed the investigations of GaAs-AlAs, GaSb-InSb and amorphous Si-Ge are the only detailed studies reported to date. So far, the sophistications of gradient energy effects and strain effects have not been included in the analyses applied to semiconductors. It appears from existing work that interdiffusion coefficients have a marked composition dependence and that therefore non-linear effects may be important. In any case, the elements of theory covered in Section 2 apply equally to semiconductors and it is to be hoped that more studies on artificially-layered semiconductors will be undertaken. Such studies, because of their ability to yield thermodynamic information from the wavelength-dependence of the interdiffusivity, and because of their unmatched sensitivity, may be of great interest in integrated circuit manufacturing.

6. CONCLUSIONS

The theory of interdiffusion in materials with a one-dimensional composition modulation has been extensively developed. Gradient energy and strain effects arise at small modulation wavelengths and analyses have been developed for diffusion both in a continuum and on a discrete lattice. There still exists a need, however, for an analysis capable of dealing with modulations of very short wavelength in amorphous or polycrystalline systems. By taking several parameters to be independent of composition a linearized diffusion equation is obtained. This enables the harmonic components of a general periodic modulation to be analyzed independently, and can be applied when the modulation amplitude is small. When the composition dependence of parameters is taken into account the interdiffusion kinetics have to be analyzed by numerical modeling.

A composition modulation gives rise to satellite peaks about the Bragg peaks and zero-order beam in x-ray diffraction. The measurement of satellite intensities is the most powerful way of monitoring changes in the composition modulation. The satellite positions give the modulation wavelength. The intensities of satellites about the zero order beam give the amplitude of the modulation. In crystalline materials the intensities of satellites about the Bragg peaks can be used to derive the degree of coherency of the layers.

The great majority of the experimental work has been on compositionally modulated fcc metals. Gradient energy effects have been observed and can be correlated with the thermodynamics of the binary solutions. In some alloys a good match between theory and experiment is obtained when only nearest neighbor atomic interactions are assumed; in other cases it is necessary to take longer range interactions into account. It is possible to measure very low interdiffusivities and the values are in agreement with Arrhenius extraploation of bulk diffusivity measurements at much higher temperatures. In some systems the degree of coherency changes sharply with modulation wavelength. The change is manifested in the interdiffusivity, and these strain effects can be explained quantitatively. Both homogenizing and phase-separating systems have been examined, and in the latter coherent spinodal temperatures have been estimated. Some sharp maxima or minima in interdiffusivity have been observed at small modulation wavelengths; these have been tentatively attributed to Brillouin zone effects.

The modulated film technique has also proved useful in the measurement of very low diffusivities in amorphous metals and in semiconductors. In general, the analyses applied in these cases have not been as sophisticated as those used for crystalline metals. Nevertheless, gradient energy effects and structural relaxation effects in the interdiffusivity have been detected in amorphous metals and non-linear diffusion effects have been detected in semiconductors. The potential for further work in these areas is great.

ACKNOWLEDGEMENTS

The author acknowledges with pleasure many useful discussions with F. Spaepen and R. C. Cammarata. At Harvard University, the author's research in this area was supported by the Office of Naval Research (Contract No. N00014-77-C-0002) and by the National Science Foundation (Contract No. DMR80-20247). At Cambridge University, current research is supported by the International Business Machines Corporation, by the Science and Engineering Research Council, and by the Office of Naval Research. The author is grateful to Prof. D. Hull for the provision of laboratory facilities.

REFERENCES

1. DuMond, J. and Youtz, J. P., An X-ray method of determining rates of diffusion in the solid state, *J. Appl. Phys.* **11**, 357-365 (1940).
2. Cook, H. E. and Hilliard, J. E., Effect of gradient energy on diffusion in gold-silver alloys, *J. Appl. Phys.* **40**, 2191-2198 (1969).
3. Eltoukhy, A. H. and Greene, J. E., Ion-bombardment-enhanced diffusion during the growth of sputtered superlattice thin films, *Appl. Phys. Lett.* **33**, 343-345 (1978).
4. Rosenblum, M. P., Spaepen, F. and Turnbull, D., Diffusion and structural relaxation in compositionally modulated amorphous metal films, *Appl. Phys. Lett.* **37**, 184-186 (1980).
5. Fleming, R. M., McWhan, D. B., Gossard, A. C., Wiegmann, W. and Logan, R. A., X-ray diffraction study of interdiffusion and growth in $(GaAs)_n(AlAs)_m$ multilayers, *J. Appl. Phys.* **51**, 357-363 (1980).
6. Abeles, B. and Tiedje, T., Amorphous semiconductor superlattices, *Phys. Rev. Lett.* **51**, 2003-2006 (1983).
7. Döhler, G. H., Künzel, H. and Ploog, K., Tunable absorption coefficient in GaAs doping superlattices, *Phys. Rev. B* **25**, 2616-2626 (1982).
8. Schuller, I. K., New class of layered materials, *Phys. Rev. Lett.* **44**, 1597-1600 (1980).
9. McWhan, D. B., Gurvitch, M., Rowell, J. M. and Walker, L. R., Structure and coherence of NbAl multilayer films, *J. Appl. Phys.* **54**, 3886-3891 (1983).
10. Yang, W. M. C., Tsakalakos, T. and Hilliard, J. E., Enhanced elastic modulus in composition-modulated gold-nickel and copper-palladium foils, *J. Appl. Phys.* **48**, 876-879 (1977).
11. Testardi, L. R., Willens, R. H., Krause, J. T., McWhan, D. B., and Nakahara, S., Enhanced elastic moduli in Cu-Ni films with compositional modulation, *J. Appl. Phys.* **52**, 510-511 (1981).
12. Zheng, J. Q., Ketterson, J. B., Falco, C. M. and Schuller, I. K., The magnetization and Curie temperature of compositionally modulated Cu/Ni films, *J. Appl. Phys.* **53**, 3150-3155 (1982).
13. Ruggiero, S. T., Barbee, T. W. and Beasley, M. R., Superconductivity in quasi-two-dimensional layered composites, *Phys. Rev. Lett.* **45**, 1299-1302 (1980).

14. Willens, R. H. Kornblit, A., Testardi, L. R. and Nakahara, S., Melting of Pb as it approaches a two-dimensional solid, *Phys. Rev. B* **25**, 290-296 (1982).
15. Spiller, E., Segmüller, A. and Haelbich, R.-P., The fabrication of multilayer x-ray mirrors, *Ann. New York Acad. Sci.* **342**, 188-200 (1980).
16. Spiller, E., in: *Handbook on Synchrotron Radiation* (E. E. Koch, ed.) Vol. 1, pp. 1091-1129, New York: North Holland (1983).
17. Döhler, G. H., Electron-hole subbands at the GaSb/InAs interface, *Surf. Sci.* **98**, 108-116 (1980).
18. Hillert, M., A theory of nucleation for solid metallic solutions, Sc.D. Thesis, Massachusetts Institute of Technology (1956).
19. Hillert, M., A solid-solution model for inhomogeneous systems, *Acta Metall.* **9**, 525-535 (1961).
20. Cahn, J. W. and Hilliard, J. E., Free energy of a nonuniform system, I. Interfacial free energy, *J. Chem. Phys.* **28**, 258-267 (1958).
21. Cahn, J. W., On spinodal decomposition, *Acta Metall.* **9**, 795-801 (1961).
22. Cahn, J. W., Spinodal decomposition, *Trans. Met. Soc. A.I.M.E.* **242**, 166-180 (1968).
23. Cook, H. E., De Fontaine, D. and Hilliard, J. E., A model for diffusion on cubic lattices and its application to the early stages of ordering, *Acta Metall.* **17**, 765-773 (1969).
24. Greer, A. L. and Spaepen, F., in: *Synthetic Modulated Structures* (L. Chang and B. C. Giessen, eds.), pp. 419-486, New York: York: Academic Press (1985).
25. Yamauchi, H. and Hilliard, J. E., A comment on Cahn's diffusion equation, *Scripta Metall.* **6**, 909-914 (1972).
26. Stephenson, G. B., Spinodal decomposition in amorphous systems, *J. Non-Cryst. Solids* **66**, 393-427 (1984).
27. Cahn, J. W., On spinodal decomposition in cubic crystals, *Acta Metall.* **10**, 179-183 (1962).
28. Cook, H. E. and De Fontaine, D., On the elastic free energy of solid solutions – I. Microscopic theory, *Acta Metall.* **17**, 915-924 (1969).
29. Cook, H. E. and De Fontaine, E., On the elastic free energy of solid solutions – II. Influence of the effective modulus on precipitation from solution and the order-disorder reaction, *Acta Metall.* **19**, 607-616 (1971).
30. Born, M. and Huang, K., *Dynamical theory of crystal lattices,* London: Oxford University Press (1954).

31. Tsakalakos, T., Interdiffusion and enhanced elastic modulus effect in composition-modulated copper-nickel thin foils, Ph.D. Thesis, Northwestern University (1977).
32. De Fontaine, D., A computer simulation of the evolution of coherent composition variations in solid solutions, Ph.D. Thesis, Northwestern University (1967).
33. Tsakalakos, T., Composition modulated materials: New materials for studying stability and critical phenomena in solid solutions, *Thin Solid Films* **86**, 79-90 (1981).
34. Purdes, A., Structure of composition modulated copper-nickel and gold-silver thin films, Ph.D. Thesis, Northwestern University (1976).
35. Murakami, M., De Fontaine, D., Sanchez, J. M. and Fodor, J., Diffusion in AgAuPd thin film microcouples, *Acta Metall.* **22** 709-719 (1974).
36. Murakami, M., De Fontaine, D., Sanchez, J. M. and Fodor, J., Ternary diffusion in multilayer Ag-Au-Cu thin films, *Thin Solid Films* **25**, 465-482 (1975).
37. Röll, K. and Reill, W., A study of interdiffusion in multilayer Cu/Ni films by Auger electron depth profiling, *Thin Solid Films* **89**, 221-224 (1982).
38. Petroff, P. M., Transmission electron microscopy of interfaces in III-V compound semiconductors, *J. Vac. Sci. Technol.* **14**, 973-978 (1977).
39. Preston, G. D., The diffraction of x-rays by age-hardening aluminum copper alloys, *Proc. Roy. Soc. London A* **167**, 526-538 (1938).
40. De Fontaine, D., in: *Local Atomic Arrangements Studied by X-ray Diffraction* (J. B. Cohen and J. E. Hilliard, eds.), New York: Gordon and Breach (1966).
41. Guinier, A., *X-ray Diffraction*, San Francisco: W. H. Freeman and Co. (1963).
42. Pomerantz, M. and Segmüller, High resolution x-ray diffraction from small numbers of Langmuir-Blodgett layers of manganese stearate, *Thin Solid Films* **68**, 33-45 (1980).
43. Segmüller, A., Small-angle interferences of x-rays reflected from periodic and near-periodic multilayers, *AIP Conf. Proc.* **53**, 78-80 (1979).
44. Lee, P., X-ray diffraction in multilayers, *Optics Commun.* **37**, 159-164 (1981).
45. Darken, L. S., Diffusion, mobility and their interrelation through free energy in binary metallic systems, *Trans. A.I.M.E.* **175**, 184-201 (1948).
46. Hilliard, J. E., in: *Phase Transformations*, pp. 497-560, Metals Park, Ohio: American Society for Metals (1970).

47. Eisen, F. H. and Birchenall, C. E., Self-diffusion in indium antimonide and gallium antimonide, *Acta Metall.* **5**, 265-274 (1957).
48. Goldstein, B., Diffusion in compound semiconductors, *Phys. Rev.* **121**, 1305-1311 (1961).
49. Palfrey, H. D., Brown, M., and Willoughby, A. F. W., Self-diffusion of gallium in gallium arsenide, *J. Electrochem. Soc.* **128**, 2224-2228 (1981).
50. Luborsky, F. E. and Bacon, F., in: *Proceedings of the Fourth International Conference on Rapidly Quenched Metals* (T. Masumoto and K. Suzuki, eds.) pp. 561-566, Sendai: Japan Institute of Metals (1982).
51. Cahn, R. W., Evetts, J. E. Patterson, J., Somekh, R. E., and Kenway, Jackson, C., Direct measurement by secondary-ion mass spectrometry of self-diffusion of boron in $Fe_{40}Ni_{40}B_{20}$ glass, *J. Mater. Sci.* **15**, 702-710 (1980).
52. Gupta, D., Tu, K. N., and Asai, K. W., Diffusion in the amorphous phase of Pd-19 at.% Si metallic alloy, *Phys. Rev. Lett.* **5**, 796-803 (1975).
53. Chen, H. S., Kimerling, L. C., Poate, J. M. and Brown, W. L., Diffusion in a Pd-Cu-Si metallic glass, *Appl. Phys. Lett.* **32** 461-463 (1978).
54. Birac, C., and Lesueur, D., Diffusion atomique du lithium dans l'alliage métallique amorphe Pd-Si, *Phys. Stat. Solidi* **A36**, 247-251 (1976).
55. Langer, J. S., Theory of spinodal decomposition in alloys, *Ann. Physics* **65**, 53-86 (1971).
56. Spaepen, F., in: *Layered Structures, Epitaxy and Interfaces* (J. M. Gibson and L. R. Dawson, eds.) pp. 295-306, Pittsburgh, Materials Research Society (1985).
57. Graham, L. D. and Kraft, R. W. Coarsening of eutectic microstructures at elevated temperatures, *Trans. Met. Soc. A.I.M.E.* **236**, 94-102 (1966).
58. Dinklage, J. and Frerichs, R., X-ray diffraction and diffusion in metal film layered structures, *J. Appl. Phys.* **34**, 2633-2635 (1963).
59. Cook, H. E., Interdiffusion in silver-gold solid solutions at low temperatures, Ph.D. Thesis, Northwestern University (1966).
60. Johnson, W. A., Diffusion experiments on a gold-silver alloy by chemical and radioactive tracer methods, *Trans. A.I.M.E.* **147**, 331-347 (1942).
61. Philofsky, E. M., Interdiffusion in copper-palladium: A verification of a new diffusion equation, Ph.D. Thesis, Northwestern University (1968).

62. Philofsky, E. M. and Hilliard, J. E., Effect of coherency strains on diffusion in copper-palladium alloys, *J. Appl. Phys.* **40**, 2198-2205 (1969).
63. Yang, W. M.-C., Spinodal decomposition in gold-nickel alloys, Ph.D. Thesis, Northwestern University (1971).
64. Paulson, W. M., Interdiffusion in composition-modulated copper-gold and copper-nickel thin films, Ph.D. Thesis, Northwestern University (1972).
65. Paulson, W. M. and Hilliard, J. E., Interdiffusion in composition-modulated copper-gold thin films, *J. Appl. Phys.* **48**, 2117-2123 (1977).
66. Cook, H. E., Brownian motion in spinodal decomposition, *Acta Metall.* **18**, 297-306 (1970).
67. Tsakalakos, T., Spinodal decomposition in Cu-Ni alloys by artificial composition modulation technique, *Scripta Metall.* **15**, 255-258 (1981).
68. Vrijen, J. and Radelaar, S., Clustering in Cu-Ni alloys: A diffuse neutron-scattering study, *Phys. Rev. B* **17**, 409-421 (1978).
69. Henein, G. E., The elastic modulus and diffusion coefficients of silver-palladium composition-modulated foils, Ph.D. Thesis, Northwestern University (1979).
70. Manikopoulos, C. N. and Tsakalakos, T., in: *Modulated Structure Materials* (T. Tsakalakos, ed.) pp. 439-453, Dordrecht: Martinus Nijhoff (1984).
71. Tsakalakos, T., and Dugan, M. P., in: *Modulated Structure Materials* (T. Tsakalakos, ed.) pp. 425-436, Dordrecht: Martinus Nijhoff (1984).
72. Luborsky, F. E. (ed.), *Amorphous Metallic Alloys*, London: Butterworths (1983).
73. Taub, A. I. and Spaepen, F., The kinetics of structural relaxation of a metallic glass, *Acta Metall.* **28**, 1781-1788 (1980).
74. Greer, A. L., Atomic transport and structural relaxation in metallic glasses, *J. Non-Cryst. Solids* **61** & **62**: 737-748 (1984).
75. Greer, A. L., Lin, C.-J. and Spaepen, F., in: *Proceedings of the Fourth International Conference on Rapidly Quenched Metals* (T. Masumoto and K. Suzuki, eds.) pp. 567-572, Sendai: Japan Institute of Metals (1982).
76. Cammarata, R. C. and Greer, A. L., Interdiffusion studies in metallic glasses using compositionally modulated thin films, *J. Non-Cryst. Solids* **61** & **62**: 889-894 (1984).
77. Hansen, M. and Anderko, K., *Constitution of Binary Alloys,* New York: McGraw-Hill (1958).
78. Cammarata, R. C., unpublished results.
79. Somekh, R. E., and Greer, A. L., unpublished results.

80. Nicolet, M.-A., Suni, I. and Finetti, M., Amorphous metallic alloys in semiconductor contact metallizations, *Solid State Tech.* 129-133 (December 1983).
81. Schwarz, R. B. and Johnson, W. L., Formation of an amorphous alloy by solid-state reaction of the pure polycrystalline metals, *Phys. Rev. Lett.* **51**, 415-418 (1983).
82. Clemens, B. M., Johnson, W. L. and Schwarz, R. B., Amorphous zirconium-nickel films formed by solid state reactions, *J. Non-Cryst. Solids* **61** & **62**, 817-822 (1984).
83. Schröder, H., Samwer, K. and Köster, U., Micromechanism for metallic glass formation by solid state reactions, *Phys. Rev. lett.* **54**, 197-200 (1985).
84. Chang, L. L. and Koma, A., Interdiffusion between GaAs and AlAs, *Appl. Phys. Lett.* **29**, 138-141 (1976).
85. Eltoukhy, A. H. and Greene, J. E., Compositionally modulated sputtered InSb/GaSb superlattices: Crystal growth and interlayer diffusion, *J. Appl. Phys.* **50**, 505-517 (1979).
86. Eltoukhy, A. H. and Greene, J. E., Diffusion enhancement due to low-energy ion bombardment during sputter etching and deposition, *J. Appl. Phys.* **51**, 4444-4452 (1980).
87. Eltoukhy, A. H., Zilko, J. L, Wickersham, C. E. and Greene, J. E., Interlayer diffusion in InSb/GaSb superlattice structure grown by multitarget rf sputtering, *Appl. Phys. Lett.* **31**, 156-158 (1977).
88. Prokes, S. M. and Spaepen, F., Interdiffusion in Si/Ge amorphous multilayer films, *Appl. Phys. Lett.* **47**, 234-236 (1985).

4

Diffusion and Growth in Oxide Films

Alan Atkinson

1. INTRODUCTION

Oxide films play a wide variety of roles in modern technologies. In some cases they have useful properties in their own right. Examples of this type are dielectric films in electronics (such as SiO_2 in MOS transistors) or oxide film sensors which respond to some aspect of their environment (such as a particular gaseous impurity). In this type of application diffusion in the oxide may be important during the formation of the film (e.g. by oxidation of a substrate), during the operation of the device (by controlling defect structure and response time) or during its eventual degradation.

In other cases the oxide film may be an unwanted nuisance, as in the soldering of components or corrosion of materials. In the latter instance, the corrosion may be reduced to acceptable proportions by employing other oxides as films which offer protection against rapid corrosion. In these cases diffusion in the oxide controls the rate of film thickening, its composition and the efficacy of protective films or treatments.

In addition diffusion controls other properties of oxides which also have relevance to thin film technology such as sintering and creep.

Here we consider some aspects of diffusion in oxides, and the relationship between them and the growth of oxide films by thermal oxidation of a metal (or semiconductor) at elevated temperature. First, some techniques which may be used to study diffusion in oxides, and particularly in thin films, are mentioned and then the current understanding of diffusion processes in oxides is briefly reviewed. The role which these diffusion processes play in the growth of oxide films by thermal oxidation is then explored and finally we mention some strategies which may be used to encourage the formation of more protective oxide films for the particular application of corrosion minimization. Emphasis is placed upon describing the current level of understanding of the underlying principles and mechanisms rather than on specific technological applications.

2. EXPERIMENTAL TECHNIQUES

In most experimental techniques for studying diffusion in oxides the distribution of diffusant, as a function of depth within the oxide, is analyzed in terms of a solution to the diffusion equations under the appropriate boundary conditions for the diffusion anneal (usually an infinitely thin source or constant surface concentration of diffusant, ideally). The available experimental techniques for measuring diffusion coefficients differ principally in the method which is used to determine the depth distribution of the diffusant. The ones which are mentioned below have a sufficiently small depth resolution that they can be used for diffusion studies in thin films.

Secondary Ion Mass Spectroscopy (SIMS) Sectioning

In SIMS an area (typically 500 μm$\times$500 μm) of the specimen is eroded by an incident ion beam and the sputtered fragments are analyzed by mass spectroscopy. It is the depth-profiling technique which probably has the greatest potential for diffusion studies in oxides. Amongst its advantages are good depth resolution (e.g. better than 5 nm at depths of up to 0.1 μm), high sensitivity, imaging capability and ability to distinguish between stable isotopes of the same element. It is this last characteristic which has been most widely exploited to date for studying oxygen self-diffusion in oxides using the ^{18}O stable isotope. The use of SIMS for this purpose has recently been described by Kilner et al. (1) and an example of a penetration profile of ^{18}O in yttrium iron garnet (YIG) obtained using this technique is shown in Fig. 1(a). The smallest diffusion coefficient which can be measured is $\sim 10^{-19}$ cm^{2}s^{-1}.

The imaging capability of SIMS is particularly valuable when using specimens in which diffusion is expected to be inhomogeneous. For example, a series of SIMS images, taken using mass 18 at different depths below the initial surface of a polycrystalline NiO film after annealing in $^{18}O_2$, have shown the preferential diffusion of oxygen along NiO grain boundaries (2).

The greatest source of error in SIMS comes from converting ion dose into depth within the specimen. This can be done satisfactorily using optical techniques or surface profiling on flat surfaces, but is very unreliable on rough surfaces.

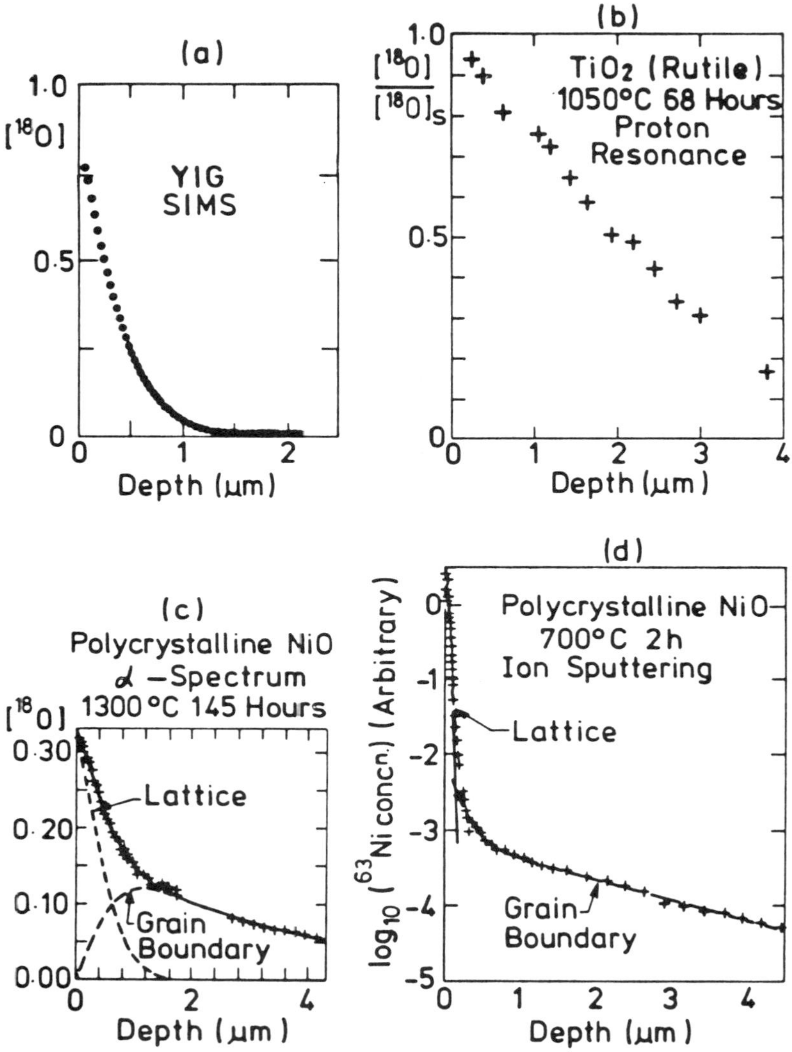

Fig. 1. Examples of diffusion penetration profiles in oxides determined by different techniques.
(a) ^{18}O in yttrium iron garnet crystal by SIMS (1).
(b) ^{18}O in TiO_2 (rutile) crystal using $^{18}O(p,\alpha)^{15}N$ resonant nuclear reaction at 1.763 MeV (4).
(c) ^{18}O in polycrystalline NiO film using α spectrometry of the $^{18}O(p,\alpha)^{15}N$ reaction at 820 keV (2).
(d) ^{63}Ni in polycrystalline NiO film using ion sputtering (10).

Nuclear Reaction Depth Profiling

In SIMS the function of the primary ion beam is to sputter material from the specimen. When the incident ions have much greater energy (typically 1 MeV rather than several keV) the nuclear reactions between the ions and the specimen can be used to analyze the specimen in a largely non-destructive way. A wide variety of depth-profiling techniques based on such nuclear reactions is now available for use in diffusion studies. Most of the techniques are specific to a particular nucleus and therefore cannot be described in general terms. In studies of diffusion in oxides, however, techniques for oxygen self-diffusion measurements are of general importance and serve to illustrate the ways in which nuclear reactions can be used.

The nuclear reaction $^{18}O(p,\alpha)^{15}N$ has been used by several groups for depth-profiling ^{18}O tracer because of its high sensitivity and lack of interference from competing reactions. The reaction may be used for depth profiling in two different ways as described by Amsel et al. (3). In the first method, use is made of the strong resonances which occur in this reaction at incident energies of 1.763 MeV and 0.629 MeV. In the experiment, the energy of the primary proton beam is gradually increased above the resonant energy and the yield of emitted α-particles measured as the resonance is driven to greater depth below the surface (due to energy loss of the proton beam in the specimen). Derry, Lees and Calvert (4) have used this method to measure ^{18}O diffusion in TiO_2 (rutile) and an example of one of their penetration profiles is shown in Fig. 1(b). In the second method the incident proton energy is maintained constant at a value where the cross section is high and has no sharp resonances. The depth information is extracted from an analysis of the energy distribution of emergent α-particles. An example of a penetration profile of ^{18}O in a polycrystalline NiO film obtained using this technique (2) is shown in Fig. 1(c).

One disadvantage of the nuclear techniques is that the observed yield curve, or energy spectrum, must be deconvoluted to obtain the required depth distribution. On the other hand the non-destructive nature of the techniques, and the fact that they do not rely on a sputtering rate for depth information, can make them more attractive than SIMS in certain applications; for example when using polycrystalline specimens or specimens with rough surfaces. The depth resolution of nuclear techniques tends not to be as good as SIMS and the lower limit of measurable diffusion coefficient is typically $10^{-18}-10^{-17}$ cm^2s^{-1} (2).

Sputter-Sectioning of Radioactive Tracers

Methods using radioactive tracers are usually preferable to those using stable isotopes because of the greater sensitivity of radiotracer detection techniques. This is particularly advantageous for measurements of self-diffusion of metal ions in oxides, because stable isotopes are difficult to obtain and natural background concentrations are inconveniently high. Consequently, the well-established techniques for tracer diffusion studies, using sectioning of the specimen, have been extended into the thin film range by the development of microsectioning techniques based on sputtering either using argon ion beams or a radio frequency (RF) argon plasma. The ion beam technique was originally developed for studying diffusion in metals by Mehrer et al. (5) and was subsequently applied to oxides by Mundy and Rothman (6). The RF technique was also originally developed for studies of metals by Gupta and Tsui (7) and subsequently applied to oxides by Atkinson and Taylor (8)(9). In all cases, radioactive material is sputtered from the specimen onto a collector which is then counted. An example of a diffusion profile (10) of ^{63}Ni tracer in a polycrystalline NiO film is shown in Fig. 1(d). The smallest diffusion coefficient which can be measured using this technique is $\sim 10^{-19}$ cm^2s^{-1} (8) and is equivalent to SIMS in this respect.

Chemical Dissolution

Liquid chemical dissolution offers an alternative microsectioning technique to those based on ion sputtering in some cases. However, the development of a suitable solvent composition, which will remove material reproducibly and uniformly rather than etch, is so difficult that the technique has only limited application. Suitable solvents are available for amorphous materials and therefore the technique has been used successfully to study diffusion in oxide glasses (11). The lower limit of diffusion coefficients which can be measured using sputtering techniques is determined by the incident ions mixing the components of the solid to a depth of about 5 nm. Since this effect is not present in chemical dissolution, the technique is capable of even better depth resolution (at least in amorphous materials) and diffusion coefficients as low as 10^{-21} cm^2s^{-1} have been claimed (12).

Gaseous Exchange

In the measurement of ^{18}O tracer diffusion by SIMS or nuclear reaction analysis of the solid, the ^{18}O is usually introduced into the specimen by exchange of ^{16}O, in the solid, for ^{18}O, in the surrounding gas, in a diffusion anneal. In principle it is possible to deduce the solid state diffusion coefficient from the exchange kinetics, measured by monitoring the isotopic composition of the gas as a function of time during the diffusion anneal. Diffusion studies in oxides using this technique have been reported by Oishi and co-workers (e.g. Refs. 13 and 14). The technique has the advantage of simplicity over those involving subsequent analysis of the solid specimen, but considerable errors are possible if there is exchange with the materials used in the diffusion apparatus, or if the rate of exchange at the specimen surface is slow by comparison with the diffusion kinetics. The lowest diffusion coefficients which have been measured using this technique are $\sim 10^{-16}$ cm^2s^{-1}, which is considerably greater than may be attained by analysis of the solid.

Diffusion in Growing Oxide Films

Of the examples given so far, only those experiments on polycrystalline NiO have been carried out using oxide in the form of a film: in this case a double-layered film of total thickness ~50 μm produced by completely oxidizing a 25 μm thick Ni foil (10). However, there are many examples to be found in the literature in which the movement of tracers has been studied in an oxide film whilst it is still growing, i.e. the tracer is present in the growing film, but the analysis is carried out afterwards. The attraction of such experiments is that they can provide a rapid indication of which species are mobile during the growth process (e.g. metal ion or oxygen ion) and which transport path they take (e.g. lattice diffusion or short-circuit diffusion). It is therefore appropriate to consider the interpretation of these experiments in more detail. There are two basic experiments of this type; one using metal tracers and the other using oxygen tracers.

The experiment with metal tracers was first carried out by Bardeen et al. (15) and consists of labelling the surface atoms of the metal by depositing radiotracer atoms prior to oxidation. After oxidation the depth distribution of tracer in the film is determined. The distribution is a function of the mobile species and its transport path, as summarized in Fig. 2. When the film grows by lattice diffusion of the metal ions the tracer distributions can be expressed analytically as

$$\frac{C}{C_s} = u^{\alpha} \exp\left[\alpha(1-u)\right] \qquad (1)$$

when the dominant point defects are cation vacancies (15) and

$$\frac{C}{C_s} = \left(\frac{1-u+\beta}{\beta}\right)^{\alpha\beta} \exp\left[\alpha(u-1)\right] \qquad (2)$$

when the point defects are cation interstitials (16). In these expressions C_s is the tracer concentration of the oxide/gas interface (u = 0),

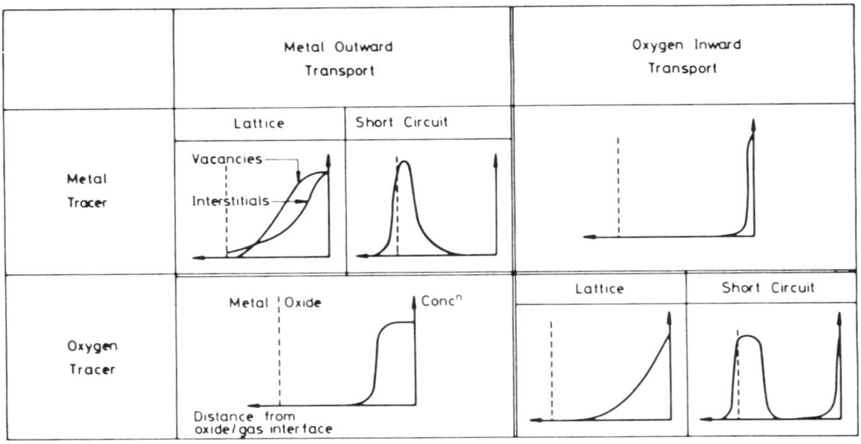

Fig. 2. Distributions of tracer atoms expected in oxide films growing by different transport processes. In the experiment with metal tracer, a thin film of tracer is deposited on the metal prior to film growth. In the experiment with oxygen tracer the film is grown sequentially first in ^{16}O and then in ^{18}O labelled gas.

u is the ratio of distance from this interface to the total oxide thickness, $\alpha = (1+|z|)/f$ where z is the effective number of electronic charges on the defect and $(1+\beta)/\beta$ is the ratio of the tracer diffusion coefficient at the metal/oxide interface to that at the oxide gas interface (i.e. $\beta \ll 1$ for interstitials).

In principle, this experiment can identify the point defect and its effective charge, but in practice this is beyond the precision of the experimental techniques. The expected distributions from lattice and short-circuit diffusion of cations are qualitatively very different and therefore these processes can be distinguished provided that $(Dt)^{1/2} < \ell$, where D is the diffusion coefficient in the oxide lattice, t is the oxidation time and ℓ is the distance between short-circuit paths (i.e. grain size in the case of grain boundary diffusion). If this condition is not satisfied then atoms in the short-circuit paths have time to mix with those in the lattice and the resulting tracer distribution is indistinguishable from that of lattice diffusion alone.

The experiment using oxygen tracers consists of first growing a film in oxygen of normal isotopic composition (99.759% ^{16}O) followed by growth in oxygen enriched in ^{18}O. Interpretation is more complicated than for the metal tracers since the ^{18}O may be incorporated in the film either by the growth of new oxide or by exchange and diffusion. Mitchell et al. (17) have demonstrated how SIMS may be used to distinguish between these two sources of incorporation. They argue that in sputtered fragments containing two oxygen atoms (e.g. MO_2^-) the two oxygen atoms will both be ^{18}O if ^{18}O was incorporated by growth, but one will be ^{18}O and the other ^{16}O if incorporated by diffusion. The ^{18}O distributions in films grown sequentially in $^{16}O_2$ and $^{18}O_2$ are illustrated in Fig. 2. They can be used to distinguish between oxygen and metal transport as the dominant growth process and between lattice and short-circuit diffusion of oxygen.

The combination of experiments using the metal and oxygen tracers is therefore a useful way of probing diffusion processes in growing oxide films. However, it should be noted that the tracer profiles cannot be used to obtain equilibrium diffusion coefficients, even though they may approximate to standard solutions of the diffusion equations. The reason for this is that diffusion is occurring in a strong gradient of chemical potential which, in turn, generates a large unidirectional flux of point defects (i.e. that responsible for film growth). Any diffusion coefficients obtained in this way can be regarded as only order of magnitude estimates at best.

3. LATTICE DIFFUSION IN OXIDES

Point Defects in Oxides

The atomistic understanding of lattice diffusion in oxides is based upon point defects and the interactions between them. The defects in the crystal lattice (e.g. vacant sites or interstitial ions), are responsible for atomic diffusion (and ionic conductivity) and those in the electronic system (e.g. electrons in the conduction band or holes in the valence band) for electronic conductivity. Most oxides are usefully regarded as ionic solids, and therefore these defects carry effective charges (with respect to the perfect lattice) and the dominant interactions between them are coulombic.

Most oxides can deviate from their exact conventional stoichiometric composition and the extent and direction of this non-stoichiometry is very important in determining the general behavior of an oxide. In Table 1 several oxides have been classified according to the extent and direction of their non-stoichiometry, x. The non-stoichiometric formulae have been written such that x is always positive and +x indicates interstitial defects and −x, vacancies. The classifications of 'large', 'small' and 'negligible' are somewhat arbitrary and are approximately as follows: 'large' is when the deviation exceeds ~1%, 'negligible' is when the deviation defies measurement (usually less than about one part in 10^4) and 'small' is neither 'large' nor 'negligible'. Not only are the classifications approximate, but in addition any one oxide may be considered to span more than one classification depending on temperature, oxygen activity or doping. Nevertheless, the general characteristics of those oxides with large deviations from stoichiometry are rapid diffusion of the non-stoichiometric species (e.g. Mn in $Mn_{1-x}O$) and a tendency for the simple point defects to associate into larger defect clusters. On the other hand, those with negligible deviation from stoichiometry exhibit very slow diffusion, which is also extremely sensitive to doping by small concentrations of foreign ions. Some oxides show deviations to either side of the stoichiometric composition (e.g. $UO_{2\pm x}$, $Fe_{3\pm x}O_4$) depending on temperature and oxygen activity.

Table 1: Non-stoichiometry in some binary oxides

Degree of non-stoichiometry	Cation deficiency	Anion deficiency
'Large'	UO_{2+x} $Mn_{1-x}O$, $Fe_{1-x}O$	UO_{2-x}
'Small'	$Co_{1-x}O$, $Cu_{2-x}O$ $Fe_{3-x}O_4$ $Ni_{1-x}O$ $Cr_{2-x}O_3$	ZrO_{2-x} $Fe_{3+x}O_4$ Fe_2O_{3-x} $Cr_{2+x}O_3$ (?)
Negligible	Al_2O_3, MgO, SiO_2	

In a pure stoichiometric oxide, MO (such as MgO) point defects may be created by Frenkel reactions at any part of the lattice, e.g.

$$M_M^x \rightarrow V_M'' + M_i^{\bullet\bullet} , \qquad (3a)$$

$$O_O^x \rightarrow V_O^{\bullet\bullet} + O_i'' , \qquad (3b)$$

or by the Schottky reaction at internal discontinuities (such as dislocations)

$$\text{null} \rightarrow V_M'' + V_O^{\bullet\bullet} . \qquad (4)$$

(The notation of Kröger and Vink is used to describe the point defects.) It is now believed that in real oxides showing negligible deviation from stoichiometry, these reactions are subordinate to doping by foreign ions even in the purest samples which have been used. For example, a foreign trivalent impurity, F, in a divalent host can generate point defects by reactions of the type

$$F_2O_3 \rightarrow 2F_M^{\bullet} + 3O_O^x + V_M'' . \qquad (5)$$

In pure non-stoichiometric oxides the non-stoichiometry is the result of reaction with oxygen gas at external surfaces. Atomic point defects are injected and some metal ions effectively change valency (i.e. defects are created in the electronic system). An example of such a reaction is

$$\tfrac{1}{2}O_2 \rightarrow V_M'' + O_O^x + 2\dot{h} \qquad (6)$$

(in which the valence band hole may be regarded as a M^{3+} ion) and is common in the transition metal monoxides. The participation of O_2 gas in these reactions results in the point defect concentrations depending on oxygen activity (partial pressure in atmospheres, for a perfect gas). In general, defect concentration is proportional to $a(O_2)^n$ where n is characteristic of the defect-forming reaction. Thus, if we apply the law of mass action and the principle of electrical neutrality to equation (6) we deduce that $n = 1/6$ for the concentration of both metal ion vacancies and electron holes in this particular case. Studies of deviation from stoichiometry, electronic conductivity and diffusion as a function of oxygen activity are therefore favored techniques for investigating and identifying point defects in oxides.

Self-Diffusion

The tracer self-diffusion coefficient, D*, depends on the concentration of point defects and the rate at which it can jump to any neighboring site, w. (The asterisk is used to denote tracer diffusion.) For the special case of diffusion by vacancies in a bcc or fcc lattice of lattice constant a_o, D* is given by (18)

$$D^* = a_o^2 f w[V] \qquad (7)$$

where [V] is the fraction of vacant sites on the sublattice being considered. Although motion of the defect is random, that of the tracer is not. This is accounted for by the correlation factor, $f(f<1)$, and has been tabulated for various mechanisms by Le Claire (18). The jump frequency is thermally activated and may be expressed as

$$w = \nu \exp(\Delta S_m/k) \exp(-\Delta H_m/kT) \qquad (8)$$

where $\Delta H_m - T\Delta S_m$ is the Gibbs free energy barrier of the jump process and ν is the average vibration frequency of an ion, and is usually assumed to be approximately equal to the Debye frequency. The activation energy for self-diffusion has therefore a contribution from motion of the defect, ΔH_m, and another from the concentration of defects. This latter contribution depends on the energetics of the reactions by which the defects are created. For the specific example of reaction (6) it may be shown that the activation energy for the concentration of cation vacancies is $\Delta H_6/3$, where ΔH_6 is the enthalpy change of reaction (6).

Cation Self-Diffusion. In recent years cation self-diffusion in the transition metal oxides has been studied by several groups. The transition metal monoxides with the rocksalt structure are examples in which the metal ions are much more mobile than the oxygen ions. The self-diffusion coefficient of the metal ions in FeO, MnO, CoO and NiO are shown as a function of $a(O_2)$ if Fig. 3. For MnO, CoO and NiO, $D^*(M)$ increases as $a(O_2)$ increases, as expected for diffusion by

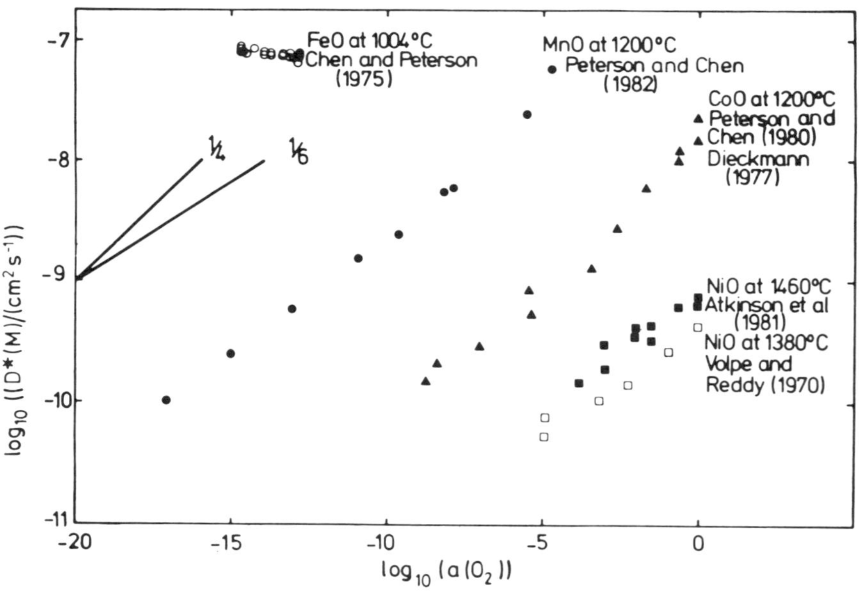

Fig. 3. The tracer self-diffusion coefficient for lattice diffusion of metal ions, in transition metal oxides having the rocksalt structure, as a function of oxygen activity. The solid lines indicate the slopes expected when either singly or doubly charged cation vacancies are the majority defects (19)-(24).

a vacancy mechanism. More detailed interpretation of tracer diffusion and other properties has revealed that the defect structure of these oxides is more complicated than expressed by the simple basic reaction (6) and that association between defects is a common occurrence. In CoO and NiO this appears to be limited to the association of doubly-charged vacancies with holes to give some singly-charged vacancies (19)(25). In MnO and FeO the association of defects is more extensive, and in FeO is such that at 1000°C increasing $a(O_2)$ increases the deviation from stoichiometry, but not the diffusion coefficient (Fig. 3). This is due to the formation of relatively immobile defect clusters, each incorporating several vacancies (26).

Magnetite, Fe_3O_4, is an oxide in which the point defect properties of the cation sublattice are particularly well-understood. In Fe_3O_4 cation self-diffusion takes place by a vacancy mechanism at high $a(O_2)$ and by an interstitialcy mechanism at low $a(O_2)$ (27). This is related to the ability of Fe_3O_4 to exhibit deviations on both sides of stoichiometry as indicated in Table 1. There is some evidence that Cr_2O_3 shows a similar behavior (28).

An example of cation diffusion in an oxide in which cation diffusion is much slower than oxygen diffusion is provided by UO_2. Matzke (29) has concluded that in UO_{2+x}, U diffuses by a vacancy mechanism and in UO_{2-x} by an interstitial defect. This behavior is therefore also similar to that found in Fe_3O_4.

Oxygen Ion Self-Diffusion. UO_2 is probably the most widely-studied example of an oxide in which oxygen diffusion is more rapid than cation self-diffusion. The measurements for UO_{2+x} have been summarized by Brietung (30) and are shown in Fig. 4. These data are presented as functions of T at constant x and therefore there is no contribution from the concentration of defects to the activation energy at large x and/or low T. The activation energy (0.52 eV) is ΔH_m for the interstitial defect O_i''. At the stoichiometric composition (or at high T) Frenkel disorder on the oxygen sublattice is dominant (reaction 3b) and the activation energy is greater by an amount equal to half the enthalpy change of reaction (3b).

An example in which oxygen ions are less mobile than the cations is provided by $Cu_{2-x}O$. Measurements of $D^*(O)$ as a function of $a(O_2)$ have shown (31) that oxygen diffusion probably involves a singly charged interstitial defect, O_i'. As in UO_{2+x}, cation diffusion is by a vacancy mechanism (32).

Diffusion in Doped Oxides. Doping by foreign ions may either be a deliberate means of controlling defect structure, or an unavoidable consequence of trace impurities. The effect of a substitutional dopant is mainly dependent on its charge state in the host lattice compared with that of the ion it replaces.

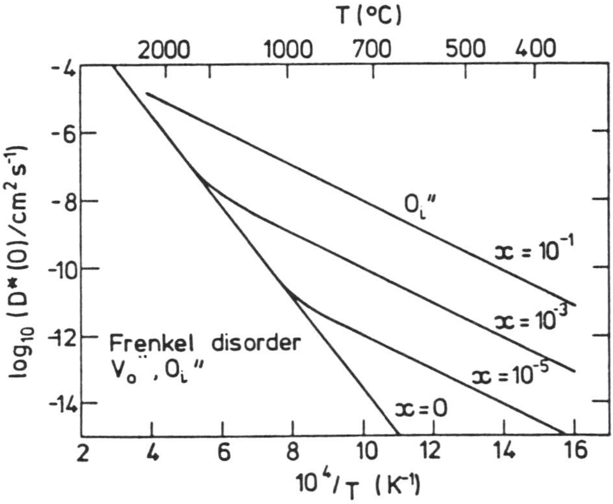

Fig. 4. Oxygen tracer self-diffusion in UO_{2+x} as a function of temperature and composition (30).

Heterovalent doping. When the dopant has a different valency from the host ion the dopant has a large influence on defect properties because of the requirement for electrical neutrality. An example of this type of doping effect is given by equation (5) and has been examined experimentally for Ni diffusion in Al-doped NiO (23). In this particular case, Al doping increases D*(Ni) as would be expected qualitatively from equation (5), but the quantitative predictions resulting from equation (5) are not confirmed (see Fig. 5). This is because the substitutional dopant ions ($Al^{\bullet}_{Ni}$) and cation vacancies have opposite effective charges and are strongly attracted ($\Delta H_A = -1.2$ eV) to give relatively immobile defect pairs. At sufficiently low temperature, depending on doping level, a second phase ($NiAl_2O_4$ spinel) precipitates. In the stoichiometric oxides, heterovalent doping has been used to probe their characteristic point defects. As an example of this approach, Reddy and Cooper (33) measured oxygen diffusion in Al_2O_3 doped with Mg^{2+} and Ti^{4+} and from the response concluded that oxygen diffusion in Al_2O_3 is by a vacancy mechanism.

Homovalent doping. When the solute (dopant) ion has the same charge as the solvent (host) ion the influence of the dopant is not primarily through the electrical neutrality condition and the resulting effect is much weaker than that of heterovalent doping. Homovalent doping effects have been mainly studied in systems which exhibit extensive mutual solid solubility. One such example is Co-doped NiO studied by Chen and Peterson (34). They concluded that in this material the main effect of the Co is to increase the concentration of cation vacancies and that solute-vacancy interactions are negligible.

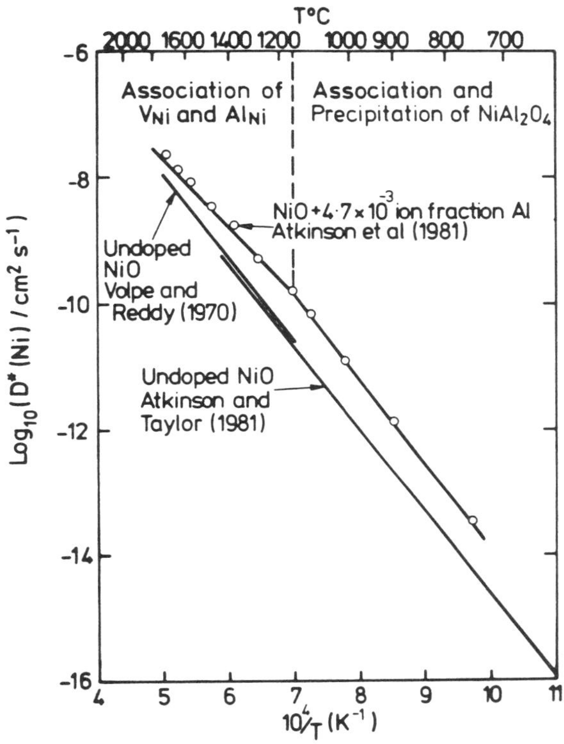

Fig. 5. Nickel tracer self-diffusion in pure and Al-doped NiO (23).

Solute Diffusion

Tracer Solute Diffusion. There are two main regimes in which a tracer solute ion may diffuse in a crystal at equilibrium, depending on the total concentration of solute with respect to the concentration of native (or intrinsic) point defects.

Dilute limit. When the solute is only present as a negligibly small concentration of diffusing tracer, it will not perturb the point defect population of the pure oxide. Diffusion of solute tracer atoms differs from diffusion of solvent tracer atoms in several respects. First there may be an interaction between a solute atom and a point defect which modifies the chance of a defect being next to a tracer atom. This interaction is characterized by an association enthalpy, ΔH_A, in the same way as described for Al-doping of NiO previously ($\Delta H_A < 0$ for an attractive interaction). Secondly, the frequency of a point defect-solute ion interchange jump, w_2, will be different from that of the solvent tracer, w_o. Thirdly, the correlation factor for the solute, f_2, is also different from that of the solvent and is in general a complex function of all the jump frequencies in the vicinity of the solute (35). In the case of a vacancy mechanism, the solute tracer diffusion coefficient, D_2^*, may be written

$$D_2^*(\text{dilute}) = a_o^2 w_2 f_2[V] \exp(-\Delta H_A/kT) \qquad (9)$$

which may be compared with equation (7) for the solvent tracer. Peterson (36) has reviewed data for diffusion of transition metal solute ions (Cr^{3+}, Ni^{2+}, Co^{2+}, Fe^{2+}) in NiO and CoO. He points out that rather surprisingly the largest ions (Fe^{2+}) have the largest values of D_2^* and the lowest activation energies. However, this trend is not universal and is not found for non-transition metal ions. For example Ca^{2+}, a much larger ion, diffuses only slowly in NiO and has a large activation energy of 3.4 eV (37). Wuensch (38) has presented a similar survey for the diffusion of Be^{2+}, Mg^{2+}, Ca^{2+}, Sr^{2+} and Ba^{2+} in MgO and finds that in this series the larger the ion the lower its diffusion coefficient. It is evident that there is a complex interplay of ionic characteristics (size, charge, polarizability) in determining solute tracer diffusion.

Non-dilute. When the concentration of unlabeled solute is large enough to influence the point defect populations in the crystal, and is uniform, the diffusion process is more complicated. As with tracer self-diffusion the effect is greatest when the dopant ion has a different valency from the host and an example of this type is ^{51}Cr in Cr-doped MgO studied by Weber et al (39). For this particular case it can be shown that diffusion occurs by an impurity-vacancy pair and

$$D^*(\text{Cr})(\text{non-dilute}) \simeq 0.5 a_o^2 w_2[\text{Cr}] \exp(-\Delta H_A/kT) \qquad (10)$$

where [Cr] is the total cation fraction of Cr solute. This is similar to equation (9) for the dilute limit but with [V] being replaced by 0.5[Cr], as might be expected from the simple doping reaction (equation (5)). In this particular case ΔH_A is -0.9 eV and the activation energy for w_2 is 1.73 eV.

Amorphous Oxides

Diffusion in an amorphous material is more difficult to characterize than in a crystalline one, because of the lack of regular long range structure and the potential diversity of sites and transitions between them. We would thus not necessarily expect to observe Arrhenius behavior of diffusion coefficients. Nevertheless, the data do usually fit an Arrhenius equation (for example see the monograph by Frischat (40)) which indicates that, for a given diffusing species, only one type of jump is likely to be rate-controlling. The simplest oxide glass is amorphous SiO_2 (a-SiO_2) and is probably also the most important in the context of thin films.

Diffusion coefficients for some species in a-SiO_2 are shown in Fig. 6 and may be divided into four main groups. The first group are those which are part of the glass-forming network. In a-SiO_2 the network is built of corner-sharing SiO_4 tetrahedra and, therefore, in this group are Si, O and ions which can substitute for either Si or O in the network, such as Al^{3+}, Fe^{3+} and, probably, Ni^{2+}. The network-forming group have the lowest diffusion coefficients and highest activation energies (>2 eV) and the mechanism of diffusion probably involves some kind of network defect which is analogous to a point defect in a crystalline lattice. In the second group are those species which have negligible interaction with the network such as rare gas atoms and some molecules such as N_2. The solubility sites for these species are believed to be within rings, comprising 5, 6 or 7 SiO_4 tetrahedral members, which exist in the glass network. The solubilities tend to be low, but the diffusion coefficients of such species may be high. In the third group are ions, usually monovalent or divalent cations, which do not enter the glass network but cause some modification of the network in order to compensate for the additional charge. These species can also exhibit large diffusion coefficients and small activation energies (<1 eV). In the fourth group are those species which may enter the ring interstitial sites, as do the rare gases, but also have significant interactions with the network. This interaction may be an exchange of oxygen (as occurs with O_2) or a chemical reaction (as occurs with H_2O to form bound OH groups). Under such circumstances it is possible to define several different diffusion coefficients and this has led to some confusion in the literature. For example, in the case of H_2O we have the diffusion coefficient of the H_2O interstitial model itself, $D(H_2O)$, the apparent diffusion coefficient of OH groups, $D_{app}(OH)$, and the tracer diffusion coefficient of network oxygen in the presence of water vapor, $D^*(O,H_2O)$. If all the chemical and exchange reactions are rapid then it may be shown that

$$D_{app}(OH) \bullet [OH] = 2D(H_2O) \bullet [H_2O] \quad (11)$$

and

$$D^*(O,H_2O) = 0.5D(H_2O) \bullet [H_2O] \quad (12)$$

(square brackets indicate mole fraction in SiO_2).

It is appropriate in this article on diffusion in films to mention in this context the work of Pfeffer and Ohring (43) who measured $D^*(O,H_2O)$ in a-SiO_2 films grown on Si as a function of external water vapor pressure. They found, in accord with equation (12), that $D^*(O,H_2O)$ is directly proportional to water vapor pressure which demonstrates that network oxygen diffuses by exchange with interstitial H_2O molecules.

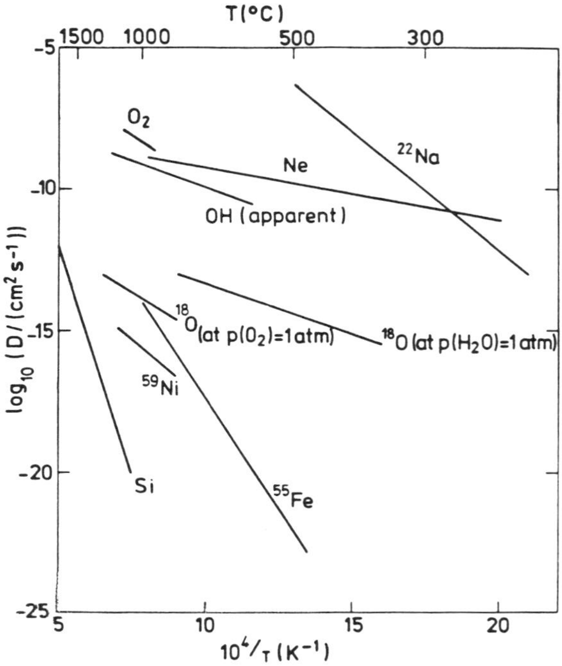

Fig. 6. Diffusion of various species in amorphous SiO_2 (12)(40)-(46).

Theoretical Estimates of Diffusion Parameters

During the past ten years or so the capability for theoretically modelling point defect and diffusion behavior in oxides has been steadily increasing in both accuracy and scope. The estimates are currently most reliable in the more ionic oxides (such as MgO) and the parameters which are most readily calculated are internal energies for the formation, interaction and migration of point defects. Comparison may thus be made between theoretical and experimental activation energies. Table 2 gives theoretical energies for point defect processes in NiO and these have been combined to yield activation energies which are compared with experimental values in Table 3. The agreement is, typically, to within a few tenths of an electron volt and strongly suggests that oxygen diffusion in NiO takes place by an interstitialcy mechanism. However, it does appear that there is a significant disagreement between theoretical and experimental estimates for migration of the Ni vacancy. Mackrodt and co-workers (48)

Table 2: Computed point defect energies for NiO

Process	Energy (eV) Sangster & Rowell (47)	Mackrodt (48)
$\frac{1}{2}O_2 \rightarrow V_{Ni}'' + 2\dot{h} + O_O^x$	–	2.2
null $\rightarrow V_{Ni}'' + V_O^{\bullet\bullet}$	7.37	7.0
$Ni_{Ni}^x \rightarrow V_{Ni}'' + Ni_i^{\bullet\bullet}$	12.09	12.6
$O_O^x \rightarrow V_O^{\bullet\bullet} + O_i''$	11.29	13.0
V_{Ni}'' jump	2.44	2.12
$Ni_i^{\bullet\bullet}$ jump	0.96	1.08
$V_O^{\bullet\bullet}$ jump	2.33	2.24
O_i'' jump	0.89	1.27

Table 3: Comparison between experimental and theoretical activation energies for NiO

Process	Q_{Meas}(eV)	Q_{theory}(eV)
Deviation from stoichometry	0.98 (ref. 23)	0.73
Chemical diffusion (V_{Ni} diffusion)	1.58 (ref. 23)	2.12-2.44
D*(Ni)	2.56	2.85-3.17 (vacancy)
D*(O)	5.6 (ref. 49)	5.2-5.57 (interstitialcy) 9.5-9.6 (vacancy)

have also calculated the activation energy for various solutes diffusing in MgO. These show good agreement with experiment, except for Mn^{2+} diffusion.

Theoretical estimation of the diffusion rate itself is more difficult because defect entropies (e.g. in equation (8)) are also required. Such calculations are in their infancy at present (50).

4. SHORT-CIRCUIT DIFFUSION

So far we have only discussed diffusion by point defects in an otherwise perfect crystal lattice. It is now apparent, however, that diffusion processes in thin films tend to be dominated by fast diffusion along short-circuit pathways such as dislocations, grain boundaries and surfaces. Since measurements of such diffusion are much more difficult than of lattice diffusion, the volume of data and quality of understanding is correspondingly lower. In the case of oxides, what data were available to 1982 have been reviewed by the author (51). The main conclusions are summarized here, together with references to more recent data. The experimental characterization of these short-circuit processes assumes that fast diffusion occurs only within a uniform region localized near the short-circuit pathway. In the case of grain boundary or surface diffusion, this is a slab of width, δ, whilst for a dislocation it is a cylindrical pipe of diameter, d.

Dislocation Diffusion

As far as the author is aware there are no measurements of diffusion along isolated dislocations in oxides. Instead, the dislocations are usually in the form of arrays in low angle grain boundaries, even in nominally single crystal samples. Nevertheless, it is possible to extract from suitable diffusion experiments, the diffusion coefficient in the cylindrical pipe and the pipe diameter and this has been done for Ni diffusion in NiO (52) and Cr in Cr_2O_3 (28). The data for NiO are shown in Fig. 7 compared with other diffusion coefficients in NiO. In both materials the diameter of the fast diffusion pipe is estimated to be approximately 1 nm, which indicates that very rapid diffusion is confined to the core of the dislocation. The dependence of dislocation diffusion coefficient on oxygen activity is observed to be similar to that of lattice diffusion in both oxides. This suggests that point defects, analogous to lattice point defects, are responsible for diffusion in the dislocation.

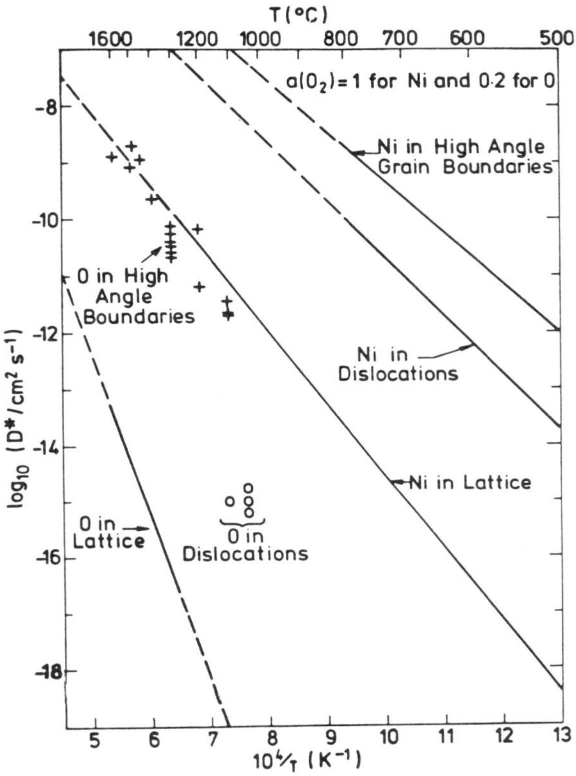

Fig. 7. Ni and O tracer self diffusion in the lattice, dislocations and grain boundaries in NiO (2)(10)(49)(52).

Grain Boundary Diffusion

In polycrystalline material the majority of grain boundaries are high angle boundaries of random misorientation, rather than the arrays of lattice dislocations which constitute low angle boundaries. Data are now available for both Ni (10) and O (2) diffusion in NiO high angle boundaries, present in films formed by the oxidation of Ni metal. These data are presented in Fig. 7 from which it can be seen that diffusion in high angle boundaries has a faster rate and lower activation energy than diffusion in a dislocation. Nevertheless, the observed dependence on oxygen activity and the measured width of the high diffusivity region (again ~1 nm) both indicate that the mechanisms are similar. It is also noteworthy that diffusion of both Ni and O is enhanced in the boundary. These data provide probably the most complete characterization of diffusion pathways in any oxide.

Surface Diffusion

Data for diffusion on oxide surfaces are even rarer than for grain boundary and dislocation diffusion. An example of direct measurement using a tracer technique is for U diffusion on UO_2 surfaces (53).

Theoretical Studies of Dislocations, Grain Boundaries and Surfaces

The techniques which were originally developed for estimating bulk lattice properties have recently been extended to deal with dislocations, grain boundaries and surfaces, which are potential pathways for fast diffusion in oxide films.

Dislocations. Rabier and Puls (54) have carried out simulations of oxygen diffusion along dislocations in MgO. They concluded that oxygen diffuses by a vacancy mechanism in the dislocation core and that the activation energy is 75% of that of bulk diffusion (both assumed to be intrinsic and stoichiometric). The lower activation energy is mainly the result of the binding energy (i.e. segregation) of vacancies to the dislocation core. The experimental measurements for Ni in NiO dislocations give an activation energy which is 78% of that for bulk diffusion; in good agreement with that predicted by theory. However, this agreement is probably fortuitous because NiO is significantly non-stoichiometric. Rabier and Puls also concluded that dislocation diffusion by an interstitial mechanism was less likely than by a vacancy mechanism.

Grain Boundaries. Extensive atomic simulations of the properties of NiO grain boundaries have been reported by Duffy and Tasker (55)(56). They concluded that Ni diffusion in the boundary occurs by a vacancy mechanism and that the migration energy of the vacancy in the boundary is 78% of that in the bulk lattice. To determine the activation energy for tracer diffusion the activation energy for the concentration of vacancies must also be included and this will further reduce the ratio of activation energies. Thus it is likely that the simulations will predict a ratio which is in reasonable agreement with the experimentally measured value of 70% (10).

5. OXIDE FILM GROWTH BY THERMAL OXIDATION

Theory of Oxide Film Growth

In this section we consider the growth of uniform single phase oxide films by thermal oxidation of a planar elemental substrate; usually a metal. A different approach to describing the growth process is taken, depending on whether the film is considered to be thick or thin. As a general rule, films thicker than 1 μm are in the thick film region. In addition, the thin films may be further subdivided (e.g. as 'very thin' and 'thin') according to criteria based on the assumptions which are believed to be valid in the mechanism. For example, the term 'very thin' is often applied to conditions in which the electric field in the film is sufficiently high to give non-ohmic ionic conductivity.

Thick Films. The theory of growth of thick films is based on the properties of bulk oxides. The growth process itself involves the transport of ionic and electronic defects across the film, driven by concentration gradients and an internal electric field. In most cases transport across the film is rate-controlling. Under these conditions the film thickness, X, increases parabolically according to the equation

$$X^2 = k_p t \qquad (13)$$

where k_p is the parabolic rate constant. Wagner (57) assumed that every point in the film was in local chemical equilibrium and thereby obtained an expression for k_p in terms of the transport properties of the oxide. In most oxides of interest the electronic conductivity is much greater than the ionic conductivity and therefore the slow ion transport controls the growth rate. It is then convenient to express k_p in terms of the tracer diffusion coefficients of metal and oxygen ions in the oxide;

$$k_p = \int_{a(O_2)_I}^{a(O_2)_{II}} \left(y \frac{D^*(M)}{f_M} + \frac{D^*(O)}{f_O} \right) d(\ln a(O_2)) \,. \qquad (14)$$

In this expression the subscripts I and II denote the metal/oxide and oxide/gas interfaces respectively, y is the oxygen to metal atom ratio in the oxide and f_M and f_O are correlation factors for diffusion. In order to calculate k_p from equation (14), it is necessary to know either $D^*(M)$ or $D^*(O)$ (whichever is the greater) as a function of oxygen activity. This is a phenomenological approach to understanding film growth since it does not require knowledge of the diffusion process in atomistic terms. Nevertheless, the required data of D^* as a function of $a(O_2)$ can themselves be interpreted atomistically, as described earlier.

An alternative approach is to formulate the equations for transport across the film in terms of atomistic point defects themselves. If local chemical equilibrium between the defects is assumed then the result is, necessarily, identical to that of the phenomenological approach. If, on the other hand, one assumes local electrical neutrality instead of local chemical equilibrium then the result is, in principle, different (58). However, since bulk equilibria also imply electrical neutrality, there is no practical difference between the two approaches.

There are many conditions which must be satisfied for the theory to be valid, but the most important are that the electrical field should be small and the majority of the film should be electrically neutral. These conditions may be expressed as

$$E << \frac{kT}{a_o e} \tag{15}$$

and

$$X >> L_D = \left(\frac{\varepsilon \varepsilon_o kT}{e^2 C_d} \right)^{1/2} . \tag{16}$$

Since the electric field is $\simeq 1/X\ Vm^{-1}$, the first condition leads to the conclusion that $X \gtrsim 10$ nm for the electric field to be sufficiently small. In the second condition, L_D is the Deby-Hückel screening length and C_d is the total number of elementary charges per unit volume resulting from charged defects at equilibrium and is not usually known. However, for an oxide containing 10^{-6} mole fraction of defects at 500°C, L_D is about 10 nm. It can therefore be concluded that the lower limit of the thick film domain for most oxides is 10-100 nm.

Thin Films. When the film is thin it is necessary to be able to include the consequences of high electric fields and large departures from local electrical neutrality. Cabrera and Mott (59) developed a theory which is expected to be valid in the extreme limit of very thin films. The basic assumption in the theory is that electrons can easily pass from the metal through the film to ionize adsorbed oxygen at the oxide/gas interface (the theory does not deal with the initial formation of the film, but only its subsequent thickening). This process is assumed to proceed to equilibrium within the electronic system, so that the Fermi level in the adsorbed layer is equal to that in the metal as illustrated in the electron energy level diagrams in Fig. 8. The resulting uniform electric field (assuming negligible space charge within the film) then drives the ions across the film to provide growth. In the original theory the film was assumed to be so thin that the electrons could tunnel through it to ionize adsorbed oxygen. Fromhold (Ref. 60, Chapter 10) has shown that this is feasible only in films less than 3 nm in thickness. However, it now appears that the concept of electronic equilibrium across the film has wider validity. Fromhold (Ref. 60, Chapter 11) has demonstrated that thermionic emission is likely to provide a 'virtual electronic equilibrium' in some oxide films up to 30 nm in thickness. Furthermore, Young and Dignam (61) have pointed out that measurements of electronic conductivity of many oxide films indicate that their interfaces are ohmic and that their electronic conductivities are high enough to maintain electronic equilibrium. They argue that this is most likely due to electronic 'short-circuit' paths (e.g., grain boundaries) in most thin films. It can be concluded that the Cabrera and Mott theory is likely to have wider validity than originally proposed. The original theory was modified by Grimley (62) to explicitly include the equilibria of oxygen species adsorbed at the oxide/gas interface and then further developed by Young and Dignam (61). An excellent summary may be found in the review of Smeltzer and Young (63) in which the wide variety of possible behavior is described. Here we give the results that are appropriate to one particular case, where the species which are adsorbed at the oxide/gas interface are O^{2-} ions.

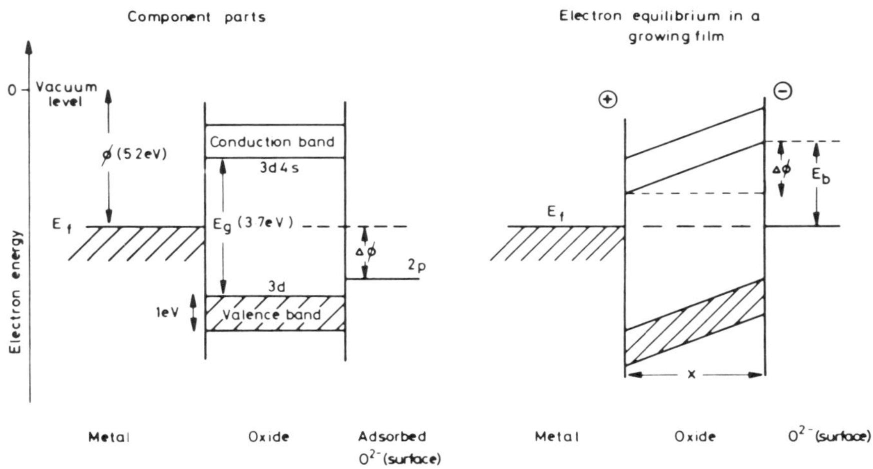

Fig. 8. Electron energy level diagrams illustrating how, in the theory of Cabrera and Mott, the transfer of electrons to adsorbed oxygen generates a uniform electric field. The energies and band labels are for Ni and NiO.

Defects injected at the metal/oxide interface. When the oxide defects are metal interstitials or oxygen vacancies the film growth rate may be written in the form

$$\frac{dX}{dt} = \frac{2D_{CM}}{a_o} \sinh\left(\frac{X_1}{X}\right). \tag{17}$$

In this expression

$$X_1 = \frac{qa_o\Delta\phi}{2kT} \tag{18}$$

where q is the charge on the diffusing ion and $\Delta\phi$ is the magnitude of the electrical potential difference across the film (oxide/gas interface negative with respect to the metal). X_1 therefore, represents the boundary between a high electric field and a low electric field. The equivalent of a diffusion coefficient in the Cabrera and Mott theory is given by

$$D_{CM} = a_o^2\nu \exp(-W/kT) \tag{19}$$

which may be compared with equations (7) and (8). The atomistic parameters relate to jumps near the interface as illustrated in Fig. 9 and W is the activation energy for injection of a point defect into the oxide film. The potential difference, $\Delta\phi$, is dependent on the adsorption equilibria and, for low surface coverage by O^{2-} ions, is given by

$$\Delta\phi \simeq -\frac{\Delta G^\circ}{2e} + \frac{kT}{2e} \ln\left(\frac{4e^2 N_s a(O_2)^{1/2} X}{kT\varepsilon\varepsilon_o}\right) \tag{20}$$

where ΔG° is the standard free energy change of the adsorption reaction

$$\tfrac{1}{2}O_2 + 2e(\text{metal}) \rightarrow O^{2-}(\text{surface}) \tag{21}$$

and N_s is the number of adsorption sites per unit surface area of oxide ($\simeq a_o^{-2}$, if all sites are available). Cabrera and Mott also drew attention to the significance of a transition temperature, $T_1 = W/40k$. This is based on a fairly arbitrary lower limit of growth rate (10^{-15} ms^{-1}) below which the film may be considered to have stopped growing for all practical purposes. If $T<T_1$ then this limiting rate is reached in the high field region ($X<X_1$) and the film grows in only a logarithmic way to reach an effective limiting thickness. If $T>T_1$ then the film thickens through the logarithmic high field region and into the low field region. Here the kinetics are parabolic with a rate constant

$$k_{p,CM} = 4D_{CM}X_1/a_o \tag{22}$$

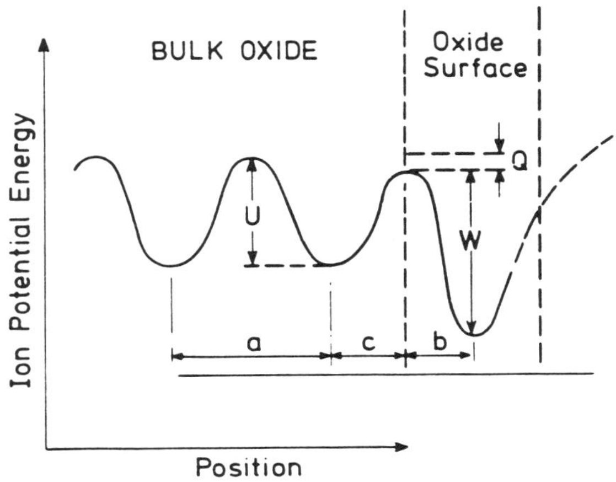

Fig. 9. Free energy diagram for the injection of a cation vacancy at the oxide/gas interface and its motion through the oxide. In the text it is assumed that $Q = 0$ and that $b = c = a_o/2$.

which is different from the thick film rate constant (eq. 14). For an oxide into which defects are injected at the metal/oxide interface, the thin film parabolic rate constant is expected to be approximately 20 times greater (i.e. by X_1/a_o) than the thick film rate constant if D* is equated with D_{CM}. To obtain a smooth transition between thin and thick films it is necessary to include the significant effects of space charge and diffusion in concentration gradients (60,61).

Defects injected at the oxide/gas interface. When the oxide defects are cation vacancies or oxygen interstitials the film growth kinetics are different since defect injection can occur only near the adsorbed oxygen ions, whereas all metal atoms are potential sources of defect injection at the metal/oxide interface. As a result the film growth rate is given by an equation of the form

$$\frac{dX}{dt} = \frac{2A_1}{X} \sinh\left(\frac{X_1}{X}\right) \tag{23}$$

with

$$A_1 = \frac{5\varepsilon\varepsilon_o \Delta\phi D_{CM}}{2eN_s a_o} \tag{24}$$

for the particular case of injection by O^{2-} surface ions. Other possibilities are summarized by Smeltzer and Young (63). In high fields ($X<X_1$) eq. (23) predicts approximately logarithmic kinetics and, in low fields, cubic kinetics. However, Smeltzer and Young (63) point out the futility of trying to distinguish between different mechanisms on the basis of their slightly different kinetics. It is more important that the atomistic parameters should take on reasonable values.

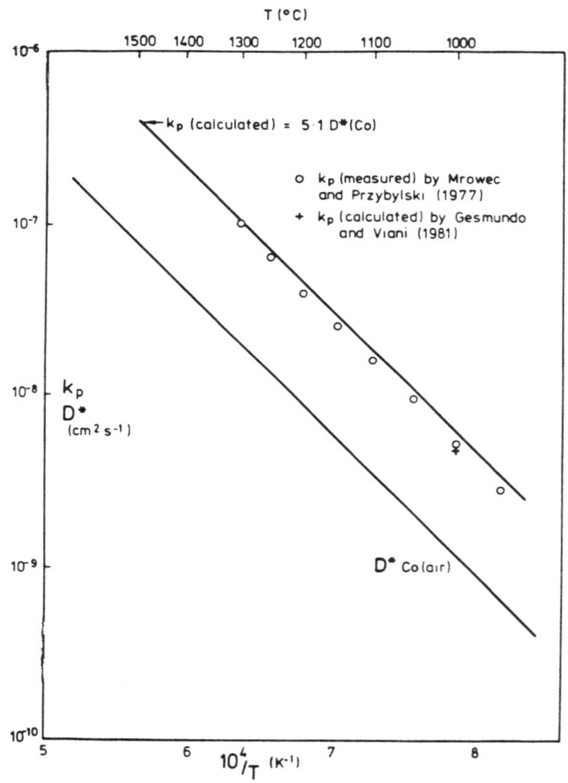

Fig. 10. The parabolic rate constant measured for the growth of thick CoO films on Co in air, compared with that calculated from theory and the measured tracer diffusion coefficient of Co in CoO (64)(65).

Oxide Films Growing by Outward Cation Transport

We will now discuss the growth of films of some oxides in the light of these theories of film growth and the transport properties of the oxides.

Growth of CoO. CoO films grow very quickly and provide a good test of thick film growth theory. For this oxide integration of eq. (14) gives $k_p = 5.1D^*(Co)$, where both k_p and D^* are measured at the same oxygen activity in the gas. Figure 10 shows the comparison between measured k_p and that calculated from eq. (14). The good agreement demonstrates the essential validity of Wagner's theory for thick films. This is further supported by the observed dependence of k_p on $a(O_2)$ also being as predicted by eq. (14). The other rate constant shown in Fig. 10 was calculated using a point defect description of film growth (65) and gives a slightly different result because of some uncertainty about the details of point defects in CoO.

Growth of NiO. NiO films grow much more slowly than CoO films and are much more representative of corrosion-resistant protective films. Furthermore, they may be grown over a very wide range of temperature and thickness and therefore offer an opportunity to study the relationship between thick and thin films. Proceeding as for CoO, evaluation of eq. (14) predicts $k_p = 6.4D^*(Ni)$ with k_p and D^* at the oxygen activity of the gas. It is now clear that k_p predicted on this basis is many orders of magnitude lower than actually observed, as illustrated in Fig. 11. Several workers have suggested that the fast oxidation rates are the result of fast diffusion of Ni along NiO grain boundaries (63). For thick films, Wagner's theory may be modified (69) to include grain boundary diffusion in parallel with lattice diffusion. In the case of NiO the result is (68)

$$k_p = 6.4\left(D^* + 2\frac{(D'\delta)^*}{g}\right) = 6.4D_{eff} \qquad (25)$$

where g is the grain size in the film. The parabolic rate constant evaluated from eq. (25) is shown compared with the corresponding measured values in Fig. 11. These experiments cover a wide range of film thickness (5 nm –30 μm) and grain size (10 nm –4.5 μm). The agreement between theory and experiment is good, even for the thin films, and the dependence on oxygen activity is also as expected from the measured dependence of $(D'\delta)^*$ on $a(O_2)$. The mean activation energy for k_p is ~1.5 eV which is somewhat less than 1.78 eV measured for $(D'\delta)^*$ because the grain size in the oxide is smaller at lower temperature. The dominant role of outward Ni diffusion along oxide grain boundaries is confirmed by Ni and O tracer experiments in the growing films (72).

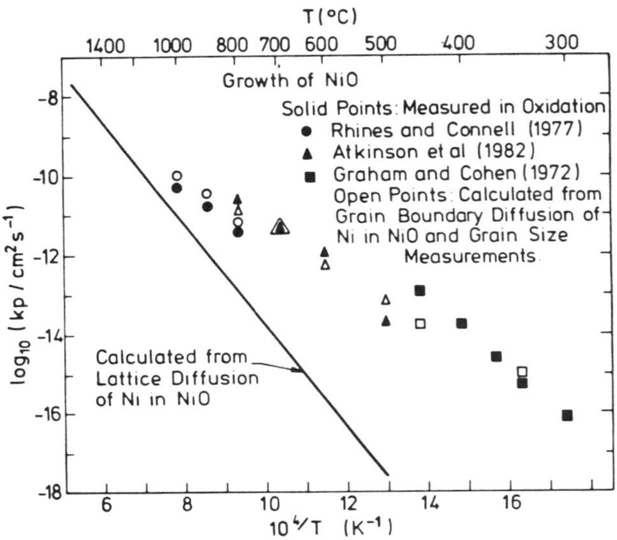

Fig. 11. The parabolic rate constant for the growth of thick NiO films on Ni at $a(O_2) = 1$. The measured values are compared with those expected from tracer diffusion measurements of Ni in the NiO lattice and in NiO grain boundaries (66)(67)(68). (The data of Graham and Cohen have been adjusted by assuming $k_p \propto a(O_2)^{1/6}$.)

Graham and Cohen (66) observed that the growth kinetics at temperatures below about 350°C were approximately logarithmic. Since NiO contains Ni vacancies at high temperatures, eq. (23) is expected to describe kinetics in the logarithmic region. In Fig. 12 the data of Graham and Cohen have been re-analyzed on the basis of eqs. (23) and (24). The magnitude of $\Delta\phi$ which fits the data is approximately 2V. Whilst this is a reasonable value, it is not possible to compare it with that expected from eq. (20) since $\Delta G°$ for the adsorption reaction is not known through independent means. We can however compare D_{CM} with the effective diffusion coefficient derived from the parabolic region (eq. (25)) which, in turn, is known to be dominated by grain boundary diffusion. At 340°C and $a(O_2) = 6.6 \times 10^{-6}$, $D_{eff} = 11 \times 10^{-22}$ m^2s^{-1} from the parabolic region and $D_{CM} = 3.2 \times 10^{-22}$ m^2s^{-1} from the logarithmic region. The agreement between the two is surprisingly close considering the approximations involved. The activation energy for D_{CM} from the two temperatures is 1.9 eV and W, deduced from eq. (19) at both 340 and 200°C is also

1.9 eV (assuming a_o = 0.418 nm and ν = 10^{13} s^{-1}). The general agreement amongst the thin film growth parameters is acceptable and the approximate equality of D_{CM} and D_{eff} indicates that it is injection of cation vacancies into oxide grain boundaries which is the rate-controlling process in thin film growth (whereas their diffusion in the boundary controls thick film growth). This conclusion is consistent with the observation that even the thinnest films are micro-crystalline (66).

Although this description of NiO film growth is broadly self-consistent, there are some observations in the thin film domain which are not in accord with this interpretation. First, film growth appears to enter a parabolic region even when $X<X_1$ and the field is still high. Second, it has been reported that in the logarithmic region the growth rate is proportional to oxygen pressure (but at much lower oxygen pressures, (70)(71)), which cannot be explained by eq. (20), or even other likely surface adsorption processes (63).

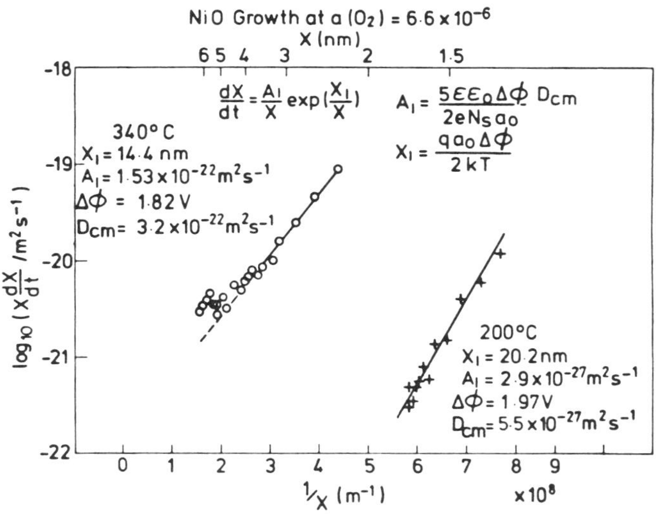

Fig. 12. The growth kinetics of thin films of NiO on Ni (66) interpreted according to the theory of Cabrera and Mott.

Growth of Other oxides by Cation Transport. The growth of these oxides is very similar to that described above for NiO. Oxygen tracer experiments in the growing films have been carried out for Cr_2O_3 (17) and Fe_3O_4 (16) and confirm that oxygen is not mobile (at least when the metals are of high purity). The behavior of cation tracers in growing Fe_3O_4 (16) is very similar to that observed in growing NiO (72) and indicates predominantly grain boundary diffusion in the oxide. Just as in NiO, the thick film parabolic rate constant predicted from lattice diffusion using eq. (14), is much lower than observed (more so for Cr_2O_3 (28) than Fe_3O_4 (9)). This behavior also indicates that faster diffusion along oxide grain boundaries is controlling the growth rate.

The growth of thin films of Cr_2O_3 is similar in many ways to the growth of thin NiO films (73)(74). A transition from approximately logarithmic kinetics to logarithmic followed by parabolic kinetics was observed at about 400°C and was accompanied by a change in microstructure. The activation energy for D_{CM} is 1.9 eV and, for the low temperature parabolic rate constant, 1.6 eV. These activation energies are similar to those reported for thick film growth and are much smaller than the 6.0 eV measured for lattice diffusion (28)(,75). As in NiO, it appears that oxide grain boundary properties control the growth rate of these oxides over wide ranges of temperature.

Oxide Films Growing by Inward Oxygen Transport

The growth of SiO_2 films is a good example of growth of this type. SiO_2 also differs significantly from the examples considered so far in that it is amorphous and therefore its growth is not controlled by grain boundary diffusion. SiO_2 growth is described by a linear-parabolic kinetic equation (76)

$$X^2 + AX = k_p(t + t_o) \qquad (26)$$

in which the linear rate constant A, characterizes the reaction at the Si/SiO_2 interface (which is sufficiently slow that it cannot be ignored in the case of SiO_2). The parabolic rate constant for transport by interstitially dissolved O_2 molecules (mole fraction $[O_2]$ at the SiO_2/O_2 interface) can be expressed as

$$k_p = 2D^*(O) = 2D(O_2)[O_2] \qquad (27)$$

provided that exchange between O_2 and network oxygen is rapid. These equalities are found to be true when transport data from bulk a-SiO_2 are used (77). However, experiments using ^{18}O tracers in growing films (78) have conclusively shown that the exchange reaction is not rapid and that most of the tracer passes straight through the film to form new oxide at the Si/SiO_2 interface. It was concluded that the small amount of exchange which was observed was confined to the outer surface layers of the film. It would therefore appear that the tracer diffusion measurements in bulk a-SiO_2 (equilibrated in O_2) presented in Fig. 6 are spurious. They are possibly the result of H_2O impurity in the O_2 gas assisting the exchange between interstitial O_2 and the network.

When SiO_2 grows on Si in steam, the water vapor reacts with the oxide to form hydroxyl groups (see eqs. (11) and (12)) and the rate constant is given by

$$k_p = 2D^*(O,H_2O) = 0.5D_{app}(OH)[OH] \tag{28}$$

provided that exchange between interstitially dissolved H_2O molecules and network oxygen is rapid. Equation (28) is also found to be true using data measured on bulk a-SiO_2 (77) but, in contrast to oxidation in O_2, ^{18}O tracer experiments in the growing films (79) show that rapid exchange with network oxygen does occur in the film (for temperatures above 600°C) when $H_2{}^{18}O$ is used instead of $^{18}O_2$.

6. GROWTH OF HIGHLY PROTECTIVE FILMS

Oxidation of Alloys

The oxides Al_2O_3, Cr_2O_3 and SiO_2 have very low diffusion coefficients and are the ones which are preferred for thin, highly protective (corrosion resistant) films. They may be deposited onto a substrate by some convenient technique (e.g. sputtering or chemical vapor deposition) or they may be thermally grown by adding the required element (Al, Cr, Si) to the substrate as an alloy and then oxidizing thermally. In this latter approach it is often necessary, for other technological reasons, to add the minimum quantity of the alloying element, B, which will enable a protective film of its oxide, BO_y, to form and to remain stable.

The process which is in competition with formation of a protective film of BO_y is internal oxidation of B to form internal oxide inclusions in the metal by inward oxygen diffusion through the alloy. Rapp (80) has given a detailed analysis of this phenomenon and has shown that internal oxidation is suppressed, and an external film of BO_y will form, if the original atom fraction of B in the alloy exceeds a critical amount which, for most systems, is given by

$$[B]^o_{int} \simeq \left(\frac{0.3\Omega_a}{\Omega_{ox}} \; \frac{\pi D_a(O)[O]_a}{2yD_a(M)} \right)^{1/2} . \tag{29}$$

In this expression Ω_a and Ω_{ox} are the volumes of alloy and oxide per atom of B, $D_a(M)$ and $D_a(O)$ are the coefficients for metal interdiffusion and oxygen diffusion in the alloy at its outer surface. This equation only strictly applies under conditions where the other element in the substrate cannot be oxidized. If the conditions are such that both elements can be oxidized then some of the internally oxidized B will be incorporated into an external film containing both oxides.

The criterion which determines the continued stability of an external film of BO_y on the alloy surface is that the supply of B to the film by diffusion in the alloy must be capable of matching the rate at which B is oxidized by continuing growth of the film, characterized by a rate constant k_p. This leads to a second critical atom fraction of B in the alloy given by (81)

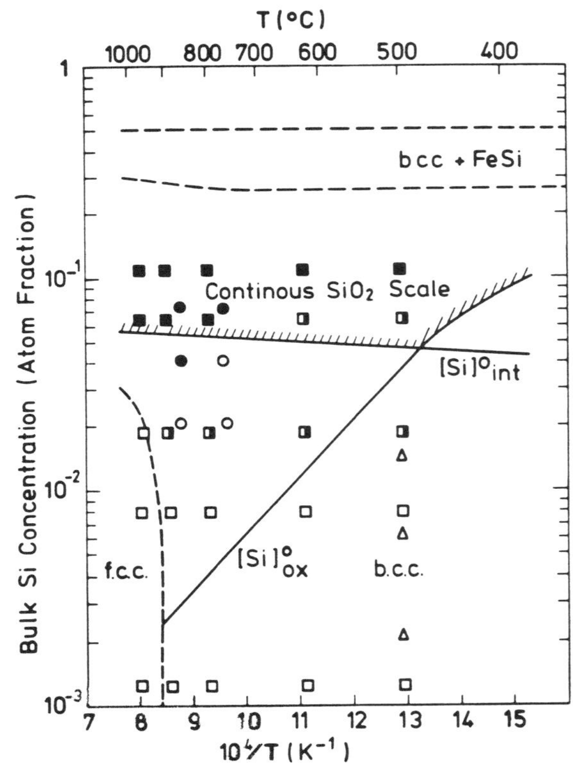

Fig. 13. The calculated critical bulk Si concentrations for SiO_2 film formation on bcc Fe-Si alloys. The experimental observations are shown as open symbols (no SiO_2 film), solid symbols (SiO_2 film) or half-solid symbols (not conclusive) (82).

$$[B]^{o}_{ox} \simeq \frac{\Omega_a}{\Omega_{ox}} \left(\frac{\pi k_p}{4D_a(M)} \right)^{1/2} . \tag{30}$$

Therefore, in order for the film of BO_y to form, the atom fraction of B in the bulk alloy must exceed both the critical values $[B]_{int}{}^{o}$ and $[B]_{ox}{}^{o}$.

Figure 13 shows a comparison between the critical concentration for SiO_2 formation of Fe-Si alloys calculated using equations (29) and (30) and experimental observations (82). The feathered line separates those alloys upon which it is predicted that a continuous film of SiO_2 will form and those upon which it is not. The predictions are in reasonable agreement with observations and also indicate that it is internal oxidation of Si which is the limiting factor over most of the temperature range.

Protection by Incorporated Coatings

Films of the highly protective oxides Al_2O_3, Cr_2O_3 and SiO_2 may be either deposited or grown on alloys and fulfil their protective function by acting as a continuous, low-diffusivity barrier to the transport of species. An alternative approach to protective film formation is to add an element (in the substrate by alloying, or as a surface coating) which will not form a continuous film of its own oxide, but will be incorporated into the oxide film, which forms during subsequent oxidation, and modifies its properties in a beneficial way (e.g. by improving its adherence to the substrate or by reducing diffusion rates). It has been know for about fifty years that certain reactive trace elements (Si, Ce, Y and rare earths) can act in this way. Many speculative ideas have been advanced to explain these effects (83) and amongst these is the suggestion that the reactive elements segregate to oxide grain boundaries and block grain boundary diffusion with a consequent reduction in form growth rate.

Chadwick and Taylor (84)(85) have studied how thin, discontinuous coatings of CeO_2 (~0.1 μm) applied to the surface of Ni influence the subsequent growth of NiO films. The parabolic rate constant for film growth on coated Ni is much less than on untreated Ni in the temperature range 600-800°C, as shown in Fig. 14. Detailed examination of the films grown on coated Ni revealed that they consisted of three layers; an outer thick columnar layer of NiO, an inner layer, also of NiO, and a thin middle layer which contained a fine dispersion of CeO_2 particles in NiO of very small grain size. Additional experiments showed conclusively that it is this middle, two-phase region which is responsible for slowing down the rate of NiO growth even though the grain size in this region is very small. All the kinetic observations can be rationalized by assuming that Ni transport along

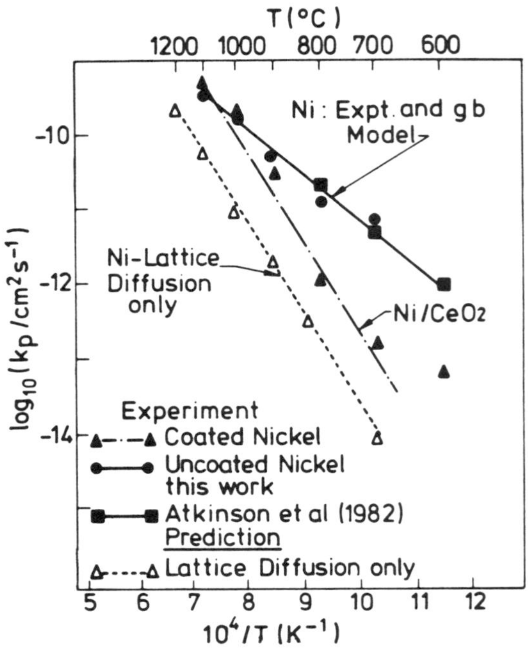

Fig. 14. The effect of a CeO_2 coating on the parabolic rate constant for the oxidation of Ni (84).

NiO grain boundaries in this two-phase region has been blocked by the incorporated coating. Chadwick and Taylor carried out Ni tracer diffusion measurements in the grain boundaries in the outer, columnar part of the film. Other measurements indicated that some Ce was present at these boundaries at low concentration (10^{-3} to 10^{-2} atom fraction). The diffusion measurements showed that diffusion in these boundaries is not very different to diffusion in pure NiO boundaries and that this low level of Ce is not sufficient to block transport along them. The possibility remains that blocking of boundary diffusion occurs only at the much higher boundary concentrations of Ce ions which are present in the two-phase region of the film. Nevertheless, the origin of this important effect is still not clear.

7. CONCLUDING REMARKS

In recent years the general level of understanding of diffusion processes in oxides has improved considerably, thanks to improvements in both experimental and theoretical capability. This progress has also led to a better appreciation of the relationships between transport in oxides and oxide film growth. No doubt this progress will continue in future studies of those areas in which understanding is still far from satisfactory.

REFERENCES

1. Kilner, J. A., Steel, B. C. H., and Ilkov, L., Oxygen self-diffusion studies using negative-ion SIMS, *Solid State Ionics*, **12**, 89-97 (1984).
2. Atkinson, A., Pummery, F. C. W., and Monty, C., Diffusion of ^{18}O tracer in NiO grain boundaries, *Proceedings of 3rd International Conference on Transport in Non-Stoichiometric Compounds* (G. Simkovich and V. S. Stubican, Eds.) pp. 359-370, Plenum Publishing Co., 1985.
3. Amsel, G., Nadai, J. P., D'Artemare, E., David, D., Girard, E., and Moulin, J., Microanalysis by the direct observation of nuclear reactions using a 2 MeV van de Graaff, *Nuclear Instruements and Methods*, **92**, 481-498 (1971).
4. Derry, D. J., Lees, D. G., and Clavert, J. M., A study of oxygen self-diffusion in the c-direction of rutile using a nuclear technique, *J. Phys. Chem. Solids* **42**, 57-64 (1981).
5. Mehrer, H., Maier, K., Hettich, G., Mayer, H. J., and Rein, G., Measurement of small diffusion coefficients using ion beam sputtering as a microsectioning technique, *J. Nucl. Mater.* **69**, 545-548 (1978).
6. Mundy, J. N., and Rothman, S. J., An apparatus for microsectioning diffusion samples by sputtering, *J. Vac. Sci. Technol. A* **1**, 74-76 (1983).
7. Gupta, D., and Tsui, R. T. C., A universial microsectioning technique for diffusion, *Appl. Phys. Lett.*, **17**, 294-297 (1970).
8. Atkinson, A., and Taylor, R. I., The evaluation of RF sputtering as a microsectioning technique for tracer diffusion studies in oxides, *Thin Solid Films*, **46**, 291-298 (1977).
9. Atkinson, A., O'Dwyer, M. L., and Taylor, R. I., ^{55}Fe diffusion in magnetite crystals at 500°C and its relevance to the oxidation of iron, *J. Mater. Sci.*, **18**, 2371-2379 (1983).
10. Atkinson, A., and Taylor, R. I., The diffusion of ^{63}Ni along grain boundaries in nickel oxide, *Philos. Mag. A*, **43**, 979-998 (1981).

11. Frischat, G. H., and Remmers, K., Measurement of low diffusion coefficient values in different glasses, in *The Physics of Non-Crystalline Solids* (G. H. Frischat, ed.) pp. 447-452, Trans Tech Publications, Aedermannsdorf, Switzerland (1977).
12. Atkinson, A., and Gardner, J. W., The diffusion of Fe^{3+} in amorphous SiO_2 and the protective properties of SiO_2 layers, *Corros. Sci.*, **21**, 49-58 (1981).
13. Oishi, Y., and Kingery, W. D., Self-diffusion of oxygen in single crystal and polycrystalline aluminum oxide, *J. Chem. Phys.*, **33**, 480-486 (1960).
14. Ando, K., Oishi, Y., Hase, H., and Kitazawa, K., Oxygen self-diffusion in single crystal Y_2O_3, *J. Amer. Ceram. Soc.* (Communications), C222-C223 (1983).
15. Bardeen, J., Brattain, W. H., and Shockley, W., Investigation of oxidation of copper by use of radioactive Cu tracer, *J. Chem. Phys.*, **14**, 714-721 (1946).
16. Atkinson, A., and Taylor, R. I., ^{55}Fe and ^{18}O tracers in magnetite scales growing on Fe and dilute Fe-Si alloys, *High Temperatures – High Pressures*, **14**, 571-580 (1982).
17. Mitchell, D. F., Hussey, R. J., and Graham, M. J., Analysis of oxygen isotope interfaces using negative molecular ion SIMS, *J. Vac. Sci. Technol. A*, **1**, 1006-1008 (1983).
18. LeClaire, A. D., Diffusion, in *Treatise on Solid State Chemistry Vol. 4* (N. B. Hannay, ed.) pp. 1-59, Plenum Press, New York (1976).
19. Peterson, N. L., and Chen, W. K., Isotope effect for cation diffusion in CoO, *J. Phys. (Paris) Colloque C-6*, **41**, 319-322 (1980).
20. Chen, W. K., and Peterson, N. L., Effect of deviation from stoichiometry on cation self-diffusion and isotope effect in wüstite, $Fe_{1-x}O$, *J. Phys. Chem. Solids,* **36**, 1097-1103 (1975).
21. Peterson, N. L., and Chen, W. K., Cation self-diffusion and the isotope effect in MnO, *J. Phys. Chem. Solids*, **43**, 29-38 (1982).
22. Dieckmann, R., Cobaltous oxide point defect structure and non-stoichiometry, electrical conductivity, cobalt tracer diffusion, *Z. Phys. Chem. N.F.*, **107**, 189-210 (1977).
23. Atkinson, A., Hughes, A. E., and Hammou, A., The self-diffusion of Ni in undoped and Al-doped NiO single crystals, *Philos. Mag. A*, **43**, 1071-1091 (1981).
24. Volpe, M. L., and Reddy, J., Cation self-diffusion and semiconductivity in NiO, *J. Chem. Phys.*, **53**, 1117-1125 (1970).
25. Farhi, R., and Petot-Ervas, G., Thermodynamic study of point defects in single crystalline NiO: Analysis of experimental results, *J. Phys. Chem. Solids*, **39**, 1175-1179 (1978).
26. Peterson, N. L., Point defects and diffusion mechanisms in monoxides of iron-group metals, *Diffusion and Defect Data*, **36**, 1-26 (1984).

27. Dieckmann, R., and Schmalzried, H., Defects and cation diffusion in magnetite, *Ber. Bunsenges Phys. Chem.*, **81**, 344-347 (1977).
28. Atkinson, A., and Taylor, R. I., Diffusion of ^{51}Cr tracer in Cr_2O_3 and the growth of Cr_2O_3 films, to be published in *Proceedings of 3rd International Conference on Transport in Non-Stoichiometric Compounds* (G. Simkovich and V. S. Stubican, Eds.) pp. 285-295, Plenum Publishing Co., 1985.
29. Matzke, Hj., Diffusion processes and surface effect in non-stoichiometric nuclear fuel oxides UO_2 and $(U,Pu)O_2$, *J. Nucl. Mater.*, **114**, 121-130 (1983).
30. Breitung, W., Oxygen self and chemical diffusion coefficients in $UO_{2\pm x}$, *J. Nucl. Mater.*, **74**, 10-18 (1978).
31. Perinet, F., Barbezat, S., and Monty, C., New investigation of oxygen self-diffusion in Cu_2O, *J. Phys. (Paris) Colloque C-6,* 315-318 (1980).
32. Peterson, N. L., and Wiley, C. L, Diffusion and point defects in Cu_2O, *J. Phys. Chem. Solids*, **45**, 281-294 (1984).
33. Reddy, K. P. R., and Cooper, A. R., Oxygen diffusion in sapphire, *J. Amer. Ceram. Soc.*, **65**, 643-638 (1982).
34. Chen, W. K., and Peterson, N. L., Correlation effects and effects of cobalt additions on cobalt and nickel diffusion in NiO, *J. Phys. Chem. Solids*, **33**, 881-892 (1982).
35. Lidiard, A. B., Ionic conductivity, *Handbuch der Physik*, **20**, 246-349 (1957).
36. Peterson, N. L., Impurity diffusion in transition metal oxides, *Solid State Ionics*, **12**, 201-215 (1984).
37. Monty, C., Diffusion in stoichiometric and non-stoichiometric cubic oxides, *Radiation Effects*, **74**, 29-55 (1983).
38. Wuensch, B. J., Diffusion in stoichiometric close-packed oxides, in *Mass Transport in Solids* (F. Bénière and C. R. A. Catlow, eds.) NATO ASI Series B: Physics, Vol. 97, pp. 353-376 (1983).
39. Weber, G. W., Bitler, W. R., and Stubican, V. S., Diffusion of ^{51}Cr in Cr-doped MgO, *J. Phys. Chem. Solids*, **41**, 1355-1359 (1980).
40. Frischat, G. H., *Ionic Diffusion in Oxide Glasses*, Aedermannsdorf, Switzerland; Trans Tech S.A. (1975).
41. Norton, F. J., Permeation of gaseous oxygen through vitreous silica, *Nature* 191: 701 (1961).
42. Moulson, A. J., and Roberts, J. P., Water in silica glass, *Trans Faraday Soc.*, **57**, 1208-1216 (1961).
43. Pfeffer, R., and Ohring, M., Network oxygen exchange during water diffusion in SiO_2, *J. Appl. Phys.*, **52**, 777-784 (1981).
44. Williams, E. L., Diffusion of oxygen in fused silica, *J. Amer. Ceram. Soc.*, **48**, 190-194 (1965).

45. Ghoshtagore, R. N., Diffusion of nickel in amorphous silicon dioxide and silicon nitride films, *J. Appl. Phys.*, **40**, 4374-4376 (1969).
46. Schaeffer, H. A., Oxygen and silicon diffusion-controlled processes in vitreous silica, *J. Non Cryst. Solids*, **38**, 545-550 (1980).
47. Sangster, M. J. L., and Rowell, D. K., Calculation of defect energies and volumes in some oxides, *Philos. Mag. A*, **44**, 613-624 (1981).
48. Mackrodt, W. C., Defect energetics and their relation to non-stoichiometry in oxides, *Solid State Ionics,* **12**, 175-188 (1984).
49. Dubois, C., Monty, C., and Philibert, J., Oxygen self-diffusion in NiO single crystals, *Philos. Mag. A*, **46**, 419-433 (1982).
50. Sangster, M. J. L., and Stoneham, A. M., Calculation of absolute diffusion rates in oxides, *J. Phys. C: Solid State Phys.*, **17**, 6093-6104 (1984).
51. Atkinson, A., Diffusion along grain boundaries and dislocations in oxides, alkali halides and carbides, *Solid State Ionics*, **12**, 309-320 (1984).
52. Atkinson, A., and Taylor, R. I., The diffusion of Ni in the bulk and along dislocations in NiO single crystals, *Philos. Mag. A* **39**, 581-595 (1979).
53. Marlowe, M. O., and Kaznoff, A. I., Tracer Study of the surface diffusivity of UO_2, *J. Nucl. Mater.*, **25**, 328-333 (1986).
54. Rabier, J., and Puls, M. P., Caculations of pipe diffusion coefficients in MgO, *Rad. Effects,* **75**, 193-201 (1983).
55. Duffy, D. M., and Tasker, P. W., Computer simulation of ⟨001⟩ tilt grain boundaries in NiO, *Philos. Mag. A*, **47**, 817-825 (1983).
56. Tasker, P. W., The calculated properties of grain boundaries in NiO, *J. Phys.* (Paris) Colloque C-4, **46**, 185-195 (1985).
57. Wagner, C., Beitrag zur Theorie des Anlaufvorgangs, *Z. Phys. Chem.*, **B21**, 25-41 (1933).
58. Fromhold, A. T., Fundamental theory of the growth of thick oxide films on metals, *J. Phys. Soc. Japan*, **48**, 2022-2030 (1980).
59. Cabrera, N., and Mott, N. F., Theory of the oxidation of metals, *Rep. Prog. Phys. Vol. 12*, **12**, 163-184 (1949).
60. Fromhold, A. T., *Theory of Metal Oxidation Vol. 1*, North Holland, Amsterdam (1976).
61. Young, D. J., and Dignam, M. J., Metal oxidation – II. Kinetics in the thin and very thin film regions under conditions of electron equilibrium, *J. Phys. Chem. Solids,* **34**, 1235-1250 (1973).
62. Grimley, T. B., Oxidation of metals, in *Chemistry of the Solid State,* pp. 336-366, Butterworths, London (1955).

63. Smetlzer, W. W., and Young, D. J., Oxidation properties of transition metals, *Prog. Sol. State Chem.*, **10**, 17-54 (1975).
64. Mrowec, S., and Przybylski, K., Oxidation of cobalt at high temperature, *Oxidation of Metals,* **11**, 365-403 (1977).
65. Gesmundo, F., and Viani, F., Application of Wagner's theory to the parabolic growth of oxides containing different kinds of defects. 1. Pure oxides, *J. Electrochem. Soc.*, **128**, 460-469 (1981).
66. Graham, M. J., and Cohen, M., On the mechanism of low-temperature oxidation (23-450°C) of polycrystalline Ni, *J. Electrochem. Soc.*, **119**, 879-882 (1972).
67. Rhines, F. N., and Connell, R. G., Role of grain growth in oxidation of Ni, *J. Electrochem. Soc.*, **124**, 1122-1128 (1977).
68. Atkinson, A., Taylor, R. I., and Hughes, A. E., A quantitative demonstration of the grain boundary diffusion mechanism for the oxidation of metals, *Philos. Mag. A*, **45**, 823-833 (1982).
69. Perrow, J. M., Smeltzer, W. W., and Embury, J. D., The role of structural defects in the growth of nickel oxide films, *Acta Metall.*, **16**, 1209-1218 (1968).
70. Mitchell, D. F., Sewell, P. B., and Cohen, M., A kinetic study of the initial oxidation of the Ni(110) surface by RHEED and X-ray emission, *Surface Sci.*, **69**, 310-322 (1977).
71. Mitchell, D. F., and Graham, M. J., A kinetic study of the initial oxidation of Ni(111) and (211) surfaces by RHEED and X-ray emission, *Surface Sci.*, **114**, 546-562 (1982).
72. Atkinson, A., Taylor, R. I., and Goode, P. D., Transport processes in the oxidation of Ni studied using tracers in growing NiO scales, *Oxidation of Metals,* **13**, 519-543 (1979).
73. Young, D. J., and Cohen, M., Oxidation behavior of Cr between 300 and 600°C, *J. Electrochem. Soc.*, **124**, 769-774 (1977).
74. Young, D. J., and Cohen, M., Pressure effects in the oxidation of Cr, *J. Electrochem. Soc.*, **124**, 775-779 (1977).
75. Hoshino, K., and Peterson, N. L., Cation self-diffusion in Cr_2O_3, *J. Amer. Ceram. Soc.*, (Communications) **66**, C202-203 (1983).
76. Deal, B. E., and Grove, A. S., General relationship for the thermal oxidation of silicon, *J. Appl. Phys.*, **36**, 3770-3778 (1965).
77. Atkinson, A., Transport processes during the growth of oxide films at elevated temperature, *Rev. Mod. Phys.* **57**, 437-470 (1985).
78. Rochet, F., Agius, B., and Rigo, S., An ^{18}O study of the oxidation mechanism of silicon in dry oxygen, *J. Electrochem. Soc.*, **131**, 914-923 (1984).
79. Rigo, S., Rochet, F., Agius, B., and Straboni, A., An ^{18}O study of cooperative diffusion and chemical reaction during thermal treatments of silica films in water vapour, *J. Electrochem. Soc.*, **129**, 867-876 (1982).

80. Rapp, R. A., Kinetics, microstructures and mechanism of internal oxidation – its effect and prevention in high temperature alloy oxidation, *Corrosion*, **21**, 382-400 (1965).
81. Wagner, C., Formation of composite scales consisting of oxides of different metals, *J. Electrochem. Soc.*, **103**, 627-633 (1956).
82. Atkinson, A., A theoretical analysis of the oxidation of Fe-Si alloys, *Corros. Sci.*, **22**, 87-102 (1982).
83. Whittle, D. P., and Stringer, J., Improvement in properties: additives in oxidation resistance, *Phil. Trans. Roy. Soc. (London)*, **A295**, 309-329 (1980).
84. Chadwick, A. T., and Taylor, R. I., Trace element effects in oxidation: the influence of Ce on grain boundary transport in NiO, *Solid State Ionics*, **12**, 343-351 (1984).
85. Chadwick, A. T., and Taylor, R. I., The reactive element effect in high temperature oxidation studied using sol-coated Ni, in *International Congress on Metallic Corrosion, Toronto:* 381-388 (1984).

5

Diffusion-Induced Grain Boundary Migration in Thin Films

Carol A. Handwerker

1. INTRODUCTION

Nature is not always as simple as we are led to believe. Just when it is imagined that we had laws well in hand to predict how a material behaves, something new is discovered to force us to reevaluate those "laws" (1). Diffusion-induced grain boundary migration (DIGM) is such a case in point. Before DIGM was recognized, it was generally accepted that intermixing in polycrystalline materials could be described as occurring only by a combination of lattice diffusion, grain boundary diffusion, and dislocation pipe diffusion (2). It was also accepted that at temperatures less than about half of the melting point, grain boundary diffusion would be expected to dominate the diffusion process with lateral "leakage" of the diffusing species from the grain boundary into the adjacent lattice by lattice diffusion. These grain boundary diffusivities are measured using averaging techniques where thin layers of material from the entire sample surface are analyzed. The individual lattice, grain boundary, and dislocation diffusivities are then calculated from the measured composition profile, which is due to diffusion along all paths. The mathematics of atom transport by multiple paths has been developed for a number of simple geometries, including several configurations applicable to thin films. In many cases, this description of atom transport is valid and the lattice diffusivities calculated from studies on polycrystals are reasonable extrapolations of diffusivities measured at higher temperatures.

However, in 1976, several researchers recognized that major discrepancies existed between lattice diffusion results from solute diffusion into single crystals and from solute diffusion into thin film polycrystals (2,3). In particular, the calculated lattice diffusivites from experiments with polycrystals were orders of magnitude larger than with single crystals for several systems, including Au-Pd and Au-Ag. This effect is seen in the Rutherford backscattering spectra for polycrystalline and single crystal thin films of Au(80nm) and Ag(100nm) annealed at 150°C, shown in Figure 1 (4). As pointed out by Baglin

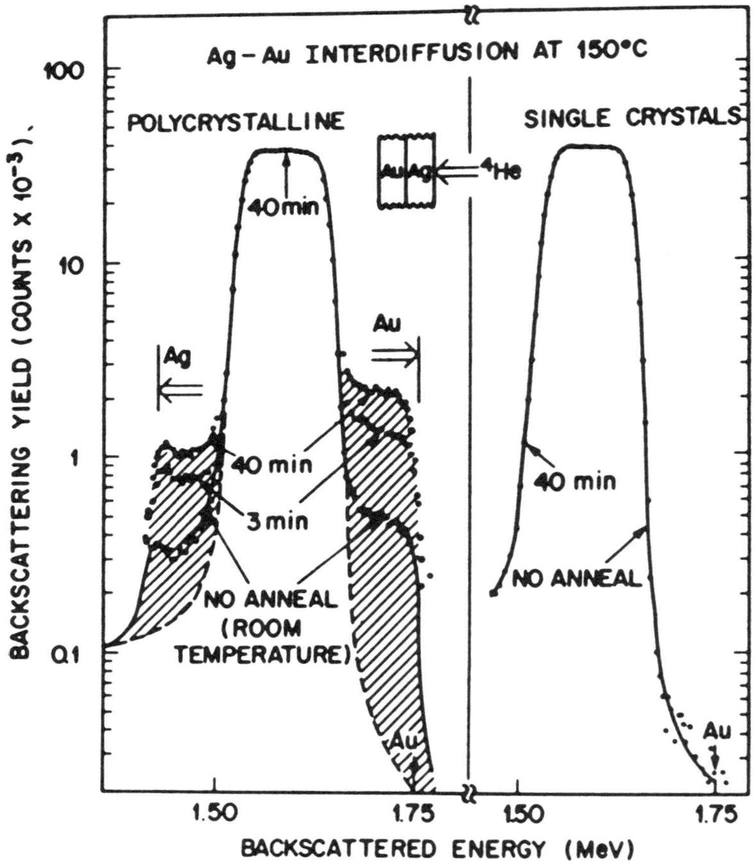

Fig. 1. Backscattering spectra from Ag(100 nm) and Au(80 nm) couples for 1.9 MeV ^{4}He. Film thicknesses were chosen so that, for beam incident on Ag, complete overlap of the Au and Ag peaks occurs. Interdiffusion then results in the diffused Au moving to the higher-energy side and Ag to the lower-energy side. Position of surface Au is indicated by arrows. As-deposited spectra without annealing are shown by smooth curve (4).

and Poate (3), the presence of grain boundaries in the Ag-Au thin film diffusion couple produced concentration profiles radically different from those in the single crystal diffusion couples. In the Au and the Ag polycrystalline films, the solute had diffused completely through the films after 40 minutes at 150°C. The solute concentrations are constant at approximately 10 atomic% solute from the original Ag-Au interface to the outer surfaces of the films. No interdiffusion was observed in single crystal Ag-Au diffusion couples annealed under the same conditions. The lack of interdiffusion in the single crystal couples is in agreement with the predicted lattice penetration of approximately 0.3nm after 40 min. at 150°C based on other Ag-Au interdiffusion experiments. However, when lattice diffusivities are calculated by the Whipple analysis using the concentration profiles shown in Figure 1, the lattice diffusivities are orders of magnitude higher than for single crystal diffusion. In order to reconcile these results with the standard grain boundary-lattice diffusion model, a "defect-enhanced lattice diffusivity" was postulated to occur in thin films but the physical significance of such a parameter was questioned. Although it is widely known that thin films are highly defective, containing high densities of short circuit paths for diffusion, the only defects which could be identified in these samples with anomalously high lattice diffusivities were high angle grain boundaries. Reviews of interdiffusion in thin films by Gupta, Campbell, and Ho (2) and by Baglin and Poate (3) describe the basic techniques for measurements of grain boundary and lattice diffusion. They also discuss how these discrepancies among lattice diffusivities in Au-Ag, Au-Pd, Al-Cu, and Au-Cu could not be explained based on the analyses of short circuit diffusion paths alone.

As a result of pioneering papers by den Broeder (5), Tu (6), and Hillert and Purdy (7) on interdiffusion at low temperatures, it has become recognized that these discrepancies are a result of a fundamental change in mechanism for solid solution formation in polycrystalline materials. In all three studies, it was determined that either alloying or dealloying of a polycrystal could have occurred only by a combination of grain boundary diffusion and grain boundary migration. Direct microstructural evidence of concomitant grain boundary migration and alloying was obtained both by den Broeder (Cr-W) and by Hillert and Purdy (Fe-Zn). In these studies, large grain-sized, well-annealed polycrystalline foils were exposed to a solute source and the solute diffused down the grain boundaries. As a result of diffusion, grain boundaries migrated, sometimes away from their centers of curvature, and solid solutions formed in the regions swept by the moving grain boundaries.

A schematic of the polycrystalline foil geometry most often used in DIGM experiments is shown in Figure 2. The foil is preannealed to

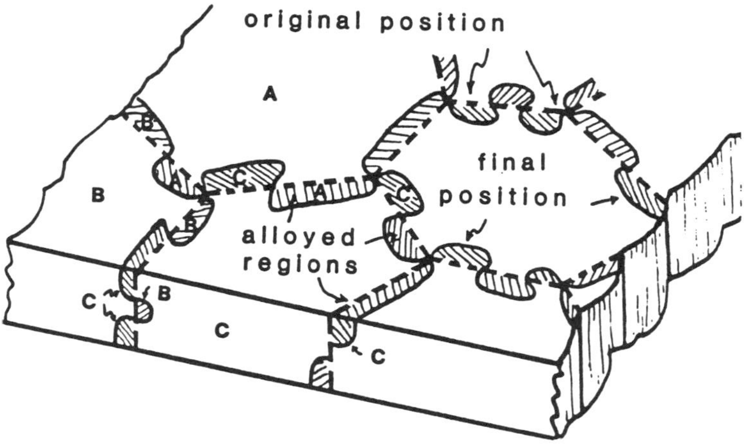

Fig. 2. Schematic of microstructure of polycrystalline foil before and after DIGM. Dotted lines refer to the initial boundary positions. Solid lines indicate the positions of the boundaries after DIGM. Volume swept by the moving boundary is alloyed. Grains with the same capital letter are of the same orientation.

"stabilize" the microstructure, that is, to promote recrystallization and grain growth of the microstructure before the lower temperature diffusion anneal. On exposure of the foil to a solute source, the grain boundaries are induced to migrate, often away from their centers of curvature, and form solid solutions in the regions through which the grain boundaries have migrated. A typical microstructure of alloyed regions in Fe foil exposed to Zn vapor are shown in Figure 3.

The difference in solute concentration between the grain interiors and the regions swept by the moving boundaries is large. Using x-ray diffraction, Tu observed that polycrystalline films of AgPd in contact with Pb at 200°C formed Pb_2Pd at the interface and the AgPd film became dealloyed. As shown in Figure 4, the change in composition of the AgPd solid solution film was accompanied by a gradual decrease in the intensity but not the position of the 200 reflection for AgPd and by the appearance of a 200 reflection for Ag containing less than 1 at.% Pd. Interdiffusion between Ag and Pd at 200°C and 450°C showed similar features with the growth of an AgPd alloy by diffusion of Pd into the Ag film.

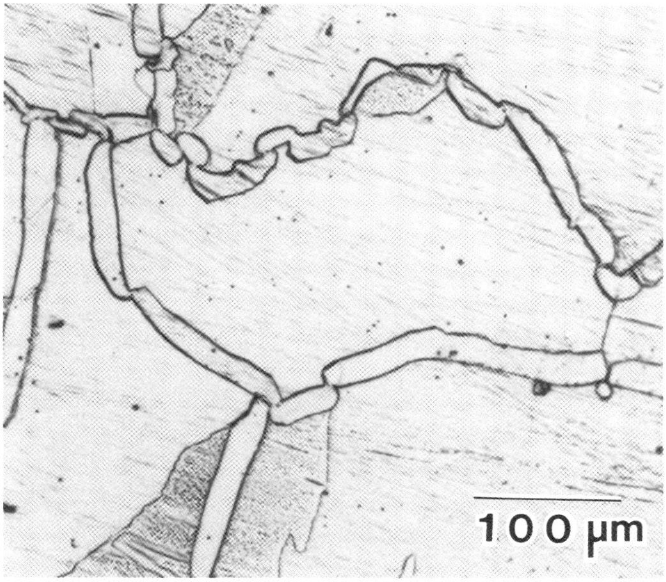

a

Fig. 3. Diffusion-induced grain boundary migration in Fe(Zn). Foils of Fe were exposed to Zn vapor from an alloy source. (a) and (b) Source: 90% Fe, 10% Zn; 550°C; 7 h.; and (c) 85% Fe, 15% Zn; 600°C; 3 h. (54).

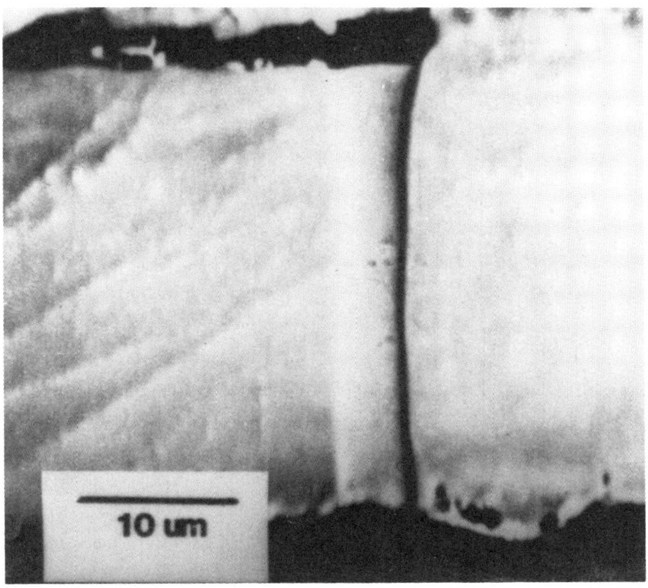

b

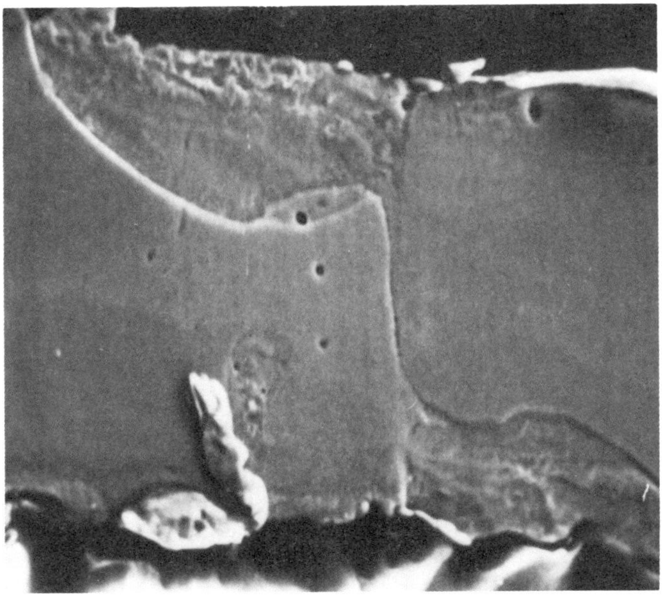

C

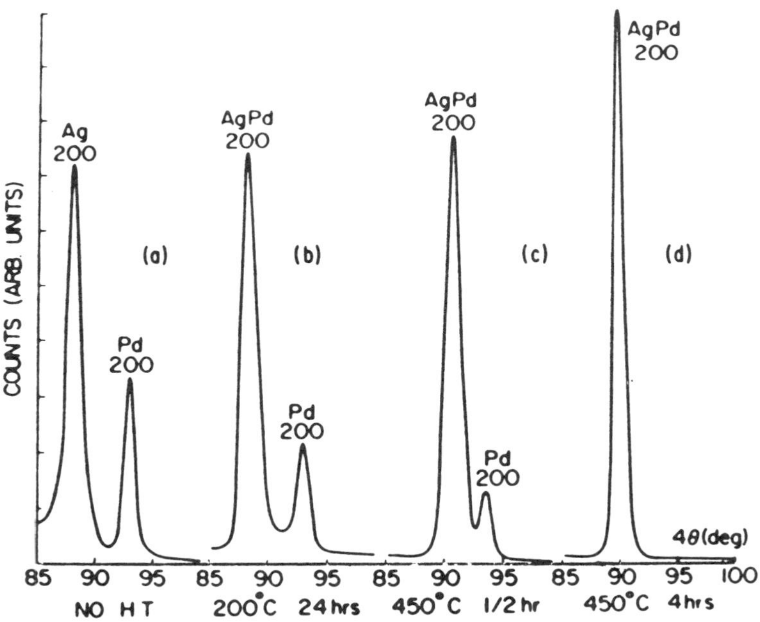

Fig. 4. Sequential x-ray spectra of various stages of homogenization of bimetallic thin films of Ag and Pd. (a) No heat treatment. (b) 24 h. at 200°C. (c) 0.5 h. at 450°C. (d) 4 h. at 450°C (6).

This low temperature grain boundary migration phenomenon, DIGM, is observed both during alloying, as in the case of Pd diffusing into Ag and during dealloying, as in the case of Pd diffusing out of a polycrystalline thin film of Ag-Pd solid solution. DIGM occurs in both symmetrical and asymmetrical grain boundaries and in systems with positive and negative deviations from ideal solution thermodynamics.

Equally surprising was the discovery that solute diffusion can produce new grains of solid solution alloy which nucleate in the original matrix grains in regions which were essentially single crystals before exposure to the solute source. New grains were observed both in Cu single crystals exposed to Zn by Doo and Balluffi (8) and in the diffusion zone between thin metallic films by Matthews and coworkers (9-11). This phenomenon, known as diffusion-induced recrystallization (DIR), can lead to complete surface alloying by formation of new grains. The possibility of nucleating and growing solid solution grains in a single crystal has serious ramifications for the use of lattice diffusion data to predict the stability of metal-metal interfaces.

The discoveries of DIGM and DIR were startling on several accounts. First, the phenomena were unexpected and the driving force for grain boundary migration during solute diffusion was not immediately obvious (12). Second, these results have caused a reevaluation of diffusion data derived from models of diffusion in stationary boundaries and in the lattice at low temperatures. And last, these phenomena have called into question the analyses of the chemical stability of multicomponent polycrystalline systems which have relied on traditional models of grain boundary and lattice diffusion. These analyses have been used in the design of critical systems, as diverse as thin film contacts and multicomponent nuclear waste storage systems, where different materials are in contact.

As with many other "new" phenomena, these effects were observed many years before any attention was drawn to them. In 1938, Rhines and Montgomery (13) described exactly these microstructures and composition discontinuities produced by grain boundary motion when Zn diffused into a Cu bicrystal forming a solid solution at 775°C. Rhines and Mehl (14) also observed the formation of new grains on the surfaces of Cu single crystals when the samples were exposed to a source of Zn vapor. In all probability there have been many diffusion experiments in thin film and bulk diffusion couples where DIGM and DIR were seen. However, these experiments had "failed" because the grain boundaries had migrated during the diffusion anneal and had, thus, gone unreported.

The most important question which has been asked about the DIGM process is why it occurs at all, that is, what is the driving force which couples the motion of the grain boundary to diffusion of the solute. The issue of driving forces in DIGM has been a topic of considerable controversy. The question of driving force for DIGM is of primary importance in the study of DIGM because we are not able to predict under what conditions DIGM will occur and to control the extent of DIGM unless we know why it occurs.

In this chapter the physical characteristics of DIGM are described for both thin film geometries and for bulk specimens. This description of the morphologies, the temperature and time regimes, and the characteristics of systems exhibiting DIGM will aid in identifying DIGM and in defining the conditions under which DIGM is likely to occur.

We then discuss the question of driving forces for boundary migration, first, in terms of whether a particular mechanism postulated for DIGM can be a driving force for migration under some circumstances and, second, in terms of their agreement with specific experimental results for DIGM. The critical step in identifying the driving force for DIGM was the observation that this anomalous migration and alloying occur even when the grain boundary is replaced by a liquid film (15-22). Establishing the relationship between DIGM and liquid film migration (LFM) has led to the conclusion that the driving force for boundary migration is a result of lattice diffusion of solute into adjacent grains, whether the penetration of the solute is as little as several layers adjacent to the grain boundary or extends micrometers into the lattice. Diffusion of solute into the lattice creates strain in the diffusion zone when the lattice parameter is a function of composition. When the diffusion layer remains coherent with the unalloyed matrix, the driving force is a result of the difference in coherency strain energy on opposite sides of an interface, regardless of whether the interface is a liquid film or a grain boundary (23). The thermodynamics of solid solution formation at an interface between a coherently strained solid and a liquid is reviewed for the case of LFM and the velocity expressions for the strain-energy-driven motion of liquid films are also presented.

When the liquid film is replaced by a grain boundary, the thermodynamic and kinetic descriptions of migration cease to be straightforward. Although the velocity expressions for LFM have been applied qualitatively to DIGM with reasonable success, a quantitative description of DIGM does not yet exist. Additional complexities for DIGM that do not occur in LFM are the effects of: the relative rates of grain boundary and lattice diffusion, the relative diffusion rates of the diffusing species in the lattice and along the grain boundary, the efficiency of the grain boundaries as sources and sinks for vacancies, and the presence of other driving forces for grain boundary migration. In spite of these complexities, a simple mechanism diagram has been developed which describes all possible low temperature and high temperature diffusion regimes over which DIGM and DIR can occur.

The identification of DIGM in thin films is not always as simple as in well-annealed bulk polycrystalline materials. Thin films can contain very high densities of defects, including grain boundaries, dislocations, twins, vacancies, and pores, and the film can be in a state of biaxial strain. Recrystallization and grain growth can occur in these materials in the absence of solute diffusion. If grain boundaries are moving due to one of these processes and solute is diffusing by grain boundary diffusion, solute will be left in the regions swept by the moving boundaries even if DIGM is not occurring. It is important to identify which process is driving grain boundary motion in order both to predict when intermixing by DIGM will occur in thin films and to control the extent of intermixing by the proper sequence of thin film layers in contact.

2. MORPHOLOGY AND KINETICS OF DIGM, DIR, AND LIQUID FILM MIGRATION

The system characteristics, the morphologies, and the kinetics of DIGM, DIR, and liquid film migration are introduced in this chapter without reference to the driving force which causes the interface to move. In later sections on migration of liquid films and grain boundaries, experimental results are presented and discussed in greater detail in light of specific predictions of theories for DIGM.

In this chapter, the following conventions are used to describe systems exhibiting DIGM: (1) "xxx(yyy)", where xxx is the original solvent phase and yyy is the species involved in the alloy formation or dealloying; (2) "yyy/xxx", where xxx is the substrate and yyy is the alloy formed by either epitaxial growth or interface motion. For example, Au(Ag) means that Ag diffuses into originally pure Au, 50% Au, 50% Ag(Ag) means that Ag diffuses into a solid solution of 50% Au, 50% Ag, and 20% Ag, 80% Au/Au means that a 20% Ag, 80% Au alloy layer was grown epitaxially on pure Au.

Systems Exhibiting DIGM and DIR

The systems in which DIGM or DIR have been confirmed are listed in Table 1(a). All these systems form substitutional solid solutions. Although DIGM has been reported in Ni(C)(68) with the formation of a solid solution by interstitial C, these results are controversial (71).

Although it was previously suggested that lattice diffusion must be "essentially frozen out" for DIGM to occur (8), there is evidence in several systems that DIGM can occur when there is significant diffusion from the grain boundary into the adjacent lattice (13,29,49). The

Table 1(a). Solvent (Solute) Systems exhibiting DIGM and/or DIR

System	Temperature Range	Maximum Time[a]	Sample Geometry[b]	Alloying or Dealloying	η^c	Ref.
Ag(Au)	Room	1.87×10^3 h (6.74×10^6 s)	TF	alloying	-1.9×10^{-3}	4
	150°C	40 m. (2.40×10^3 s)	TF	alloying		4
	200°C	24 h. (8.64×10^4 s)	TF	alloying		25,26
Ag(Pd)	200°C,250°C	48 h. (1.73×10^5 s)	TF	dealloying	−0.049	27
	200°C	24 h. (8.64×10^4 s)	TF	alloying		6
	450°C	4 h. (1.44×10^4 s)	TF	alloying		6
	200°C	1.33 h. (4.80×10^3 s)	TF	dealloying		6
	160°C-200°C	4 h. (1.44×10^4 s)	TF	alloying		26
Ag(Zn)	350°C	89 h. (3.20×10^5 s)	F	alloying	−0.049	28
Al(Zn)	160°-190°C	120 h. (4.32×10^5 s)	W	alloying	−0.045	29
	220°C	72 h. (2.59×10^5 s)	F	alloying		30
Au(Ag)	150°-300°C	200 h. (7.20×10^5 s)	TF	alloying	1.89×10^{-3}	31,32,33
	200°C	24 h (8.64×10^4 s)	TF	alloying		34
	175°-375°C	72 h (2.59×10^5 s)	F	alloying		35
Au(Cu)	160°-220°C	500 h. (1.80×10^6 s)	TF	alloying	−0.104	36
	150°C	96 h. (3.46×10^5 s)	TF	alloying		31,32,33
	150°C,300°C	30 h. (1.08×10^5 s)	TF	alloying		37
Au(Pd)	–	200 s.	TF	alloying	−0.046	10
	175°-300°C	200 h. (7.20×10^5 s)	TF	alloying		38
	200°-389°C	2.40×10^3 s	TF	alloying		39
Cu(Al)	300°-500°C	24 h. (8.64×10^4 s)	TF	alloying	0.066	40,41
Cu(Cd)	360°-450°C	96 h. (3.46×10^5 s)	F	alloying	0.231	42
Cu(In)	412°,450°C	16 h. (5.76×10^4 s)	F	alloying	0.259	42
Cu(Ni)	450°-600°C	8 h. (2.88×10^4 s)	TF	alloying	−0.026	43
	450°-520°C	–	TF	alloying		44,45,46
	200°-500°C	4 h. (1.44×10^4 s)	P	alloying		47,48
	350°-900°C	6 h. (2.16×10^4 s)	F			49,50
Cu(Sb)	457°C	264 h. (9.50×10^5 s)	F	alloying	0.295	42
Cu(Zn)	250°-500°C	10 h. (3.60×10^4 s)	F	alloying	0.061	28
	150°-450°C	2.78×10^3 h. (1.00×10^7 s)	B	alloying		51
	400°C	96 h. (3.46×10^5 s)	F	alloying		52
	200°-550°C	1.25×10^3 h. (4.49×10^6 s)	B	dealloying		53

Fe(Zn)	545°-600°C	4 h. (1.44×10^4 s)	TF,F	alloying and dealloying	0.054	7
	500°-600°C	20 h. (7.20×10^4 s)	F	alloying		54
	460°-650°C	245 h. (8.81×10^5 s)	F	alloying		55
	550°-630°C	48 h. (1.73×10^5 s)	F,B	alloying		56,57
	600°C	–	F	alloying		58
Mo(Ni)	1250°-1400°C	2 h. (7.20×10^3 s)	P	alloying	−0.053	16
	800°-1290°C	20 h. (7.20×10^4 s)	F	alloying		18,59
	1380°-1510°C	8 h. (2.88×10^4 s)	PC	alloying and dealloying		19,60
Ni(Cu)	350°-900°C	6 h. (2.16×10^4 s)	B	alloying	0.023	49,50
Ni(Fe)	550°C	384 h. (1.38×10^6 s)	F	alloying	0.033	61
Ni(Pd)	400°-500°C	2 h. (7.20×10^3 s)	TF	alloying	0.134	11
Pd(Cu)	Room	4.67 h. (1.68×10^4 s)	TF	alloying	−0.071	62
W(Cr)	1250°-1500°C	96 h. (3.46×10^5 s)	W	alloying	−0.070	5
W(Pd)	200°-650°C	16 h. (5.76×10^4 s)	TF	dealloying		63
Non-metallic systems						
$CaCO_3$(Sr)	650°-800°C	192 h. (6.91×10^5 s)	BC	alloying		64
$CaCO_3$(Ba)	650°-800°C	192 h. (6.91×10^5 s)	BC	alloying		64
Ge(Al)	500°C	–	TF	alloying		65
Ge(Au)	600°C	–	TF	alloying		65
Ge(Cu)	150°-400°C	–	TF	alloying		65
Ge(Sn)	440°-600°C	–	TF	alloying		65

Table 1(b). Solvent (Solute) Systems exhibiting Liquid Film Migration

System	Temperature Range	Maximum Time[a]	Sample Geometry[b]	Alloying or Dealloying	η^c	Ref.
Mo(Ni)	1345°,1400°C	2 h. (7.20×10^3 s)	LPS	alloying	−0.054	16
	1390°-1510°C	8 h. (2.88×10^4 s)	LPS	alloying and dealloying		19,60
W(Ni)	1670°C	6 h. (2.16×10^4 s)	LPS	alloying	−0.091	15
W(Ni,Fe)	1550°C	1 h. (3.60×10^3 s)	LPS	alloying		20
$MgO\text{-}V_2O_5$(NiO)	1300°C	52 m (3.12×10^3 s)	LPS	alloying	$-4.\times10^{-3}$	67
$MgO\text{-}V_2O_5$(CoO)		20 m (1.20×10^3 s)	LP	alloying	0.011	67

[a] Maximum annealing time reported. – indicates no specific time cited in reference.
[b] Sample geometries include: TF – thin films; W – bamboo wire; B – bulk samples; BC – bicrystal; P – sintered powders; PC – particle-particle contact forming grain boundary in liquid-containing system; LPS – liquid-solid samples prepared by liquid phase sintering.
[c] $\eta = (1/a_o)(da/dC)\times100\%$, where a is the lattice parameter and C is the composition assuming that a is linear in C. Lattice parameter measurements from Ref. (24). Most systems listed here deviate significantly from this relationship. Therefore, these values are only rough estimates of the change in lattice parameter with composition.

ranges of temperature and time over which DIGM occurs are frequently designated by the Harrison diffusion categories (69): Type-B kinetics (grain boundary and lattice diffusion) and Type-C kinetics (grain boundary diffusion only). These diffusion regimes where DIGM is observed are illustrated by the examples of lattice penetration depths expected in systems showing DIGM in Table 2.

Table 2. Approximations of penetration distance indicating the contribution of lattice diffusion to the mixing process based on solute lattice diffusion data (70) extrapolated from higher temperatures.

Solvent (Solute)	Temperature(°C)	Time(s)	Penetration Depth by Lattice Diffusion(m)	Temperature Range(°C)	Boundary Velocities (m/s)	Ref.
Cu(Ni)	550	1.8×10^{3}	2×10^{-8}	550	1×10^{-7}	43
	900	3.6×10^{3}	5×10^{-6}	900	5×10^{-9}	50
Ni(Cu)	550	7.2×10^{2}	2×10^{-9}	550	1×10^{-7}	43
	900	3.6×10^{3}	8×10^{-6}	900	5×10^{-9}	50
Au(Ag)	150	2.4×10^{3}	5×10^{-12}	150	8×10^{-14} 6×10^{-12}	32
	200	1.4×10^{6}	2×10^{-9}	200	-	34
Ag(Au)	150	2.4×10^{3}	6×10^{-17}	150	-	4
	200	1.4×10^{6}	7×10^{-11}	200	-	25,26

DIGM is a widespread phenomenon in thin films as evident from Table 1. Most experiments in which DIGM has been confirmed have been performed on thin films. In the following sections, the physical and chemical characteristics of DIGM in metals, ceramics, and semiconductors are described, using results from both thin film and bulk systems.

Grain Morphology of DIGM

As shown schematically in Figure 2, there are several morphological characteristics which are common to all systems exhibiting DIGM. The original positions of the grain boundaries are indicated by the dotted lines. As a result of exposure to a solute source, either through the vapor phase or by contact with a liquid or solid, solute diffuses along the grain boundaries and the grain boundaries migrate to the new positions indicated by the solid lines. A solid solution has formed in the regions between the initial and final positions of the grain boundaries. The morphology of the grain boundaries is often referred to as "S-shaped" or "sinusoidal". During normal grain growth (grain boundary migration as a result of capillary forces), grain boundary migration occurs only towards the center of curvature of the boundary. In contrast to the normal grain growth, boundaries move by DIGM into both adjacent grains. Total grain boundary area often is increased and migration is frequently observed to occur away from the center of curvature.

The extent of alloy formation as a function of the distance from the surface (or from the solute source) is illustrated in perpendicular sections through Fe foils alloyed with Zn, Figures 3(b) and 3(c). At low temperature boundaries migrate through the entire film thickness. As the temperature is increased boundaries migrate only near the free surfaces and DIGM becomes a "surface effect" (54,55).

Crystallographic Orientation of Alloyed Regions

The alloyed regions have been reported to form epitaxially on one of the two adjacent grains as indicated by the letters A, B, and C in Figure 2b (7,32,33,52). When the change in lattice parameter with composition is small, such as in the case of Al(Zn) (29) and Mo(Ni) (18,59,73), the strain due to the change in lattice parameter with composition is accommodated elastically and no misfit dislocations are observed at the position of the original grain boundary. When the change in lattice parameter with composition is large, as in Fe(Zn) and Au(Cu), diffuse arrays of misfit dislocations have been observed at the initial position of the boundary.

Table 3. Critical Layer Thickness, L_c, for Coherency Loss based on Equation from Ref. (72) compared with Observations of Dislocations at Initial Boundary Position

Compositions of Layer/Substrate (Reference)	L_c(m)	Dislocations ?
98.04% Mo, 1.96% Ni/98.25% Mo, 1.75%Ni (19)	1.6×10^{-5}	-
99.15% Mo, 0.85% Ni/Mo (59)	1.4×10^{-6}	no (18,59)
95% Cu, 5% Ni/Cu (49)	2.1×10^{-8}	-
92% Cu, 8% Zn/90% Cu, 10% Zn (53)	7.1×10^{-8}	-
93% Fe, 7% Zn/Fe (7)	1.7×10^{-8}	yes (7)
75% Au, 25% Cu/Au (32,33)	2.4×10^{-9}	yes (32,33)

The growth of the alloy at the original grain boundary position can be modeled as the epitaxial growth of a film with a different lattice parameter than the substrate. When the strain energy of an epitaxial layer with a different lattice parameter from the substrate is greater than the energy required to form misfit dislocations, dislocations can form at the substrate-epitaxial layer interface. Calculations based on the model of Jesser and Kuhlmann-Wilsdorf (72) are presented in Table 3 for comparison with observations of misfit dislocations at the initial boundary position. These observations of alloyed region thickness and misfit dislocations at the initial position of the interface are in good qualitative agreement with these calculations. Isolated dislocations in the growing grains in Au(Cu) (32,33) and Fe(Zn) (7) are also seen lying in a plane perpendicular to the moving interface. In Figure 5, dislocations in regions of Au-Cu alloy in a Au thin foil appear to extend from the wall of misfit dislocations at the initial boundary position to the migrating high angle grain boundary (32,33). It is not known whether these isolated dislocations are simply extensions of the

Fig. 5. Dislocations extend from original position of grain boundary to the final boundary position in alloyed regions of Au(Cu) [111] tilt boundary annealed at 150°C. (Courtesy of J. D. Pan.)

wall of misfit dislocations or are formed by a more complex process during migration (7,32).

From the results of an annealing experiment by Grovenor (34) on thin Au(Cu) foils conducted in-situ by use of a TEM hot stage, the crystallographic relationship between the original grain and the alloyed region is not necessarily as simple as indicated by Figure 2 and Table 3. Arrays of dislocations were left behind at the original boundary position during the migration of 32° and 43.5° (001) tilt boundaries. However, the boundaries were observed to move in discrete jumps with a low angle grain boundary left behind at each intermediate position of the grain boundaries. In contrast to the expected array of misfit dislocations, the dislocation arrays in Au(Cu) produce a substantial misorientation between the grains. Intermediate boundary positions as indicated by surface relief and etching contrast have also been observed in bulk samples of Cu(Zn) and Fe(Zn) (53,54). Small misorientations between the original grain and the alloyed region in Cu(Zn) also indicate that the array of dislocations separating these regions is not composed entirely of misfit dislocations (74).

After prolonged annealing, the morphology deviates from the simple "S-shape" shown in Figures 2 and 3. The segments with the same orientation coalesce and continue to migrate forming a grain structure with orientations A-B-A-B as shown schematically in Figure 6. The coalescence of grains is shown in a Fe-Zn microstructure in Figure 7. In Fe(Zn) at 550-600°C migration proceeds until the surface is completely covered with an Fe(Zn) alloy and the grain size has been reduced by a combination of DIGM and the nucleation of new grains on the sample surface (54,55). In other systems (28) and in Fe(Zn) (55) at other temperatures, migration stops before the entire surface has alloyed. After further annealing, boundaries sometimes are seen to reverse the direction of motion and move towards rather than away from their centers of curvature, migrating through the alloyed

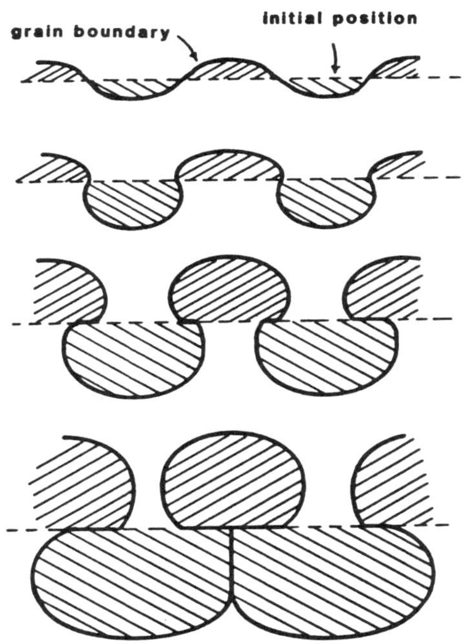

Fig. 6. Schematic of different stages in coalescence of migrating grains during DIGM. Grains of the same orientation have identical cross-hatching.

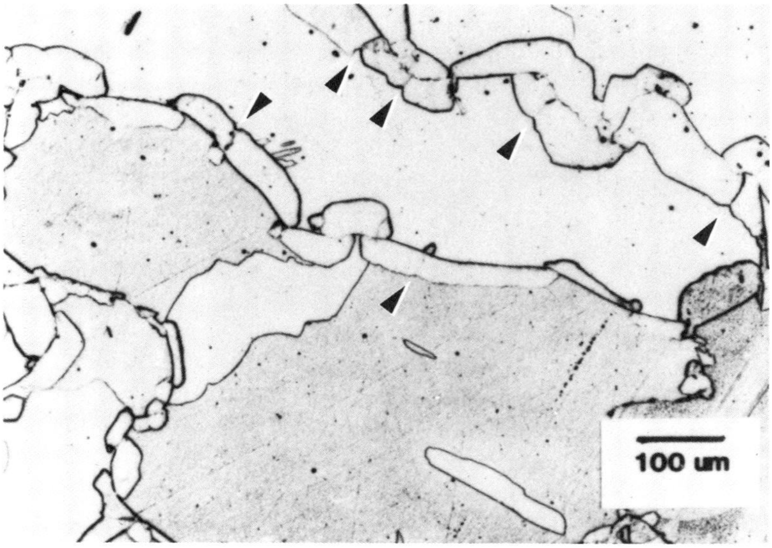

Fig. 7. Grain coalescence in Fe(Zn). Arrows indicate junction between two grains migrating in the same direction. Source: 90% Fe, 10% Zn; 550°C; 20 h.

regions back to the position of the original interface (55,58). These changes in migration direction are referred to as "reversals".

Facetting has been reported for some grain boundaries undergoing DIGM in Au(Cu) (32,33), Mo(Ni) (18,59,73), and Cu(Zn) (74,75,76). There is no apparent correlation between grain boundary facetting and grain misorientation or migration velocity (32,33).

Surface Relief

Surface relief has frequently been observed to correspond to the alloyed regions. Whiskers and hillocks are formed on the surface when Zn supplied through the vapor phase diffuses into Cu or Fe (8,52,58,65). The amount of surface relief depends on several factors, including the difference in the diffusivities between the solute and the matrix atoms (76), the change in lattice parameter with composition, and the amount of expansion that can occur in the plane of the sample. For thin bicrystals where the boundary remains planar through the entire film thickness during migration, the change in the volume caused by DIGM can be accommodated by uniform contraction or expansion in the plane of the foil. Because of the complex grain geometry of polycrystalline films, non-uniform contraction or expansion leads to changes in stress in the film. For semi-infinite samples where DIGM occurs only at the surface, the unalloyed material constrains the sample dimensions and the change in volume due to alloy formation must be accommodated by surface relief and by elastic and plastic strains. The amount of surface relief accompanying DIGM in massive $CaCO_3$ bicrystals has been measured by interferometry (65). The amount of material above the foil surface is found to correspond to the volume of Ba or Sr solute incorporated into the region swept out by DIGM.

Kinetics of Migration

As shown in Figure 3, the migration rate is not uniform from boundary to boundary. There are broad distributions in lengths of the boundary segments and in the maximum migration distances of the segments (32). Since grain boundary diffusivity and mobility depend strongly on the misorientation of the grain boundary, DIGM also depends on misorientation. Low angle grain boundaries with misorientation angles less than about 15° in Au(Cu) (32-34), Au(Ag) (32,33), and Cu(Zn) (76) do not show DIGM. The maximum migration distance in [111] tilt boundaries in Au(Cu) (32-34) and [001] tilt boundaries in Cu(Zn) (76) increased with increasing misorientation. Special boundaries, such as the Σ = 5, 17, 25, and 29 symmetrical (001) tilt boundaries in Cu(Zn), show low migration rates relative to random high angle grain boundaries (76).

Incubation times have sometimes been reported (28,54,55). The migration rates for boundaries with similar misorientations in the same specimen are significantly different (32). The distribution in migration velocities spans a much broader range for different experimental studies performed on the same system. Velocities measured in the Au(Cu) system at 150°C vary from 8×10^{-6}m/s (34) to 8×10^{-13} m/s (32,33). Migration rates computed from an averaging of the migration distances, from the surface area of the alloyed regions, or from the maximum migration distance, do not follow a simple kinetic law (6,28,54,55). The migration velocity is approximately constant at short times (Fe(Zn) (54), AgPd(Pd) (6)) and decreases with longer diffusion time (Au(Cu) (32), Mo(Ni) (18,59)). Typical migration velocities applying at short times are listed in Table 2. Migration frequently ceases before the entire sample is alloyed by DIGM (28).

A major experimental problem in determining the kinetics of DIGM is the establishment of a constant source throughout the duration of the experiment. This is a particularly serious problem when a polycrystalline film is annealed with a solid solution source material which supplies the solute through the vapor phase. While the polycrystalline film is undergoing DIGM by alloying with a solute from the vapor phase, the source material must dealloy by DIGM to supply solute to the film. The changing rate of DIGM in the source will, thus, affect the rate of DIGM in the film.

Source problems can also occur in bilayer couples. In most interdiffusion experiments, one layer contracts while the other layer expands during interdiffusion and DIGM. As a result of the stress due to differential volume changes, the bilayer could possibly delaminate at the interface. Solute depletion and void formation could also occur at the intersection of grain boundaries with the source film. If the solute must then be supplied to the grain boundary by surface diffusion, slow surface diffusion of the solute can limit migration. Migration in Au(Cu) where the solute is deposited as discrete islands on the sample surface is faster in the regions in contact with the islands than in the regions between the islands (34).

The sample geometry should also be taken into account when the results from different studies are compared. The physical models of diffusion are different for thin films than for semi-infinite solids. For a semi-infinite solid containing a planar boundary, diffusion from the free surface is described by Whipple's original equations (2). Penetration from both sides of a thin foil will lead to overlapping concentration profiles and higher concentrations at any position in the boundary than for diffusion from a single direction. When solute diffusion occurs from a single surface in thin films the solute can either exit the other side of the film and spread by surface diffusion or be reflected at the other surface to produce overlapping diffusion fields (2). Since DIGM depends on the concentration of solute in the grain boundary, the diffusion geometry affects the kinetics of DIGM.

The migration of grain boundaries by DIGM has many similarities to the migration of grain boundaries during discontinuous precipitation (7,77). In discontinuous precipitation, a phase, α_1, which had been stable at a different temperature, decomposes into two phases, α_2 and β. α_1 and α_2 are the same phase but of different compositions. The α_2 and β phases form heterogeneously as parallel lamellae at a grain boundary of the initial solid solution (78). The interface between the α_1 and α_2 is a grain boundary which moves into the parent grain, α_1, and the grain boundary serves as the diffusion path for the redistribution of the solute from the α_1 to the α_2 and β phases. The similarities between DIGM and discontinuous precipitation are especially pronounced at the early stages of discontinuous precipitation with a high driving force (7) and throughout the complete process at low driving forces (77).

The motion of the grain boundary joining two lamellae of β has been modelled by Cahn (78) based on the assumptions that the sideways transport to the β phase is by grain boundary diffusion and that the velocity is proportional to the free energy decrease due to precipitation. The analysis of the transport along the grain boundary has been applied to DIGM in thin foils when the grain boundary is perpendicular to the foil surfaces. Both surfaces of the foil are exposed to the same solute source, and migration occurs through the entire foil thickness. Such a microstructure is seen in Figure 3(b) for Fe(Zn). In the case of DIGM in Fe(Zn), the free surfaces in contact with the solute source are parallel solute sources whereas the β-lamellae in discontinuous precipitation serve as sinks for the solute. The relationship between composition, grain boundary diffusion, and migration velocity, v, is of the following form for a planar interface:

$$\frac{x - x_1}{x_0 - x_1} = \frac{\cosh \frac{z\sqrt{a}}{S}}{\cosh \frac{\sqrt{a}}{2}} \qquad (1)$$

where $a = vS^2/KD_b\delta$, x_1 is the solute content of the parent grain, x_0 is the solute content of the growing lamellae at the sides of the grain boundary, z is the coordinate in the direction of diffusion in the boundary, S is the spacing between lamellae, K is the equilibrium distribution coefficient of the solute between the boundary and the growing grain, D_b is the grain boundary diffusion coefficient and δ is the width of the grain boundary. Hillert and Purdy used this equation to fit the experimentally observed composition profile along a boundary in Fe(Zn) at 580°C when $S = 4.5\times10^{-5}$m, $x_o = 0.06$, and $x_1 = 0$, as shown in Figure 8. The parameter a which best fits these data was $a = 4.4$. Using the experimentally determined migration rate, $v = 7\times10^{-10}$ m/s, the a value yields $KD_b\delta = 3.2\times10^{-19}m^3/s$. It was noted that this value is four orders of magnitude larger than the grain boundary diffusivity observed in diffusion studies.

Compositions of Alloyed Regions

To gain insight into the relationship between alloy composition left behind the moving grain boundary and migration velocity, it is important to know the alloy composition as a function of position and time in the regions alloyed by DIGM. This complete set of informa-

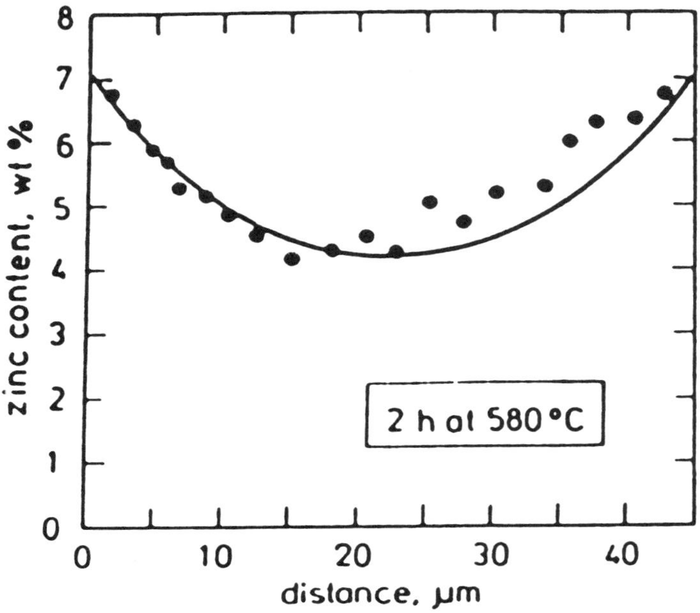

Fig. 8. Experimental concentration profile of Zn from one side of a thin foil to the other compared with concentration profile calculated from Eq. (1). Source: 11.2 wt.%Zn; 580°C; 3 h. Reprinted with permission of Acta Metallurgica, Vol. 26, Mats Hillert and Gary R. Purdy, "Chemically Induced Grain Boundary Motion," 1978, Pergamon Press.

tion – composition, position, and velocity – has not been determined for any system. Limited information on compositions in alloyed regions has been obtained from techniques which average over the entire specimen and from energy dispersive x-ray spectrometry, scanning transmission electron microscopy or wavelength dispersive x-ray spectrometry, which average over a small interaction volume.

In thin film couples, two techniques which average over many grains, Rutherford backscattering and glancing-angle Seeman-Bohlin x-ray diffraction, have shown that the compositions of the alloyed regions are uniform through the film thickness. The x-ray diffraction studies of Ag(Pd) (6), Cu(Au) (36), and Au(Cu) (36) show the gradual decrease with time in intensity of the 200 peak for the solute film accompanied by the appearance of a sharp 200 peak corresponding to an alloy with a well-defined solute concentration. For diffusion of Pd from a AgPd alloy, the AgPd 200 reflection decreased in intensity with time while a 200 reflection corresponding to pure Ag appeared. Since these techniques provide average composition values over many grains, they do not provide information of sufficient detail to test theory. Data with better spatial resolution is required to analyze the relationship among kinetics, boundary properties, and composition in and near the alloyed region.

Early attempts to measure the composition as a function of distance from the initial boundary position consisted of step scans using STEM or electron microprobe in a direction normal to the initial boundary plane. These experiments in Au(Cu) (32,33), Au(Ag) (32,33), and Cu(Zn) (51) indicated that the composition in the alloyed regions is far from uniform and that the profile does not change in a systematic way from the initial position to the final boundary position. Recent developments in image processing of electron microprobe data (79,80,81) allow quantitative analyses of entire boundaries. From these digitized concentration maps, concentration profiles across any region of interest can be easily obtained. A map of Zn concentration, Figure 9(b), corresponds to the boundary in Figure 9(a). The line drawn perpendicular to the grain boundary in Figure 9(b) is the location of the concentration profile shown in the lower left-hand corner. The concentration maps of Fe(Zn) alloyed at 550° and 600°C indicate that the profiles measured on surfaces perpendicular to the free surface are almost flat except for small Zn-rich precipitates occurring at the initial position of the grain boundary (82). Precipitates identified as Fe-Zn carbides have previously been observed on the surfaces of Fe(Zn) undergoing DIGM at these temperatures (55). The approximately flat concentration profiles in Fe(Zn) are consistent with the kinetics and morphology of DIGM at these temperatures. There is a short incubation time for the start of DIGM and DIR and migration occurs until the entire surface is covered by the Fe(Zn) alloy. The velocities and composition profiles in Fe(Zn) at these temperatures are the most uniform that have been observed in a DIGM system. In contrast, composition profiles in Cu(Zn) show concentrations varying from background levels to as high as 4.5 wt.% Zn in a single DIGM region (83).

An, as yet, unexplained complication in Cu(Zn) (51,53) and Cu(As) (84) is that oxygen is frequently detected throughout the entire specimen at concentrations higher than the solubility limit of oxygen in Cu. This result suggests that precipitation of an oxygen-containing phase has occurred during the diffusion anneal. Furthermore, precipitation seemed to be enhanced near grain boundaries where DIGM was occurring and the precipitates were larger in the regions swept by the grain boundaries than in the unswept matrix.

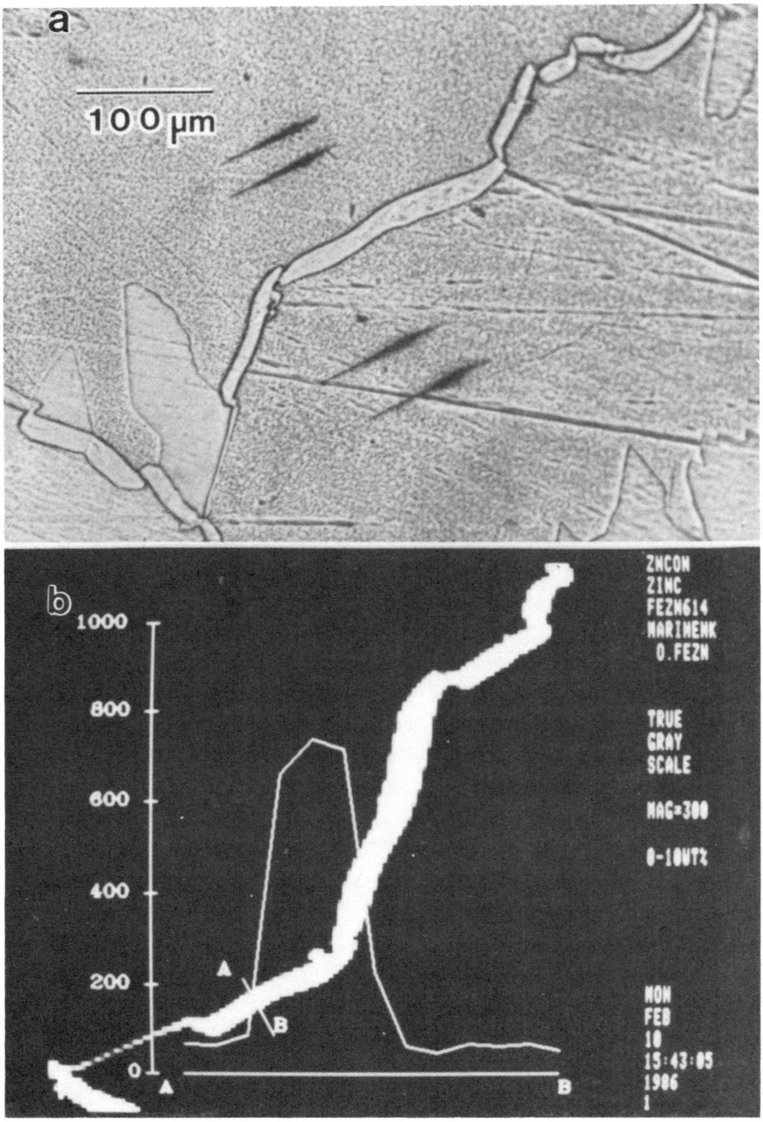

Fig. 9. (a) Microstructure of DIGM grains in Fe(Zn). Source: 90% Fe, 10% Zn; 550°C; 7 h. (b) Concentration map of Zn for region shown in (a) obtained by electron microprobe analysis. Concentration profile for line AB (82).

Nucleation of New Grains – Diffusion-Induced Recrystallization

When the source concentration is high and at the higher temperatures where DIGM occurs, solid solution grains have been observed to nucleate in the matrix grains. This process, known as diffusion-induced recrystallization (DIR), is not restricted to materials which are initially polycrystalline. New grains can nucleate during interdiffusion at the interface between single crystal epitaxial layers of different materials as well as at the surface of single crystals exposed to a vapor or liquid solute source. As with DIGM, migration of these high angle grain boundaries is accompanied by alloy formation. In addition, migration of these high angle grain boundaries through pre-existing dislocation arrays leads to the formation of almost defect-free grains and to a change in misorientation across the grain boundaries as a result of the absorption of these dislocations (44-46). The specific change in misorientation is dependent on the efficiency with which the boundaries absorb the dislocations. In Figure 10, a DIR grain in Cu-Ni (44) is migrating into the original epitaxial layer of Cu which contained a high dislocation density. The DIR grain is bent due to the

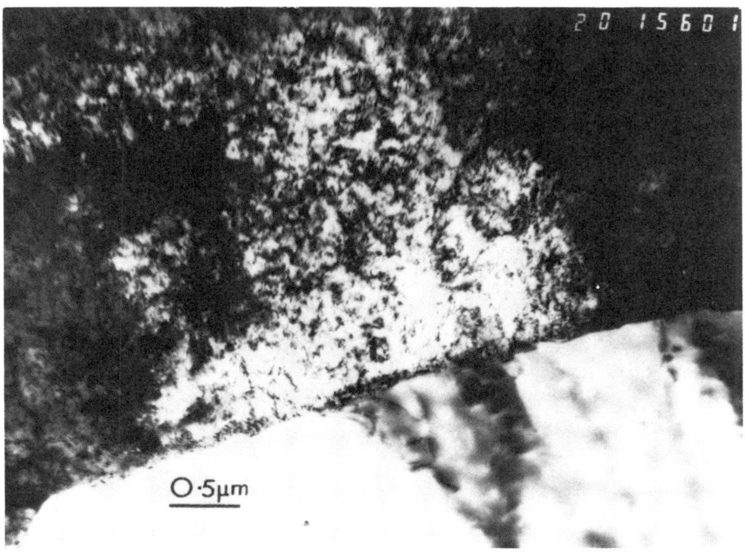

Fig. 10. TEM micrograph of new grain formed at the interface between Cu and Ni in a thin bilayer diffusion couple grows into Cu containing a high dislocation density. (Courtesy of C. R. M. Grovenor.)

change in composition, and hence, lattice parameter from the base to the periphery of the spherical cap. The dislocation content of the DIR grain is extremely low (10). As the misorientation of the grain boundary changes with the absorption of dislocations, the velocity of the boundary changes (44,45). Grains formed by DIR move in a jerky manner. Boundaries migrate quickly to a new position where they remain temporarily. When the boundaries begin to migrate again, a mobile boundary nucleates by a twinning process (44) from the original grain boundary leaving a low angle boundary at the intermediate position. An in-situ TEM study of DIR grains in polycrystalline Cu-Au bilayers shows similar grain morphologies, bend contours, twinning-velocity relationships, and dislocation density differences between the DIR grains and the matrix (34). In a region containing a high density of DIR grains, Grovenor observed that most of the grains had either ⟨110⟩ or ⟨100⟩ directions lying normal to the film plane when the original film orientation was ⟨100⟩. The orientations and morphologies of DIR grains in Au(Cu) are seen in Figure 11.

In Au-Pd bilayer films produced by physical vapor deposition using NaCl substrates (10), pores were observed to impede the motion of DIR boundaries. The pores were produced during the deposition and annealing processes before diffusion occurred. The pinning of boundaries by pores or particles has not been reported in any other investigation. It has also been reported that DIR grains do not occur in sandwich-like structures (A-B-A) although DIR grains form at the interface of A-B bilayers of the same materials (85). In polycrystalline specimens, the fine-grained surface layer usually advances into the interior of the material with a slightly higher rate along the initial grain boundaries (28,55).

At high temperatures, a new phenomenon occurs in the layer containing DIR grains. Li and Hillert (55) determined that a peculiar kind of abnormal grain growth occurs in Fe(Zn) at temperatures above 650°C with the large matrix grains growing back to the surface and completely consuming the fine-grained DIR structure. At temperatures above 720°C, this reaction was so rapid that the fine-grained surface was never observed. However, a non-uniform distribution of zinc indicated that there had been a fine-grained surface layer.

Grovenor and Smith (66) have postulated that DIGM also occurs in bilayer films of Cu/Ge, Sn/Ge, Al/Ge, and Au/Ge. Upon heating, rapid grain growth occurred in the regions in contact with a eutectic (Sn/Ge, Al/Ge, and Au/Ge) or an intermediate phase (Cu/Ge). It is possible that the process leading to this rapid grain growth is an initial reduction in grain size by DIR followed by normal grain growth. It is also possible that this grain coarsening occurred by the motion of a liquid film.

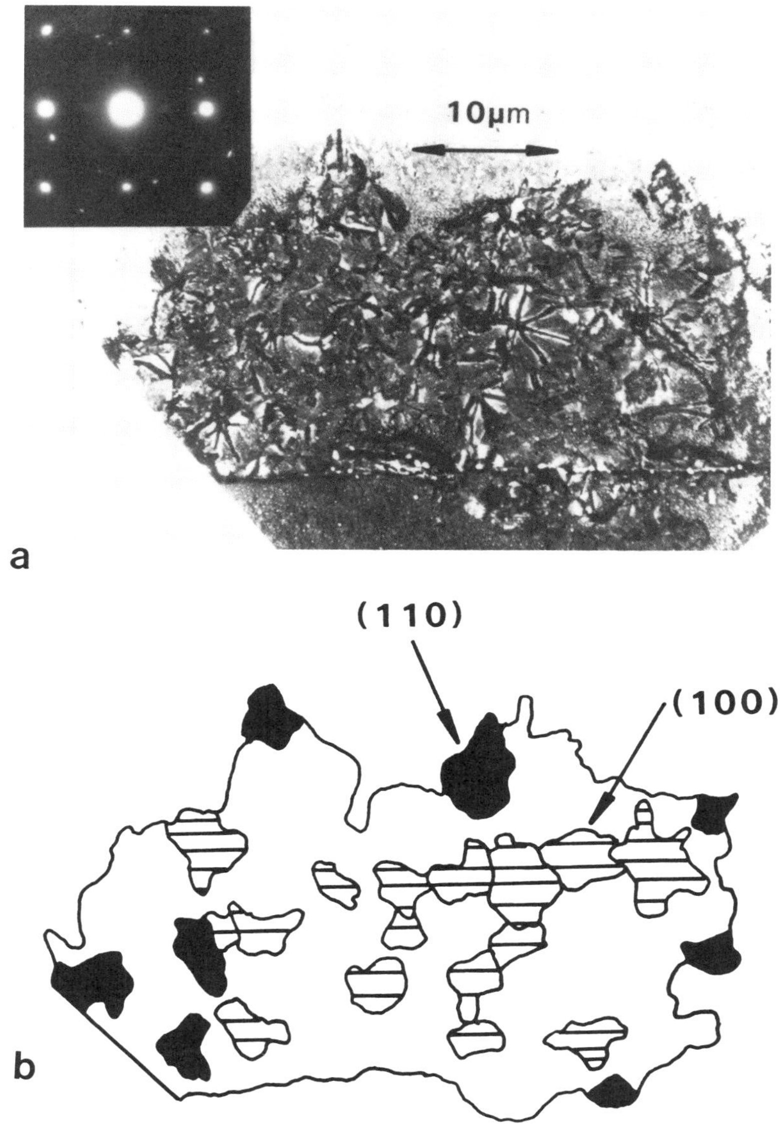

Fig. 11. (a) A micrograph showing a large group of DIR grains in Au(Cu) nucleated after a few minutes at 300°C. This is an example where the 18 degree tilt boundary plays no part in the nucleation process. (b) A schematic diagram of the DIR grains indicating the orientation of the DIR grains. The original orientation was (100). Reprinted with permission of Acta Metallurgica, Vol. 33, C. R. M. Grovenor, "Diffusion Induced Grain Boundary Migration in Thin Gold/Copper Films", 1985, Pergamon Press.

Liquid Film Migration

Migration and solid solution formation also occur when liquid films are present instead of grain boundaries. This liquid film analogue to DIGM is important in identifying general driving forces for DIGM since many proposed driving forces require that a grain boundary be present. A typical microstructure of an equilibrated, partially liquid Mo-Ni alloy formed by coarsening is shown in Figure 12(a). When the grain structure is subsequently annealed at a temperature different from the coarsening temperature, liquid film migration (LFM) occurs causing the formation of the Mo-Ni solid solution in equilibrium at the annealing temperature and an increase in the total interfacial area, Figure 12(b). Similar microstructures occur when a substitutional solute, such as W, Co, or Fe, is added to the system. The essential features which make LFM analogous to DIGM are present: interface motion against curvature and solid solution formation in the regions swept by the moving liquid films. This analogy between LFM and DIGM was noted previously by Song, Ahn, Baik, and Yoon (19,20) and by Hillert (23). Of greatest importance is that, when grain boundaries as well as liquid films are present in partially liquid Mo-Ni alloys, DIGM and LFM occur simultaneously, with the same morphologies, and at approximately the same rate. This result leads to the conclusion that the driving force for the two processes is the same.

The characteristics of systems which undergo LFM are listed in Table 1(b) and are similar to those listed for DIGM. As in DIGM, the solid solution regions formed under the new annealing conditions deposit epitaxially on the original grains and the alloyed regions have a very low dislocation density compared with the matrix. The lack of misfit dislocations at the interface (Table 3) between the original grain and the epitaxial deposit in Mo(Ni) is in good qualitative agreement with the calculated critical thickness for maintenance of coherency. In contrast to DIGM and DIR, the composition of the alloyed regions in Mo(Ni) is the equilibrium solid composition (19).

The discovery that LFM and DIGM occur in the same sample and at about the same rate indicates that a grain boundary is no longer a necessary feature of diffusion-induced interface migration. The common factors which remain from a comparison among DIGM, DIR, and LFM are (1) the systems are not at chemical equilibrium and diffusion and interface migration occur to establish equilibrium; (2) interface migration occurs away from the centers of curvature of the interfaces; (3) alloyed regions are formed in the regions swept by the moving interfaces; (4) the dislocation densities in the alloyed regions are low; and (5) lattice diffusion is slow relative to diffusion along short circuit paths at the temperatures and times where these effects are observed.

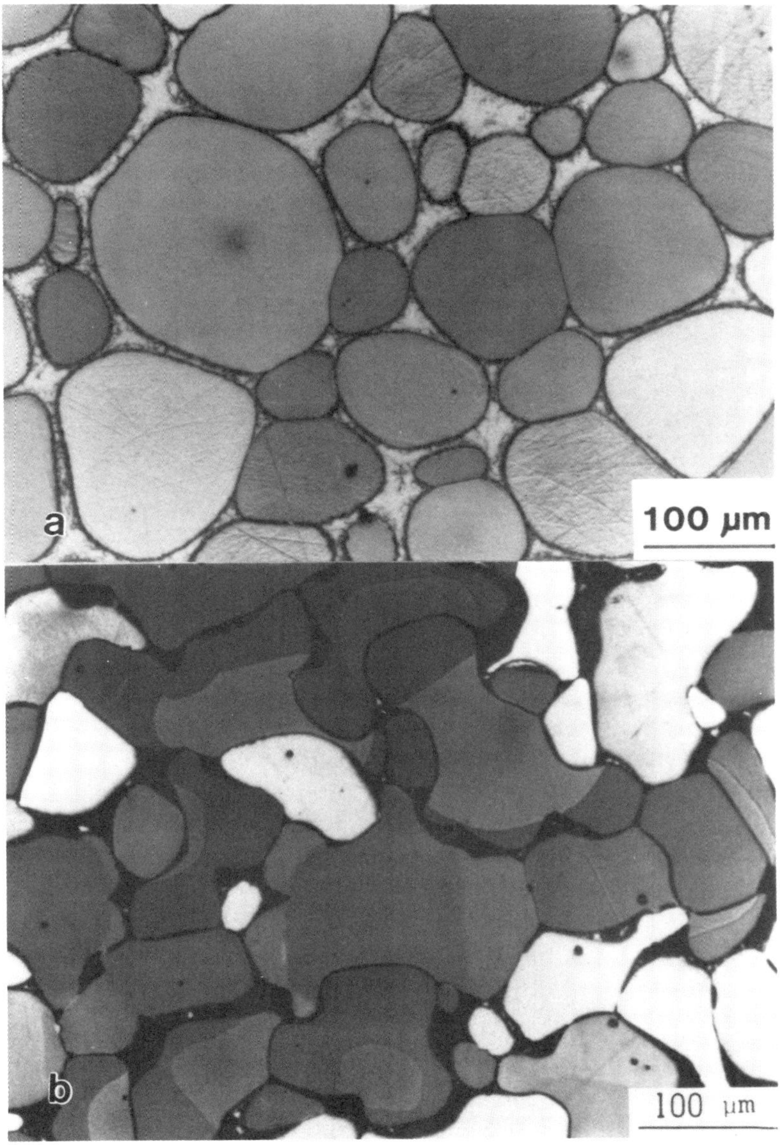

Fig. 12. Optical micrographs of 95%Mo-5%Ni alloys: (a) Microstructure produced by liquid phase sintering of fine powders; (b) Liquid film migration observed after annealing alloy in (a) at a lower temperature. (Courtesy of Y.-J. Baik and D. N. Yoon (60).)

3. DRIVING FORCE FOR GRAIN BOUNDARY/INTERFACE MOTION

Distinctions Between Free Energy Changes and Driving Forces

The key to the understanding of diffusion-induced phenomena, such as DIGM, DIR, and LFM, has been the identification of a driving force which couples boundary motion and diffusion. When solute diffusion occurs along a boundary at the same time that the boundary is moving, solute can be incorporated into the grains as the boundary moves, regardless of the driving force for boundary motion. In most cases, the motion of the boundary and diffusion of the solute are independent processes. For DIGM and LFM boundary motion is coupled to diffusion of the solute. A common misconception about grain boundary migration in general is that any free energy decrease resulting from boundary motion provides a driving force for the boundary to move (68,86,87). In the case of DIGM, a free energy decrease is a *necessary* condition for migration but it is not a *sufficient* condition to cause motion of the grain boundary. A force is required to drive interface motion. The distinction between free energy changes and driving forces are illustrated here using the examples of grain growth and isotope mixing. The proposed models of driving forces for DIGM and LFM are then analyzed as to whether they can lead to boundary migration under the experimental conditions where DIGM is observed.

Let us first consider a 2-D grain microstructure consisting of hexagonal grains of a single size, G_1, with an isotropic grain boundary energy, γ_g. Under these conditions, all grains have straight sides. If the microstructure were to coarsen to an array of straight-sided, hexagonal grains of $G_2 > G_1$, the grain boundary area would decrease with a concomitant decrease in the free energy of the system of $\gamma_g dA$. In grain growth the driving force is proportional to the curvature, κ, to some power. For the case of straight-sided grains, κ is zero and no migration can occur although the free energy of the system would decrease by a reduction in the grain size.

Balluffi and Cahn (8) used the case of mixing between ideal species to illustrate that a free energy decrease is necessary, but not sufficient, to cause boundary migration during diffusion. They considered the case of two species which are identical except for a difference in labelling. This could correspond to the cases of isotope self-diffusion or to interdiffusion in an ideal diffusion couple. During interdiffusion between two species that are ideal and differ only in labelling, mixing of the species occurs by random walk diffusion, that is, the motion of the species is identical and diffusion is identical to that in a homogeneous alloy where the atoms are unlabeled. Free energy, which in this case corresponds to the entropy of mixing, is dissipated by the diffusion process because the species are labelled. If

DIGM occurred, mixing of the two species would be enhanced and the rate of energy dissipation would be increased by the motion of the grain boundary. For diffusion to induce grain boundary motion during interdiffusion of ideal species, motion of the grain boundary would have to be the same in all experiments which have the same arrangement of atoms but have the atoms labelled differently. Furthermore, the same motion of the grain boundary would also have to occur in homogeneous alloys where no diffusion is occurring since ideal diffusion is purely by random walk. Motion of the grain boundaries in homogeneous alloys under the time and temperature conditions where DIGM is observed is due only to random thermal fluctuations ($\propto kT$). In this case of ideal mixing, free energy is dissipated by diffusion but diffusion cannot cause the boundaries to move.

Balluffi and Cahn examined some of the ways in which real systems differ from ideal systems to find ways in which diffusion could couple to grain boundary migration. The determined that the excess part of ΔG_m, the enthalpy of mixing ΔH_m, could not be the coupling between diffusion and migration. If ΔH_m were to cause boundary migration in some undefined way, it was reasoned that materials with ΔH_m of different signs should behave differently. A negative ΔH_m (ordering) would then be associated with alloying (mixing) and a positive ΔH_m (segregation of like species) with de-alloying (de-mixing). By examination of the systems which exhibit DIGM (Table 1), systems with positive ΔH_m (Fe-Zn) and with negative ΔH_m (Cu-Zn) are found which exhibit DIGM on both alloying and de-alloying (8). Therefore, neither ideal solid solution behavior, nor deviations from ideality, could cause DIGM even though free energy is dissipated by the interdiffusion process and by DIGM.

Diffusional Transport in Polycrystals with Migrating Grain Boundaries

When grain boundary diffusion of a solute and grain boundary migration are occurring simultaneously, grain boundary migration leads to a mixing of the solute into the lattice as the grain boundary migrates regardless of the driving force for migration (88). The driving force for migration (such as curvature or prior cold work) can be unaffected by the presence of the solute but the solute will be incorporated into the lattice making it appear that the migration is coupled to grain boundary diffusion when they are, in fact, independent processes.

The major theoretical question in DIGM has been why the boundary moves rather than why the solute is incorporated into the lattice behind the moving boundary. In many, but not all, DIGM experiments, curvature and prior cold work had been minimized by annealing the samples at high temperature before the diffusion anneal. All other driving forces which could lead to net boundary motion, such as temperature gradients, were also eliminated but the boundaries still moved in response to solute diffusion and sometimes against curvature.

Driving Force for DIGM, DIR, and LFM

The theories of the driving force for DIGM fall into four categories: (1) free energy of mixing (68,87,89,); (2) energy changes of the grain boundary with solute additions (55,90,91); (3) grain boundary Kirkendall effect leading to the climb of grain boundary dislocations (8,92,93); and (4) the coherency strain due to lattice diffusion ahead of the moving interface (23,28,94,95). As discussed above, the free energy decrease due to intermixing is not a driving force for boundary migration. All of the other driving forces postulated for DIGM can lead to interface motion under some circumstances. As we describe below, experimental evidence rules out many of these as general driving forces for DIGM.

Changes in the energy of the grain boundary with solute additions can lead to migration of the interface. However, the theories based on this concept require that the composition be discontinuous across migrating interface and hence, that there be no diffusion from the grain boundary into the adjacent lattice. Significant lattice diffusion ahead of the migrating interfaces in Al(Zn) (29) and Cu(Ni) (49) removes the effect of grain boundary energy changes with solute additions from consideration as a general law for DIGM. Furthermore, the grain boundary energy theories require that there be a grain boundary for the effect to occur. The theories based on a grain boundary Kirkendall effect also require that the migrating interface must be a grain boundary. Liquid film migration precludes the need for grain boundary structures or unequal diffusivities of the solute and solvent in the grain boundary. The coherency strain mechanism is a driving force which is qualitatively consistent with most, but not all, experimental observations.

If the lattice parameter is a function of composition, most compositional inhomogeneities lead to a stress field. When a polycrystalline film is exposed to a solute source, grain boundary diffusion and some lattice diffusion of the solute occur. The regions in which lattice diffusion has occurred adjacent to the grain boundary will be stressed if the lattice parameter changes with composition. When the diffusion layer is thin compared to the principal radii of curvature of the surface of the solid, the stress depends only on the difference in lattice parameter between the composition produced by lattice diffusion, $\tilde{C}^s$, and the initial composition C_o, and for elastic anisotropy, the surface normal, $\hat{n}$. This situation is shown schematically in Figure 13 for two grains separated by a liquid film. The two grains of composition, C_o, are initially in equilibrium with the surrounding liquid of composition, C^ℓ. When a solute which causes a change in lattice parameter is added to the system, a stressed diffusion layer with a surface composition of $\tilde{C}^s$ is formed. If the stresses in the adjacent grains are different, such

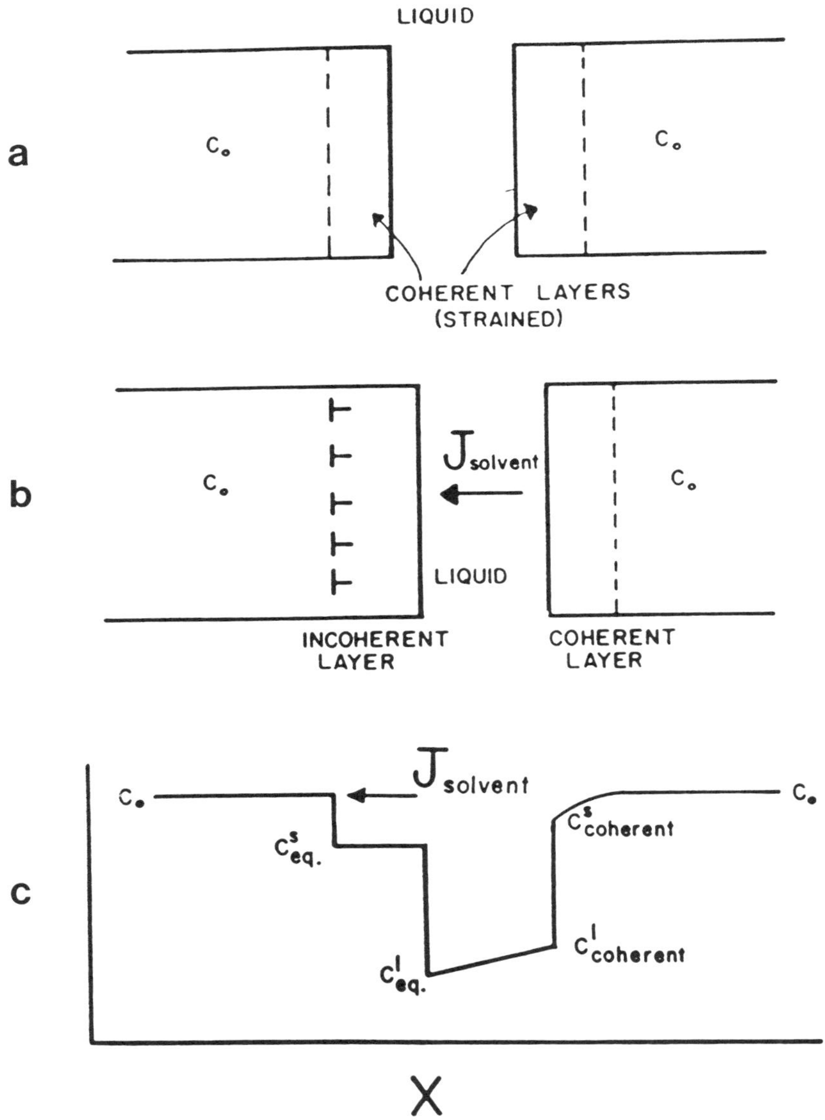

Fig. 13. Schematic of LFM with coherency loss occurring in one grain. (a) Diffusion of a solute from the liquid into the solid leads to the formation of a strained layer at the surface. (b) Stress relaxation occurs in one grain. The solubility of the stressed solid in the liquid is different from the solubility of the unstressed solid. (c) Concentration profile arising from this solubility difference.

that the liquids in equilibrium with these two grains are of different compositions, a concentration gradient will exist in the liquid film separating the grains. This concentration difference across the liquid film leads to diffusion across the film according to Fick's first law. One grain dissolves and the other grain grows as a result of this concentration difference. It has been demonstrated that dissolution and growth under these conditions can be formulated as a boundary value problem for diffusion across the liquid film. The boundary conditions for the diffusion equations are calculated from the thermodynamic theory of the solubility of stressed solids and the velocity is proportional to the concentration gradient (21-23).

Stresses also result from lattice diffusion when the liquid film is replaced by a grain boundary. As a result of the differences in stress across the grain boundary, the grain boundary can move. The theory for DIGM based on the coherency strain mechanism is similar to, but less tractable than the LFM case, as discussed below. Although an exact kinetic expression does not exist for DIGM, we use the expressions derived for liquid films to predict various qualitative features of DIGM and to compare these predictions with experimental observations.

Additional driving forces for migration occur when lattice diffusion produces significant diffusion ahead of the grain boundary and in the surface adjacent to the solute source. In all real systems, the diffusivities of two chemical species are not equal. This unequal lattice diffusion of the solute and the solvent atoms leads to the formation of dislocations in the regions where diffusion occurs. This effect is known as the "Kirkendall effect". The strain energy associated with these dislocations can be orders of magnitude greater than the strain energy associated with an elastically stressed coherent diffusion layer. Migration of pre-existing grain boundaries and nucleation and growth of new dislocation-free grains is identical to the growth of dislocation-free grains in a matrix deformed as a result of an external stress. In this case, however, the deformation is diffusion-induced. In the last section, a simple mechanism map is presented which describes the conditions for initiation and continued migration of grain boundaries driven by coherency strain and by diffusion-induced deformation.

4. EFFECT OF COHERENCY STRAIN ON INTERFACE MIGRATION

The concept of coherency strains affecting thermodynamic equilibria is not new. It has been suggested that coherency strains affect discontinuous precipitation (cellular precipitation) (94-96), spinodal decomposition (97-99), solute redistribution during surface melting and regrowth (100), nucleation and growth (101-103), massive transformations (104), and late stage phase separation processes, such as, Ostwald ripening (105). In the case of DIGM and LFM, coherency strains develop as a result of lattice diffusion adjacent to the grain boundary in DIGM and in the surface layer of the solid in LFM (23). The presence of coherency strains creates conditions of constrained equilibrium at the interface. When two solid surfaces are separated by a liquid film (LFM), the compositions at the solid-liquid interfaces become the boundary conditions for the one-dimensional problem of diffusion through a liquid layer. When the coherency strain energies in two adjacent grains are different, a concentration gradient will exist in the liquid. Diffusion will occur across the liquid film and the film will migrate by dissolution of one grain and reprecipitation onto the other grain. The velocity of the liquid film resulting from the dissolution and reprecipitation process is:

$$V = \frac{D_{liq} V_m}{C^{\ell} - C^{s}} \frac{\Delta \tilde{C}^{\ell}}{\lambda} \tag{2}$$

where D_{liq} is the diffusivity through the liquid film, V_m is the molar volume, $\Delta \tilde{C}^{\ell}$ is the difference in equilibrium liquid composition across the liquid film of thickness λ, and the term $(C^{\ell} - C^{s})$ accounts for conservation of mass at the interface. This mechanism, suggested by Hillert in 1983 (23), directly couples diffusion of the solute to migration of the interface.

Hillert suggested that the model for migration of liquid films can be used quantitatively for DIGM if a grain boundary can be described as a thin film of another phase constrained to a fixed, small, but non-zero, thickness. As we discuss below, shrinking the liquid film to grain boundary thickness is not sufficient to model the thermodynamics and the kinetics of DIGM.

Model for Coherency Strain Energy

Cahn (97) has calculated the strain energy for a coherent surface layer of composition C^s, and lattice parameter, a, on a solid of composition, C_o, and lattice parameter, a_o, where $a \neq a_o$. This was done by calculating the work require to elastically deform a slice of material so it can be added coherently to a solid with a different lattice parameter. If the composition change is normal to the surface (the x-direction), the surface layer must be subjected to a strain of:

$$\delta = (a - a_o)/a_o \tag{3}$$

in the plane of the surface (the y and z directions). The deformation required to produce this deformation is done in two steps: first, the surface layer is deformed hydrostatically to produce the required strains in the y and z directions and, second, the stress in the x direction is allowed to relax reversibly while the sides of the surface layer parallel to the x direction are clamped. This deformation is, thus, a plane strain. For cubic crystals, the coherency strain energy is:

$$f_{coherent} = Y(\hat{n})\delta^2 \tag{4}$$

where $Y(\hat{n})$ is an orientation-dependent elastic modulus and δ is the strain. For elastic isotropy,

$$Y = E/(1-\nu) \tag{5}$$

where E is Young's modulus and ν is Poisson's ratio. If stress relaxation occurs, the magnitude of δ will be reduced. In particular, δ equals zero if the diffusion layer is completely unstressed because of plastic flow, because the grain is compositionally homogeneous, or because the lattice parameter does not vary with composition. When the change in lattice parameter with composition is constant (Vegard's law),

$$f_{coherent} = Y(\hat{n})\eta^2(C^s - C_o)^2 \tag{6}$$

where $\eta = (da/dC)/a(C_o)$.

The variation of Y with orientation can be large, even for cubic crystals. Hilliard (99) derived a general expression for $Y(\hat{n})$ in cubic crystals:

$$Y(\hat{n}) = 1/2(c_{11} + 2c_{12})\left\{3 - \frac{(c_{11}+c_{12})}{(c_{12}+c_{44})} \times \left[\frac{3(\alpha-1)^2\theta + 2\beta(\alpha-1)\phi + \beta^2}{(\alpha-1)^2(\alpha+2)\theta + \beta(\alpha^2-1)\phi + \beta^2(\alpha+\beta)}\right]\right\} \tag{7}$$

where $\theta = l^2m^2n^2$, $\phi = l^2m^2 + m^2n^2 + l^2n^2$, $\alpha = (c_{11} - c_{44})/(c_{12} + c_{44})$, $\beta = c_{44}/(c_{12} + c_{44})$, and l, m, and n are the directional cosines of $\hat{n}$. The directions of the maxima in Y in cubic systems are $\langle 100 \rangle$ when $2c_{44}/(c_{11} - c_{12}) < 1$ and $\langle 111 \rangle$ when $2c_{44}/(c_{11} - c_{12}) > 1$. The values of Y for low index directions are listed in Table 4 for some cubic crystals which undergo DIGM or LFM.

Table 4. Orientation-dependent Elastic Coefficients $Y(\hat{n})$ for Selected Cubic Materials using Data from Reference (114)

	$Y(\hat{n})$ (10^{11} Pa)			
	Y(100)	Y(111)	Y(111)/Y(100)	$2c_{44}/(c_{11}-c_{12})$
Au	0.81	1.92	2.38	2.88
Cu	1.15	2.62	2.27	3.19
Fe	2.06	3.51	1.71	2.34
GaAs	2.04	1.69	0.83	0.93
KCl	0.38	0.32	0.84	0.37
MgO	3.23	3.98	1.23	1.54
Mo	4.59	3.95	0.86	0.78
Si	1.82	2.13	1.17	1.65
W	5.68	5.71	1.00	1.01

Migration of Liquid Films

Larché and Cahn have recently derived the conditions for equilibrium between binary liquids and planar, non-hydrostatically stressed solid solutions by the simultaneous solution of two equations for the concentrations of each phase (106). These thermodynamic expressions have been modified to include the effects of curved interfaces on interfacial equilibria. On the basis of the calculated composition gradient in the liquid film between two stressed and curved solids, the kinetics of liquid film migration have been treated as a boundary-value diffusion problem (21,22).

For a curved and stressed solid in equilibrium with a liquid, the free energy of the solid is:

$$f_{total} = f_{chemical} + f_{coherent} + f_{curvature} \tag{8}$$

where $f_{chemical}$ is the chemical free energy, $f_{curvature}$ is $\sigma\kappa$, σ is the isotropic solid-liquid interfacial energy, κ is $(1/r_1+1/r_2)$, and r_1 and r_2 are the signed principal radii of curvature of the interface, and $f_{coherent}$ is Eq. (5). Equation (8) comes from a set of reversible steps. From Eq. (8), along with the equation for the free energy of the liquid phase, the equilibrium concentrations of the solid and liquid in equilibrium can be derived as outlined below.

For dilute solutions, the equilibrium conditions for a stressed and curved solid in contact with a liquid are:

$$\begin{aligned} \mu_1^{o,s}-\mu_2^{o,s} + RT\ln\frac{\tilde{\gamma}_1^s\tilde{C}^s}{\tilde{\gamma}_2^s(1-\tilde{C}^s)} + 2V_m Y(\hat{n})\eta^2(\tilde{C}^s-C_o) \\ = \mu_1^{o,\ell}-\mu_2^{o,\ell} + RT\ln\frac{\gamma_1^\ell\tilde{C}^\ell}{\gamma_2^\ell(1-\tilde{C}^\ell)} \end{aligned} \tag{9a}$$

and

$$\mu_2^{o,s} + RT \ln \tilde{\gamma}_2^{s}(1-\tilde{C}^s) + V_m Y(\hat{n})\eta^2(C_o^2-(\tilde{C}^s)^2) + 2\sigma\kappa = \mu_2^{o,\ell} + RT \ln \gamma_2^{\ell}(1-\tilde{C}^{\ell}) \tag{9b}$$

For an unstressed and planar solid in contact with its liquid, the equilibrium conditions are:

$$\mu_1^{o,s} - \mu_2^{o,s} + RT \ln \frac{\gamma_1^s C^s}{\gamma_2^s(1-C^s)} = \mu_1^{o,\ell} - \mu_2^{o,\ell} + RT \ln \frac{\gamma_1^{\ell} C^{\ell}}{\gamma_2^{\ell}(1-C^{\ell})} \tag{10a}$$

and

$$\mu_2^{o,s} + RT \ln \gamma_2^s(1 - C^s) = \mu_2^{o,\ell} + RT \ln \gamma_2^{\ell}(1-C^{\ell}) \tag{10b}$$

where the components are labelled 1 and 2 as in Figure 14, T is the temperature, $\mu_i^{o,s}$ and $\mu_i^{o,\ell}$ are the standard chemical potentials and γ_i^s and γ_i^{ℓ} are the activity coefficients of the ith component in the solid and liquid, C^s and C^{ℓ} are the mole fractions of component 1 in the solid and liquid, V_m is the molar volume of the solid, κ is the mean curvature, σ is the isotropic solid-liquid interfacial energy, and ~ indicates the quantities for the stressed solid and liquid.

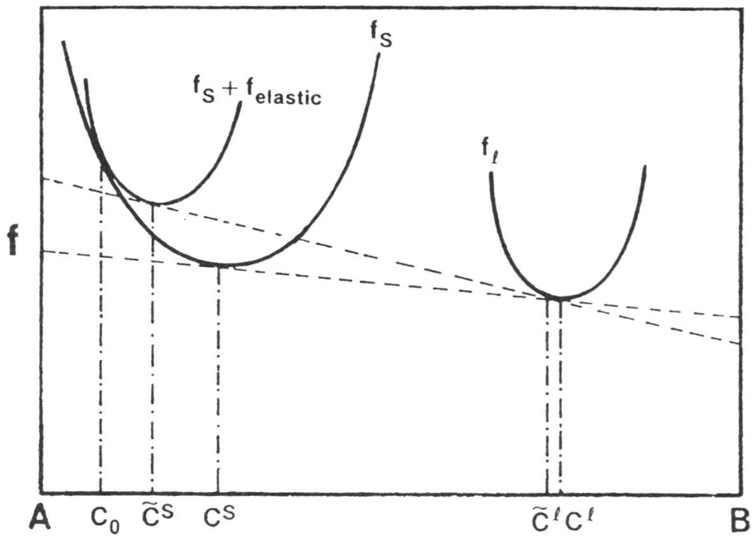

Fig. 14. Common tangent constructions for an unstressed planar solid in contact with its equilibrium liquid and for a coherently stressed and curved solid in contact with its equilibrium liquid (21,22).

These equations were derived from the basic laws of thermodynamics. They imply, however, a simple graphical construction on a free energy diagram for the solutions of Eqs. (9) and (10) for C^s and C^ℓ and for $\tilde{C}^s$ and $\tilde{C}^\ell$. The tangent to the free energy curve for the solid must be parallel to that of its equilibrium liquid as stated in Eqs. (9a) and (10a). Furthermore, the tangents to the solid and liquid must go through a common point as stated in Eqs. (9b) and (10b). These two conditions together define the common tangents shown schematically in Figure 14.

It is worthwhile to examine explicitly the implications of incorporating δ and $f_{coherent}$ (Eqs. (4) and (6)) into the conditions for coherent equilibrium, Eqs. (9a) and (10b). Equation (6) states that the strain in the alloyed region is dependent only on the difference in lattice parameters between the surface composition $\tilde{C}^s$ and the initial matrix material, C_o. Thus, the composition profile and the thickness of the diffusion layer have no effect on the strain of the surface layer in contact with the liquid. Furthermore, Eqs. (9a)-(9b) imply that local equilibrium is maintained at the solid-liquid interface.

The concentration of a liquid in equilibrium with a stressed and curved solid was found to be

$$\tilde{C}^\ell = C^\ell + \left\{ \left[\frac{Y(\hat{n})\eta^2 V_m \dfrac{RT}{C^s(1-C^s)}\left[1+\dfrac{\partial \ln \gamma^s}{\partial \ln C^s}\right]}{2Y(\hat{n})\eta^2 V_m + \dfrac{RT}{C^s(1-C^s)}\left[1+\dfrac{\partial \ln \gamma^s}{\partial \ln C^s}\right]} \right] (C^s-C_o)^2 + 2\sigma\kappa V_m \right\} \times \frac{C^\ell(1-C^\ell)}{RT(C^\ell-C^s)}\left[1+\frac{\partial \ln \gamma^\ell}{\partial \ln C^\ell}\right]^{-1} \qquad (11)$$

in terms of the concentrations for an unstressed planar solid C^s in equilibrium with a liquid of composition C^ℓ. The equilibrium liquid composition depends on the local crystallographic orientation as defined by the normal to the interface, $\hat{n}$, the local interface curvature, κ, and on the composition C_o of the underlying matrix.

The kinetics of liquid film migration are then treated as a boundary value problem for diffusion through the liquid using Eq. (11) to provide the boundary conditions for the flux equation Eq. (2). Equation (11) is simplified significantly by assuming that the small strain limit and Henrian solution behavior apply. For two solid grains, A and B, of the same phase and the same initial composition but having different elastic coefficients, $Y(\hat{n}_A)$ and $Y(\hat{n}_B)$, and different curvatures, κ_A and κ_B, the velocity of the liquid film becomes:

$$V = \frac{D_{liq}V_m}{\lambda RT}[(Y(\hat{n}_A)-Y(\hat{n}_B))\eta^2(C^s-C_o)^2 + 2\sigma(\kappa_A-\kappa_B)]\frac{C^\ell(1-C^\ell)}{(C^\ell-C^s)^2} \tag{12}$$

where λ is the liquid film thickness and D_{liq} is the diffusivity through the liquid film. When the stress in grain B is completely relaxed, the coherent strain energy goes to zero in that grain and the velocity of the liquid film becomes:

$$V = \frac{D_{liq}V_m}{\lambda RT}[Y(\hat{n}_a)\eta^2(C^s-C_o)^2 + 2\sigma(\kappa_A-\kappa_B)]\frac{C^\ell(1-C^\ell)}{(C^\ell-C^s)^2} \tag{13}$$

It is important to note that these expressions for the migration velocity contain no unmeasurable materials parameters. The parameters are all thermodynamic, elastic, crystallographic, or diffusive and can be measured independent of the migration process, except for interface curvature, grain orientation, the strain, and the liquid film thickness. When all of these parameters are known, velocities can be calculated to compare quantitatively with measured velocities. The interface curvature changes slowly with time and can be reliably measured from samples quenched to room temperature. The liquid film thickness should be measured during LFM. However, the direct observation of the liquid film is difficult due to the high temperatures and the thinness of the liquid film. When the film thickness cannot be measured in-situ, the film thickness can be controlled by the use of inert spacers or, as a last resort, must be measured on quenched samples. The interlocking grain structures produced by LFM limit rearrangement during quenching thereby justifying, somewhat, the use of measurements from quenched samples. The grain orientations can be determined by TEM or selected area electron channeling in the SEM. The state of strain can be inferred from the composition and the dislocation structure in the alloyed regions.

Migration Regimes for Liquid Films

In describing the initiation of film motion, it has been assumed that the solid is initially homogeneous and stress-free and is surrounded by a compositionally homogeneous liquid. When the conditions are changed, either by changing the temperature or by adding solute to the system, the compositions of the liquid and the solid will no longer be the equilibrium compositions. There will be an initial transient after which the liquid compositions will be in equilibrium with the stressed surface layers of the solid, as described by Eq. (11). If the values of C^ℓ are different for two adjacent surfaces, a concentration gradient will exist in the liquid and the film will migrate according to Eq. (12). The direction of migration will depend on both the differences in $Y(\hat{n})$ and κ in the two adjacent grains. When κ is negligible, the direction of migration is into the grain with larger $Y(\hat{n})$.

The diffusion layer thickness, L, in the growing grain increases by both lattice diffusion and migration according to:

$$L = \sqrt{D_\ell t} + Vt \tag{14}$$

where D_ℓ is the lattice diffusivity of the solute in the solid and t is time, which at long times reduces to L = Vt. For the dissolving grain, L can be approximated as:

$$L = \sqrt{D_\ell t}\left(\frac{\sqrt{D_\ell t}}{\sqrt{D_\ell t} + Vt}\right) \tag{15}$$

which for $\sqrt{D_\ell t} \gg Vt$, $L = \sqrt{D_\ell t}$ and for $\sqrt{D_\ell t} \ll Vt$, $L = D_\ell/V$.

At a critical thickness of the alloyed zone, L_c, the stresses due to coherency can relax by plastic flow. The critical layer thickness occurs when the coherency strain energy exceeds the energy needed to create the dislocations needed to relax the stress. The coherency strain energy is proportional to the thickness and the energy needed to create the misfit dislocations is, to a first approximation, independent of the layer thickness. To determine when coherency will be lost in the growing and dissolving grains, we have considered the two mechanisms by which the coherently strained region may grow: by lattice diffusion and by liquid film migration. We can use Eqs. (14) and (15) to describe the loss of coherency in one solid. If $\sqrt{D_\ell t} \gg Vt$, the values of L in both grains will be determined by $\sqrt{D_\ell t}$. Since L_c is inversely proportional to $Y(\hat{n})$, coherency loss will first occur in the grain with the larger value of $Y(\hat{n})$. In this case, migration will occur first in the direction with the larger values of $Y(\hat{n})$ and will reverse directions when coherency is lost. If $\sqrt{D_l t} \ll Vt$, migration will occur into the grain with the larger $Y(\hat{n})$. Coherency loss will occur in the growing grain, the grain with the lower $Y(\hat{n})$, and migration will continue into the grain with the larger $Y(\hat{n})$. After coherency is lost in one layer, the velocity of film migration will be described by Eq. (13).

Stress Relaxation Mechanisms

Stress relaxation in one grain at L_c, as described above, occurs uniformly. However, the stresses due to coherency strain can also relax nonuniformly in the alloyed regions by the interactions of the stress field with pre-existing dislocations or by the multiplication of dislocations from pre-existing dislocation sources. During stress relaxation, the velocity will depend on the local stress difference across the interface. In the special case of closely matched values of $Y(\hat{n})$, little motion would occur until stress relaxation begins. If stress relaxation occurs non-uniformly, film migration would begin in the vicinity of each site of stress relaxation leading to the local bowing of the interface.

Migration Velocities and Morphology of the Migrating Interface

Equations (12) and (13) give migration velocities which depend on the local stress differences across the interface. For an interface with no stress relief in either grain, the local velocity is determined by the local value of $Y(\hat{n}_A)-Y(\hat{n}_B)$ and, hence, on the misorientation between the grains. After coherency loss, the growing grain is stress free while the dissolving grain remains stressed. Under these conditions, the local velocity is determined only by the anisotropy of $Y(\hat{n})$ of the dissolving grain. The growth rate is independent of the orientation of the growing grain or of the grain misorientation.

From this relationship between growth rate and the orientation of the dissolving grain, shapes of the films between the dissolving and growing grains have been predicted as a function of the elastic anisotropy (107). The shapes which evolve in this analysis are analogous to those developing during crystal growth where the directions of fastest growth become corners in the final shape and the directions of slowest growth become planes. These shapes have been derived using Frank's general theory of crystal habit formation (108,109) based on the assumption that the rate of growth depends on the crystallographic orientation of the dissolving grain alone. The predicted habits for KCl and Mo and for Fe and Co, shown in Figure 15, represent the two extremes in shape and orientation for cubic materials. For Co and Fe, where $2(c_{44})/(c_{11}-c_{12})>1$, the shape is cuboidal with curved faces parallel to the (001) and with sharp edges. For KCl and Mo, where $2(c_{44})/(c_{11}-c_{12})<1$, the shape is an octahedron with curved faces parallel to the (111) and with sharp edges for KCl but not for Mo. For an elastically isotropic material, that is, for $2(c_{44})/(c_{11}-c_{12}) = 1$, the equilibrium shape is a sphere. The dependence of shape on orientation can be tested by measuring the shapes of growing grains and their orientations relative to the crystallographic orientation of the dissolving grains.

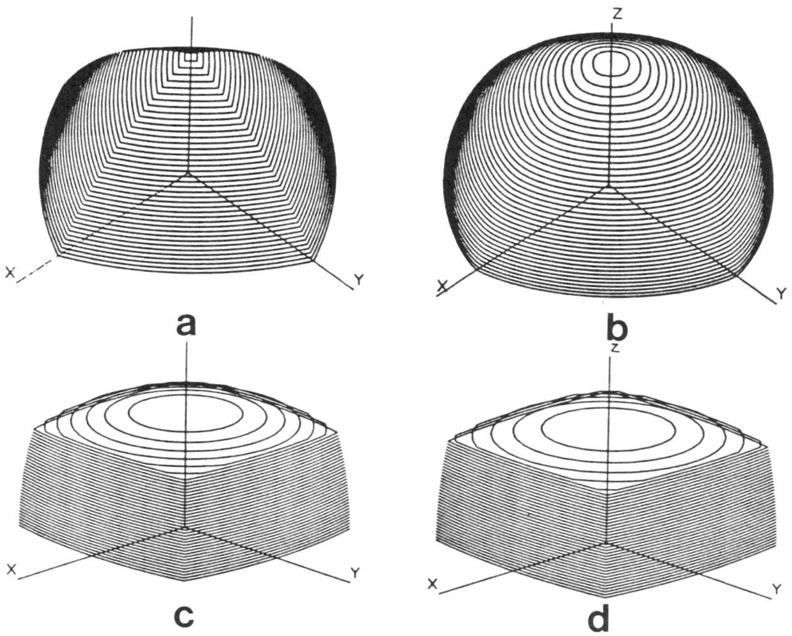

Fig. 15. Shapes of dissolving grains with a coherently stressed diffusion layer ahead of the migrating interface. Shapes calculated assuming that the migration rate is proportional to the elastic energy of the dissolving grain alone. (a) KCl; (b) Mo; (c) Fe; and (d) Co (107).

Controlled Variation of Strain

The critical test of the coherency stress mechanism as the driving force for LFM is the observation of how the velocity changes with changes in the coherency strain energy while ΔG_m is held nearly constant. As seen by examination of Eqs. (12) and (13), the migration rate is proportional to the square of the strain. No migration should occur when the strain is zero regardless of the magnitude of the ΔG_m.

For a binary system that obeys Vegard's law, the coherency strain energy and ΔG_m cannot be varied independently. They both depend on $(C^s - C_o)^2$ so that at zero strain, ΔG_m is also zero. Under this condition of zero strain, the system is at equilibrium ($C^s = C_o$) and no migration is expected. For binary systems that do not obey Vegard's law, it is possible under special conditions to have the experimental situation where the coherency strain energy and ΔG_m do not go to zero simultaneously.

In multicomponent systems, the strain can be varied readily by changing the concentrations of two or more species simultaneously. When Vegard's law applies for ternary solid solution, changing the concentrations of two components leads to

$$\delta = \eta_1(C_1^s - C_1) + \eta_2(C_2^s - C_2) \tag{16}$$

where $\eta_1 = (1/a_o)(da/dC_1)$ for constant $C_2{}^s$ and $\eta_2 = (1/a_o)(da/dC_2)$ for constant $C_1{}^s$. From Eq. (16) it is seen that δ can be varied from positive to negative by changing $(C_1{}^s - C_1)$ and $C_2{}^s - C_2)$ regardless of the signs of η_1 and η_2. Of particular importance is that δ can be zero for many different compositions under conditions that keep ΔG_m greater than zero. Under these conditions, no migration induced by coherency stresses is expected even though mixing of the three components will occur by diffusion.

Tests of the Coherency Strain Theory for Liquid Film Migration

The critical tests of the coherency strain model for liquid film migration have been performed by Yoon and co-workers at the Korea Advanced Institute of Science and Technology (KAIST), Seoul, Korea, and the present author and co-workers at the National Bureau of Standards (NBS), Gaithersburg, MD (21,22). In this section we compare the experimental results with each quantitative prediction of the model as presented above. The critical tests determine whether: (1) the velocity of the liquid films is described by Eq. (13) when coherence is lost in the growing grain; (2) the velocity varies as δ^2; (3) the velocity goes to zero as δ goes to zero even when the free energy of mixing is greater than zero; (4) the crystallographic orientation of the dissolving grain determines the shape of the dissolution front for a stress-free grain growing into a stressed matrix; (5) curvature can either increase or retard the motion of the liquid film according to Eqs. (12) and (13).

The systems used in these quantitative tests are binary Mo-Ni and Co-Cu alloys, and ternary and quaternary Mo-Ni-x-y alloys. The Mo-Ni system is ideal for liquid film migration experiments since Baik and Yoon demonstrated that changing the temperature will significantly change the equilibrium solid and liquid compositions and will induce migration of films in previously equilibrated samples. In this case, the change in Ni concentration of the solid and the liquid occurs at all liquid-solid interfaces throughout the system. Hence, migration of the film does not require diffusion parallel to the interface to replenish solute in the film as the film migrates. The experimental conditions approximate the one-dimensional model of diffusion in a solid and liquid undergoing LFM It is also important to note that the physical model for a coherent surface layer described above allows relaxation of the stress perpendicular to but not in the plane of the interface. In the case of solid grains separated by liquid films, the grains are free to expand or contract perpendicular to the interface.

Migration Velocities. Baik and Yoon (19) manipulated the strain parameter in the Mo-Ni system by equilibrating samples at 1510°C and lowering or raising the temperature to induce migration. The change in Ni concentration of the solid and, hence, the strain increased as the temperature difference increased. The migration distance measured as a function of time and the migration rate measured as a function of strain are in excellent agreement with the coherency strain model. A steady state velocity was found as indicated by the linear increase in the migration distance with time and the migration rate, V, was found to be a parabolic function of ΔC (Fig. 16) and, hence, parabolic in δ, as predicted by Eq. (13).

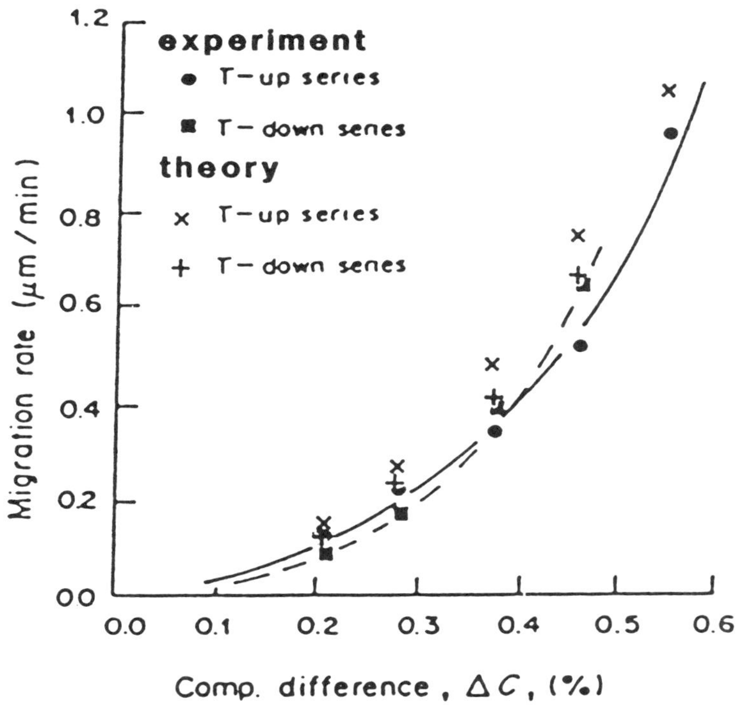

Fig. 16. Measured migration rate versus the estimated composition difference ΔC between the initial and the final solid solutions from Ref. (19) compared with calculated migration rates based on estimated ΔC and Eq. (12).

The maximum migration rate can be calculated if the strain in the adjacent grains is known. There are limited data available on the state of coherency of the growing and the dissolving grains. In a TEM investigation of migration in Mo-Ni by Hofmann-Amtenbrink and co-workers (18,59,73), coherency was not lost in growing grains for migration distances of the order of several micrometers. The calculations of critical layer thickness for coherency loss in Mo-Ni are shown in Table 3. The migration distances at short times are of the same order of magnitude as the calculated values of L_c; the migration distances at longer times are greater than L_c by up to a factor of eight. At short times and low strains, the velocity should be described by Eq. (12) and at longer times the velocity should be described by Eq. (13). In addition to the uncertainty in the state of strain, there are other uncertainties in the data required to use Eq. (13). There are no data on D_{liq} for Mo-Ni liquids and measured by Baik and Yoon are averages from all migrating interfaces. Because of these uncertainties, we cannot compare calculated and λ measured migration velocity for a particular misorientation and film thickness. However, an upper limit for the migration velocity can be estimated by using Eq. (13), for which it is assumed that the growing grain is strain free. Calculations were performed using solubility data (110), the maximum value of the elastic coefficient for Mo, a film thickness of 1.4×10^{-6} m (19), and D_{liq} of $10^{-9} m^2/s$. The calculated migration velocity, shown by the dashed line in Fig. 16 is in remarkable agreement with the measured values. Because of the uncertainties in the parameters used in the calculations, the agreement can only be regarded as fortuitous.

Controlled Variation of Strain. The condition of zero coherency strain was achieved by Kang (111), Rhee (112), Baik (60), and Yoon (113) in liquid phase sintered Mo-Ni alloys during formation of ternary and quaternary solid solutions with W, Co, and Fe. By changing the relative amounts of Ni and one other alloying element, the strain and, hence, the velocity could be controlled. As predicted by Eqs. (12) and (13), the velocity of the liquid film goes to zero as δ goes to zero.

The results for the zero strain condition are important for two reasons. First, these results confirm that coherency strain is the driving force for liquid film migration. Second, they also demonstrate unequivocally that having a negative ΔG_m is a necessary but not a sufficient condition for LFM to occur (21,22).

Morphology of Migrating Interfaces. Calculations of the shape of the migrating interface in the Mo-Ni system have been performed based solely on the crystallographic orientation of the dissolving grain. The degree of anisotropy can be estimated from the ratio of the elastic coefficients, $2c_{44}/(c_{11}-c_{12})$. For Mo, this parameter is 0.78, indicating that Mo is almost isotropic (114). The shape of the dissolving interface for Mo, shown in Fig. 12, reflects this isotropy. In contrast to the shape of the KCl interface ($2c_{44}/(c_{11}-c_{12}) = 0.19$) (114), the shape of a dissolving Mo grain is close to spherical with no sharp edges or corners. It is, therefore, not surprising that the migrating interfaces in Mo-Ni are generally observed to be smoothly curved.

Baik and Yoon (113) performed additional experiments in the Co-Cu system where $2c_{44}/(c_{11}-c_{12}) = 3.5$ (115), to determine if sharp corners exist in a system that is more anisotropic than Mo. The calculated and observed shapes of Co(Cu) in a Co-Cu liquid are shown in Figs. 15 and 17, respectively. It is also frequently observed that two grains will grow with parallel growth fronts into a single grain regardless of the orientations of the growing grains. Parallel growth fronts indicate that the crystallography of the growing grain does not affect the shape and the orientation of the dissolving interface.

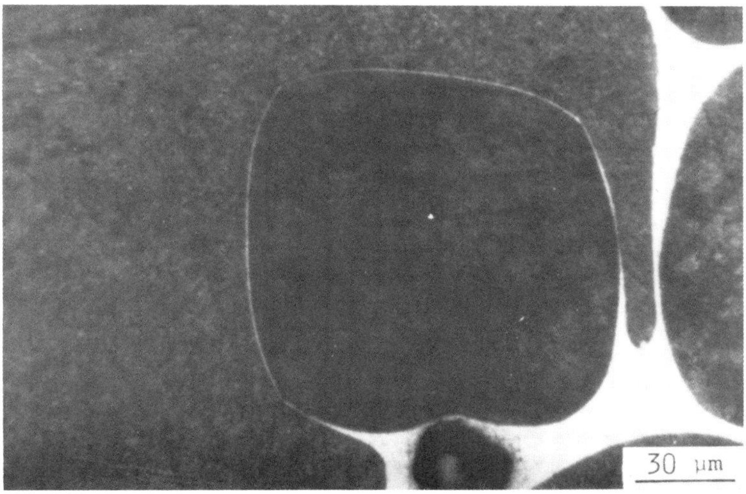

Fig. 17. Growth of stress-free Co(Cu) grains into stressed matrix. Alloys of 80% Co-20% Cu were annealed at 1300°C for 20 h. The samples were subsequently annealed at 1150°C for 3 h. and LFM occurred. (Courtesy of Y.-J. Baik and D. N. Yoon (60).)

Other Driving Forces for Boundary Migration. In the model for coherency stress, the term in the velocity expressions, Eqs. (12) and (13) that includes the coherency strain energy is a driving force for boundary motion. As also seen in these expressions, curvature can change the total driving force on the boundary. The interfaces in LFM are sometimes observed to move away from their center of curvature ($\kappa_A - \kappa_B < 0$, for B growing into A) and, hence, migration due to coherency strain can be opposed by curvature. A limit can be placed on driving force due to coherency strain by calculating the retarding force due to curvature. For the LFM geometry, $\kappa_A = -\kappa_B = \kappa$ and the curvature term becomes $(4\sigma_s V_m \kappa)$. Baik and Yoon (19) have compared the two terms for a sample sintered at 1400°C and reannealed at 1510°C, where $\Delta C = 0.0045$. Under these conditions, $f_{coherent} = 0.23$J/mole and $(4\sigma_s V_m \kappa) = 0.61$J/mole, assuming $\sigma_s = 1.3 J/m^2$ and $\kappa = 1.25 \times 10^4 m^{-1}$. Because of the uncertainties in the data used in these calculations, it can be concluded only that the two terms are the same order of magnitude. Under these conditions, it is expected that the net velocity of the interface is close to zero.

There is strong evidence for the mitigating effect of curvature on boundary motion due to coherency strains. In the W-Fe-Ni system (20), in the Mo-Ni system (21), and in the MgO-CoO-V_2O_5 system (67), boundaries have been observed to reverse directions. Alloying first occurs with interfaces moving away from their centers of curvature. The interfaces then move back due to curvature through the alloyed region. We believe that coherency is lost in the dissolving grain as the boundary slows down due to the effect of curvature. When coherency is lost in the dissolving grain, curvature becomes the only driving force on the interface and the direction of motion reverses.

Liquid Film Migration in Ceramics

The test of a general physical model like the coherency strain model is the ability to use the model for predicting the behavior of materials. The coherency strain model has been used with great success to choose candidate systems in which LFM had not been observed and to predict the conditions for LFM. Of special interest was to determine if the phenomenon could be observed in a ceramic system. Although LFM had been observed in metal alloy systems, attempts to find this effect in a ceramic system had not been successful prior to the development of this theory (116). The MgO-V_2O_5-NiO and the MgO-V_2O_5-CoO systems (67) were selected because of the extensive solid solutions formed between MgO and NiO and between MgO and CoO and the lattice parameter change with composition and the grain boundary and lattice diffusivities of CoO and NiO into MgO were well documented. Equations (12) and (13) were used to calculate the times and temperatures for LFM to occur at temperatures where lattice diffusion is appreciable. Samples of submicron MgO and V_2O_5 were annealed at 1350°C for 24 hours to produce large, compositionally homogeneous, almost spherical solid particles in a liquid matrix. The average size of the solid particles was approximately 4×10^{-5}m. When NiO-V_2O_5 or CoO-V_2O_5 liquid was added to the system at temperatures lower than the initial equilibration temperature, LFM occurred with morphologies identical to those in the Mo-Ni system. This was the first unequivocal observation of LFM in a non-metallic system.

5. EFFECT OF DIFFUSION-INDUCED STRAIN ON GRAIN BOUNDARY MIGRATION

The theoretical description and the experimental tests for coherency strain as the driving force for liquid film migration were summarized in the previous section in order to have a simple starting point from which DIGM and DIR can be analyzed. The same physics apply in these three processes whether the interface is a liquid film or a grain boundary and whether the sample is a bulk polycrystal or a thin film bicrystal. However, the reformulation of the thermodynamics and kinetics for DIGM and DIR requires major modifications to the model for LFM. As a result, the thermodynamic and kinetic expressions which were generated straightforwardly for LFM have not yet been derived for DIGM or DIR.

In this section, experimental results are used to qualitatively test the coherency strain model, as derived for LFM, as a mechanism for DIGM. The differences between DIGM and LFM are described and several of the complications in the full three-dimensional diffusion problem for DIGM are outlined. From an analysis of the three-dimensional diffusion problem as a function of temperature and time, the diffusion-induced nature of DIGM and DIR can be understood. The coherency strain is shown to be only one driving force for migration resulting from diffusion into the lattice adjacent to the grain boundary. The other driving force for grain boundary migration is seen to arise from the high dislocation densities generated when there are unequal diffusivities of solute and solvent during interdiffusion.

Distinctions Between Grain Boundaries and Liquid Films

The effect of diffusion-induced strain on boundary migration can be applied qualitatively when liquid films are replaced by grain boundaries. However, shrinking the liquid layer to grain boundary dimensions is not sufficient to model DIGM. The coexistence of a liquid with a solid removes a thermodynamic degree of freedom. In a binary two-phase system at fixed temperature, the phase compositions are determined even if the solid is stressed due to coherency strains. In contrast, grain boundaries exist in solids over the entire range of compositions where the solid is stable. Establishing interfacial compositions is an essential part of solving the boundary value diffusion problem for grain boundaries in an analogous way we solved the problem for the case of liquid films (21,22).

A kinetic description of grain boundary motion due to coherency strain is also more complex than the corresponding theory for the motion of liquid films. The liquid film is a source or sink for composition changes and diffusion of solute parallel to the liquid film is not required for migration of the film. Thus, the diffusion problem for LFM is one-dimensional. In contrast, the diffusion geometry in DIGM is three-dimensional. Diffusion into the lattice and along the grain boundary will play important roles in DIGM.

Qualitative Test of the Coherency Strain Theory for DIGM

Qualitatively the coherency strain theory provides a mechanism by which the motion of the boundary is coupled to grain boundary diffusion of the solute. The coherency strain is created in the lattice adjacent to the grain boundary by diffusion of solute supplied by grain boundary diffusion. Thus, migration of the boundary by a coherency strain mechanism requires grain boundary diffusion of the solute.

Several of the critical tests of the coherency strain model for liquid film migration can be applied qualitatively to DIGM. These tests determine whether: (1) the velocity varies as δ^2; (2) the velocity goes to zero as δ goes to zero even when the free energy of mixing is greater than zero; (3) the crystallographic orientation of the dissolving grain determines the shape of the dissolution front for a stress-free grain growing into a stressed matrix; and (4) curvature can either increase or retard the motion of the grain boundary.

In liquid-solid Mo-Ni alloys, low angle grain boundaries will exist between some particles even when most particle contacts contain a liquid film. Baik and Yoon (19) determined that some low angle grain boundaries move in response to a temperature change at approximately the same rate as the liquid films. The average migration distance of these grain boundaries also appeared to increase with the driving force, δ. (The migration rates for Mo-Ni liquid films, shown in Fig. 16, were proportional to δ^2.) However, the rate of grain boundary motion varied from boundary to boundary in the same sample. Some boundaries did not migrate at all. Since DIGM requires grain boundary diffusion to supply and redistribute solute along the interface, slow grain boundary diffusion in these low angle grain boundaries may limit motion by DIGM. It is, therefore, not surprising that the migration rates by DIGM were not uniform.

The effect of grain boundary diffusion rates on DIGM is seen more strikingly in the experiments by Rhee (112), Baik (60), and Yoon (113) to determine the zero strain condition for Mo-Ni-x ternary and quaternary alloys. In these experiments, the equilibrium compositions of solid and liquid were changed by a change in temperature and by addition of a solute. The experiments were designed so that the lattice parameter of original binary solid was matched to the lattice parameter of the new ternary solid solution. They observed that, under conditions where the migration rate of liquid films went to zero, sections of grain boundaries farthest from the solute source would

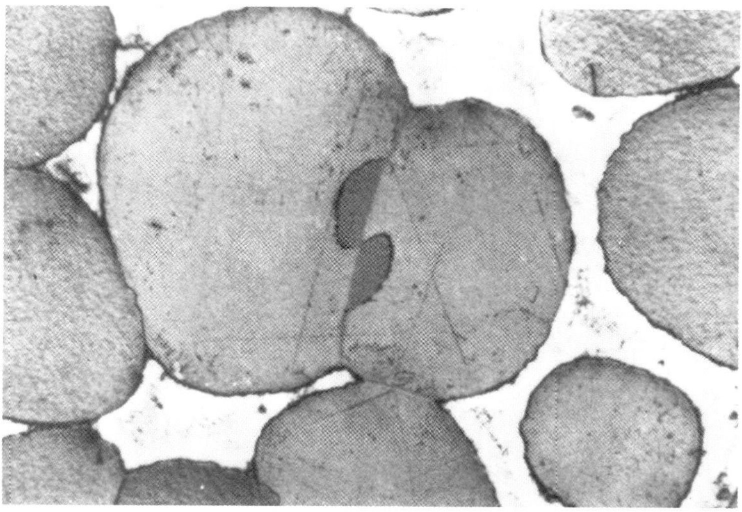

Fig. 18. Grain boundary formed during liquid phase sintering of 90%Mo-10%Ni alloy at 1520°C for 20 h. Subsequently annealed at 1400°C for 1 h. with a solution containing 1.2% Fe. DIGM occurred only in the center of the grain boundary not in contact with the Fe. (Courtesy of Y.-J. Baik and D. N. Yoon (60).)

migrate due to the temperature change. The sections nearest the solute source, which would be affected by both the temperature change and solute additions, would be stationary. These results illustrate the effect of slow grain boundary diffusion on migration rates in DIGM. A typical microstructure of the migrating grain boundary in the center of the grain boundary is shown in Fig. 18 for Mo-Ni-Fe.

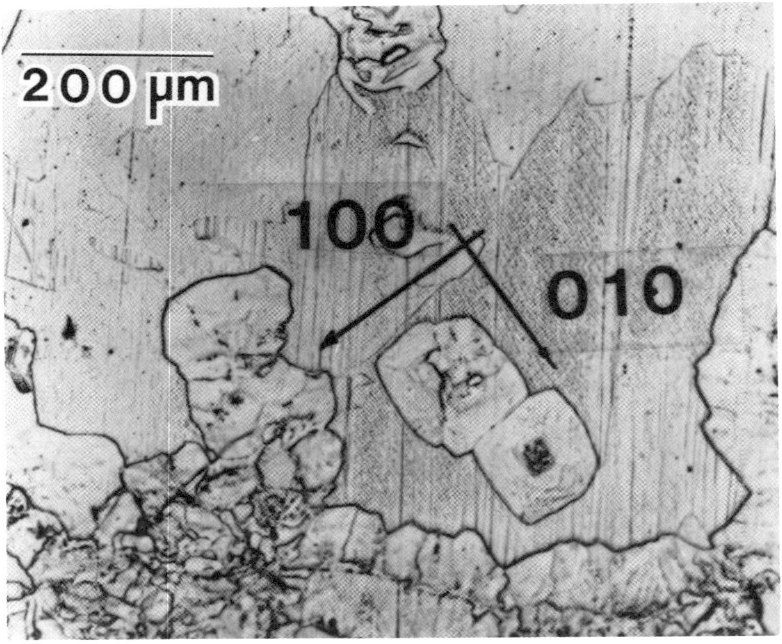

Fig. 19. New grains formed by DIR and grain boundaries moved by DIGM in Fe(Zn) are aligned with the [100] of the original grain. Source: 85% Fe, 15% Zn; 600°C; 20 h. Orientations measured by selected area electron channeling (107).

To qualitatively test the effect of elastic anisotropy on the shapes and orientations of the dissolving grain, we require the condition that a stress-free grain grow into a stressed matrix. The new grains which grow by DIR meet this criterion. These grains have been observed to be almost dislocation-free while the matrix contains a high density of defects, including dislocations. In Fe(Zn), DIR grains and the regions alloyed by DIGM at 600°C formed curved faces aligned with the [001] of the dissolving grains. As shown in Fig. 19, three DIR grains are misoriented with respect to each other as indicated by the dihedral angles but they are aligned with the [001] of the dissolving grain as determined by selected area electron channeling. The alloyed regions formed by DIGM do not have the same orientation as the dissolving grains of the DIR grains but they, too, formed curved faces aligned with [001] of the dissolving grain. This result agrees with the shape and orientation of the dissolution front for Fe, Fig. 15, based on the growth rate being a function only of $Y(\hat{n})$ of the dissolving grain (107).

These results on the orientation relationship between the dislocation-free growing grain and the stressed matrix are not general. In Au(Cu) (34), the newly nucleated grains have irregular boundaries. In Cu(Zn), a pre-existing grain boundary migrates with curved faces and sharp corners with the orientation of the faces parallel to low index planes in the growing grain and to high index planes with respect to the dissolving grain (74).

Critical Values of D_ℓ/V.

A major question about the application of coherency strain theory to DIGM is how much lattice diffusion is required to induce the boundary motion. We can address this question, first, by examining the kinetic data for systems where measurable lattice diffusion occurs and, second, by examining the conditions for coherency loss when lattice diffusion occurs.

It was initially suggested that DIGM could only occur when lattice diffusion was "frozen out", that is, when the composition changed discontinuously across the grain boundary (8). Both den Broeder and Nakahara (49) and Tashiro and Purdy (29) established that this is not a necessary condition for DIGM. Significant lattice diffusion was observed during DIGM in Ni(Cu) at 900°C (49) and in Al(Zn) (29) at 160°C-190°C. As seen in Eq. (14), the thickness of the diffusion zone in front of a moving boundary reduces to D_ℓ/V when $\sqrt{D_\ell t} << Vt$. For Ni diffusing into Cu at 900°C, $L = 1.6\times10^{-6}$ m using D_ℓ data from (70). For Zn diffusing into Al, Tashiro and Purdy calculated the following values of D_ℓ/V: 0.7×10^{-6}m (166°C); 1.5×10^{-6}m (178°C); and 2.8×10^{-6}m (190°C). The calculated values are the same order of magnitude as those observed by etching differences (Al(Zn)) and by EDS (Cu(Ni)). In addition, den Broeder and Nakahara determined the temperature dependence of migration in terms of $V = V_0 \exp(-Q/RT)$ over a wide range of temperature. From 900°C to 350°C, the velocity is linear on a plot of log V against 1/T. Although the calculated value of D_ℓ/V at 350°C is 5.4×10^{-14}m, this linearity suggests that the mechanism for migration remains the same over the entire temperature range.

The primary objection to the coherency strain model for DIGM is the widespread observation of DIGM at low temperatures where lattice diffusion is slow (53). At these low temperatures, D_ℓ/V is estimated to be less than 1nm. For example, DIGM occurs in Cu(Ni) at 350°C where D_ℓ/V was calculated to be 5.4×10^{-14}m as well as at 900°C where D_ℓ/V is 1.6×10^{-6}m. Lattice diffusion in the first layers adjacent to the grain boundary may be required for significant penetration to occur by grain boundary diffusion. If diffusion occurs at all in the lattice, even in the first few atomic planes parallel to the grain boundary plane, a thin, coherently strained layer will form which meets the criteria of the coherency strain model. This implies that some small amount of diffusion will occur into the lattice as long as grain boundary diffusion can operate. While there have been no direct measurements with sufficient resolution to detect diffusion in the layers adjacent to the grain boundary, molecular dynamics calculations of grain boundary diffusion in α-Fe by Kwok and co-workers (117) do show vacancy diffusion from the grain boundary into the region adjacent to the interface. The question still remains about how much diffusion or alternately, how much strain, is necessary to move grain boundaries.

By examining Eq. (13), we can describe qualitatively the limiting values of diffusion required from the coherency strain model. The composition at the grain boundary is not constrained by thermodynamics to be at a single composition. The condition for motion of the grain boundary which must apply is that there must be a net force on the grain boundary. When the driving force due to coherency strain is high enough to oppose any other driving force, the grain boundary will move according to the coherency strain driving force. This implies that DIGM should occur with a decreasing penetration depth as the ratio of lattice diffusivity to grain boundary diffusivity increase. As this ratio increases, there is a decrease in the time and penetration depth at which the critical condition of strain are reached. Blendell and co-workers (54) noted that DIGM in Fe(Zn) at 550°C led to migration of boundaries through the entire foil thickness but DIGM at 600°C led to migration of the boundaries only near the foil surfaces. This effect of temperature on extent of migration through the foil thickness is shown in Fig. 3.

As seen in Eq. (13), $f_{elastic}$ must be greater than $f_{curvature}$ for coherency strain to be the driving force for DIGM, that is, $\Delta C = (4\sigma_{gb}\kappa/Y\eta^2)^{1/2}$. For Fe(Zn), $\kappa = 3.3\times10^5 m^{-1}$ (54), $\eta = 5\times10^{-2}$, $\sigma_{gb} = 0.7\ J/m^2$ (grain boundary energy for γ-Fe used as estimate for σ_{gb} for α-Fe (118)), $Y = 3.7\times10^{11}$ Pa, $\Delta C > 3\times10^{-2}$. For 90Cu 10Zn(Cu), $\kappa = 2\times10^5 m^{-1}$ (53), $\eta = 6\times10^{-2}$, $\sigma_{gb} = 0.625\ J/m^2$ (118), $Y = 2.5\times10^{11}$ Pa, $\Delta C > 2\times10^{-2}$. This relationship between elastic energy and curvature establishes a lower limit on the concentration in the region adjacent to the grain boundary. However, this relationship does not give us any insight into the thickness of the diffusion layer required for DIGM. From the model presented in the previous section, the elastic energy of a layer adjacent to the interface is dependent only on the difference in lattice parameter between that layer and the bulk. A diffusion monolayer with a specific ΔC yields the same strain at the grain boundary as a diffusion layer of a micrometer with the same ΔC.

From Eq. (13) and the value of ΔC calculated above, we can derive the upper limit on the amount of lattice diffusion which still allows the coherency strain mechanism to operate. From Eq. (15), the coherent layer ahead of the interface can only exist when $L_c < D_\ell/V$. When $D_\ell(T)$, $V(T)$, and L_c are known, we can calculate the temperature at which coherency loss is expected. Above this temperature it might be expected that either DIGM would stop or that the mechanism for migration would change. For 90Cu10Zn(Cu), $D_{Zn} = 0.73 \times 10^{-4}$ exp($-$17lkJ/RT) (m^2/s) (119), V = 3.1×10^{-5} exp($-$76 kJ/RT) (m/s) (53), and L_c = 71 nm using $\Delta C = 2 \times 10^{-2}$ calculated above and Eq. (36) in (72), the calculated transition temperature is 386°C. In 90Cu10Zn(Cu), den Broeder observed that DIGM occurred between 200°C and 550°C, but that nucleation and growth of alloyed grains (DIR) at the surface in contact with the solute source did not occur below 375°C but did occur above 400°C. The appearance of DIR above the calculated temperature where lattice diffusion becomes significant reflects a change in migration mechanism, as discussed below.

Features of Grain Boundaries Required to Quantitatively Model DIGM

The migration of liquid films can realistically be treated as a one-dimensional diffusion problem. However, in DIGM, the grain boundary has properties which make DIGM more complex than LFM. To account for the complexities that the grain boundary introduces, the following processes have been considered: (1) unequal grain boundary diffusivities of the solute and solvent; (2) unequal lattice diffusivities of the solute and solvent in the lattice adjacent to the grain boundary and from the surface in contact with the solute source; (3) vacancy creation and annihilation process at grain boundaries; (4) stresses developing as a result of geometrical constraints; (5) the effect of grain boundary structure on grain boundary diffusion and migration; (6) the interaction of pre-existing lattice dislocations with the stress field around the alloyed region; and (7) the effect of other driving forces acting simultaneously with DIGM. We will discuss each of these separately and then define regimes of temperature and time where specific combinations of these effects will dominate.

Unequal Grain Boundary Diffusivities. Balluffi and Cahn (8) and Smith and King (92,93) (BCSK) proposed that an unequal flux of the solute and solvent atoms along the grain boundary cause the climb of grain boundary dislocations. The divergence in the flux of atoms requires that vacancies in the grain boundary be either created or destroyed by the climb of grain boundary dislocations. The climbing dislocations cause the boundary to move by the motion of grain boundary steps which are intrinsically associated with grain boundary dislocations. The operation of this mechanism requires the continuous generation of grain boundary dislocation line length. Recent experiments on the motion of grain boundaries in Au indicate that the climb and generation of grain boundary dislocations may be a physically necessary process for boundary migration. In an in-situ TEM study of grain boundary migration in Au, Babcock (120) demonstrated that grain boundaries move by the motion of mobile grain boundary dislocations. In this study, it was also shown that grain boundaries are composed of arrays of secondary grain boundary dislocations, only some of which contribute to the boundary motion, and that sources and sinks of grain boundary dislocations operate in the boundary which can aid grain boundary motion. Since a boundary in a pure material moves by the motion of grain boundary dislocations, it is likely that DIGM boundaries also move by the same mechanism. When there is a divergence of the atom flux caused by unequal diffusivities in the boundary, it also is likely that this will lead to motion of the boundary by dislocation climb. The only way of distinguishing if this effect is a major contributing factor to DIGM is to have a grain boundary Kirkendall effect under conditions when the coherency strain is zero. This situation is difficult to obtain experimentally. It is likely, however, that the effect of a grain boundary Kirkendall effect is minor since the same grain boundary morphology ("S-shaped" boundaries) is observed in the absence of solute diffusion when the driving force is a stress difference across the grain boundary due to local differences in dislocation densities (122).

Unequal Lattice Diffusivities of the Solute and Solvent Species. Unequal diffusion coefficients of the chemical species in the lattice will affect interdiffusion processes in several ways (9). Tashiro and Purdy (29) considered the case of DIGM with both grain boundary diffusion and diffusion from the grain boundary into the adjacent lattice. The unequal lattice diffusivities lead to the creation or annihilation of vacancies in the lattice by the climb of dislocations and to a net flow of vacancies normal to the grain boundary if the vacancy concentration is in local equilibrium. The grain boundary would act as a source/sink for this vacancy flux, again by the climb of grain boundary dislocations.

This unequal flux would also lead to additional elastic and plastic strains in the diffusion layer adjacent to the boundary that are orders of magnitude higher than the coherency strain resulting from lattice parameter changes alone. A lattice Kirkendall effect would then require a more complex description of the strain and the strain energy in the diffusion zone.

At higher temperatures, there will be an increase in lattice diffusion into the surface in contact with the solute source. If the diffusivities of the chemical species are different, the diffusion zone in the surface will be strained due to both the change in lattice parameter with composition and the change in volume due to the climb of dislocations required to maintain a equilibrium vacancy concentration. As in the coherency strain model presented above, the volume changes are constrained by the unalloyed region below the diffusion zone. For A diffusing into B, the stresses in the diffusion zone will be tensile for $D_\ell^B > D_\ell^A$ and compressive for $D_\ell^A > D_\ell^B$.

The stress fields can lead to extensive plastic deformation by climb and slip (9,14) and at higher temperatures, can lead to substructure formation, recrystallization, and grain growth (9,121). We believe that the extensive plastic flow and the elastically accommodated strains associated with the combined effect of a lattice parameter mismatch and a lattice Kirkendall effect lead to stresses which are high enough to cause recrystallization in the diffusion zone at the surface in contact with the solute source. As the stress-free grains grow into the stressed matrix, grain boundary diffusion distributes solute into the volume swept by the migrating boundary. den Broeder (53) has noted that DIR in Ni(Cu) was preceded by formation of a high density dislocation network in the matrix in contact with the solute source. Initial results from a TEM study in Cu(Zn) by Hackney and coworkers (74) suggest that DIR occurs in the surface layers in Cu(Zn) at 360°C as a result of substantial plastic deformation generated by a lattice Kirkendall effect. Twinning, another mechanism for stress relief, is also observed within the surface grains in Cu(Zn) annealed over a wide range of temperatures (9,28,51,53,76).

Grain Boundaries as Sources/Sinks for Vacancies. The operation of a grain boundary Kirkendall effect as described originally (8,92) requires that the net flux of vacancies and chemical species be zero. This requirement means that the grain boundary is a perfect source/sink of vacancies and that all vacancies are accommodated by climb of grain boundary dislocations. As discussed recently by Jahn and King (123), the motion of the grain boundary may be sufficiently fast compared with diffusion along the boundary to cause a supersaturation of vacancies along the grain boundary. This supersaturation of vacancies may lead to vacancy deposition in the lattice swept by the moving grain boundary. They also discussed the effect of significant lattice diffusion on the vacancy deposition rate. It was postulated that vacancy deposition will occur when the velocity of the boundary is faster than the rate at which vacancies can diffuse back to the moving grain boundary. When $Vt < \sqrt{D_\ell t}$, vacancies can always diffuse back to the migrating boundary.

We suggest additional complications when there is significant lattice diffusion with unequal lattice diffusivities of the chemical species. A lattice Kirkendall effect can provide a net vacancy flux normal to the grain boundary. This vacancy flux from the lattice will change the vacancy concentration in the grain boundary and, thus, can change both diffusion along and migration of the grain boundary. As a result of this increased vacancy flux, grain boundary diffusion could be faster in a system undergoing chemical mixing by grain boundary and lattice diffusion. If the increased vacancy concentration due to lattice diffusion cannot be compensated by climb of grain boundary dislocation as in the model by Jahn and King (123), a net deposition of vacancies will occur in the lattice swept by the migrating boundary.

Jahn and King (123) also reported that stacking fault tetrahedra were observed in regions swept by DIGM boundaries in Cu(Zn). TEM specimens of recrystallized polycrystalline Cu foil were prepared by jet polishing and subsequently annealed for 30 min. with brass filings (11% Zn). Stacking fault tetrahedra were seen near the edge of a DIGM specimen in a region which had apparently been swept by a moving grain boundary. From the presence of stacking fault tetrahedra in Cu(Zn), a system with a low stacking fault energy, they concluded that a vacancy supersaturation existed behind the migrating boundary and ascribe this supersaturation to a grain boundary Kirkendall effect. As we suggest above, a vacancy supersaturation can be a result of several processes, which involve either grain boundary or lattice Kirkendall effects.

Geometrical Constraints. In DIGM there is a net accumulation of atoms in the polycrystalline foils when the boundaries migrate. Shewmon noted that the accumulation of solute will be accommodated by surface relief with no change in stress in the plane of the foil when the boundary diffusivities of the chemical species are equal. If the boundary diffusivities of the chemical species are different, there will be a net accumulation or depletion of material not accommodated by surface relief and stresses will build up in the foil. For example, consider the case of B diffusing into A from vapor source of B with $D_b{}^B > D_b{}^A$ and the vapor pressure of A being much less than the vapor pressure of B. Under these conditions, the flux of B into the boundary is greater than the flux of A and the volume of the alloyed regions will increase putting the surface into compression. Shewmon illustrated this effect by referring to differences in migration rates with sample thickness (76). In a thick sample, only the grain boundaries near the surface migrate while the same grain boundaries in the interior do not migrate. In thinner samples, grain boundaries migrate through the entire foil thickness and at rates faster than boundaries in the thicker samples. Shewmon postulated that the difference in volume between the alloyed and unalloyed bulk in the interior leads to a stress which slows grain boundary diffusion and migration.

To determine the magnitude of this effect, several other geometrical factors must be considered. By examining typical DIGM microstructures, it becomes obvious that there is considerable unalloyed material distributed throughout the bulk. Large unalloyed regions are encircled by alloyed regions, as shown in Fig. 3(a). Unequal boundary or lattice diffusivities of the chemical species will lead to a stress in these regions whether the boundary moves as a unit through the entire sample thickness or only near the surface. Thus, thick samples would not necessarily show significantly different migraton rates than thinner samples. Migration will be affected by this stress build-up regardless of the sample dimensions.

An alternate explanation for the increase in migrating rate with decreasing sample thickness is increased solute concentrations in the boundary in the thinner samples due to diffusion from both foil surfaces. Diffusion along a grain boundary from both foil surfaces will lead to overlapping diffusion fields and higher local concentrations. Migration due to diffusion-induced strains can begin sooner and occur at a faster rate with a higher solute concentration in the grain boundary.

Grain Boundary Structure Effects in DIGM. An obvious condition for DIGM is that grain boundary diffusion must be fast enough to produce significant penetration into the grain boundary. The observation of migration rates varying from grain boundary to grain boundary would seem, at first analysis, to be directly related to the differences in grain boundary diffusivities from boundary to boundary, that is, that some critical concentration of solute is required in the grain boundary before migration can begin. Results of Ostertag on grain boundary diffusion in Ni(In) (124) and of Shen and King on DIGM in [001] tilt boundaries in Cu(Zn) (76) indicate that the effect of grain boundary structure on migration is more complex.

Boundary migration was observed in most Ni bicrystals (approximately 20 out of 24) during diffusion anneals using In as the solute. Using optical microscopy, the boundaries in the other four bicrystals appeared to remain stationary. Either grain boundary diffusion produced measurable alloying by diffusion into stationary boundaries or migration and alloying occurred by DIGM but at a much slower rate than for other boundaries so that migration was not detectable by optical microscopy. In the study of DIGM in Cu(Zn), low Σ boundaries were observed to migrate more slowly than random high angle grain boundaries, as shown in Fig. 20. When the low Σ boundaries did migrate, the Zn concentrations in the regions swept by these moving grain boundaries were higher than for random high angle boundaries. This difference in concentration indicates that the solute content in the special grain boundary may be higher than originally expected, approaching that of the random high angle grain boundary, and that a critical concentration of solute in the grain boundary is not an adequate criterion for the initiation of migration. The lower migration rates of special boundaries may indicate: lower mobilities for special boundaries; changes in grain boundary energy with boundary plane for these special boundaries; or changes in the magnitude of the grain boundary Kirkendall effect with grain boundary structure.

Relative Rates of Grain Boundary and Lattice Diffusion. For a diffusion-induced strain to occur adjacent to the grain boundary, lattice diffusion must provide significant penetration into the lattice from the boundary. The rate at which this penetration occurs for a

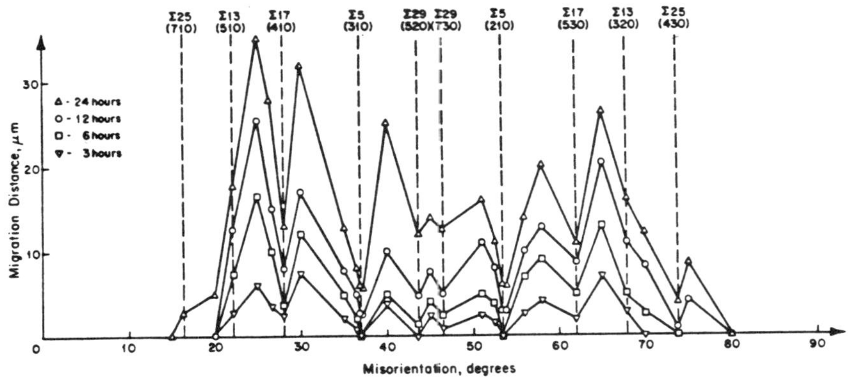

Fig. 20. Migration distance (measured on the free surface) versus misorientation angle for [100], initially symmetric tilt boundaries in Cu exposed to Zn vapor for various times. Each point on the graph corresponds to a single bicrystal specimen. (Courtesy of F.-S. Chen and A. H. King (76).)

given distance along the grain boundary from the solute source depends on the relative rates of boundary and lattice diffusion. The activation energy for boundary diffusion is less than that for lattice diffusion so that grain boundary diffusion increases relative to lattice diffusion as the temperature decreases. Grovenor (34) has analyzed the penetration depth as a function of temperature and distance from the solute source by numerically integrating the approximate Fisher solution for infinite source conditions. For Au(Cu) at 150°C, almost all of the solute diffusing along the boundary from the top surface reaches a depth of 1×10^{-6}m whereas only 55% of the solute reaches 10 micrometers from the foil surface. At 300°C, only 40% of the solute flux reaches 1×10^{-6}m and almost no solute reaches a depth of 1×10^{-5}m. If diffusion into the lattice is required to initiate DIGM, this required amount of diffusion will occur at shorter distances from the solute source as the temperature increases. This effect of temperature on the depth at which alloying by DIGM occurs is seen in Fig. 3 for Fe(Zn) at 550°C and 600°C.

Stress Relief Mechanisms. Coherency stresses in the lattice adjacent to the grain boundary and stresses caused by unequal diffusivities of the chemical species will generate a stress field ahead of the migrating boundary. Lattice dislocations can be attracted to the interface where they will cause local stress relief. This stress relief can be transitory; these lattice dislocations which impinge on the grain boundary then become short circuit diffusion paths for solute diffusion. Dislocation pipe diffusion and diffusion into the lattice from these dislocations can ultimately create higher local stresses than existed before impingement of the dislocations. Dislocations in the vicinity of the grain boundary may explain the "S-shape" of DIGM boundaries. If dislocation pipe diffusion is fast compared with migration, the region containing the dislocations will become stressed and migration will occur towards these regions of higher stress.

Other Driving Forces. Curvature, prior cold work, and any other driving force for boundary migration can act simultaneously with DIGM. If one of these driving forces is present in a DIGM experiment, a kinetic analysis of migration must consider its effect. For the case of LFM, the effect of curvature was straightforwardly included in the coherency strain model. The compositions of solid and liquid in equilibrium at a curved interface are different from the equilibrium compositions at a planar surface. For the case of DIGM, it is not obvious how to include these other driving forces with the complex effects described above into a physically meaningful kinetic model.

Mechanism Maps for DIGM and DIR

Although the features of the kinetic model have only been described qualitatively, a velocity-diffusion diagram for DIGM and DIR has been constructed which contains many of the basic elements of the coherency strain theory for grain boundaries and includes the effect of plastic deformation induced by a lattice Kirkendall effect (125). This mechanism diagram, based on the general diagram constructed by Cahn and Balluffi (88) to describe diffusion processes in stationary and migrating grain boundaries, represents a new approach to quantifying the phenomena of DIGM and DIR. Here, we describe, first, the diffusion and migration regimes defined for general migration and diffusion by Cahn and Balluffi. Next, we describe the additional regimes which arise in DIGM and DIR as the relative amounts of lattice and boundary diffusion change. We then use this diagram to describe the diffusion and velocity regimes for a grain boundary during a diffusion anneal.

The DIGM and DIR diagram, shown in Figure 21(a), has axes of log(velocity • time) and log (lattice diffusivity • time). In this figure, V is velocity, t is time, D_ℓ is the lattice diffusivity, λ is the grain boundary width, S is the grain size, L_c is the critical layer thickness for coherency loss, and λ' is the minimum penetration distance into the lattice for coherency strain to exist. The term λ' is defined below. In this diagram, grain boundary diffusion is assumed to produce a constant composition in the grain boundary. The parameters which change in this diagram are the penetration distance from the grain boundary into the lattice, as described by $\sqrt{D_\ell t}$, and the distance which the grain boundary has migrated, as described by Vt.

From this basic diagram, the effect of lattice diffusion on the amount of penetration from the grain boundary into the adjacent lattice can be understood for stationary boundaries by referring to the region when $Vt < \lambda$. When the boundaries migrate ($Vt > \lambda$), the amount of penetration from the grain boundary into the lattice depends on the relative rates of lattice diffusion, which moves the diffusing species into the lattice ahead of the grain boundary, and grain boundary migration, which "dissolves" the diffusion zone ahead of the grain boundary and "reprecipitates" this material as the boundary moves.

Cahn and Balluffi first considered the case of stationary grain boundaries, that is, when $Vt < \lambda$. With fast grain boundary diffusion in stationary grain boundaries, there are three regimes on this diagram corresponding to the definitions of Harrison Type-A, -B, and -C diffusion. For $D_\ell t < \lambda^2$, penetration into the lattice is less than the λ and, hence, grain boundary diffusion dominates (Harrison Type-C). This situation will occur at short times at high temperatures where lattice diffusion is fast compared with grain boundary diffusion and at increasingly longer times as the lattice diffusivity and temperature decrease. For $D_\ell t > S^2$, the penetration of the grain by lattice diffusion exceeds the grain boundary spacing. Each diffusing species will have diffused through several grain boundaries in addition to diffusing through the lattice between the grain boundaries. This regime corresponds to Harrison Type-A behavior. At intermediate temperatures and times, such that $\lambda^2 < D_\ell t < S^2$, lattice diffusion is fast enough to produce some penetration from the grain boundary into the grain but slow enough that the diffusing species only diffuses within a single grain. This region corresponds to Harrison Type-B diffusion.

They next considered the regimes which exist when the grain boundary is moving. These regimes are based on the combined effects of migration and lattice diffusion on the composition profile either ahead of the moving interface, Eq. (15) above, or behind the moving interface, Eq. (14). When $D_\ell/V < \lambda$, little diffusion occurs in the lattice ahead of the migrating boundary. This inequality is bounded by a line of slope = 1 on the log-log plot in Fig. 21(a). This boundary intersects the line for $D_\ell t = \lambda^2$ at $Vt = \lambda$. When $\sqrt{D_\ell t} + Vt > S$, all grains have been either swept by grain boundaries ($Vt > S$) or diffusion from the lattice has occurred in several grains ($D_\ell t > S^2$) and along several grain boundaries. Under these conditions, intermixing may be complete.

Starting from this basic description of diffusion in migrating and and stationary grain boundaries, the $D_\ell t$-Vt diagram has been modified to describe the conditions for initiation and migration of grain boundaries driven by diffusion down the grain boundaries and into the adjacent lattice. Three new lines appear on the $D_\ell t$-Vt diagram for DIGM. These three lines correspond to the conditions of no penetration into the lattice adjacent to the grain boundary, coherency loss in the growing grain, and coherency loss in the dissolving grain.

In the regimes for DIGM, the existence of a coherency strain in one or both adjacent grains at the start of migration and the maintenance of these strains during migration determine the migration rate by purely a coherency strain effect. It has been suggested above that diffusion in the lattice adjacent to the grain boundary is similar to grain boundary diffusion. When grain boundary diffusion produces significant penetration into the grain boundary plane, there will be some lateral leakage from the grain boundary to give rise to coherency stresses. There will be a lower limit to the value of $D_\ell t$ for which this will hold. The parameter λ' is defined to be the minimum penetration distance which will lead to a coherency strain in the lattice adjacent to the boundary. The value for λ' is calculated based on the $\lambda' = \sqrt{D_\ell t}$ for a stationary grain boundary and $\lambda' = D_1/V$ for migrating grain boundaries. This corresponds to the solid line in Fig. 21(a). Below this solid line, no DIGM is predicted. When coherency is lost in both grains or when a lattice Kirkendall effect exists, the process which can drive migration becomes recrystallization of a plastically deformed solid by the motion of pre-existing or newly formed grain boundaries. In between the two extremes of behavior, no diffusion into the adjacent lattice and extensive diffusion into the lattice, DIGM driven by coherency strains is predicted.

In the diagram in Fig. 21(a), it was assumed that the critical layer thicknesses for coherency loss for two adjacent grains, Grain 1 and Grain 2, are different. Without loss of generality, it was assumed that $L_1 < L_2$. For a stationary boundary, $(Vt < \lambda)$, coherency loss will occur in Grain 1 when $D_\ell t = (L_1)^2$. The condition of coherency loss in only one grain corresponds to the initiation mechanism implied in Eq. (13). When migration begins without coherency loss in one grain, as described by Eq. (12), coherency loss will occur in the growing grain when $\sqrt{D_\ell t} + Vt = L_1$, Eq. (14). When $(D_\ell t) = 0$, coherency loss in the growing grain will occur when $L_1 = Vt$. The curve with the extremes of $(D_\ell t) = (L_1)^2$ and $Vt = L_1$ define the transition between migration with no coherency loss, Eq. (12), and migration with loss of coherency in the growing grain, Eq. (13).

If migration does not occur before $D_\ell t = (L_2)^2$, coherency loss will occur in Grain 2 and coherency loss will then have occurred in both grains. When the grain boundary is migrating, coherency will be

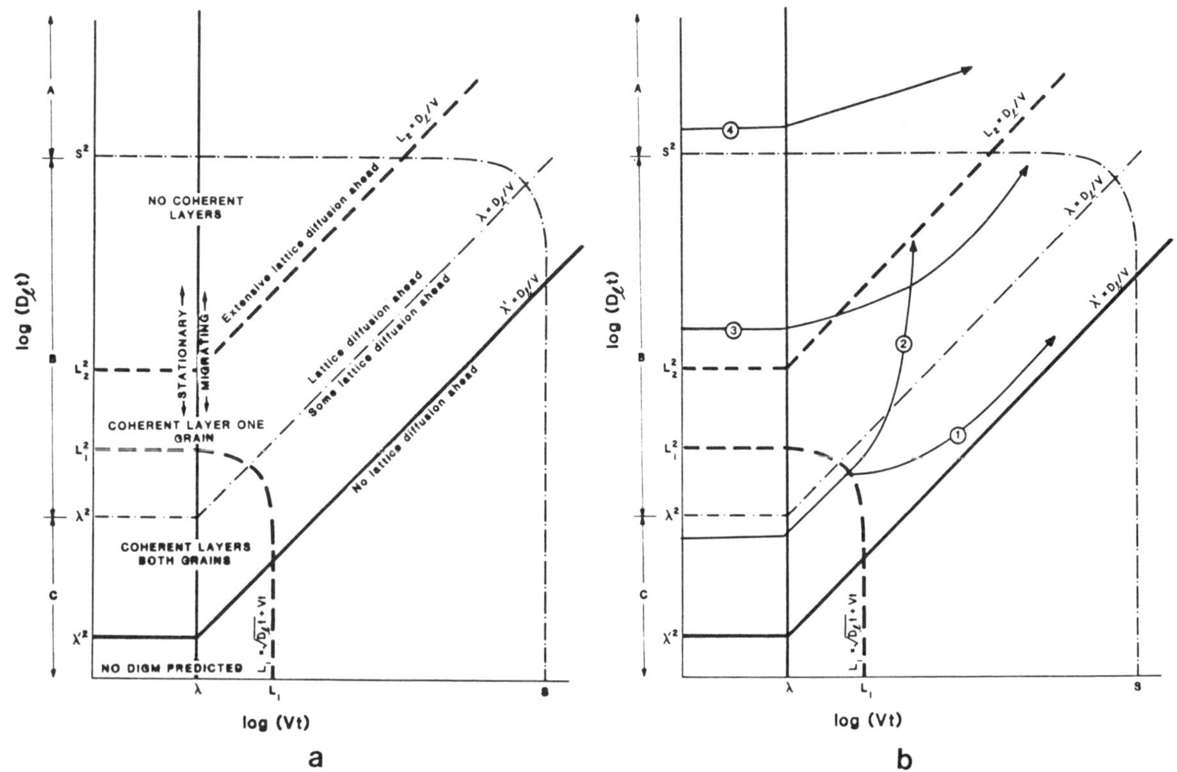

Fig. 21. Mechanism maps for DIGM and DIR (a) basic regimes (b) trajectories for various diffusion temperatures and additional driving force. Dotted lines refer to the original regimes defined in (88). Solid lines delineate additional regimes for DIGM and DIR (125).

lost if $D_\ell/V > L_2$, that is, the diffusion zone ahead of the migrating boundary exceeds L_2, as described by Eq. (15) above. In the region defined by $\sqrt{D_\ell t} + Vt > L_1$ and $D_\ell/V < L_2$, migration can occur with a coherent layer maintained ahead of the grain boundary and migration will be described by Eq. (13).

If coherency loss occurs in both grains, migration cannot be induced by differences in coherency strain energy across the grain boundary. When coherency is lost, another driving force for migration can operate. If a lattice Kirkendall effect exists, as was described above, there will be a high dislocation content adjacent to the grain boundary. Migration of the grain boundaries can occur if the dislocation density adjacent to the boundary is sufficiently high to allow local bowing out of the grain boundary (125). Recrystallization by motion of pre-existing grain boundaries can incorporate solute into the region swept by the moving grain boundary if grain boundary diffusion is significantly higher than lattice diffusion (88). This diffusion-induced recrystallization will occur in the region where $D_\ell/V > L_2$.

As the values of $D_\ell t$ increase, lattice diffusion from the free surface into the grains can produce significant solute concentrations near the grain boundaries intersecting the surface. If lattice diffusion occurs with unequal diffusivities of the diffusing species and with a change in lattice parameter with composition, the diffusion zone at the free surface can become stressed as a result of the change in lattice parameter with composition and the climb of dislocations into the lattice. When the stress is sufficiently high, recrystallization (DIR) in regions away from the grain boundaries can occur as a result of lattice diffusion by the nucleation of new grains within the stressed matrix. Solute can be incorporated in the growing grains at higher levels than in the stressed matrix if grain boundary diffusion remains significant. If lattice diffusion is sufficiently fast compared with grain boundary diffusion, no solute enrichment will occur as the boundaries migrate.

The sequence of events leading up to DIGM and DIR and the transition between different diffusion and migration regimes can be understood by following trajectories on the $D_\ell t$-Vt diagram. Consider first trajectory 1 in Fig. 21(b), which corresponds to diffusion at low temperatures where D_ℓ is low and grain boundary diffusion is significant and to grains with a difference in elastic coefficients, ΔY. DIGM will occur as soon as there is sufficient penetration into the lattice to create an elastic energy difference across the grain boundary as a result of a difference in coherency strain resulting from ΔY. The amount of coherency strain required to move the boundary depends on the presence of other driving forces, such as curvature, which could aid or suppress migration. For trajectory 1, it is assumed that only diffusion-induced driving forces are present, that penetration distance into the lattice required to initiate DIGM is small and occurs as soon as grain boundary diffusion is significant, and the amount of penetration into the lattice is small enough to ignore the effect of a lattice Kirkendall effect. In a previous section we have asserted that coherency strains adjacent to the grain boundary and a grain boundary Kirkendall effect occur approximately simultaneously and, thus, are not separable. Migration will be described by Eq. (12) as long as coherency is maintained in both grains. When coherency is lost in the growing grain at $L_1 = \sqrt{D_\ell t} + Vt$, migration will then be described by Eq. (13). Migration will continue until $S = \sqrt{D_\ell t} + Vt$, that is, until each grain has been swept by a moving grain boundary.

The effect of an opposing driving force, such as curvature, can be seen by referring to trajectory 2. The value of $D_\ell t$ at $Vt = \lambda$ is the same as for trajectory 1. In many DIGM systems, grain boundary curvature increases as the boundary migrates. This implies that the $\sigma_{gb}\kappa$ term in Eqs. (12) and (13) increase as migration continues and boundary velocity decreases. Since the velocity decreases while D_ℓ remains constant, coherency loss can occur in the dissolving grain when $D_\ell/V > L_2$. Under these conditions of coherency loss in both grains, the grain boundary will move by capillarity back to the initial position of the grain boundary after crossing the line $L_2 = D_\ell/V$.

Diffusion at higher temperatures is illustrated by trajectories 3 and 4. For trajectory 3, diffusion produces coherency loss in both grains before migration can occur. At these higher temperatures, a lattice Kirkendall effect can operate in addition to a grain boundary Kirkendall effect. The strains in the zone of lattice diffusion adjacent to the grain boundary will be a function of the lattice parameter mismatch and the magnitude of the lattice Kirkendall effect. The driving force for migration of the grain boundary is the same as for migration into a plastically deformed single component system, except in this case the deformation was produced by diffusion. Local differences in the stresses in the adjacent grains may cause the boundary to migrate with an "S-shape". Migration at this temperature can occur with significant penetration into the lattice ahead of the migrating interface. Migration can continue after the region of local deformation has been swept if a coherently stressed layer forms by diffusion ahead of the migrating boundary, that is, if the boundary velocity is such that $L_2 > D_\ell/V$.

In trajectory 4, lattice diffusion from the surface in contact with the solute source occurs concurrently with DIGM. Extensive plastic deformation can occur in the diffusion zone at the free surface where lattice diffusion has occurred. For the grain boundaries near the surface, there will be two driving forces for boundary motion at long times. Coherency stresses in the diffusion zone ahead of the boundary due to lattice parameter mismatch and plastic deformation in the lattice as a result of a lattice Kirkendall effect from the free surface and the grain boundary can both act as driving forces for migration. Within this region of extensive plastic deformation near the free surface, stress-free regions can be produced either by the migration of pre-existing grain boundaries or by the nucleation and growth of new grains. Plastic deformation and formation of surface grains (DIR) can occur in the surface diffusion zone in contact with the solute source before pre-existing grain boundaries can migrate. Solute can be incorporated at high levels into the regions swept by the moving grain boundaries if grain boundary diffusion is still fast relative to lattice diffusion regardless of the driving force for migration. The regime for DIR occurs at large values of $D_\ell t$ and may occur concurrently with migration of pre-existing grain boundaries in regions far from the free surface for which $L_2 < \sqrt{D_\ell t}$.

These time and temperature regimes for different interdiffusion processes are illustrated using data from the literature for the Cu(Zn) system. The studies described here include both the initiation and migration regimes at intermediate and high temperatures. A trajectory at a specific temperature can be followed qualitatively by looking at the microstructure as a function of distance from the solute source. At the surface in contact with the solute source, the lattice and the grain boundaries see the effect of the solute immediately. Grain boundary and lattice diffusion are required to supply solute to the regions farther away from the surface. This implies that the effective annealing time decreases from the free surface towards the interior (74,126).

The intermediate temperature experiments by Hackney and coworkers (74,126) were performed using high purity, recrystallized Cu foils annealed for 32 and 60 hrs. at 360°C with a Zn source containing 11 at.% Zn. The high temperature experiments were performed with single crystals of Cu supplied with Zn through the vapor phase (9). Five separate regimes were noted which correspond to trajectories 3 and 4 in Fig. 21(b): (1) no migration farthest from the sample surface; (2) local plastic deformation adjacent to the boundary before migration occurred followed by bowing out of the boundary; (3) DIGM boundaries showing grain boundary steps and having low curvatures; (4) DIGM boundaries with high curvatures growing into a highly deformed matrix; and (5) DIR grains nucleating at the surface in contact with the vapor source. Characteristic microstructures in these regimes are shown in Fig. 22. The density of lattice defects in the region showing no migration was very low and corresponded to that in the as-recrystallized foils. Segments of the boundary closer to the solute source showed extensive, localized plastic deformation near the grain boundary but with no apparent migration. Some segments show the same deformation characteristics but with a small scale undulation in the boundary plane. In Fig. 22(a) is shown a plastically deformed, undulated grain boundary segment adjacent to a grain boundary seg-

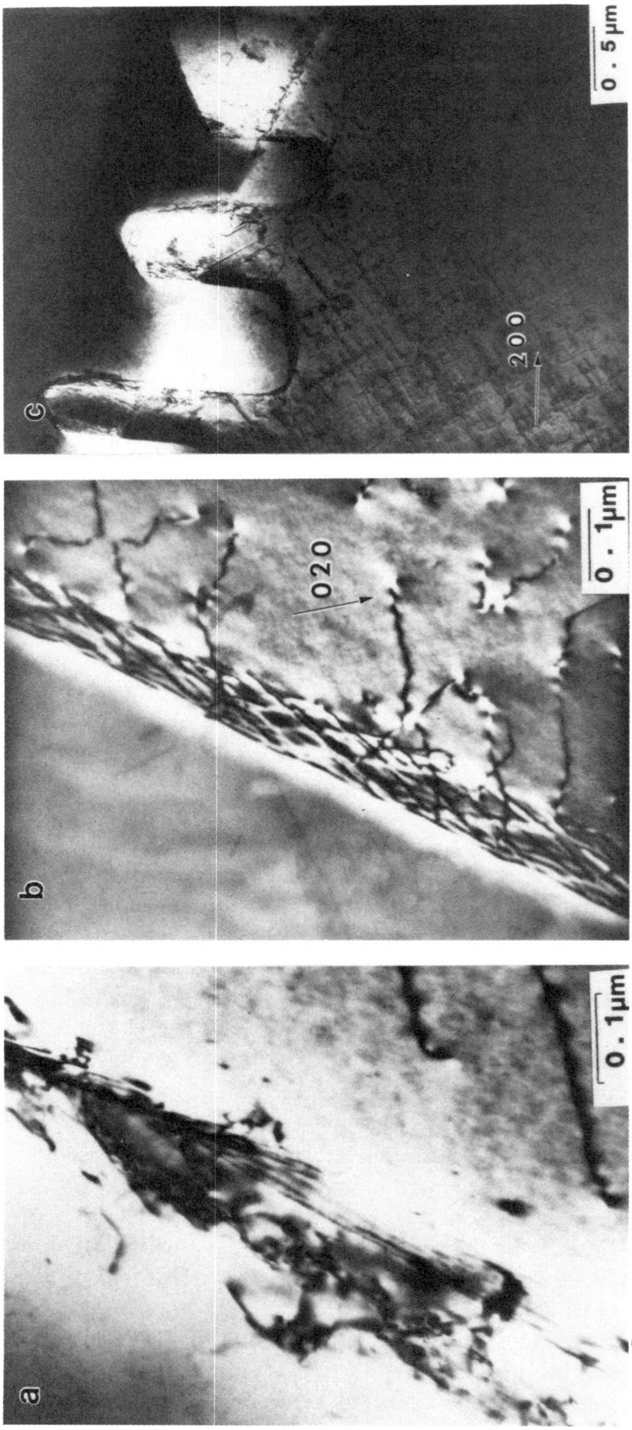

Fig. 22. TEM micrographs of boundaries in Cu(Zn) during DIGM. (a) Initiation of migration adjacent to region with local plastic deformation (126); (b) Migrated grain boundary has a high density of defects with step character in the grain boundary; (c) Sine wave form of the grain boundary caused by the migration of different segments into highly deformed lattice on both sides of the grain boundary (74).

ment which has migrated. The dislocations behind the migrated boundary appear to be arranging into a dislocation wall typically associated with the initial boundary position. Where more migration has occurred, the boundary appears to have migrated out of the region of plastic deformation. In the region showing extensive migration, Fig. 22(b), grain boundaries which migrated into unalloyed matrix contained grain boundary steps and had a near zero curvature. The matrix had a low dislocation density. Groups of isolated dislocations were occasionally seen near the interface ahead of the migrating grain boundary. The growing grain was alloyed with approximately 5% Zn. In the region immediately adjacent to the surface, Fig. 22(c), DIGM occurred into the highly deformed matrix with grain boundaries exhibiting grain boundary steps and a high boundary curvature (κ = $1.1 \times 10^7 m^{-1}$). The average Zn concentration in the alloyed region was 5.4% Zn. The matrix grains were alloyed with approximately 1.5% Zn. The dislocations in the matrix were sessile, undissociated Lomer-Cottrell dislocations, characteristic of slip on different slip systems in fcc metals. Comparisons of the coherency strain energy with the $\sigma_{gb}\kappa$ term in this case indicate that coherency strain energy cannot be the only driving force driving the motion of this boundary. In the sample annealed for 60 hrs., new dislocation-free, alloyed grains formed by DIR were observed on the surface in contact with the vapor source.

In the high temperature experiments by Doo and Balluffi (9), Cu single crystals were annealed with fine alpha brass chips at 680° and 880°C for times up to 730 h. The following regimes were found within the diffusion zone: (1) dislocation formation; (2) arrangement of dislocations into sub-boundaries; (3) recrystallization and the formation of new grains; and (4) twin formation. The dislocations formed at the lower diffusion temperature appeared to be associated with slip planes, indicating the presence of slip dislocations. The development of the sub-boundary network occurred continuously and the structures became coarse and equiaxed at the higher temperature. These morphologies would be expected after most diffusion had occurred since fewer dislocations would be then added. At the higher temperature, sub-grain formation and recovery were observed as annealing time increased. At the lower temperature, the high dislocation density persisted for long times. The formation of new grains was favored by low temperatures and long diffusion periods and by low initial crystal perfection. This nucleation and growth of new grains (DIR) can occur concurrently with DIGM for large values of $D_\ell t$, such as for trajectory 4. The presence of dislocations in regions alloyed by lattice diffusion from the surface exposed to the solute source and the formation of new grains within the highly deformed matrix are results common to experiments performed at 360°C (74,126) and 680°C (9).

The mechanism maps, shown in Fig. 21, help us to understand the DIGM and DIR phenomena by placing them in the context of diffusion, migration, and recrystallization processes which are qualitatively understood. DIGM occurs as a result of stresses introduced in the near grain boundary region, by diffusion from the grain boundary into the lattice and, in some cases, by additional diffusion from the free surfaces near the solute source into the lattice. New grains nucleate and grow by DIR in the diffusion zone near the free surface as a result of lattice diffusion with a Kirkendall effect and the change in lattice parameter with composition. The specific phenomena which occur depend on complex factors, described above, including the relative rates of boundary and lattice diffusion and the initial dislocation density in the samples. The concept of diffusion causing stress differences across a grain boundary applies regardless of whether the diffusion-induced stress is accommodated elastically or plastically or whether dislocations are generated in the diffusion zone as a result of a lattice Kirkendall effect.

By inspection of the typical DIGM and DIR microstructures, it becomes obvious that DIGM and DIR cannot be described as steady-state processes. During DIGM, the curvatures of the grain boundary segments are changing. Since curvature provides a driving force for the motion of the boundary in DIGM and DIR, the total driving force for boundary motion will change as the curvature changes. The diffusion distance from the solute source increases as the boundary segment increases in length through the film thickness leading to a decreasing supply of solute at a given depth in the foil. If the misorientation of the grain boundary changes by absorption or emission of dislocation arrays, the grain boundary diffusion and migration characteristics can also change discontinuously. In spite of the difficulty in accurately modelling these complex processes, the phenomena of DIGM and DIR can be understood qualitatively by relating them to the processes of grain boundary diffusion, lattice diffusion, grain boundary migration, and recrystallization.

Special Considerations for Thin Films

Diffusion in thin films can be significantly more complex than diffusion in well-annealed bulk single crystals and polycrystals. The characteristics of thin films which drastically affect diffusion were summarized by Balluffi and Blakely (127). These characteristics are: (1) all volume elements are close to a free surface or an internal interface; (2) defect densities can be high; (3) biaxial stress state can exist as a result of the film deposition process; (4) high impurity concentrations can be incorporated during processing; and (5) the crystal structure of the film can be metastable or disordered. For example, thin films can contain dislocation densities as high as $10^{10}-10^{11}$ cm^{-2} and can have an average grain size smaller than 10 nm (128,129).

As a result of these factors causing the "poor thermal stability" of thin films, recrystallization and grain growth as well as diffusion and diffusion-induced phenomena, such as DIGM and DIR can occur at low temperatures. For example, the high dislocation densities that pre-exist in the thin films may lead to nucleation and growth of new grains at the same time that pre-existing grain boundaries move due to both solute diffusion and the local density of pre-existing dislocations adjacent to the grain boundary. When more than one driving force is present, it is expected that the migration velocity will reflect the total local driving force. For example, boundary motion caused intermixing of Au-Ag thin films at room temperature, with measurable interdiffusion after 5 h. (4). In other experiments, where it was determined that only DIGM was occurring in Au-Ag thin film specimens, the migration velocities were significantly lower (32,33).

DIGM and DIR are important processes affecting the thermal stability of thin films. As can be seen in Table 1, the majority of experiments where DIGM was observed were performed using thin films. These diffusion-induced phenomena are widespread; DIGM has been observed in almost every binary thin film system where it has been sought.

The one example where no DIGM has yet been observed is Au(Al), a technologically important binary system (41). Extensive lattice diffusion of Al into Au was seen at 270°-450°C with no accompanying migration of grain boundaries. However, as pointed out (41), DIGM may be important in Au(Al) at temperatures lower than 270°C.

Recent results on grain boundary migratio in P-doped polycrystalline Si suggest that DIGM and DIR may be of increasing importance in VLSI technology. In many of the experiments described above, DIGM and DIR led to a degradation of materials properties. However, in a report by Chou, Wong, and Tu, a grain size increase from 10 nm to 400 nm occurred when phosphorus diffused along the grain boundaries from phosphorus-doped poly Si (130). The phosphorus was removed from the poly Si by a combination of grain boundary diffusion and grain boundary migration during formation of $TiSi_2$ at the interface between Si and Ti and the phosphorus was incorporated into the $TiSi_2$ phase. When $TiSi_2$ formed by reaction of undoped Si and Ti, no grain boundary migration occurred and the grain size remained at 10 nm. These two results suggest that DIGM may be occurring at selected grain boundaries to lead to an overall increase in grain size. An additional interesting observation in this study was that large voids occurred at the interface between the poly Si and the silicide layer during silicide formation. These voids are typically known as "Kirkendall" voids because they are due to the unequal diffusivities of solute and solvent species. In the phosphorus-doped Si, no voids were formed during silicide formation. The absence of voids at the Si-silicide interface is desired from the viewpoint of device reliability. More importantly, producing local redistribution of dopant at the scale of a single grain by DIGM may allow the production of uniquely small devices, such FET's, unattainable by more conventional VLSI technologies.

ACKNOWLEDGEMENTS

Critical discussions on DIGM and LFM with John E. Blendell, John W. Cahn, Steve A. Hackney, John R. Manning and Duk N. Yoon are gratefully acknowledged. Stimulating discussions were also held with Wolfgang A. Kaysser while the author was a visiting scientist at Max-Planck-Institut, PML, Institut für Metallforschung, Stuttgart, West Germany. Thomas W. Coyle, Ann Van Orden, and Robb Thomson are thanked for their careful reviews of the manuscript, as are Y.-J. Baik, F.-S. Chen, C. R. M. Grovenor, R. S. Hay, R. J. Jahn, Alex H. King, J.-D. Pan, G. R. Purdy and K.-N. Tu for supplying micrographs and results from unpublished research. Special thanks also go to Peter K. Schenck for the development of the computer graphics to display the shapes of the growing grains. This work was supported in part by the Office of Naval Research.

REFERENCES

1. Shechtman, D., Blech, I., Gratias, D., and Cahn, J. W., Metallic phase with long-range orientational order and no translational symmetry, *Phys. Rev. Lett.* **53** 1951-1953 (1984).
2. Gupta, D., Campbell, D. R., and Ho, P. S., in "*Thin Films-Interdiffusion and Reaction*" (J. M. Poate, K. N. Tu, and J. W. Mayer, ed.), pp. 119-160, John Wiley, New York (1978).
3. Baglin, J. E. E., and Poate, J. M., in "*Thin Films-Interdiffusion and Reaction*" (J. M. Poate, K. N. Tu, and J. W. Mayer, ed.), pp. 305-358, John Wiley, New York (1978).
4. Kirsch, R. G., Poate, J. M., and Eibschutz, M., Interdiffusion mechanisms in Ag-Au thin-film couples, *Appl. Phys. Lett.* **29**, 772-774 (1976).
5. den Broeder, F. J. A., Interface reaction and a special form of grain boundary diffusion in the Cr-W system, *Acta Metall.* **20**, 319-332 (1972).
6. Tu, K. N., Kinetics of thin-film reactions between Pb and AgPd alloys, *J. Appl. Phys.* **48**, 3400-3404 (1977).

7. Hillert, M., and Purdy, G. R., Chemically induced grain boundary migration, *Acta Met.* **26**, 333-340 (1978).
8. Balluffi, R. W., and Cahn, J. W., Mechanism for diffusion induced grain boundary migration, *Acta Metall.* **29**, 493-500 (1981).
9. Doo, V. Y., and Balluffi, R. W., Structural changes in single crystal copper-alpha brass diffusion couples, *Acta Metall.* **6**, 428-438 (1958).
10. Matthews, J. W., and Crawford, J. L., Formation of grain boundaries during diffusion between single crystal films of gold and palladium, *Phil. Mag.* **11**, 977-991 (1965).
11. Matthews, J. W., and Jesser, W. A., Formation of grain boundaries during diffusion between single-crystal films, *J. Vac. Sci. Technol.* **6**, 641-644 (1969).
12. Matthews, J. W., Some features of the behavior of misfit dislocations during diffusion, *Thin Solid Films* **25**, 199-211 (1975).
13. Rhines, F. N., and Montgomery, A. M., A new type of structure in the alpha-copper-zinc alloys, *Nature* **141**, 413 (1938).
14. Rhines, F. N., and Mehl, R. F., *Trans. Amer. Inst. Min. (Metall.) Engrs.* **128**, 185 (1938).
15. Yoon, D. N., and Huppmann, W. J., Grain growth and densification during liquid phase sintering of W-Ni, *Acta Metall.* **27**, 693-697 (1979).
16. Kaysser, W. A., and Pejovnik, S., Grain boundary migration during sintering of Mo with Ni additions, *Z. Metall.* **71**, 649-653 (1980).
17. Song, Y. D., and Yoon, D. N., The driving force for chemically induced migration of molten Ni films between W grains, *Met. Trans.* **15A**, 1503-1504 (1984).
18. Hofmann-Amtenbrink, M., Kaysser, W. A., and Petzow, G., Grain boundary migration in recrystallized Mo foils in the presence of Ni *J. Phys. C4* **46**, C4-545-C4-552 (1985).
19. Baik, Y. J., and Yoon, D. N., Migration of liquid film and grain boundary in Mo-Ni induced by a temperature change, *Acta Metall.* **33**, 1911-1917 (1985).
20. Song, Y. D., Ahn, S. T., and Yoon, D. N., Chemically induced migration of liquid film in W-Ni-Fe alloy, *Acta Metall.* **33**, 1907-1910 (1985).
21. Handwerker, C. A., Cahn, J. W., Yoon, D. N., and Blendell, J. E., in "*Diffusion in Solids: Recent Developments*", TMS/AIME Publication, Warrendale, PA (1985).
22. Yoon, D. N., Cahn, J. W., Handwerker, C. A., Blendell, J. E., and Baik, Y. J., in "*Interface Migration and Control of Microstructure*", ASM Press, Metals Park, OH (1985).
23. Hillert, M., On the driving force for diffusion induced grain boundary migration, *Scr. Metal.* **17**, 237-240 (1983).

24. Pearson, W. B., "*A Handbook of Lattice Spacings and Structures of Metals and Alloys*", Pergamon Press, Oxford (1967).
25. Tu, K. N., Mass transport along grian boundaries in silver, *J. Appl. Phys.* **43**, 1303-1304 (1972).
26. Tu, K. N., and Rosenberg, R., Mass transport in layered polycrystalline thin films, *Thin Solid Films* **13**, 163-167 (1972).
27. Tu, K. N., and Chance, D. A., Thin-film reactions of Pb with AgAu and Ag-Pd alloys, *J. Appl. Phys.* **46**, 3229-3234 (1975).
28. Li, Chongmo, and Hillert, M., Diffusion induced grain boundary migration in Cu-Zn, *Acta Metall.* **30**, 1133-1146 (1982).
29. Tashiro, Kelvin, and Purdy, Gary R., The role of volume diffusion, in DIGM, a reappraisal, *Scr. Metall.* **17**, 455-458 (1983).
30. Purdy, G. R., and Tashiro, K., in "*Interface Migration and Control of Microstructure*", ASM Publication, Metals Park, OH (1986).
31. Cahn, J. W., Pan, J. D., and Balluffi, R. W., Diffusion induced grain boundary migration, *Scripta Metall.* **13**, 503-509 (1979).
32. Pan, Ju-Don, Diffusion Induced grain boundary migration in Au-Cu and Au-Ag thin films, Ph.D. Thesis, Cornell University, May (1980).
33. Pan, J. D., and Balluffi, R. W., Diffusion induced grain boundary migration in Au/Cu and Au/Ag thin films, *Acta Metall.* **30**, 861-870.
34. Grovenor, C. R. M., Mechanisms of interdiffusion reactions during the heating of metallic layer contact systems, *Thin Solid Films* **89**, 367-375 (1982).
35. Butrymowicz, D. B., Newbury, D. E., Turnbull, D., and Cahn, J. W., Diffusion induced grain boundary migration in the Au-Ag system, *Scr. Metall.* **16**, 1005-1010 (1984).
36. Tu, K. N., and Berry, B. S., X-ray study of interdiffusion in bimetallic Cu-Au films, *J. Appl. Phys.* **43**, 3283-3290 (1972).
37. Grovenor, C. R. M., Diffusion induced grain boundary migration in thin gold/copper films, *Acta Met.* **33**, 579-586 (1985).
38. Hall, P. M., Morabito, J. M., and Poate, J. M., Diffusion mechanisms in the Pd/Au thin film system and the correlation of resistivity changes with Auger electron spectroscopy and Rutherford backscattering profiles, *Thin Solid Films* **33**, 107-134 (1976).
39. Debonte, W. J., and Poate, J. M., Low temperature interdiffusion in the Au-Pd and Au-Rh thin film systems, *Thin Solid Films* **25**, 441-448 (1975).
40. Vanderberg, J. M., den Broeder, F. J. A., and Ramm, R. A., Formation of intermetallics and grain boundary diffusion in Cu-Al thin film couples, *Thin Solid Films* **93**, 277-280 (1982).
41. den Broeder, F. J. A., Klerk, M., Vanderberg, J. M., and Hamm, R. A., A comparative study of DIGM, recrystallization, and vol-

ume diffusion during the low temperature diffusion of Al into Cu and Au, *Acta Metall.* **31**, 285-291 (1983).

42. Kim, W., Meyrick, G., and Shewmon, P. G., Diffusion induced boundary migration and discontinuous precipitation in Cu alloys, *Scripta Metall.* **17**, 1435-1440 (1983).
43. Mittemeijer, E. J., and Beers, A. M., Recrystallization and interdiffusion in thin bimetallic films, *Thin Solid Films* **65**, 125-135 (1980).
44. Grovenor, C. R. M., Smith, D. A., and Goringe, M. J., Nucleation and migration of high angle grain boundaries in bilayer foils I: Nucleation, *Thin Solid Films* **74**, 257-267 (1980).
45. Grovenor, C. R. M., Smith, D. A., and Goringe, M. J., Nucleation and migration of high angle grain boundaries in bilayer foils II: Migration, *Thin Solid Films* **74**, 269-279 (1980).
46. Grovenor, C. R. M., On the measurement of interfacial diffusion coefficients during diffusion induced grain boundary migration, *Scripta Metall.* **16**, 317-320 (1982).
47. Novikov, V. I., Trusov, L. N., Lapovok, V. N., and Geleishvili, T. P., Mechanism of low-temperature diffusion activated by a migrating boundary, *Sov. Phys. Solid State* **25**, 2128-2129 (1984).
48. Lapovok, V. N., Novikov, V. I., Svirida, S. V., Semenikhin, A. N., and Trusov, L. I., Formation of non-equilibrium vacancies due to recrystallization of ultradisperse nickel powder, *Sov. Phys. Solid State* **25**, 1063-1064 (1983).
49. den Broeder, F. J. A., and Nakahara, S., Diffusion induced grain boundary migration and recrystallization in the Cu-Ni system, *Scripta Metall.* **17**, 399-404 (1983).
50. Nakahara, S., and den Broeder, F. J. A., Diffusion-induced grain boundary migration in heated Cu-Ni alloy targets during sputtering, *Scripta Metall.* **17**, 607-610 (1983).
51. Piccone, T. J., Butrymowicz, D. B., Newbury, D. E., Manning, J. R., and Cahn, J. W., Diffusion induced grain boundary migration in the Cu-Zn system, *Scripta Metall.* **16**, 839-843 (1982).
52. Tsai, Y. S., Meyrick, G., and Shewmon, P. G., Surface relief produced by diffusion induced boundary migration in Cu-Zn, *Met. Trans. A*, **15A**, 495-499 (1984).
53. den Broeder, F. J. A., Diffusion-induced grain boundary migration and recrystallization, exemplified by the system Cu-Zn, *Thin Solid Films*, **124**, 135-148 (1985).
54. Blendell, J. E., Handwerker, C. A., Kaysser, W. A., and Petzow, G., in "*Surface and Interfaces in Ceramic and Ceramic-Metal Systems*", Materials Science Research (J. Pask and A. G. Evans, ed.) Vol. 14, pp. 217-226, Plenum Press, NY (1981).

55. Li, Chongmo, and Hillert, M., A metallographic study of diffusion induced grain boundary migration in the Fe-Zn system, *Acta Metall.* **29**, 1949-1960 (1981).
56. Yu, Z. S., and Shewmon, P. G., Diffusion driven grain boundary migration in Fe during zincification, *Met. Trans. A* **13A**, 1567-1572 (1982).
57. Yu, Z. S., and Shewmon, P. G., Size effect, internal stress, and diffusion driven boundary motion, *Met. Trans. A* **14A**, 1579-1582 (1983).
58. Meyrick, G., Iyer, V. S., and Shewmon, P. G., Morphological changes due to diffusion induced grain boundary migration, *Acta Metall.* **33**, 273-279 (1985).
59. Hofmann-Amtenbrink, M., Kaysser, W. A., and Petzow, G., Grain boundary migration in recrystallized foils in the presence of Ni, *Z. Metallkde.* **73**, 305-310 (1982).
60. Baik, Y. J., Ph.D. Thesis, Korea Advanced Institute of Science and Technology (1986).
61. Parthasarathy, T. A., and Shewmon, P. G., Vapor transport and DIGM in the Ni-Fe system, *Met. Trans.* **14A**, 2560-2563 (1983).
62. Nakahara, S., Abys, J. A., and Abys, S. M., Room-temperature diffusion induced grain boundary migration in the fine-grained Pd side of Cu-Pd diffusion couples, *Mater. Lett.* **2**, 155-159 (1983).
63. Olowolafe, J. O., Tu, K. N., and Angiello, J., Contact reaction between Si and Pd-W alloy films, *J. Appl. Phys.* **50**, 6316-6320 (1979).
64. Muster, W. J., and Willerscheid, H., Crystallographic orientation relationship in coalescing sintered tungsten spheres, *Metallogr.* **12**, 287-294 (1979).
65. Evans, B., Hay, R. S., and Shimizu, N., Diffusionally induced grain boundary migration in calcite, *Geology* **14**, 60-63 (1986).
66. Grovenor, C. R. M., and Smith, D. A., Metal induced grain growth in germanium and silicon thin films, in *Thin Films and Interfaces* II (J. E. E. Baglin, D. R. Campbell, and W. K. Chu, ed.) North Holland, New York (1984).
67. Handwerker, C. A., and Blendell, J. E., submitted to *J. Amer. Ceram. Soc.*.
68. Parthasarathy, T. A., and Shewmon, P. G., DIGM in Ni-C, *Scripta Metall.* **17**, 943-946 (1983).
69. Harrison, L. G., Influence of dislocations on diffusion kinetics in solids with particular reference to the alkali halides, *Trans. Faraday Soc.* **57**, 1191-1199 (1961).
70. Askill, J., "*Tracer Diffusion Data for Metals, Alloys, and Simple Oxides*", IFI/Plenum Publishing Co., New York (1970).
71. Purdy, G., personal communication.

72. Jesser, W. A., and Kuhlmann-Wilsdorf, D., On the theory of interfacial energy and elastic strain of epitaxial overgrowths in parallel alignment on single crystals, *Phys. Stat. Sol.* **19**, 95 (1967).
73. Hofmann-Amtenbrink, M., "*Untersuchungen zum aktivierten Sintern von Molybdan*", Doktor-Ingenieur thesis, Technischen Universitat Berlin (1985).
74. Hackney, S. A., Biancaniello, F. S., Yoon, D. N., and Handwerker, C. A., Observations on crystal defects associated with diffusion induced grain boundary migration in Cu-Zn, *Scripta Metall.*, **20** [6] 937 (1986).
75. Shewmon, P. G., Diffusion driven grain boundary migration, *Acta Metall.* **29**, 1567-1572 (1981).
76. Chen, F.-S., and King, A. H., The misorientation dependence of diffusion induced grain boundary migration, *Scripta Metall.* **20** [10] 1401 (1986).
77. Hackney, S. A., and Biancaniello, F. S., Experimental indications of a low undercooling instability in the cellular mechanism of discontinuous precipitation, submitted to *Scripta Metall.*
78. Cahn, J. W., The kinetics of cellular segregation reactions, *Acta Metall.* **7**, 18-28 (1959).
79. Myklebust, R. L., Marinenko, R. B., Newbury, D. E., and Bright, D. S., in "*EMAG '85, Inst. Phys. Conf. Ser. 78*", (G. J. Tatlock, ed.) Adam Hilger, Ltd., Bristol (1985).
80. Marinenko, R. B., Myklebust, R. L., Bright, D. S., and Newbury, D. E., in "*Microbeam Analysis – 1985*" (J. T. Armstrong, ed.) San Francisco Press, San Francisco, CA (1985).
81. Myklebust, R. L, Marinenko, R. B., Newbury, D. E., and Bright, D. S., in "*Microbeam Analysis – 1985*" (J. T. Armstrong, ed.) San Francisco Press, San Francisco, CA (1985).
82. Handwerker, C. A., Blendell, J. E., Newbury, D. E., and Marinenko, R. B., unpublished work.
83. Newbury, D. E., National Bureau of Standards, Washington, DC, personal communication.
84. Butrymowicz, D. B., Cahn, J. W., Manning, J. R., Newbury, D. E., and Piccone, T. J., in "*Character of Grain Boundaries*", (M. F. Yan and A. H. Heuer, eds.) Advances in Ceramics, Vol. 6, pp. 202-212 (1983).
85. Matthews, J. W., Preparation of thick β-cobalt films, *Phil. Mag.* **18**, 1149-1154 (1968).
86. Grant, E., Porter, A., and Ralph, B., Grain boundary migration in single-phase and particle-containing materials, *J. Mat. Sci.* **19**, 3554-3573 (1984).

87. Shewmon, P. G., Meyrick, G., Mishra, S., and Parthasarathy, T. A., Diffusion induced dislocation glide, *Scripta Metall.* **17**, 1231-1235 (1983).
88. Cahn, J. W., and Balluffi, R. W., On diffusional mass transport in migrating and stationary grain boundaries, *Scripta Metall.* **13**, 499 (1979).
89. Li, J. C. M., and Rath, B. B., A mechanism for the diffusion induced grain boundary motion, *Scripta Metall.* **19**, 689-693 (1985).
90. Meyrick, G., On the initiation of discontinuous precipitation, *Scripta Metall.* **10**, 649-653 (1976).
91. Louat, N., Pande, C. S., and Rath, B. B., On a new approach to diffusion induced grain boundary motion, *Phil. Mag. A* **51** (6) L73-77 (1985).
92. Smith, D. A., and King, A. H., On the mechanism of diffusion induced grain boundary migration, *Phil. Mag.* **A44**, 333-340 (1981).
93. King, A. H., Some problems with the grain boundary dislocation climb mechanism for diffusion induced grain boundary migration, and possible solutions, *Phil. Mag* **A48**, L39-L44 (1983).
94. Sulonen, M. S., On the driving force of discontinuous precipitation and dissolution, *Acta Metall.* **12**, 749-753 (1964).
95. Sulonen, M. S., Discontinuous mode of dissolution of a β-phase precipitate into a Cu-Cd solid solutions, *Acta Metall.* **8**, 669-676 (1960).
96. Hillert, M., On theories of growth during discontinuous precipitation, *Met. Trans. A* **3**, 2729-2741 (1972).
97. Cahn, J. W., On spinodal decomposition, *Acta Metall.* **9**, 795-801 (1961).
98. Cahn, J. W., On spinodal decomposition in cubic crystals, *Acta Metall.* **10**, 179-183 (1962).
99. Hilliard, J. E., in "*Phase Transformations*", American Society for Metals, Metals Park, OH, pp. 497-560 (1970).
100. Perepezko, J. H., and Boettinger, W. J., in "*Surface Alloying by Ion, Electron, and Laser Beams*", ASM Publication, Metals Park, OH (1986) in press.
101. Eshelby, J. D., The determination of the elastic field of an ellipsoidal inclusion, and related problems, *Proc. Roy. Soc.* **A241**, 376-396 (1957).
102. Cahn, J. W., Coherent fluctuations and nucleation in isotropic solids, *Acta Metall.* **10**, 907-913 (1962).
103. Hilliard, J. E. Nucleation within crystalline phases, *Ind. Eng. Chem.* **58**, 19-25 (1966).
104. Hillert, M., Thermodynamics of the massive transformation, *Met. Trans. A* **15A**, 411-419 (1984).

105. Voorhees, P. W., and Johnson, W. C., Interfacial equilibrium during a first-order phase transformation in solids, *J. Chem. Phys.* **84**, 5108-5121 (1985).
106. Larche, F. C., and Cahn, J. W., The interactions of composition and stress in crystalline solids, *Acta Metall.* **33**, 331-357 (1985).
107. Handwerker, C. A., Effect of coherency strain on interface migration 2. Quantitative tests, TMS/AIME Meeting, New Orleans, LA, March (1986).
108. Frank, F. C., in "*Growth and Perfection of Crystals*", (R. H. Doremus, B. W. Roberts, and D. Turnbull, eds.) John Wiley, New York, pp. 411-419 (1958).
109. Frank, F. C., On the kinematic theory of crystal growth and dissolution processes, II, *Z. Phys. Chem.* Neue Folge, **77**, 84-92 (1972).
110. Kang, S. J. L., Song, Y. D., Kaysser, W. A., and Hofmann, H., Determination of Mo solidus in the Mo-Ni system by electrolytic phase separation method, *Z. Metall.* **75**, 86-91 (1984).
111. Kang, H. K., M. S. Thesis, Korea Advanced Institute of Science and Technology (1985).
112. Rhee, W. H., M. S. Thesis, Korea Advanced Institute of Science and Technology (1985).
113. Yoon, D. N., Effect of Coherency Strain on Interface Migration 1. Overview of Theories and Experiment, presented at TMS/AIME Meeting, New Orleans, LA, March, 1986.
114. Simmons, G., and Wang, H., "*Single Crystal Elastic Constants and Calculated Aggregate Properties*", MIT Press, Cambridge, MA (1971).
115. Leamy, H. J., and Warlimont, H., The elastic behavior of Ni-Co alloys, *Phys. Stat. Sol.* **37**, 523-534 (1970).
116. Handwerker, C. A., Coble, R. L., and Blendell, J. E., in "*Character of Grain Boundaries*", (M. F. Yan and A. H. Heuer, eds.) Advances in Ceramics, Vol. 6, pp. 213-223 (1983).
117. Kwok, T., Ho, P. S., Yip, S., Balluffi, R. W., Bristowe, P. D., and Brokman, A., Evidence for vacancy mechanism in grain boundary diffusion in bcc iron: a molecular dynamics study, *Phys. Rev. Lett.* **47**, 1148-1151 (1981).
118. Murr, L. E., "*Interfacial Phenomena in Metals and Alloys*", Addison-Wesley Publishing Co., Reading, MA (1975).
119. Hino, J., Tomizuka, C., and Wert, C., Internal friction and diffusion in 31% alpha brass, *Acta Metall.* **5**, 41-49 (1957).
120. Babcock, S. E., Ph.D. Thesis, M.I.T., Cambridge, MA (1986).
121. Read-Hill, R. E., "*Physical Metallurgy Principles*", D. Van Nostrand Co., New York (1973).

122. Bailey, J. E., and Hirsch, P. B., The recrystallization process in some polycrystalline metals, *Proc. R. Soc. London*, **267**, 11-30 (1961).
123. Jahn, R. J., and King, A. H., Vacancy deposition during diffusion induced grain boundary migration, submitted to *Scripta Metall.*
124. Ostertag, C., "*SIMS – Untersuchungen zur Volumen – und Korngrenzenfremddiffusion von Indium in Kupfer*", Ph.D. Thesis, Universitat Stuttgart, 1981.
125. Handwerker, C. A., and Hackney, S. A., Mechanism diagrams for diffusion induced grain boundary migration and diffusion induced recrystallization, **20** [10] 1385 (1986).
126. Hackney, S. A., Experimental observations on the initiation of DIGM, *Scripta Metall.* **20**, [10], 1385 (1986).
127. Balluffi, R. W., and Blakely, J. M., Special aspects of diffusion in thin films, *Thin Solid Films* **25**, 363-392 (1975).
128. Chopra, K. L., "*Thin Film Phenomena*", McGraw-Hill, New York (1969).
129. Neugebauer, C. A., in "*Handbook of Thin Film Technology*" (L. I. Maissel and R. Glang, eds.) McGraw-Hill, New York (1970).
130. Chou, T. C., Wong, C. Y. and Tu, K. N., "Enhanced grain growth of polycrystalline silicon at low temperature by dopant redistribution", *Appl. Phys.*, **62**, 2722 (1988).

6

Effects of Ambients on Thin Film Interactions

Chin-An Chang

ABSTRACT

Thin film interactions have often shown dependence on the annealing ambients used. The intriguing phenomenon not only has interesting science involved, but also has great impact on thin film technologies. The ambient effects have been observed for the metal-metal systems, metal-semiconductor and metal-compound systems. Both ambient-enhanced and ambient-reduced interactions have been observed. The observed ambient effects are reviewed, along with the proposed mechanisms for such observations. A surface potential model is described which correlates quite well a vast number of systems, especially the gold covered ones. The predicted reduced interactions by oxygen from this model are especially interesting, and have been confirmed for many systems. The widely interested effect of oxygen on the formation of silicides is discussed with the Pt/Si system.

1. INTRODUCTION

Thin film metallurgies have been of increasing importance in many areas of technological significance, including semiconductor devices, surface coating, interface metallurgy, and corrosion resistance. The interactions occurring in these films have therefore great impacts on the stability and properties of the films used. Much studies have been devoted to understanding the various parameters which affect such interactions, as evident from the vast number of reports on such subjects.

Among the factors that affect thin film interactions is the ambient in which the thin film system resides during annealing or operation of the structures. The ambient effects, notably that of oxygen, have often been undesired and interfere with the properties of the thin film structures formed. It is therefore not unexpected that oxygen is notorious to thin films and is avoided whenever possible.

In this Chapter we review the various ambient effects observed for a large number of systems, covering metal-metal, metal-semiconductor and metal-inorganic compound systems. The ambient is seen to either enhance or reduce the interactions of concern. Several models are described to account for such observations. The surface potential model is described in detail, which is shown to correlate the ambient effects for many systems, especially those involving gold as the covering layer. The model further predicts that, opposite to the case where gold is the covering layer, for certain systems the interactions can be reduced by oxygen. This has been confirmed with a number of systems, with useful application of making structures stable in air for both storage and operation.

The effect of oxygen on the formation of silicides has been of increasing interest in recent years, and has a great influence on the use of silicides in VLSI. The Pt/Si system is used to illustrate the effect of oxygen on the different stages of silicide formation.

2. OBSERVED AMBIENT EFFECTS

As stated above, both enhanced and reduced interactions by the annealing ambients have been observed. In this Section we review the literature reports on such ambient effects and the different mechanisms proposed.

Enhanced Interactions

The widely known systems where the interactions are enhanced by the annealing ambients are the ones where gold is the covering metal and the outdiffusion of the underlying species can be greatly enhanced by steam, air and oxygen. This has been observed for Si (1,2), Ge (3), GaAs (4), Cr (5), Cu (6), Ni (7), Fe (8) Ti (9) and Pd (9). The Pt/Au system shows an enhanced diffusion of Au through Pt in N_2 and He than that in vacuum (10).

In the above mentioned systems where gold is the overcoating metal with a thickness of about 2000Å, outdiffusion is enhanced by steam, air and oxygen when compared with those using inert ambients or in vacuum. The Au/Si system was studied in detail (1,2) and is probably more suitable for the illustration of the effects of ambients on the interactions present. It was observed that using steam or air, outdiffusion of Si to the surface of gold and oxidation there are greatly enhanced over that using vacuum. An enhanced outdiffusion was also observed in oxygen. When the Au/Si system was first heated above the eutectic temperature of Au-Si, 363°C, further annealing in an oxidizing ambient gave no outdiffusion as observed in the unpretreated samples.

The ambient effect becomes more complicated when the Au/Si sample is compared with the Ag/Si one and those with sequential Au-Ag layers on Si. Heating the Ag/Si sample under the same conditions in oxidizing ambients showed little outdiffusion of Si and formation of Si oxide on the silver surface. However, using the sequential Au-Ag layers, oxide was observed on the silver surface of the Ag/Au/Si structure, but not on the gold surface of the Au/Ag/Si one. Similar results were also observed for Cr using sequential Au-Ag layers (11).

The observed enhanced outdiffusion through gold is often attributed to the formation of oxide on the gold surface in the oxidizing ambients. The oxide is thought to act as a sink which becomes the driving force for the enhanced outdiffusion observed. Furthermore, the outdiffusion of the underlying species and its accumulation on the surface of the overcoating metal is related to the surface energy of the system concerned (12). These will be discussed later with other suggested mechanisms for the various ambient effects observed.

Reduced Interactions

Reduced interactions by the ambients are also commonly known for thin films. The Au/Si interaction, for example, is reduced by hydrogen (1). Similarly reduced interactions by hydrogen have also been observed for the Au/Co (13) and Au/Al (14) ones. Other reduced interactions have been observed in oxygen-containing ambient, where the reduced interactions observed are attributed to possible interface oxide formation. Examples are Pd/Ti (9), Pt/Al (15), and Pt/Si (16-19). In this respect we do not consider the cases where an interface oxide is formed during film deposition process. It was reported that for many systems, depending on the reactivity of the first metal layer deposited, an oxide layer can form on the surface of the first metal film before depositing the second metal film due to poor vacuum (20). The presence of such an interface oxide was found to hinder subsequent interactions. Such cases are not considered in this Chapter for the effect of annealing ambient on thin film interactions.

The oxygen effect observed for the Pt/Si system is especially interesting. For the Pt silicide formation, Pt_2Si is first formed, followed by the second phase PtSi. It is reported that the formation rates of both Pt_2Si and PtSi are reduced by oxygen, and that an oxide layer is observed at the interface of Pt and Pt_2Si or PtSi (16-19). The reduced formation rates of these and other silicides by oxygen are often attributed to the oxide layer observed (20). In a later Section a detailed study on the Pt/Si system will be described to illustrate how the oxide layer is formed in an oxygen ambient, and the effect of this oxide layer on the formation processes of both Pt silicides.

3. MECHANISMS FOR THE AMBIENT EFFECTS

Several mechanisms have been suggested for the observed ambient effects on the thin film interactions, including the sinking effect, surface energy effect, interface oxide formation effect, and surface potential model.

Sinking Effect Model

As described above, when the annealing is carried out in oxidizing ambients, outdiffusion of the underlying species to the surface of the overcoating metal is often observed, leading to the formation of a surface oxide layer. This has often been suggested as the driving force for the enhanced outdiffusion observed in such ambients (21). The sinking effect would be consistent with the observation that a higher outdiffusion of Si is observed using steam than air, which in turn, is much higher than that using oxygen (1,2). The oxidation rate of Si is known to be higher in humid oxidizing ambients (22). It is also consistent with the fact that using reduced ambients such as hydrogen, a reduced outdiffusion is observed.

The most difficult problem associated with the sinking effect is that it cannot account for the results on Ag-Si and those using sequential Au-Ag layers described earlier (1,2). If the surface oxidation is the major driving force for the enhanced outdiffusion in these ambients, one should expect similar outdiffusion rates using Ag or other metals. In other words, little dependence on the overcoating metal should be expected. This is in clear contradiction to the observation of little outdiffusion in the Ag/Si case, and that outdiffusion is observed for the Ag/Au/Si but not for the Au/Ag/Si system. These considerations indicate that the sinking effect is partly responsible for the surface oxide formed using oxidizing ambients, but is not the major driving force for the observed ambient effects.

Surface Energy Model

The observed ambient effects on both the enhanced outdiffusion and formation of oxide layer on the surface of the overcoating metal layer have been suggested to be due to a surface energy effect (9). Due to the minimization of surface energy, it helps for the underlying species to accumulate on the surface of the overcoating metal. Accordingly, the underlying species diffuses to the surface to reduce the undesired surface energy of the system.

Disregarding quantitative consideration of the surface energy requirement, conceptually there is one major difficulty when one compares the Au/Si and Ag/Si systems with the Au/Ag/Si and Ag/Au/Si ones. The Au/Si and Au/Ag/Si systems both have the Au surface, yet Si oxide is observed on the gold surface of Au/Si but not on that of Au/Ag/Si. On the other hand, little Si oxide is observed on the Ag surface for the Ag/Si system, but is observed on that of the Ag/Au/Si one. These results indicate that some other driving force is more dominant in determining the ambient effects observed.

Interface Oxide Model

The reduced interactions in thin films in the presence of oxygen is often attributed to the formation of interface oxide which blocks the diffusion of the species involved (23). Such interface oxide layers have been observed in cases like the Pt/Si system when Pt films containing an appreciable amount of oxygen were used (16-19). A detailed account of the effect of such interface oxide layer on the formation rates of silicide will be given later for Pt/Si, which will provide insight into the formation of the oxide layer and its effect on the silicide formation rates.

The interface oxide alone cannot account for the fact that oxygen reduces the interactions in certain systems, but enhances those in the others, such as those having gold as the overcoating layer. One therefore needs to study individual cases to understand whether oxygen is retarding or promoting the outdiffusion of the species involved to understand the effects observed.

Surface Potential Model

Among the models proposed, the surface potential model has been the most extensively explored for both the mechanism involved and the predictions it makes to other systems. This model will be described in detail here and its application to other systems will also be illustrated.

As indicated above in the discussion of the sinking effect and surface energy effect, the ambient effects on the Au/Si, Ag/Si, Au/Ag/Si and Ag/Au/Si systems compose a challenging problem to any model proposed for the ambient effects observed. In addition, one has to deal with the opposing effects of oxygen for different systems, i.e., enhancing interactions for the vast number of systems where gold is the overcoating metal, but reduced interactions for many other systems. Furthermore, the model should be able to predict ambient effect for other systems which can be tested for a better understanding of the thin film interactions present. The surface potential model is proposed to deal with these and other concerns (24-26).

In all the systems where ambient effects are reported, the overcoating metal layer is always a polycrystalline film. This allows fast diffusion along the grain boundaries compared with that observed in single crystalline metals (21). Furthermore, it is reported that the ambient species can diffuse into the metal film along its grain boundaries and reach the interface (21, 27). The grain boundary diffusion is thus a crucial factor in all the ambient effects observed, and is consistent with the fact that after heating the Au/Si system to temperatures above its eutectic point, further annealing in oxidizing ambient no longer gives outdiffusion of Si as observed for the untreated samples (1). Crystallization taking place during the process apparently greatly reduces the number of grain boundaries available for the outdiffusion to take place.

In the surface potential model, two major considerations about the ambient effects are: the roles of the overcoating metal layers and the ambients present. It is clear from the described effects that the overcoating metal plays an important role in the effects observed. The bonding between the two materials at the interface is taking into concern by considering the difference in electronegativity. The ambient gas is known to alter the surface properties of the metal upon adsorption, as observed for many systems (28). Since the ambient species can diffuse into the metal layer along the grain boundaries, it can alter the surface property of the metal at the interface, similar to its effect on the surface itself. The ambient-induced surface potential change at the interface is assumed to be similar to that at the surface. Such a surface potential change is measured by the work function change induced by the ambient; an increase in work function of the metal is taken as a positive surface potential induced by the ambient (29). The change in work function of the metal grain surface at the interface is also expected to increase with coverage of the in-diffused ambient species, similar to that observed on the metal surface itself (28,30).

Based on this model, the Au-Si interaction can be described as:

$$Au_x + Si_y \rightarrow Au_x\text{-}Si + Si_{y-1} \qquad (1)$$

$$Au_x\text{-}Si \rightarrow \text{(outdiffusion of Si, or alloy formation, etc.)} \qquad (2)$$

Here, a Si atom is released from the Si lattice, Si_y, and bonds to the Au lattice, Au_x, at the interface, as Reaction (1) shows. The Si atom thus released can subsequently outdiffuse to the gold surface along the grain boundaries, through bulk diffusion, or form alloys with gold, as Reaction (2) shows. In analogy to chemical reactions, it is assumed that Reaction (1) is the rate-limiting step where the overall reaction rate can be enhanced by stabilizing the intermediate complex Au_x-Si. In other words, the reaction is enhanced by lowering the potential barrier for the formation of the activated complex involved. The electronegativities of Au and Si are 2.4 and 1.8, respectively (31), which give a partial ionic bond in the Au_x-Si complex as $Au_x\delta^-$ -$Si\delta^+$, with a partial charge of one-tenth of a unit charge. The ambient species can affect this bonding through its effect on the surface potential of Au at the interface. Steam and air are known to induce a negative surface potential on gold. For example, a decrease in work function of gold by 0.5 and 1.2 eV is observed upon exposure to air and steam, respectively (32,33). No work function data are available for gold in dry oxygen, but the resistivity measurement of gold in dry oxygen also shows an induced negative surface potential (34). The negative potential induced would make the Au_x more negative and strengthen the bonding of the $Au_x\delta^-$ -$Si\delta^+$ complex, therefore enhancing both Reactions (1) and (2). Hydrogen, on the other hand, induces a positive surface potential of Au by increasing its work function by 0.18 eV (35), which is unfavorable to Reactions (1) and (2), and should reduce the Au-Si interactions. Such an analysis is seen to be consistent with not only the ambient effects observed for the Au-Si system, but also with those of all the other systems mentioned earlier where the overcoating layer is gold. All the diffusing species of these systems have smaller electronegativities than gold, and the same ambient effects as that described for Si should also apply to these systems.

In this model, two parameters are considered: the electronegativity relation between the two species of interest, and the ambient-induced work function change of the metal through which diffusion of the other species is studied. Four combinations are expected from these two parameters, as listed in Table 1. In our discussion, the term "overcoating" layer is used in the general sense to refer to the layer through which diffusion of the other species is considered. Here, we see that all the systems containing gold as the overcoating layer discussed so far belong either to Case B when the ambient reduces the work function of gold, or Case A with an increased work function. We will refer to Table 1 frequently in our subsequent discussion of the ambient effects for various systems.

When silver is used as the overcoating metal, much less ambient effects are expected. Take the Ag/Si system, for example, the ionicity of the Ag-Si bond is less than on-tenth of that of the Au-Si due to the smaller electronegativity of silver, 1.9 (31). The same reasoning also applies to the systems when Au/Ag sequential layers are used for the overcoating metal. According to the reaction mechanism described above, the metal-Si interaction is determined by the interface reaction which involved only the metal that is in immediate contact with Si. Accordingly, significant ambient effects are expected for the Ag/Au/Si system, similar to the binary Au/Si one. On the other hand, the Au/Ag/Si system resembles that of Ag/Si, and little outdiffusion of Si is expected. Such correlations are seen to agree with observations for both Si and Cr using sequential Au-Ag layers (2,11).

It is seen that the surface potential model correlates very well the observed ambient effects for the systems with gold being the overcoating metal. Both the enhanced and reduced interactions by the ambients are accounted for. The reduced interactions reported for the Pd/Ti (9), Pt/Al (15) and Pt/Si (16-19) systems were attributed to interface oxide layer. As is shown later, the reduced interactions for the Pt systems are expected from the surface potential model without invoking the interface oxide layer. The Pd system cannot be treated at present due to a lack of the needed surface data for this metal in oxygen.

4. OBSERVATION OF OTHER AMBIENT EFFECTS FROM THE SURFACE POTENTIAL MODEL

In this Section the application of the surface potential model is described. Interesting ambient effects have been observed for the various systems, including reduced interactions for the Au systems by CO, reduced interactions by oxygen for many Pt systems and Ni and Cu systems, competing ambient effects, ambient effects on competing interactions and on preferential interactions, and effects of oxidation of the outdiffused species.

Reduced Interaction in Au Systems by CO

From the surface potential model, interactions for the Au systems, i.e., systems where the overcoating metal is gold, are reduced by ambients which induce a positive surface potential on the gold surface. Carbon monoxide, CO, is reported to increase the work function of gold by 0.9 eV (35). The interactions of the Au/Si, Au/Cr and Au/GaAs systems were studied, and all showed reduced interactions by CO (24,25,36). Figure 1 shows the results on the Au/Cr system, comparing N_2 with N_2-CO ambients. It is seen that a higher accumulation of surface Cr is observed using the nitrogen ambient. An Auger

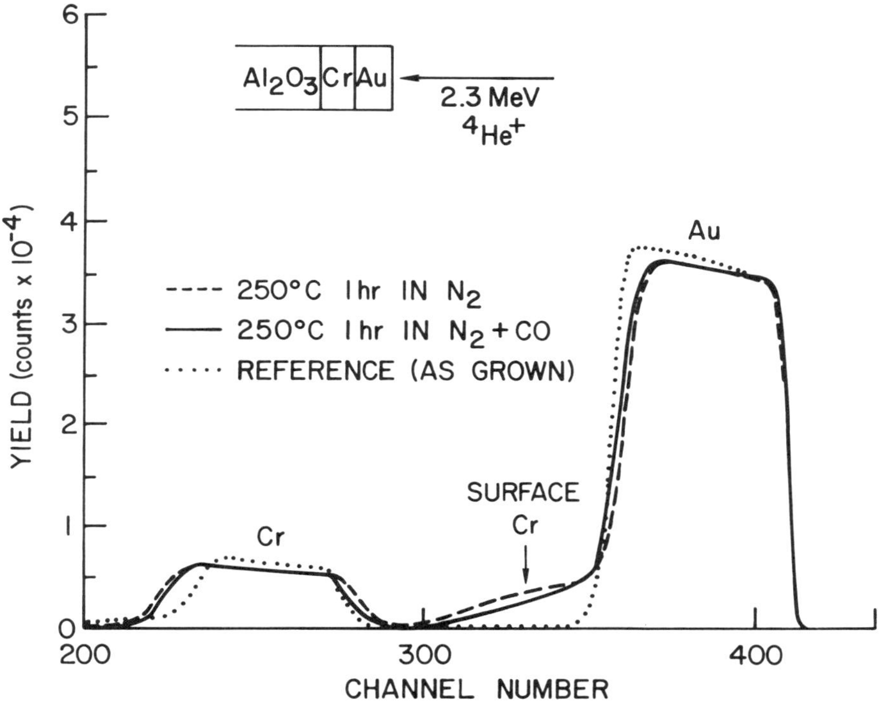

Fig. 1. Backscattering spectra showing effect of CO on the interdiffusion and outdiffusion of the Au/Cr system.

analysis of the sample annealed in nitrogen is shown in Fig. 2. The surface Cr is seen to be accompanied by oxygen, indicating oxidation of the surface Cr; the surface Cr and oxygen are removed after a 50 min sputtering. For comparison, it takes only 7 min to sputter off the Cr and oxygen for the sample annealed in N_2-CO. The result on the Au/Si system is shown in Fig. 3, showing reduced outdiffusion of Si in the presence of CO.

Further studies on the Au/Cr system using CO indicate interesting results shown in Figs. 4 and 5. Comparing samples annealed in air with those in air-CO mixtures, the reduced outdiffusion of Cr using an air-CO (5:1) mixture is similar to that observed using N_2-CO shown in Fig. 1. Using an air-CO (1:1) mixture, both the interdiffusion between Au and Cr and the outdiffusion of Cr are completely suppressed, as Fig. 5 shows. Such effects can be useful to the semicon-

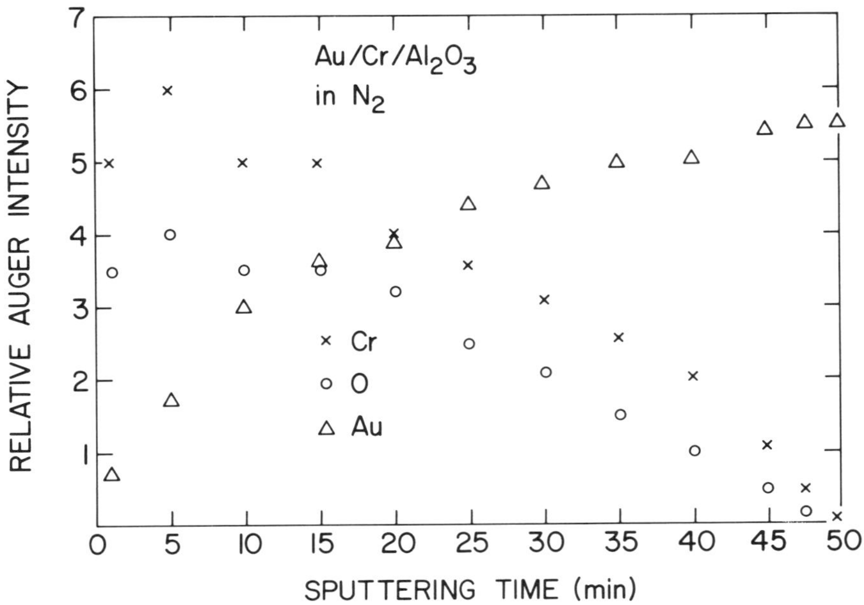

Fig. 2. Auger depth profile showing outdiffusion and surface accumulation of Cr of the Au/Cr system.

ductor technology where outdiffusion of Cr to the gold surface leaves Cr oxide, hindering subsequent bonding to Au for the various devices.

Reduced Interactions in Pt Systems by Oxygen

In this Section it is shown that reduced interactions by oxygen is expected for Pt systems, with Pt being the overcoating metal and the underlying species being less electronegative. Such reduced interactions will be technologically important for the devices which employ thin film structures. Operation of these devices are normally done in air, and stability in air becomes a vital factor for the thin film structures used.

Considering the ambient effect on the Pt systems, oxygen increases the work function of Pt by 1.2 eV (35). According to the surface potential model, interaction of a less-electronegative underlying species with Pt should be reduced by oxygen, a case A of Table 1. Take

Pt/Cr for example. The electronegativities are 2.2 and 1.6 for Pt and Cr, respectively (31). The rate-limiting step involves the $Pt_x\delta^-$ -$Cr\delta^+$ complex, similar to the Au/Si case described earlier. Oxygen makes Pt more positive, which weakens the bond strength in the complex, and is unfavorable to the subsequent release of Cr from the interface for outdiffusion, alloy formation, etc.. The interactions between Pt and Cr, and similarly between Pt and other less-electronegative species, should therefore be reduced by oxygen.

The predicted effect of oxygen has been confirmed on Pt/Si (37), Pt/Cr (37), Pt/Ni (38), Pt/Al (39), and Pt/Cu (40), where all the underlying species have smaller electronegativities than Pt. As shown in Figs. 6-10, the interactions include outdiffusion, interdiffusion and alloy formation. All interactions are reduced by oxygen when a N_2-O_2 (4:1) mixture is compared with forming gas, a N_2-H_2 mixture. The results on the outdiffusion of Cr, shown in Fig. 6 is especially interesting. Both the outdiffusion and accumulation of Cr on the Pt surface are clearly reduced by oxygen, an observation in contradiction to the surface sinking effect described earlier. For the Pt/Si system, the results show reduced formation rate by oxygen for the Pt_2Si, similar to that reported by other workers (16-19). The suggested effect of the interface oxide will be described in a later Section.

Reduced Interactions by Oxygen for the Ni and Cu Systems

In addition to the Pt systems described above, other systems with different overcoating metals also show reduced interactions by oxygen, as observed for Ni (41) and Cu (40, 42) systems.

The work functions of Ni and Cu are increased by oxygen by 1.6 and 0.68 eV, respectively (35). The interactions between these metals and a less-electronegative species should therefore be reduced by oxygen, similar to those using Pt as the overcoating metal. These also belong to Case A of Table 1. The ambient effects on the Ni/Al system are shown in Fig. 11. The electronegativities of Ni and Al are 1.8 and 1.5, respectively. Several alloys are formed at different temperatures. Figure 11 shows reduced alloy formation rate in the presence of oxygen. Pt was used in this case to protect the Ni surface from oxidation, allowing the study of the Ni-Al interactions.

Two Cu systems have been tested for the reduced interactions by oxygen, Cu/Cr and Cu/Al. The electronegativities of Cu, Cr and Al are 1.9, 1.6 and 1.5, respectively. The Cu-Cr interaction shows grain mixing and is reduced by oxygen, as shown in Fig. 12. The Cu-Al interactions involve both outdiffusion of Al to the Cu surface, where oxidation takes place, and interdiffusion between Cu and Al. The former is enhanced by oxygen, as Fig. 13 shows for the Cu/Al system, and the latter is reduced by oxygen, as Fig. 14 shows for the Pt/Cu/Al system (42). The results show interesting interactions and different ambient effects on these interactions. For the sample annealed in N_2-O_2, only the outdiffsion of Al and interdiffusion between Al and Cu are observed. For the sample annealed in forming gas, on the other hand, alloy formation between Al and Cu is observed (42). Using the Pt/Cu/Al structures, no outdiffusion of Al to the Pt surface is observed; instead, both interdiffusion between Al and Cu and alloy formation are observed both of which are reduced by oxygen. More of the ambient effects on competing interactions will be discussed later.

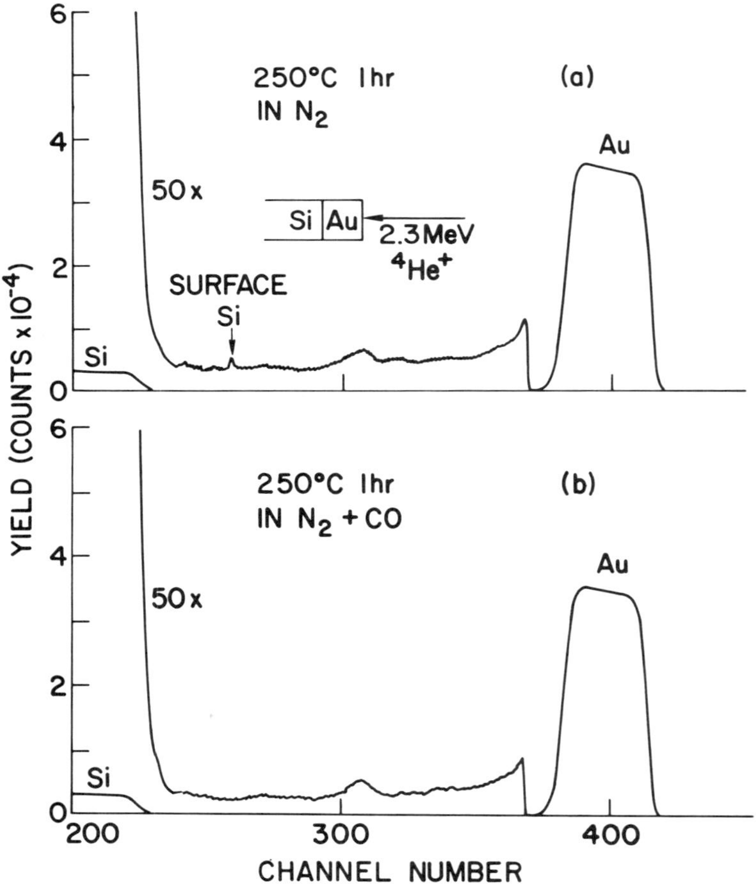

Fig. 3. Backscattering spectra showing effect of CO on the outdiffusion of Si of the Au/Si system.

Competing Ambient Effects

In the application of the surface potential model, the correlations have been made for systems sharing the same overcoating metals. This is more suitable in the treatment of the outdiffusion process and subsequent accumulation of the diffusing species on the surface of the overcoating metal. The interdiffusion, on the other hand, involves migration of both the overcoating metal and the underlying materials, and it is interesting to compare the expected ambient effects from both directions of movement. In other words, the roles of the two materials are reversed when the diffusion of the overcoating metal into the underlying one is considered. Since both diffusion processes are present for interdiffusion, such a comparison can lead to further understanding of the interactions observed.

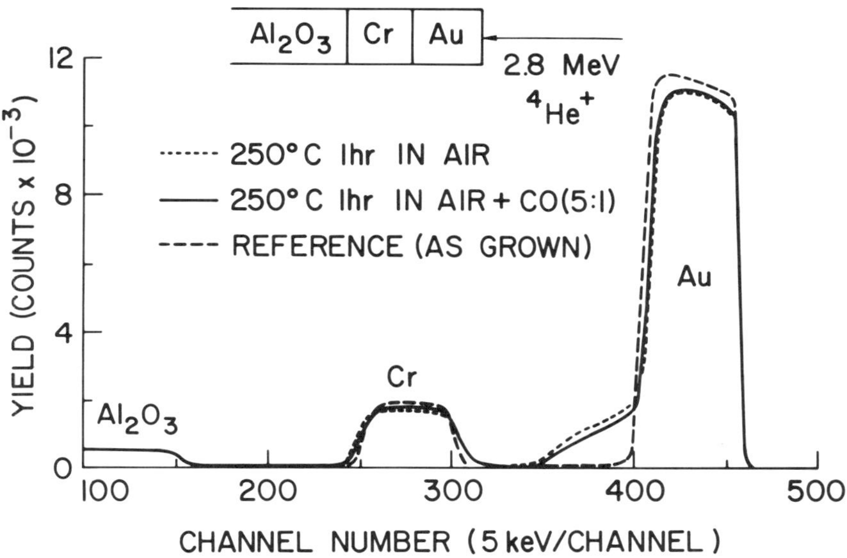

Fig. 4. Backscattering spectra showing effect of CO on the interdiffusion and outdiffusion of the Au/Cr system in air.

The observed interdiffusion is the sum of both diffusion processes described above. Outdiffusion of the underlying species could leave vacancies for the overcoating metal to diffuse in, and vice versa. The situation is simple when the ambient affects both processes in the same way. However, if opposing ambient effects are expected, i.e. one is enhanced but the other is suppressed, the competing effects can be interesting and provide further understanding of the interactions involved. Two such cases are available for the study of the competing ambients effects described, the Pt/Ni and Pt/Cu systems.

As described earlier, oxygen increases the work function of all of Pt, Ni and Cu, and the electronegativities of Ni and Cu are smaller than that of Pt. A reduced interaction by oxygen is expected in the systems involving any of these metals as the overcoating metal on a less-electronegative species, which has been confirmed for a number of systems, shown in earlier Sections. According to the surface potential model, diffusion of Pt into Ni and Cu should be enhanced by oxygen, as evident from reversing the roles of the overcoating and underlying species. This belongs to Case C of Table 1. However, a reduced overall interaction is observed for both Pt/Ni and Pt/Cu systems, indicating the dominant role played by the more electronegative species, here the Pt. The reason for this dominance is not clear, and can be important to our understanding of the interactions in thin films.

The reported enhanced diffusion of Au in Pt using nitrogen and helium than that using vacuum could be accounted for it the gases used contain residual oxygen (10). Both the diffusion of Au into Pt and of Pt into Au are enhanced by oxygen, which are Case C and Case B of Table 1, respectively. Further work is needed to test this correlation for the Pt/Au system.

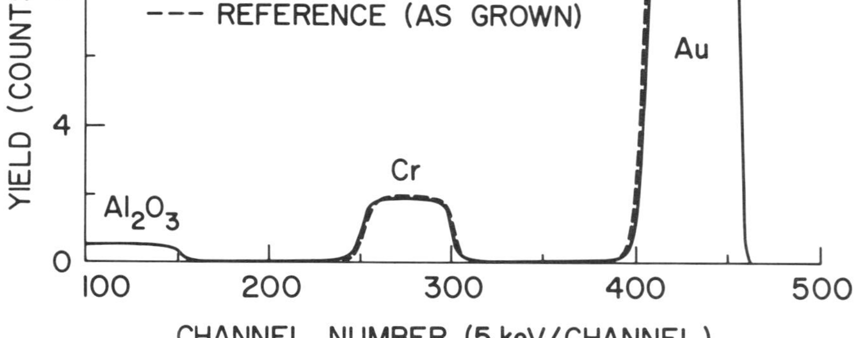

Fig. 5. Backscattering spectra showing suppressed interaction between Au and Cr by CO.

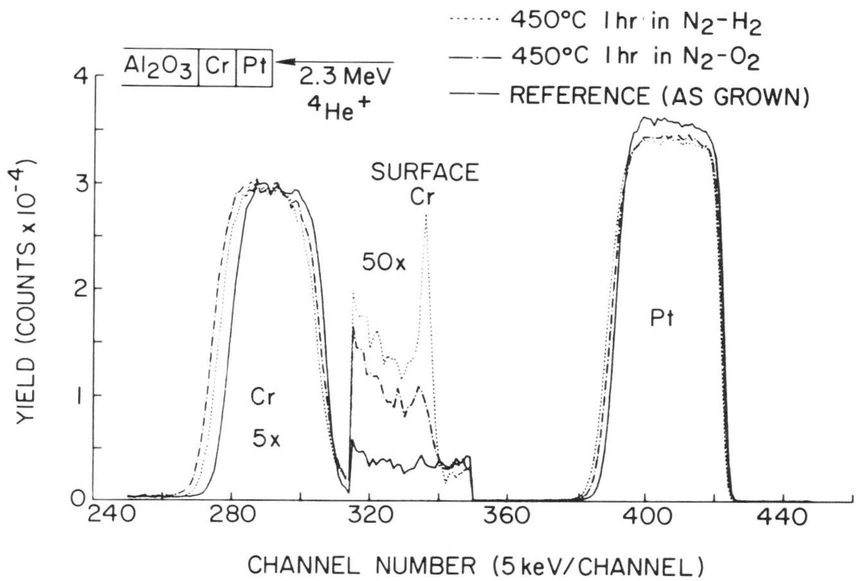

Fig. 6. Backscattering spectra showing reduced outdiffusion of Cr by oxygen in Pt/Cr.

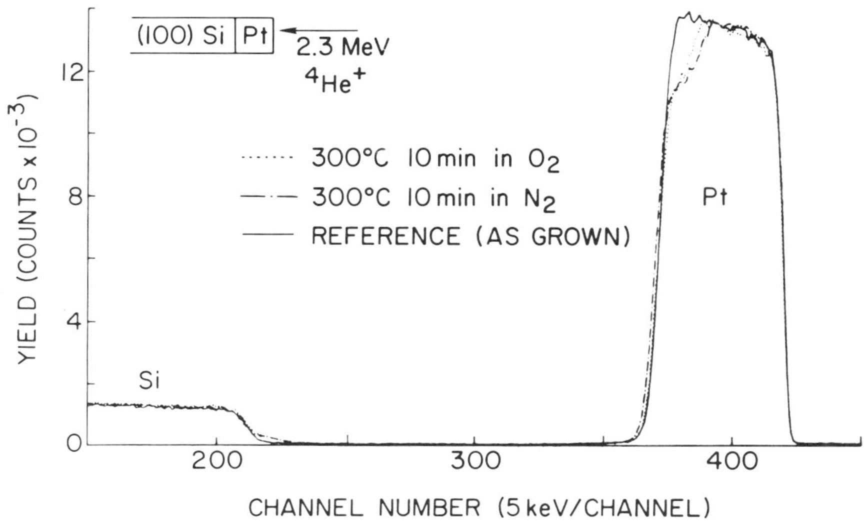

Fig. 7. Backscattering spectra showing reduced rate of forming Pt_2Si by oxygen.

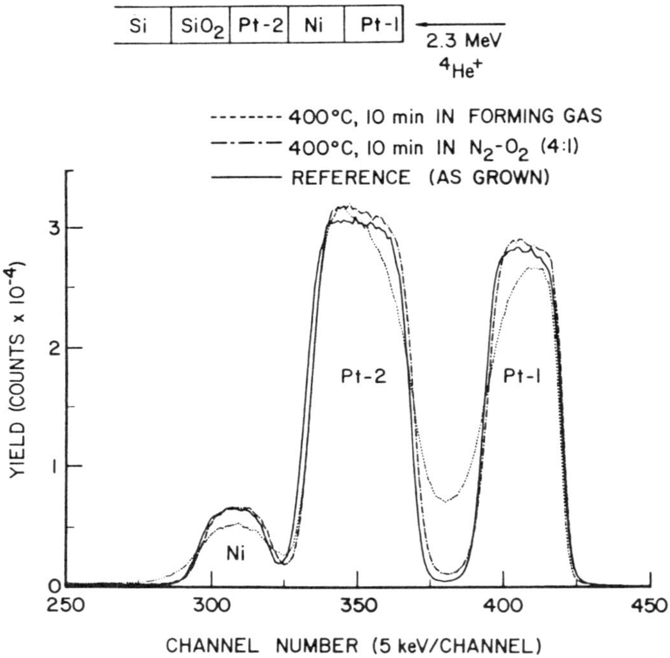

Fig. 8. Backscattering spectra showing reduced interdiffusion between Pt and Ni by oxygen.

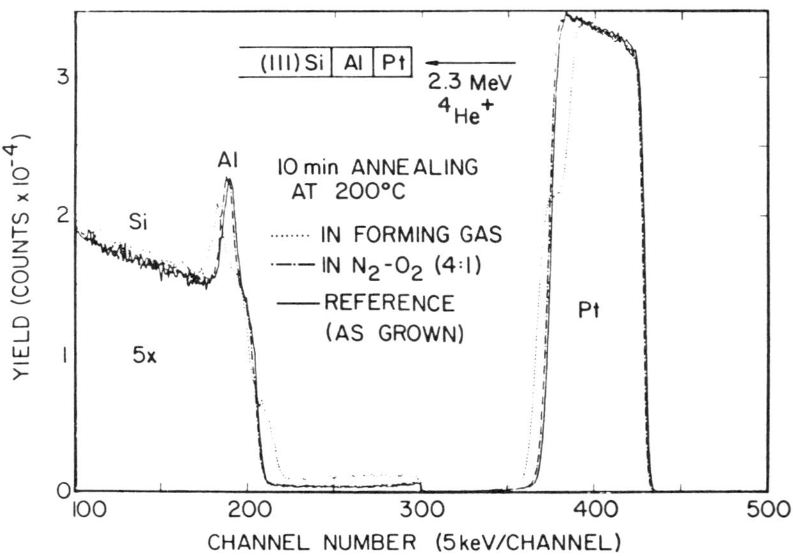

Fig. 9. Backscattering spectra showing reduced alloy formation between Al and Pt by oxygen.

Table 1. Predicted ambient dependence of diffusion as functions of the electronegativity relation between the diffusing species and the overcoating metal, and the ambient-induced work function (W.F.) change of the later.

	Overcoating metal showing	
Overcoating metal being	Increased W.F. by ambient	Decreased W.F. by ambient
More Electronegative	A Reduced diffusion	B Enhanced diffusion
Less Electronegative	C Enhanced diffusion	D Reduced diffusion

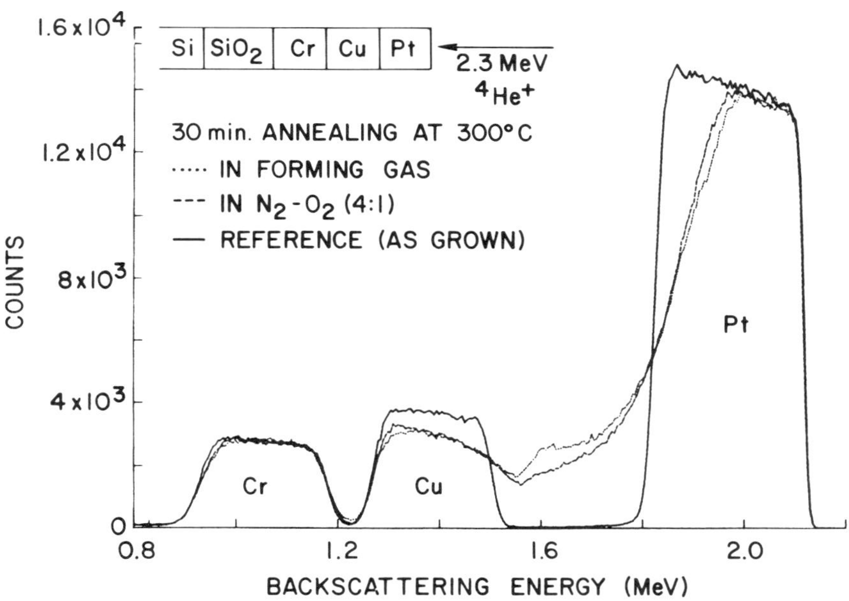

Fig. 10. Backscattering spectra showing reduced inter-diffusion between Cu and Pt by oxygen.

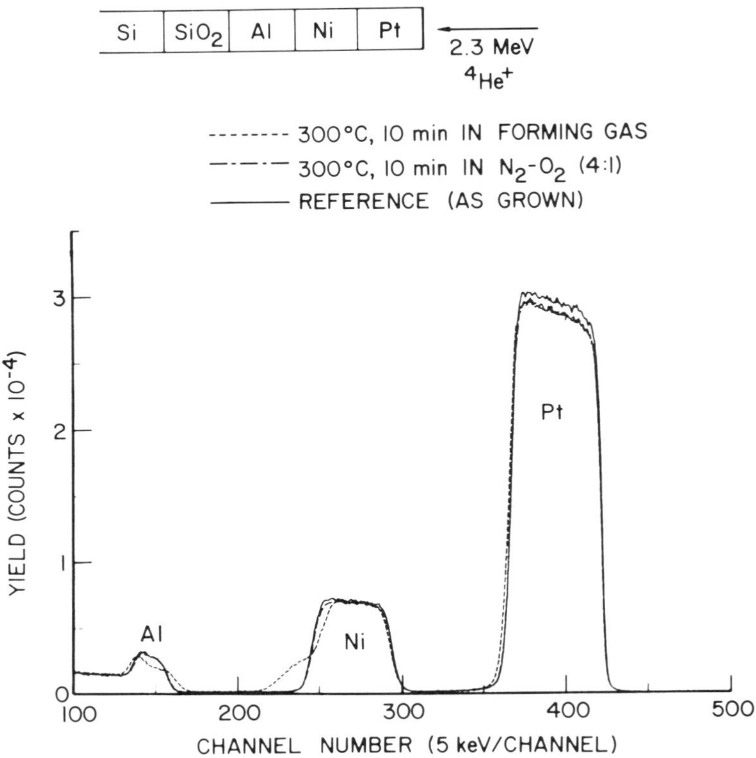

Fig. 11. Backscattering spectra showing reduced alloy formation between Al and Ni by oxygen.

Diffusion Through a Less-Electronegative Metal Layer

In most of the examples discussed so far, diffusion through a more-electronegative metal layer is studied. For the Pt/Ni and Pt/Cu cases where diffusion in both directions are present, i.e., both Cases A and C are operating, the ambient effect on the more-electronegative metal dominate. It is therefore very interesting to find structures where only Case C is operating. This allows a better comparison with the predicted ambient effect as described in Table 1. In other words, the diffusion through a less-electronegative metal in such systems is not complicated by a diffusion in the reverse direction. The recent study on the diffusion of Au and Cu via the intermediate Ni layer in the Au/Ni/Cu structure provides such an example (43). Comparison of the ambient effects on the Au/Ni/Cu structure with those on the Pt/Ni/Cu one (44) provides further insight into the roles of the initial interface interaction in determining the ambient effects observed.

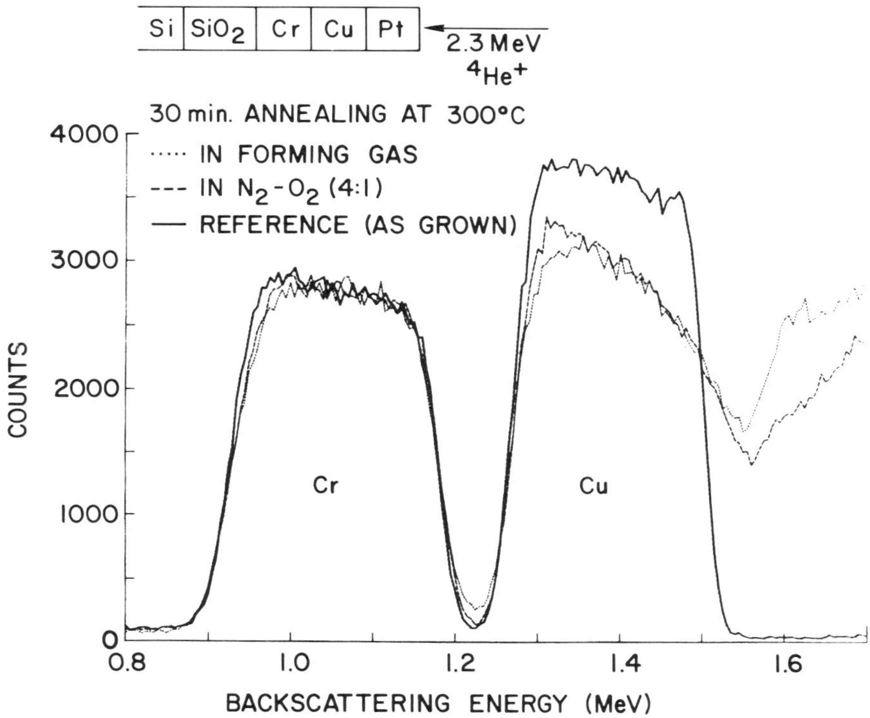

Fig. 12. Backscattering spectra showing reduced interaction between Cr and Cu by oxygen.

The Au/Ni/Cu structure shows interesting diffusion between Au and Cu across the Ni layer, yet with little interactions involving the Ni (45). Figures 14-16 illustrate this observation with different thicknesses of the Ni layer. It is seen that the distribution of Au shows a discontinuity under the Ni layer, an indication of little Au in the Ni layer. The diffusion of Au through the Ni layer into the Cu layer, and of Cu through the Ni layer, therefore, consist of examples of Case C of Table 1 if an ambient is used which increases the work function of Ni. Figures 17 and 18 show the effects of hydrogen and oxygen on such diffusion processes. Both hydrogen and oxygen increase the work function of Ni by 0.35 and 1.6 eV, respectively (35). An enhanced diffusion of both Au and Cu through the Ni layer is observed in hydrogen compared with that in a nitrogen-hydrogen mixture, in agreement with the ambient effect described in Table 1. A similarly enhanced diffusion is also observed using Ni_2-O_2 than N-H_2, consistent with the larger work function change of Ni by oxygen.

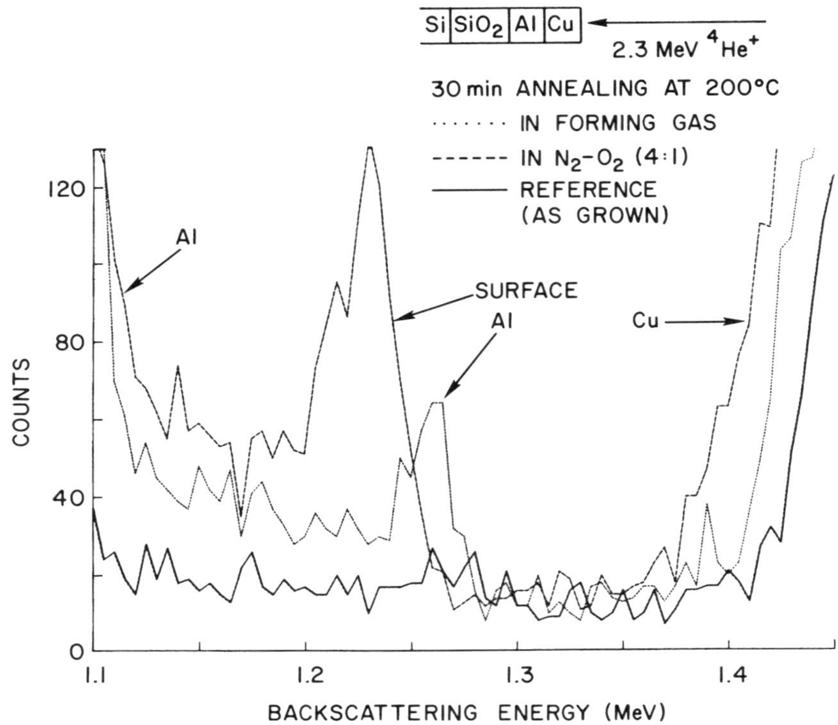

Fig. 13. Backscattering spectra showing enhanced surface accumulation of Al on Cu by oxygen for the Cu/Al system.

For the Pt/Ni/Cu structure, the interdiffusion between Pt and Ni affects the diffusion of Pt into the Cu layer, whereas the diffusion of Cu through the Ni layer resembles that in the Au/Ni/Cu structure. As Figure 19 shows, the diffusion of Pt into the Cu layer via Ni is reduced by oxygen, with a simultaneous enhanced diffusion of Cu through the Ni layer (45). The reduced Pt diffusion via Ni by oxygen is consistent with the effect of oxygen on the Pt/Ni interdiffusion described earlier. The studies on the Pt/Ni/Cu and Au/Ni/Cu structures thus provide further information as to how the ambient effect on the initial interface interaction affects the subsequent diffusion and reaction processes. The two types of interactions observed here for trilayer structures are depicted in Fig. 20. Using an A/B/C structure, the ambient effect on the diffusion of each species is determined by that on the interaction of the initial involving such a species. It is clear that Au/Ni/Cu belongs to the case of Fig. 20a, and Pt/Ni/Cu that of Fig. 20b.

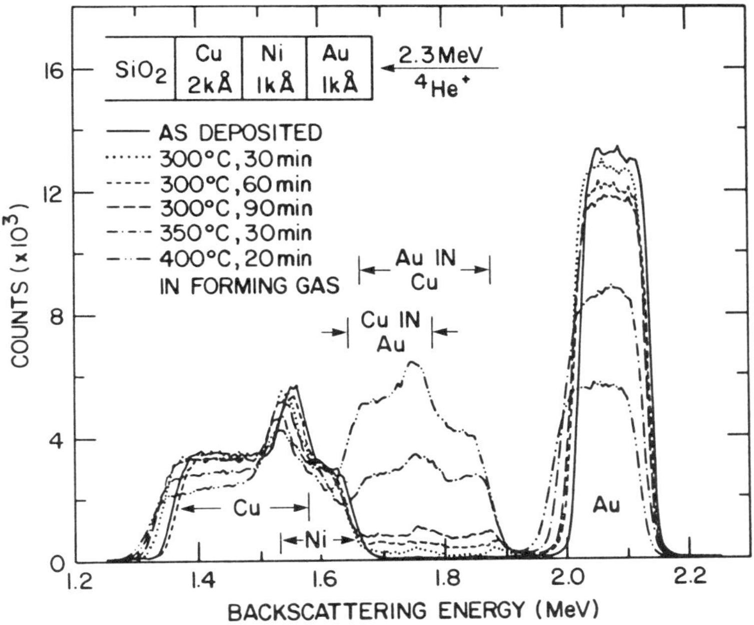

Fig. 14. Backscattering spectra of the Au/Ni/Cu structure annealed at different temperatures, showing interdiffusion between Au and Cu across the Ni layer.

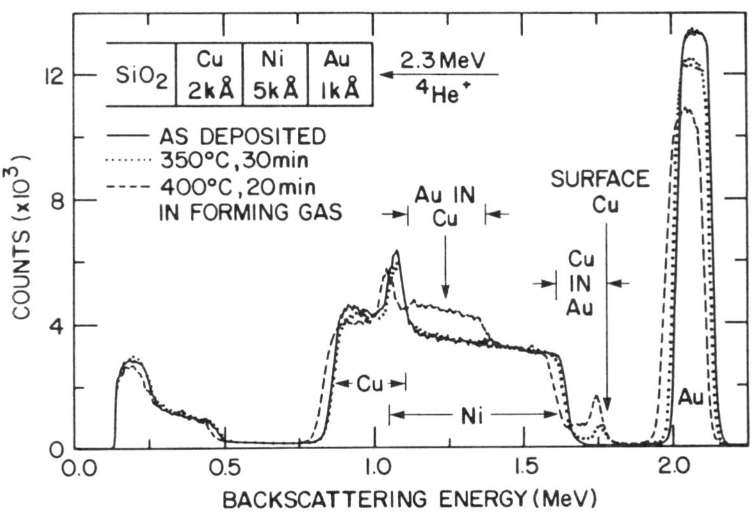

Fig. 15. Backscattering spectra of the Au/Ni/Cu structure with a Ni layer thickness of 5000Å.

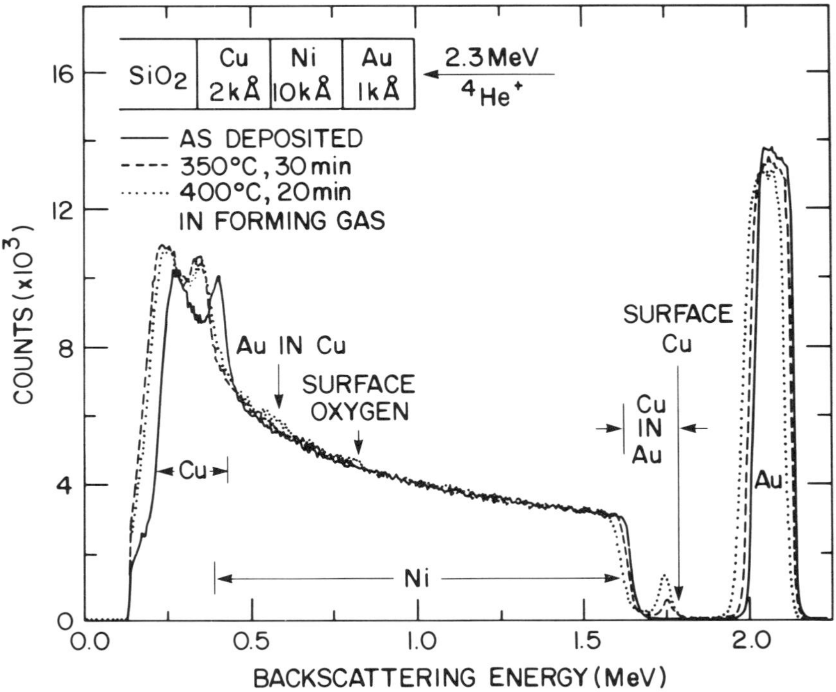

Fig. 16. Backscattering spectra of the Au/Ni/Cu structure with a Ni layer thickness of 1 μm.

Ambient Effects on Competing Interactions and on Preferential Reactions

As mentioned earlier for the Cu/Al system, interesting ambient effects on the competing thin film interactions and on preferential reaction have also been observed. An enhanced outdiffusion of Al to the Cu surface is observed in the presence of oxygen, but an alloy formation for that using forming gas (42), as Fig. 21 shows. The fact that the alloy formation is observed using forming gas but not the oxygen-containing ambient seems to indicate a faster rate of outdiffusion of Al than those of the interdiffusion and alloy formation in the presence of oxygen. When the surface accumulation of Al is suppressed using a top Pt layer, alloy formation is observed at higher temperatures in both ambients.

Using the Pt/Cu/Al structure, another interesting phenomenon has been observed: a preferential reaction between Pt and Al from the CuAl alloy formed under Pt (42), as shown in Fig. 22. The CuAl

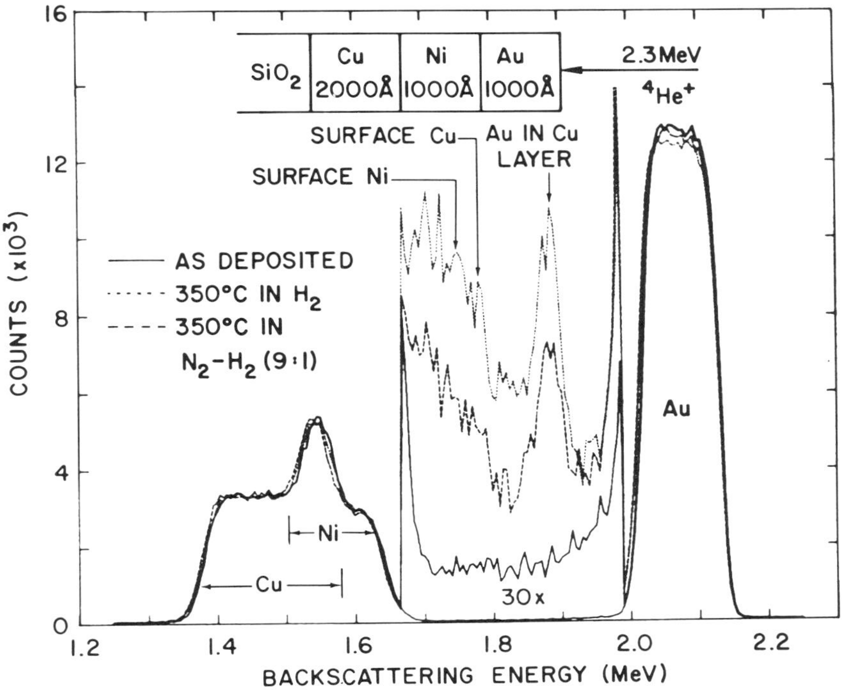

Fig. 17. Backscattering spectra of the Au/Ni/Cu structure, showing enhanced diffusion between Au and Cu across the Ni layer by hydrogen.

alloy formation has been completed after a 30 min annealing at 300°C in both forming gas and N_2-O_2 ambients, with no sign of interaction between either Cu or Al and the top Pt layer. However, Pt is found to react preferentially with Al from the underlying CuAl alloy at 350°C, forming a PtAl alloy, as shown in Fig. 22. The alloy formation between Pt and Al is seen to be reduced in oxygen, similar to that observed for the Pt/Al binary system (39).

The Pt/Cu/Al system has thus shown several competing interactions: outdiffusion of Al through Cu and surface accumulation, interdiffusion and alloy formation between Cu and Al, and preferential reaction between Pt and Al from the CuAl alloy formed. All interactions are affected by the ambients used, which provide interesting information to our understanding of the ambient effects on thin film interactions.

Oxidation Effects of the Outdiffused Species

The effect of oxidation of the outdiffused species has been discussed in the Section on sinking effect. Its contribution to the outdiffusion observed has also been compared with those from other reaction mechanisms. Several other systems also show interesting oxidation effect, for example, the Pt/Ni system (38). At lower temperatures, the interactions in this system are dominated by interface mixings and are reduced by oxygen. At temperatures above 500°C, oxidation becomes important which accounts for an enhanced outdiffusion rate of Ni using an oxygen-containing ambient.

Another interesting system is Cu/Al discussed earlier, which shows an oxygen-enhanced outdiffusion of Al. Using a Pt top layer to

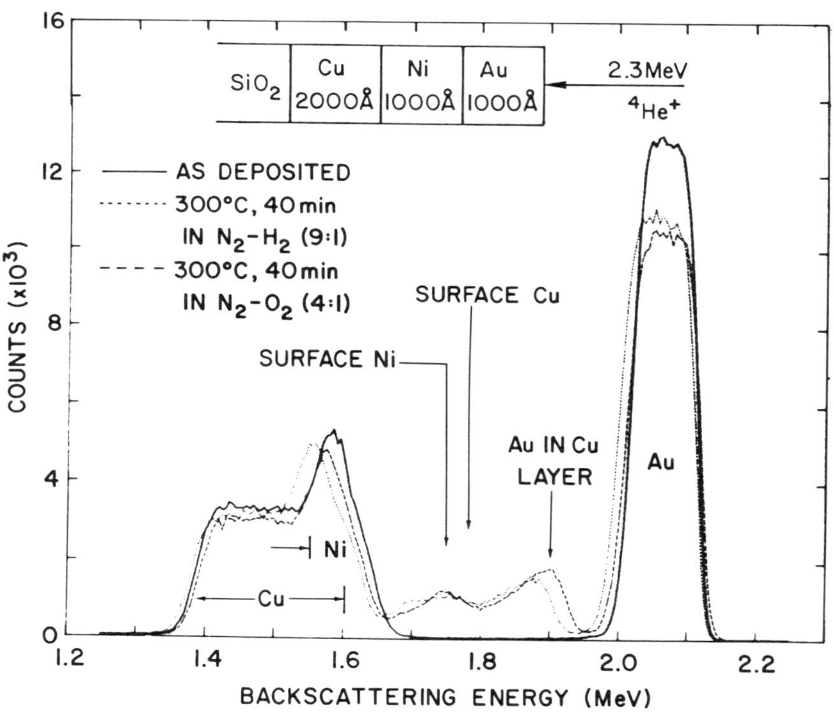

Fig. 18. Backscattering spectra of the Au/Ni/Cu structure, showing enhanced diffusion between Au and Cu across the Ni layer by oxygen.

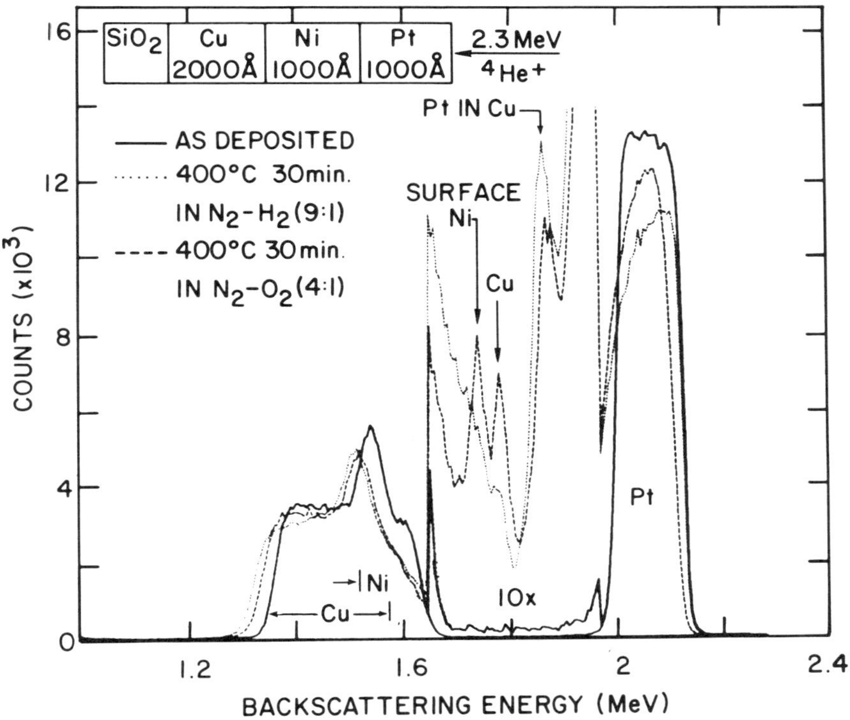

Fig. 19. Backscattering spectra of the Pt/Ni/Cu structure, showing reduced diffusion of Pt into Cu by oxygen, with an enhanced outdiffusion of Cu.

suppress such an effect, the interdiffusion between Cu and Al can be studied and is found to be reduced by oxygen.

These studies indicate that the interactions observed for various thin film systems can involve several different mechanisms. The interactions can take place at the same time or each dominates at different temperatures. It is therefore important to explore all the possible interactions to understand the system under investigation.

5. EFFECT OF OXYGEN ON THE SILICIDE FORMATION

As pointed out earlier, the formation of silicide has often been noted to be reduced by oxygen, and an interface oxide is suggested to be the reason (16-19). In these studies the Pt films contained an appreciable amount of oxygen which served as the source for the oxide layer observed. A reduced interaction by oxygen for the Pt systems has been shown to be expected from the surface potential model

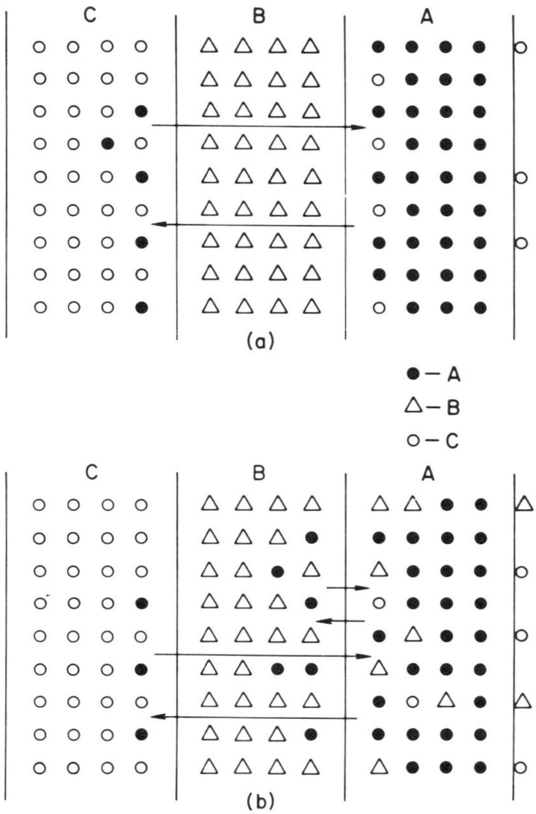

Fig. 20. Schematics of the diffusion patterns for two trilayer structures A/B/C: (a) Interdiffusion between A and C across B with little interactions involving B. (b) Interdiffusion between A and C across B in the presence of strong interdiffusion between A and B.

without invoking the oxide layer. In this Section we describe recent work using Pt films containing little oxygen to eliminate the difficulty encountered in the earlier studies, and to study the silicide formation in an oxygen ambient. With such controlled conditions, one is then able to study the true ambient effect of oxygen on the silicide formation (46).

Pt films were deposited by sputtering onto precleaned Si wafers, and annealed between 200 and 500°C in an oxygen ambient. The phases formed are identified by both x-ray diffraction and transmission electron microscopic measurements, and the composition distribution by Auger spectroscopy. The Auger depth profiles are shown in Figs.

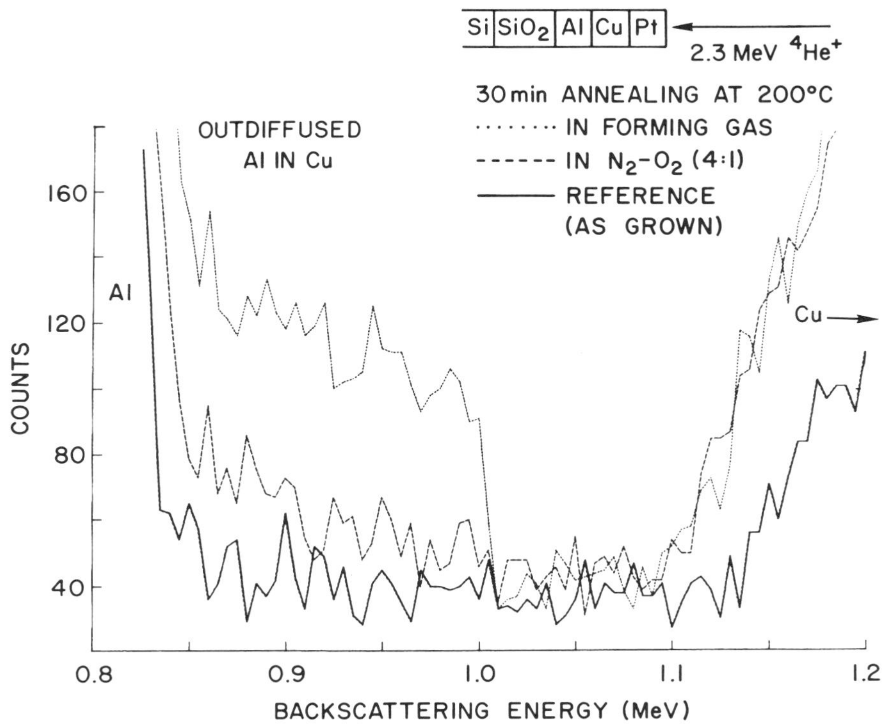

Fig. 21. Backscattering spectra showing reduced outdiffusion of Al into Cu by oxygen in the Pt/Cu/Al system.

23-25. The as-deposited Pt films contain little oxygen, and little reaction took place at 200°C. Pt_2Si and PtSi are the major products after additional annealings at 300°C and at 300 and 500°C, respectively. After such heat treatments, an oxide layer has developed, as seen in both Figs. 24 and 25. This indicates that the oxygen observed comes from the ambient used for annealing, with a higher oxygen concentration for the sample annealed at higher temperatures. The oxide is seen to locate at the Pt-Pt_2Si interface in Fig.24, and at the Pt-PtSi interface in Fig. 25, after different heat treatments. Auger analysis indicated that the oxide is of silicon oxide in nature, which is consistent with the observation of a simultaneous depletion in Pt and the pile-up of oxygen shown in the Auger depth profiles.

To further explore the origin of the oxide layer observed, a Pt/Si sample is annealed at 250°C and analyzed, as shown in Fig. 26. It shows the most interesting result that oxygen clearly comes from the ambient by diffusing into the Pt layer. The formation of the Pt_2Si phase has not yet been completed after this heat treatment, and the

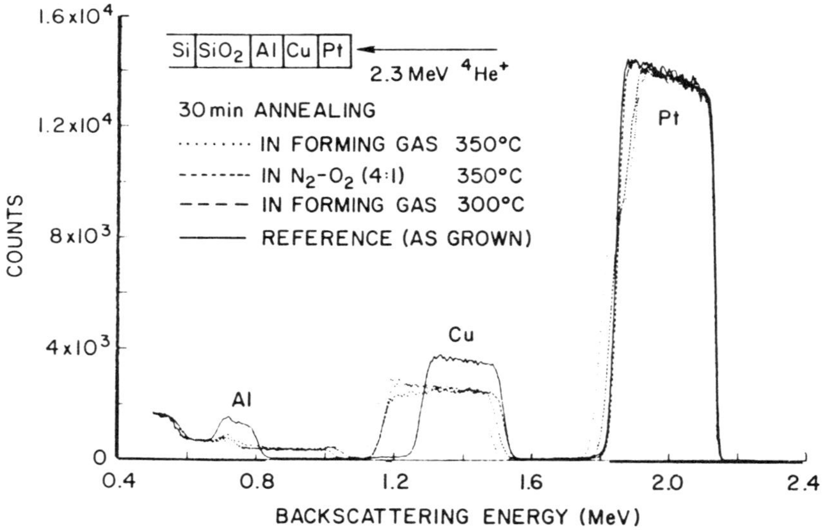

Fig. 22. Backscattering spectra showing (1) preferential interaction between Pt and Al from the CuAl alloy underneath, and (2) reduced Pt-Al interaction by oxygen.

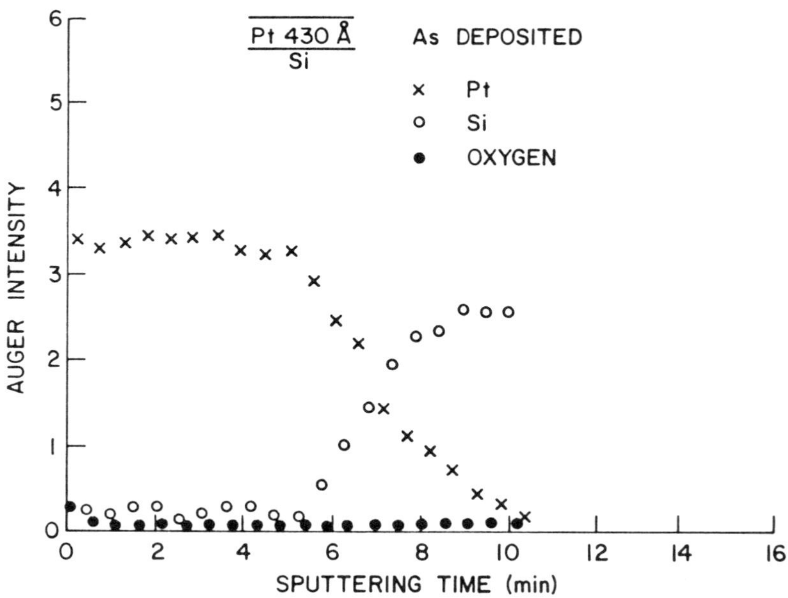

Fig. 23. Auger depth profile of the as-deposited Pt/Si sample, showing Pt film containing negligible amount of oxygen.

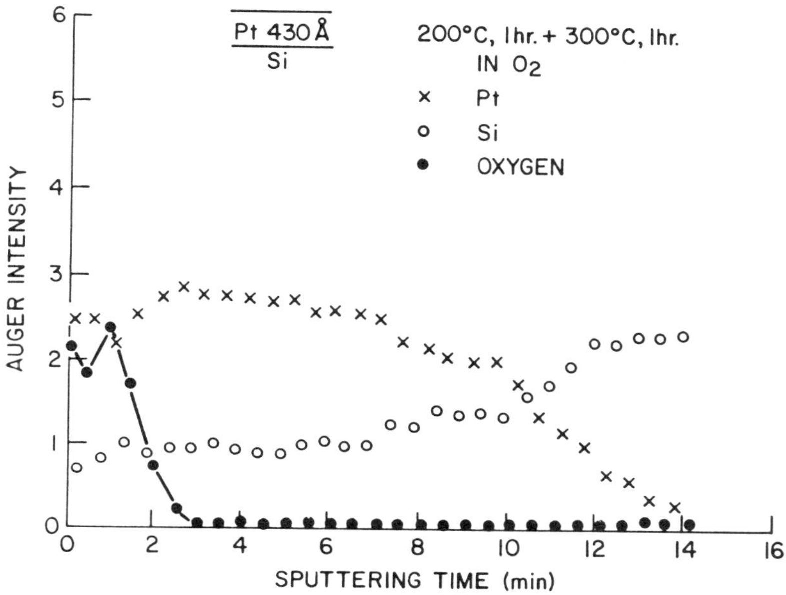

Fig. 24. Auger depth profile for the Pt/Si system after heat-treatments at 200 and 300°C in oxygen.

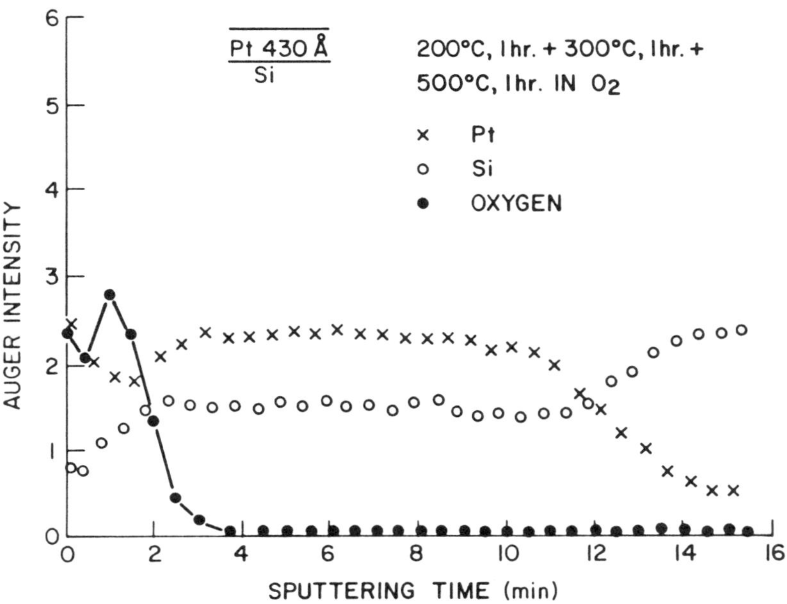

Fig. 25. Auger depth profile of the Pt/Si system after heat-treatments at 200, 300 and 500°C in oxygen.

front of the Pt_2Si phase is not in contact with the oxygen front. This provides an unambiguous evidence of the origin of the interface oxide layer observed in the present study using Pt films containing little oxygen. The results also indicate that the oxide layer observed has little effect on the formation rates of both Pt_2Si and PtSi. The oxide is seen to be confined near the surface, and comes into contact with the silicide phase only when the majority of the Pt film has been consumed for the silicide formation. Its effect lies mainly in preventing a complete formation of Pt_2Si and, later, PtSi.

The reaction sequence of the Pt silicide formation and the development of the interface oxide layer are depicted in Fig. 27. At low temperatures, Pt_2Si starts to form with little oxygen diffusion into the Pt film, as Fig. 27a shows. As the reaction proceeds, both the outward growth of Pt_2Si and inward diffusion of oxygen continue, as shown by that of the stage 2 of the Pt_2Si formation in Fig. 27a. Once the two fronts meet, oxygen reacts with the Si of Pt_2Si to form an oxide, which hinders further growth of Pt_2Si and leaves either an unreacted Pt or Pt-rich silicide film near the surface. This is shown in Fig.27b where an oxygen pile-up has developed between the surface Pt-rich layer and the underneath Pt_2Si. As further reaction proceeds, PtSi starts to form, giving distributions of the various elements as shown in Fig. 27c for different stages of reaction, and for that at the end of reaction shown in Fig. 27d. During these stages of reaction, oxygen diffusion into the film and its reaction with the Si front also continue, further enhancing the oxygen concentration and a pile-up of oxygen in the reacted film.

From the results shown, it is clear that, using Pt films with little oxygen and annealed in an oxygen ambient, the oxide layer observed has little effect on the overall formation rates of either Pt_2Si or PtSi. One therefore needs mechanisms other than the interface oxide to account for the reduced formation rates of silicide by oxygen. The surface potential model has been shown to be able to account for such ambient effects, as described for many other systems. It is therefore interesting to extend similar studies as described here for the Pt/Si system to other silicide systems to better understand the mechanism of the ambient effects observed for such systems.

The effect of oxygen described above on the formation of Pt silicides has been shown to have a significant impact on the PtSi contact metallurgy used for device fabrication. For a Pt/Si structure heated directly to 550°C, an incomplete reaction between Pt and Si is observed, with a surface oxide layer of poor protection against etching in aqua regia needed in device processing. This is observed using forming gas, nitrogen, oxygen or nitrogen-oxygen ambients (47). Using a three-temperature annealing sequence at 200-300-550°C in forming gas, a complete reaction between Pt and Si is obtained, with a

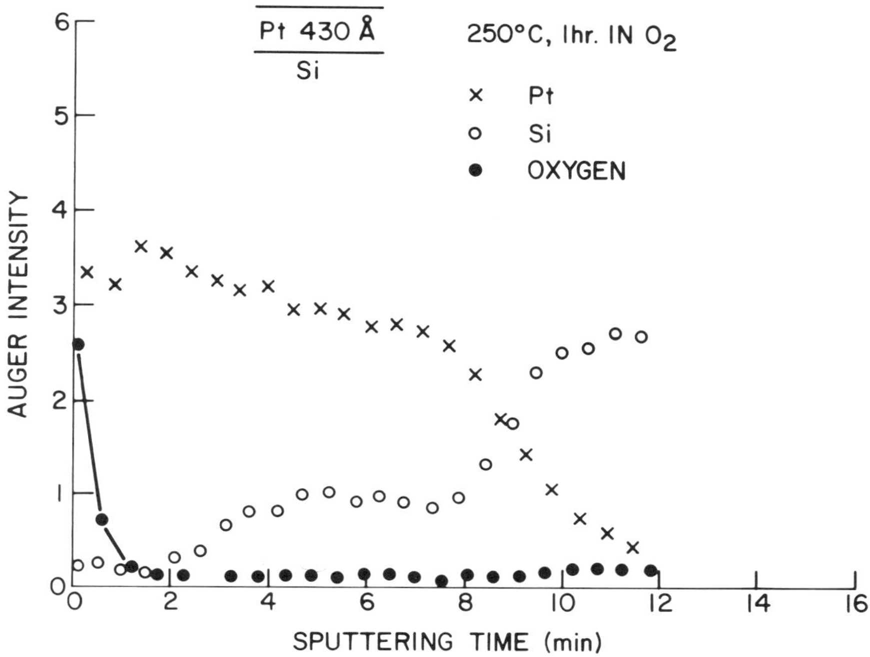

Fig. 26. Auger depth profile of the Pt/Si system after a heat-treatment at 250°C in oxygen. Notice that the oxygen front is separated from the Si front of Pt_2Si.

surface oxide layer of excellent resistance to etching (47). The Auger depth profiles of representative samples are shown in Figs. 28-30, and the x-ray diffraction patterns shown in Figs. 31-33. It is seen that using the one-temperature process, the film shows an appreciable change upon etching. The surface signals of both Pt and oxygen change appreciably upon etching. The corresponding x-ray patterns show the disappearance of the unreacted Pt signal after etching, indicating most of the unreacted Pt is confined near the surface. This is in agreement with the Auger studies, and with the results described earlier on the effect of oxygen on the formation of Pt silicides. Using the three-temperature process in forming gas, on the other hand, an excellent surface Auger profile is obtained, with high oxygen and low Pt signals at the surface, and no unreacted Pt from the x-ray diffraction measurement. Detailed studies of the formation processes of PtSi contact metallurgy have demonstrated the superior features of the three-temperature annealing sequence in forming gas as needed for device fabrication (48-52).

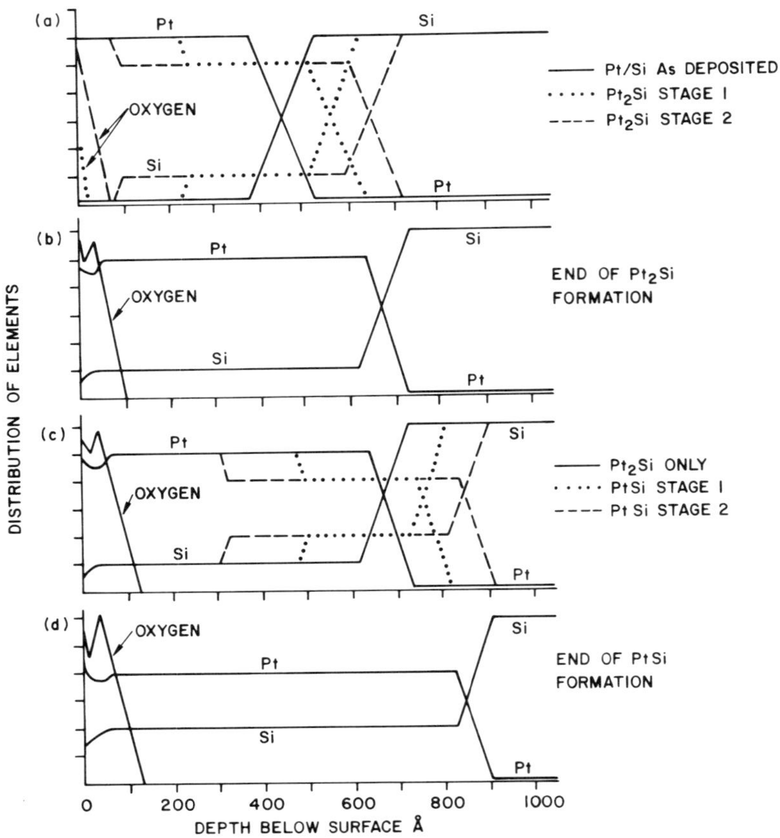

Fig. 27. Schematic description of the ambient effect of oxygen on the Pt-Si interactions.

6. APPLICATIONS AND FUTURE STUDIES

The reported ambient effects on the interactions of a large number of thin film systems have been reviewed and discussed. Several proposed mechanisms are compared. Many systems can be well correlated by the surface potential model. The predicted ambient effects from this model have also been confirmed for many systems. Several interesting interactions and the ambient effects on such interactions have also been observed.

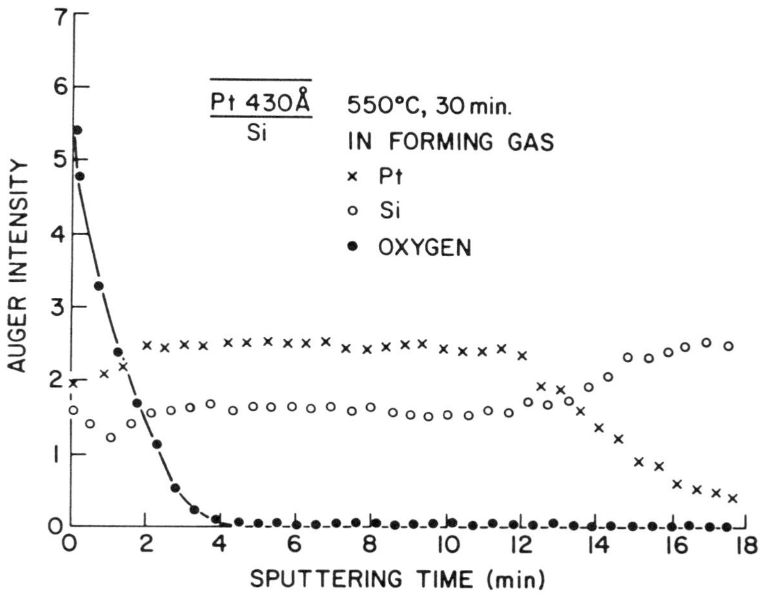

Fig. 28. Auger depth profiles of the Pt/Si sample annealed at 550°C in forming gas.

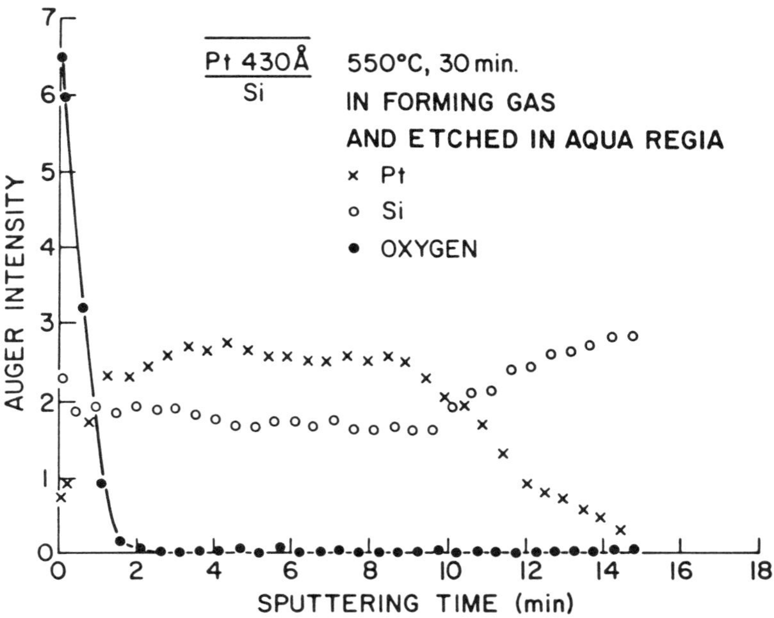

Fig. 29. Auger depth profiles of the Pt/Si sample annealed at 550°C in forming gas, and etched in aqua regia.

Among the ambient effects discussed, the reduced interactions by oxygen is interesting and can be technologically significant. For example, surface oxide due to outdiffused species is always a trouble source for the materials joining for various devices. During storage of the structures, which is often done in air, stability against such outdiffusion is a vital factor for the successful joining later. Such a stability is known to be poor when gold is used as the surface protective layer, as reported for the many Au systems reviewed in this Chapter. During subsequent operation of the devices containing thin film structures, the ambient effect is also vital to the device lifetime and performance. The electric current passing through the interface will unavoidably generate local heating, and therefore enhances the interactions there. If the interactions generated are further enhanced by the ambient, the device lifetime will be shortened. On the other hand, a proper metallurgy may by chosen to suppress such interactions, and contribute to the lifetime of the device fabricated. It is thus important to under-

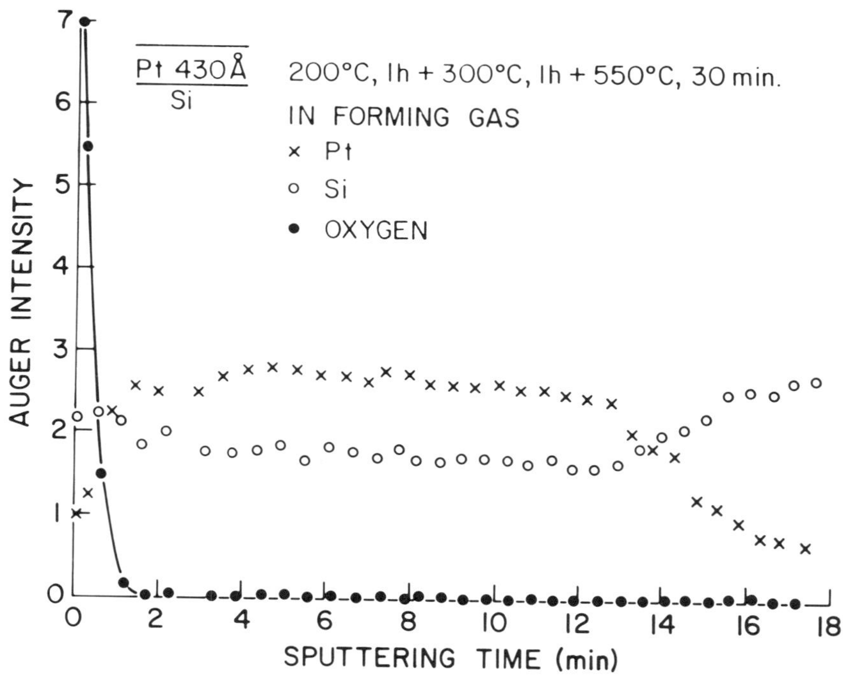

Fig. 30. Auger depth profiles of the Pt/Si sample annealed at 200-300-550°C in forming gas.

stand the ambient effects present and to choose proper metallurgies for the optimal performance and lifetime of devices.

An example is the use of Pt to reduce the undesired outdiffusion of Cu through Au in the solder wetting studies (53). An Au/Pt/Cu structure is compared with the Au/Cu one for the wetting of PbSn solder using no chemical flux. The results are shown in Fig. 34 upon heating the solder/metallurgy structures in hydrogen and forming gas to 350°C for the PbSn (95-5) solder, and to 215°C for the PbSn (37-63) solder. Wetting of the PbSn (95-5) solder on the Au/Pt/Cu metallurgy is clearly better than that on the Au/Cu one in both ambients. Backscattering and Auger analysis of the reactions in these two metal structures show the outdiffusion of Cu to the Au surface using the Au/Cu structure, but not for the Au/Pt/Cu one, as shown in Figs. 35 and 36. The spectra also show a reduced outdiffusion of Cu using hydrogen than that using forming gas for the Au/Cu structure, and a similarly reduced Pt-Cu interdiffusion by hydrogen, both in agreement with the surface potential model. For the wetting of the PbSn (37-63) solder at 215°C, little outdiffusion of Cu through Au is observed at such temperatures, and good wetting is obtained on both metallurgies.

Another example using Pt for protection in air is the Pt/GaAs structure, which has been shown to retain good Schottky characteristics after heating in air for 30 min at 310°C. The structure becomes softened after a similar heat treatment at 370°C (54).

In addition to its application to technology, an understanding of the ambient effects provides valuable information to our understanding of the thin film interactions in general. From the vast number of systems reported and discussed in this Chapter, it is clear that the ambient effects observed reveal interesting information about the interactions taking place at the very interface, which is often difficult to obtain. In other words, the ambient acts as an interface probe by catalyzing or hindering the interactions there (25). The electronegativity and factors which can affect the bonding at the interface, such as the ambient-induced surface potential changes, should be seriously considered in future studies of thin film interactions. It is clear that once the initial interface mixing is affected, many of the subsequent reactions that follow are affected similarly, including outdiffusion, interdiffusion, alloy formation, grain mixing, etc. The observed ambient effects for the many systems reported and discussed here are summarized in Table 2. It illustrates the general behavior of the various interactions observed in the same ambient for systems sharing the same overcoating metal layer.

For almost all of the systems reviewed here whose ambient effects can be correlated, the difference in electronegativity between the two interacting materials is rather large. There are many other systems

Table 2. Observed Ambient Effects on Thin Film Systems. Interactions Observed: OD-Outdiffusion; ID-Interdiffusion; AF-Alloy Formation; GM-Grain Mixing; CR-Chemical Reaction.

System	Interaction	Ambient Effect	Reference
Au/Si	OD	steam>air>O_2>H_2	1,2,58
		N_2-CO<N_2	24
Au/Ge	OD	steam>N_2	3
Au/GaAs	OD	air>vacuum	4
		N_2-CO<N_2	36
Au/Cr	OD	O_2>vacuum	5
	OD, ID	N_2-CO<N_2	24
	OD, ID	air-CO<air	25
Au/Cu	OD	air>vacuum	6
	OD/ ID	O_2>N_2	53
		H_2<N_2-H_2	
Au/Ni	OD	air>vacuum	7
		N_2-O_2>N_2-H_2	45
Au/Fe	OD	air>vacuum	8
Au/Ti	OD	air>vacuum	9
Au/Pd	OD	air>vacuum	9
Pt/Au	OD	He>N_2>vacuum	10
Ni/Au	OD	H_2>N_2-H_2	43
		O_2>N_2-H_2	

(Continued on the following page)

Table 2 (Continued)

System	Interaction	Ambient Effect	Reference
Au/Co	OD	N_2-H_2>N_2	13
Au/Al	OD	H_2<Ar	14
Pd/Ti	ID	air<vacuum	9
Pt/Al	AF	air<Ar, He, N_2-H_2	15
		N_2-O_2<N_2-H_2	39
Pt/Si	AF	O_2<vacuum	16-19
		O_2<O_2-N_2<N_2	37
Pt/Cr	OD	N_2-O_2<N_2-H_2	37
Pt/Ni	ID	N_2-O_2<N_2-H_2	38
Pt/Cu	ID	N_2-O_2<N_2-H_2	40
Ni/Al	AF	N_2-O_2<N_2-H_2	41
Ni/Cu	OD	H_2>H_2-H_2	43
		O_2>N_2-H_2	
Cu/Cr	GM	N_2-O_2<N_2-$H_2$43	
Cu/Al	OD	N_2-O_2>N_2-H_2	42
	ID, AF	N_2-O_2<N_2-H_2	42
Ag/Tl halides,	ID, CR	air>vacuum	56
Co/Tl halides,		Cl_2 H_2S	
Cu/Cs halide,		Cl_2 air	
Ag/Cs halide			

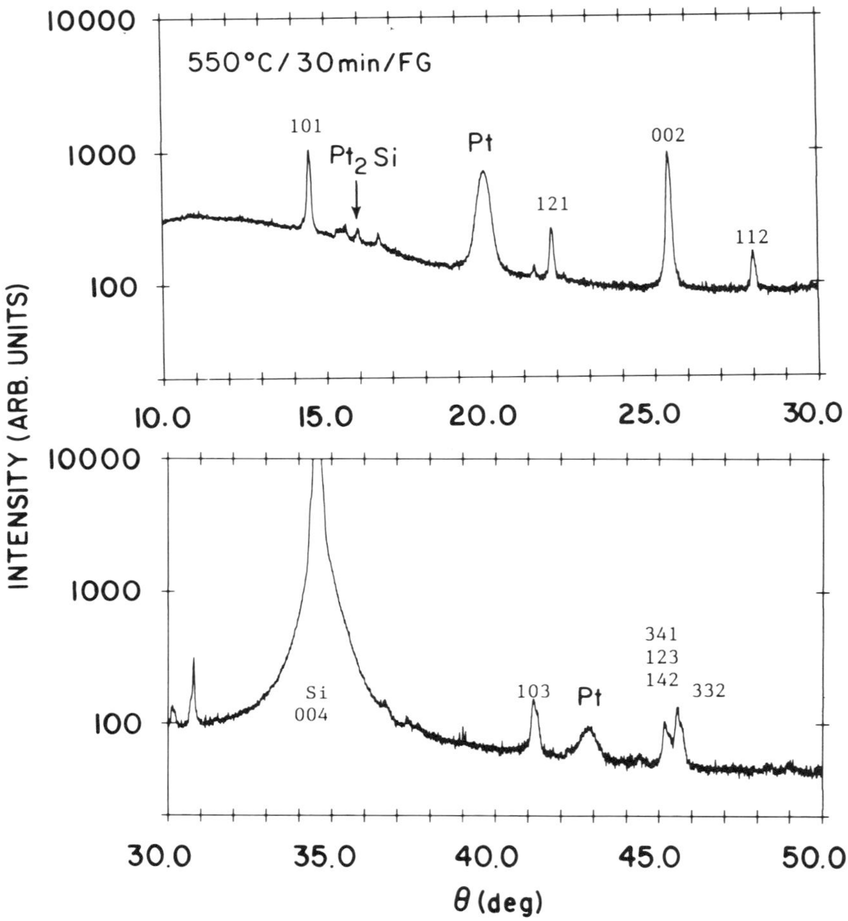

Fig. 31. X-ray diffraction patter of the Pt/Si sample annealed at 550°C in forming gas.

with much smaller difference in electronegativities. The situation resembles the chemical systems where either ionic bonding or covalent bonding can be dominating. Extension of studies to such "covalent" thin film systems to understand their ambient effects, if any, is worth pursuing.

Several mechanisms for the ambient effects observed are discussed, and good correlations have been shown with the surface potential model. The predicted ambient effects from this model have been confirmed with many systems. These observations therefore provide

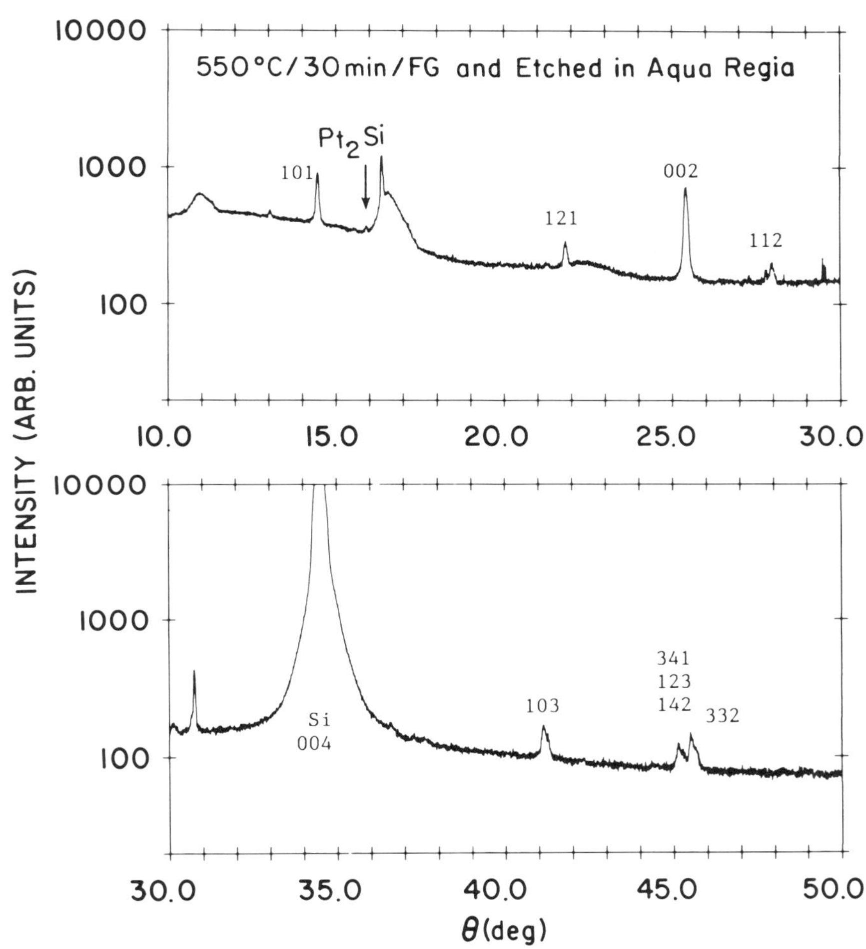

Fig. 32. X-ray diffraction pattern of the Pt/Si sample annealed at 550°C in forming gas, and etched in aqua regia.

an experimental basis for any mechanism describing the ambient effects known. In the present form of the surface potential model, only a conceptual interaction mechanism is suggested, and a more detailed one remains to be determined. The systems treated should also be compared with different studies on the surface properties. For example, a reduced work function of gold by CO has been reported recently (55), which is different from the one used earlier. More studies are therefore needed to understand the effects of CO on the systems reported.

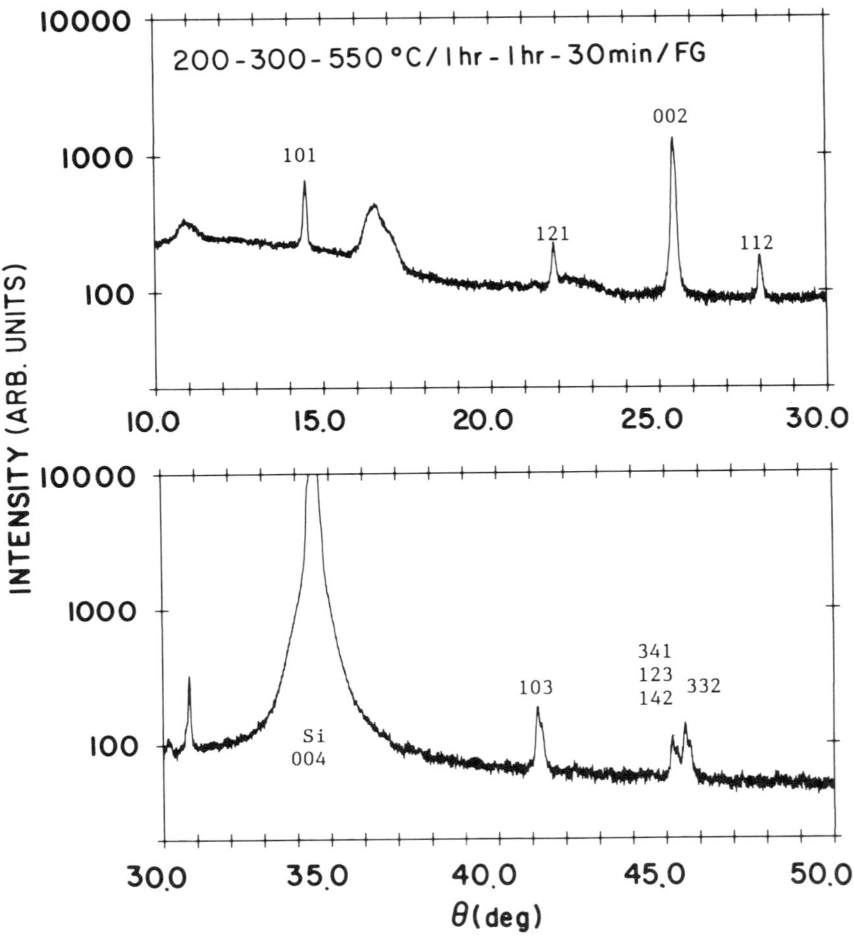

Fig. 33. X-ray diffraction pattern of the Pt/Si sample annealed at 200-30 0-500°C in forming gas.

There are systems where chemical reactions can be involved which are affected by the ambient. For example, an ambient-induced diffusion has been observed in several metal/halide systems (56). Compound formation has been suggested to be the cause for the ambient effects observed. Such systems are different from the ones discussed in this Chapter, and provide further challenges for the understanding of the ambient effects observed. For example, an interface reaction mechanism, not considered by these workers, could provide reactions that are energetically more favorable for the ambient effects observed than those considered by these workers (57).

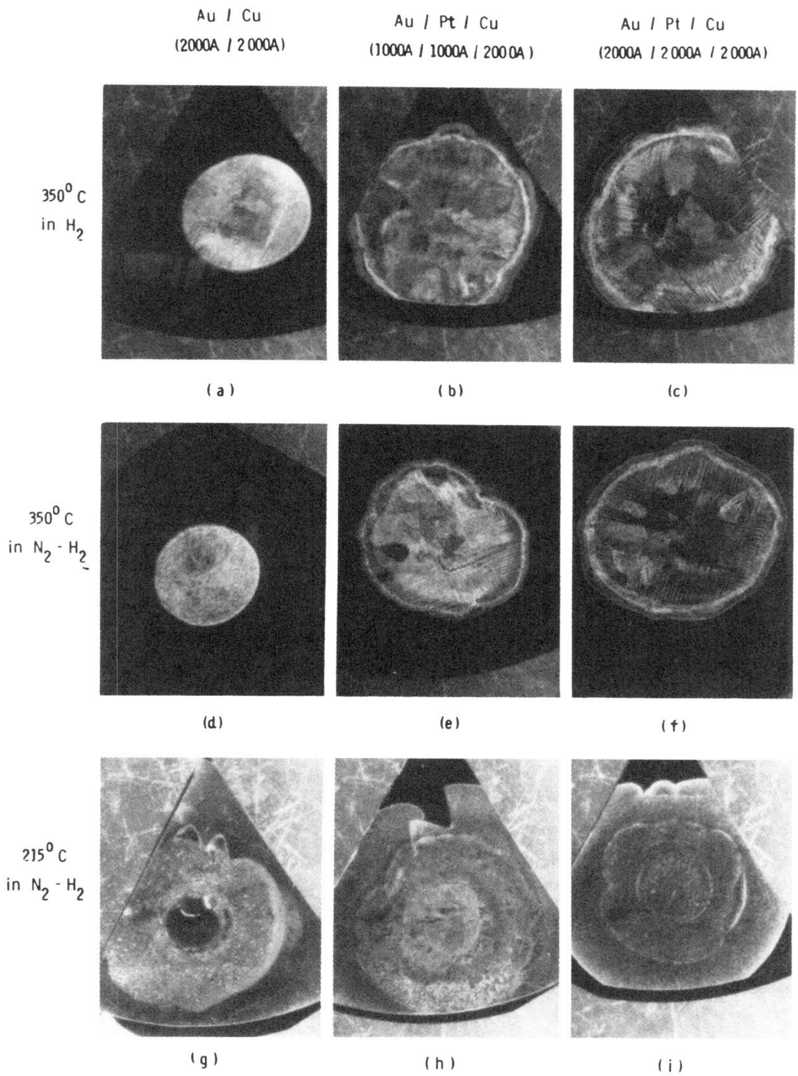

Fig. 34. Results of solder wetting on different metallurgies. a-c: wetting of PbSn (95-5) solder at 350°C in hydrogen. d-f: Wetting of PbSn (37-63) solder at 215°C in forming gas.

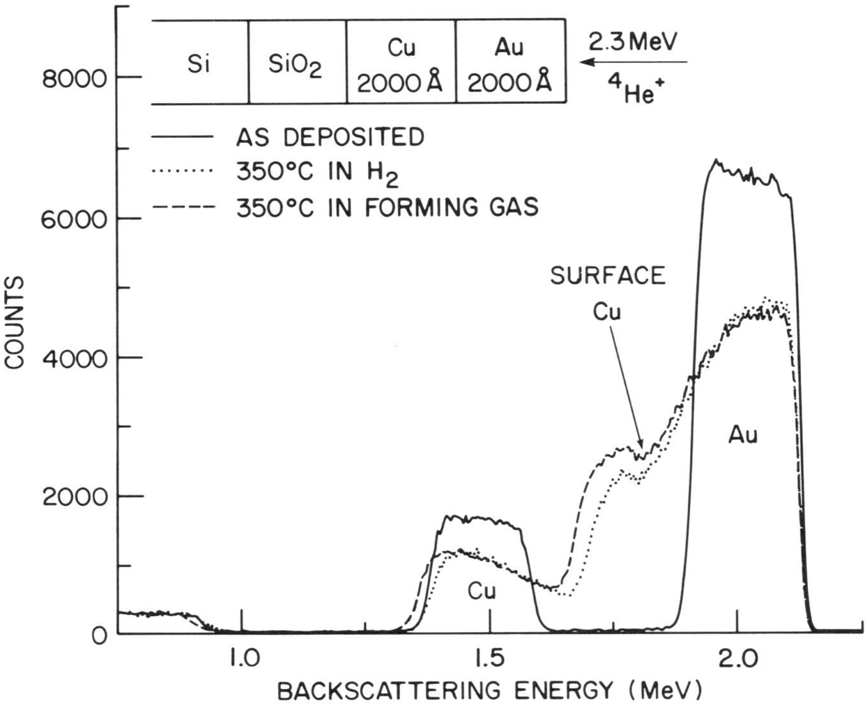

Fig. 35. Backscattering spectra of the Au/Cu samples heated to 350°C in hydrogen and forming gas.

The effects of ambients can be applied to systems with different crystal orientations and different metal film structures, and compare the combined effects on the thin film interactions. Recent studies of the Au/Si system, for example, have shown significantly stronger interactions for the Au/(100)Si system than the Au/(111)Si one upon annealing in air (58). Using different deposition techniques for gold, much enhanced interactions have been observed for the sample using sputtering than those using the electron-beam evaporation techniques. The interactions are much reduced in oxygen for all cases, with Au/(100)Si being still more reactive than Au/(111)Si, and the sputtered samples more reactive than the evaporated ones.

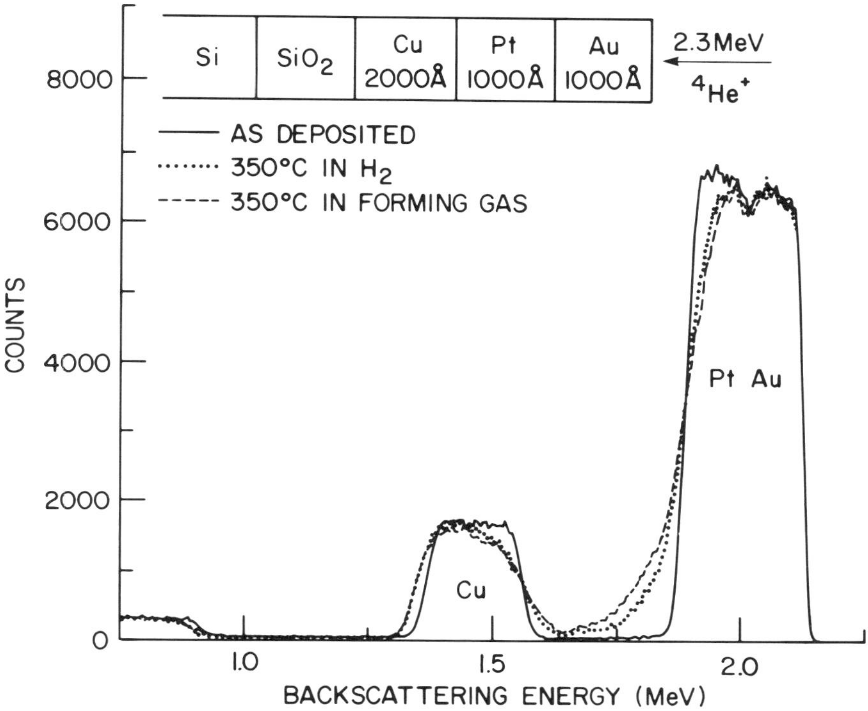

Fig. 36. Backscattering spectra of the Au/Pt/Cu samples heated to 350°C in hydrogen and forming gas.

7. SUMMARY

A large number of thin film systems, scattered in literature and seemingly unrelated, have shown interactions which depend on the annealing ambients used. The systems studied include metal-metal, metal-semiconductor, and metal-compound ones. By considering the roles played by the overcoating metal layer and the ambients, a model is proposed to explain the ambient effects observed, and is shown to correlate the observed ambient effects quite well.

Extension of the work to other systems has confirmed the predicted ambient effects for many systems, and has led to the observation of interesting new ambient effects. These studies of the ambient effects on thin film interactions not only have important applications to technology, but also provide insights into our basic understanding of the thin film interactions in general.

REFERENCES

1. A. Hiraki, E. Lugujjo and J. W. Mayer, J. Appl. Phys. **43**, 3643 (1972).
2. A. Hiraki, E. Lugujjo, M.-A. Nicolet, and J. W. Mayer, Phys. Status Solide A**7**, 401 (1971).
3. E. M. Lawson and A. J. Tavendale, J. Appl. Phys. **53**, 8771 (1982).
4. A. K. Sinha and J. M. Poate, in Thin Films-Interdiffusion and Reactions, ed. J. M. Poate, K. N. Tu and J. W. Mayer (Wiley, New York, 1978), p. 407.
5. G. C. Nelson and P. H. Holloway, Metallurgical Applications of Surface Analysis Technbiques, ASTMSTP 596 (American Society fo Testing Mterials, Philidelphia, 1976).
6. H. G. Thompkins and M. R. Pinnel, J. Appl. Phys. **47**, 3804 (1976).
7. J. R. Rairden, C. A. Neugebsuer, and R. A. Sigsbee, Metall. Trans. 2, 719 (1971); A. Barcz, A. Turos and L. Wielundki, in Ion Beam Surface Layer Analysis, O. Mayer, G. Linker and F. Kappeler, eds., Plenum Press, New York (1976), p. 407.
8. J. F. Ziegler, J. E. E. Baglin, and A. Gangulee, Appl. Phys. Lett. **24**, 36 (1974).
9. J. M. Poate, P. A. Turner, W. J. DeBonta, and J. Yahalom, J. Appl. Phys. **46**, 4275 (1975).
10. C. C. Chang and G. Quintana, Appl. Phys. Lett. **29**, 453 (1976).
11. K. D. Kang, R. R. Burgess, M. G. Coleman, and J. G. Keil, IEEE Trans. Electron Devices **16**, 356 (1969).
12. See, for example, W. J. DeBonte, J. M. Poate, C. M. ,Elliar-Smith, and R. A. Levesque, J. Appl. Phys. **46**, 4284 (1975).
13. J. E. E. Baglin and J. M. Poate, in Ref.4, p. 305.
14. Da-Yuan Shih and P. J. Ficalora, 16th Annual Proceedings Reliability Physics 1978, p.268.
15. S. P. Murarka, I. A. Blech and H. J. Levinstein, J. Appl. Phys. **47**, 5175 (1976).
16. J. Bindell, J. W. Colby, D. R. Wonsidler, J. M. Poate, D. K. Conley, and T., C. Tisone, Thin Solid Films **37**, 441 (1976).
17. C. Canali, C. Catellani, M Prudenziati, W. H. Wadlin, and C. A. Evans, Appl. Phys. Lett. **31**, 43 (1977).
18. F. Nava, S. Valeri, G. Majni, A. Cembali, G. Pignatel, and G. Queirolo, J. Appl. Phys. **52**, 6641 (1981).
19. C. A. Crider and J. M. Poate, Appl. Phys. Lett. **36**, 417 (1980).
20. R. B. Belser, J. Appl. Phys. **31**, 562 (1960).
21. See, for example, Ref.13.

22. See, for example, H. F. Wolf, Semiconductors, Wiley-Interscience, New York (1971), p. 342.
23. G. Ottaviani, J. Vac. Sci. Tech. **16**, 1112 (1979).
24. Chin-An Chang, J. Electrochem. Soc. **127**, 1331 (1980).
25. Chin-An Chang, Appl. Phys. Lett. **38**, 860 (1981).
26. Chin-An Chang, Mat. Res. Soc. Sym. Proc. Vol.25, p.111(1984).
27. J. C. M. Hwang, P. S. Ho and R. W. Ballufi, Appl. Phys. Lett. **33**, 458 (1978).
28. Chin-An Chang, Surf. Sci. **95**, L239 (1980).
29. F. C. Thompkins, in The Solid-Gas Interface, ed. E. A. Flood (Dekker, New York, 1967), Chap.25.
30. Chin-An Chang, J. Electrochem. Soc. **123**, 1245 (1976).
31. L. Pauling, The Nature of the Chemical Bond, 3rd ed. (Cornell University, New York, 1960).
32. N. A. Surplice and W. Brearly, Surf. Sci. **52**, 62 (1975).
33. R. L. Wells and R. Fort, Jr., Surf. Sci. **32**, 554 (1972).
34. K. L. Chopra, Thin Film Phenomena, McGraw Hill Co., New York (1969), p.380.
35. G. A. Somorjai, Principles of Surface Chemistry (Prentice-Hall, Englewood, Cliffs, 1972).
36. Chin-An Chang and N. Chou, J. Vac. Sci. Tech. **17**, 1358 (1980).
37. Chin-An Chang and W. K. Chu, Appl. Phys. Lett. 37, 161 (1980).
38. Chin-An Chang and W. K. Chu, J. Appl. Phys. **52**, 512 (1981).
39. Chin-An Chang, J. Appl. Phys. **52**, 4620 (1981).
40. Chin-An Chang, J. Appl. Phys. **53**, 7092 (1982).
41. Chin-An Chang, J. Appl. Phys. **53**, 7088 (1982).
42. Chin-An Chang, Appl. Phys. Lett. **44**, 310 (1984).
43. Chin-An Chang and H. L. Yeh, Appl. Phys. Lett. **49**, 1233 (1986).
44. Chin-An Chang, to be published.
45. Chin-An Chang, J. Appl. Phys. **60**, 1220 (1986); Mat. Res. Soc. Symp. Proc. Vol. 72 (1986), p. 267.
46. Chin-An Chang, J. Appl. Phys. **58**, 1412 (1985).
47. Chin-An Chang, J. Appl. Phys. **58**, 3258 (1985).
48. Chin-An Chang, J. Appl. Phys. **59**, 3116 (1986).
49. Chin-An Chang, A. Segmuller, H.-C. W. Huang, B. Cunningham, F. E. Turene, A. Sugerman, and P. A. Totta, J. Electrochem. Soc. **133**, 1256 (1986).
50. Chin-An Chang, B. Cunningham, A. Segmuller, H.-C. W. Huang, F. E. Turene, A. Sugerman, and P. A. Totta, J. Vac. Sci. Technol. B**4**, 745 (1986).

51. Chin-An Chang, A. Segmuller, H.-C. W. Huang, F. E. Turene, B. Cunningham, and P. A. Totta, J. Vac. Sci. Technol. A**4**, 841 (1986).
52. Chin-An Chang and A. Segmuller, J. Appl. Phys., in press (January, 1987).
53. Chin-An Chang and H. L. Yeh, to be published.
54. Chin-An Chang, J. Vac. Sci. Technol. **21**, 639 (1982).
55. B. E. Nieuwenhuys, Surf. Sci. **105**, 505 (1981); A. M. Bradshaw and J. Pritchard, Proc. Roy. Soc. London A**316**, 169 (1970); J. Pritchard, Trans. Faraday Soc. **59**, 437 (1963).
56. A. A. Simanovsky, S. V. Stolyarova, and G. P. Upit, Thin Solid Films, **97**, 301 (1982).
57. Chin-An Chang, (unpublished).
58. Chin-An Chang and G. Ottaviani, Appl. Phys. Lett. **44**, 901 (1984).

7

Electromigration in Metallic Thin Films

Thomas Kwok and Paul S. Ho

1. INTRODUCTION

Electromigration, or electrotransport, describes the atomic motion in a metal under the influence of an applied electric field. It is basically a diffusion phenomenon under a driving force. The phenomenon of electromigration has been observed for more than a century (1), however, systematic studies did not start until the early 1950's with the work of Seith and Wever (2). By measuring the mass transport across the phase diagram of some Hume-Rothery alloys, they observed that the direction of mass transport can be reversed and is correlated with the type of the majority charge carriers, i.e. electrons or holes, in the specific alloy phase. This observation provided the first evidence concerning the nature of the driving force for electromigration. It showed that the atomic motion is not determined solely by the electrostatic force imposed by the applied field, instead it depends on the direction of motion of the charge carriers. This prompted Seith to introduce the idea originated by Skaupy (3) of "electron wind" to account for the induced mass transport, an idea which laid the foundation for the basic understanding of electromigration. Seith and Wever introduced also the method of using the displacement of an indentation on a metal wire to measure the induced mass transport. This "marker motion" technique, now called the vacancy flux method, has become one of the standard measurements of electromigration.

The concept of the "electron wind" driving force was first formulated by Fiks (4) and Huntington and Grone (5). Working independently, these authors employed a semiclassical "ballistic" approach to treat the collision of the moving atom by the charge carriers. In a more rigorous treatment, Huntington and Grone showed that not only the initial and final states of the charge carriers have to be considered in the collision process, the spatial variation of the force experienced by the moving atom has to be taken into account as well. These ideas yielded a driving force depending on the type of defects and the atomic configuration of the jumping path.

The formulation of the driving force was a major contribution to the study of electromigration as it demonstrated the possibility of using electromigration to probe directly the interaction of mobile defects and charge carriers. This stimulated considerable interest during the 1960's to carry out experimental studies on electromigration in metals. During that period of time, the investigations were primarily basic studies and concentrated mostly on bulk materials. The interest has initially concentrated on self-electromigration in pure metals (6), which was extended later to alloys and liquid metals (7). Until that time, the applied study of electromigration was rather limited with efforts largely directed toward the purification of interstitial, particularly gaseous impurities in refractory metals (8). The interest as well as the direction of electromigration study took a drastic turn in the late 1960's when electromigration was identified as causing the failure of aluminum interconnecting lines in integrated circuits (9). This stimulated a flurry of activities in studying electromigration in thin-film conductors. Most of the studies were aimed at the rather practical problem of conductor line failure, or the "cracked stripe" problem in microelectronic integrated circuits. Instead of bulk materials, these studies were carried out in thin films prepared by evaporating metal layers onto insulating substrates and at moderate temperatures of about half the absolute melting point. Under these conditions, electromigration occurs primarily along the grain boundaries. Particularly relevant to damage formation is the flux divergence site where damages, e.g. voids or hillocks, can form because of the local imbalance of atomic flux caused by inhomogeneity of parameters controlling the diffusion process, e.g. microstructure and impurity distribution. This aspect of electromigration damage has shifted the focus of the investigations to that of understanding electromigration at structural defects in thin-film materials (10-13). This trend continued into the 1970's, particularly at the industrial laboratories.

The idea of expressing the driving force in terms of electron-ion interactions has intrigued a number of theorists to employ quantum mechanical approaches to formulate the problem starting in the 70's. Most of the theoretical studies have focused on the physical basis of the driving force, particularly regarding the dynamic screening of the moving ion and the effect of the band structure on the electron-ion scattering. A controversy has emerged from these analyses with regard to the extent of screening of the moving ion from the applied field, a problem which seemed to have resolved only recently. While the theory continues to develop, data have accumulated to show systematic trends to check the theoretical results. The interest in understanding the nature of the driving force has continued to the present time.

In the past several years, an impetus of studying electromigration in very fine conductor lines has arisen from the development of very large-scale integrated (VLSI) circuits. The conductors of interest are not only small in dimensions (with a width of about one micron and smaller), they are often assembled into a multilevel structure with certain combinations of conductors and insulators. This gives rise to electromigration problems with characteristics distinctly different from simple single-level conductor lines (14), where most of the studies have concentrated so far.

This chapter is organized into four sections. The first focuses on the studies of bulk materials. It contains a description of the phenomenological formulation based on irreversible thermodynamics and a review on the experimental results from bulk studies. The second section discusses the general aspects of electromigration in thin films including damage formation and methods for improving electromigration resistance. This is followed by the third section dealing with electromigration in fine conductor lines and the last section discussing electromigration in multilevel interconnection.

There are several comprehensive reviews on electromigration studies in metals (6,15-19). It is not the goal of this article to provide a similar comprehensive review, instead this article summarizes the current progress of the field with emphasis on the studies of thin films.

2. ELECTROMIGRATION IN BULK METALS

In this section, the formal basis for describing electromigration is introduced. First electromigration is treated as a phenomenon of forced motion in solids within the framework of irreversible thermodynamics. This is followed by discussions on the theory of the driving force, the techniques of measurement and a review on the results of electromigration in substitutional systems.

2.1 Phenomenological Description

Electromigration can be treated as a general phenomenon of forced motion in a solid with several interacting components (20). The problem has been treated within the framework of irreversible thermodynamics by Huntington in a review (6).

In a multicomponent system, the atomic motion of the ith component depends on its interaction with all the other elements in the system. This can be expressed in general by writing the atomic flux J_i as

$$\vec{J}_i = \sum_{j=1} L_{ij}\vec{X}_j \quad (i = 1, 2, ...n) \tag{1}$$

Where L_{ij} is the phenomenological coefficient correlating the flux of the ith component to the driving force X_j of the jth component. In an electric field, the driving force is derived from the gradient of the chemical potential μ_i as well as from the electrostatic potential ϕ. In case of electromigration, one of the moving components is the charge carrier. For a discussion of electromigration, it is convenient to express the flux equations separately for the motion of the atoms and the charge carriers as

$$J_i = -T^{-1}\sum_j L_{ij}\nabla(\mu_j + q_j\phi) - T^{-1}L_{ie}q_e\nabla\phi \tag{2a}$$

$$J_e = -T^{-1}\sum_j L_{ej}\nabla(\mu_j + q_j\phi) - T^{-1}L_{ee}q_e\nabla\phi \tag{2b}$$

Expressed in this way, it becomes clear that the diagonal elements L_{ii} are related to the diffusion coefficients of the ith element except for L_{ee} which involves the flow of the charge carriers. There are two types of off-diagonal elements: L_{ij}, relating to the correlation effects on the atomic jumps of the ith element due to the atomic jumps of the jth element, and L_{ie}, relating to the interaction between the ion and the electrons. The first type of these terms L_{ij} applies to both the chemical potential μ_j and the electrical potential $q_j\phi$. Therefore the correlation affects not only the diffusion process (which relates to μ_j) but also the electromigration process (which relates to $q_j\phi$). This introduces interesting correlation effects on electromigration and such an effect will be examined for dilute alloys. The second type of the off-diagonal elements constitutes the "electron-wind" force, which is usually combined with the direct electrostatic force on the moving ion, i.e. the $q_i\nabla\phi$ term, to yield the electromigration driving force. The driving force is expressed usually in terms of an effective charge Z^*, which includes the electrostatic and the electron-wind contributions and can be defined as

$$\begin{aligned} F_{eff} &\equiv |e|Z^*E \\ &= |e|(Z^e + Z^w)E \end{aligned} \tag{3}$$

where $|e|$ is the absolute value of the electronic charge and E, equal to $-\nabla\phi$, is the electric field.

At this point, not all of the nondiagonal coefficients are independent so that the Onsager relation cannot be applied to reduce the number of independent coefficients (6). Before this simplification is carried out, the mass conservation condition, i.e. $\sum_i J_i = 0$ and mechanical equilibrium condition, i.e. $\sum_i \nabla(\mu_j + q_j\phi) = 0$ should be imposed to assure an equilibrium state of the system. These conditions reduce further the number of independent coefficients L_{ij}. For a system where mass transport involves diffusion via vacancies, it is convenient to introduce the vacancy as an additional component to balance the mass transport of the moving atoms as

$$J_v = -\sum_i J_i. \tag{4}$$

Using this condition, one can transfer Eqs. 2a and 2b from the laboratory reference frame to a lattice reference frame where the atomic motion can be observed from the relative movement of the lattice planes. This approach of tracking the mass transport in electromigration is one of the commonly used methods in measurement and is called the vacancy flux method (6).

After transposing the transport equations to the lattice frame, the system is under mechanical equilibrium and the Onsager relation can be applied to yield $L_{ij} = L_{ji}$. After this simplification, a formal basis is established to treat electromigration in multi-component systems. Applications to several cases of common interest can be found in Ref. 6. In the following, we illustrate an application by considering electromigration in a dilute alloy (21,22) This example is of technological interest as well since the addition of alloying elements has been one of the most common approaches used to improve the lifetime of thin-film conductor lines.

For electromigration in a dilute alloy with solvent a and solute b, the transport equations are:

$$J_a = -(L_{aa}\nabla\mu_a + L_{ab}\nabla\mu_b)/T + (L_{aa}q_a + L_{ae}q_e + L_{ab}q_b)E/T, \tag{5a}$$

$$J_b = -(L_{bb}\nabla\mu_b + L_{ba}\nabla\mu_a)/T + (L_{ba}q_a + L_{bb}q_b + L_{be}q_e)E/T \tag{5b}$$

and

$$J_a + J_b + J_v = 0 \tag{5c}$$

These equations can be simplified by assuming that the vacancy concentration can be maintained at equilibrium by the presence of sources and sinks, so $c_a \gg c_v$. Using the Gibbs-Duhem relations and defining

$$L_{ae}q_e \equiv L_{aa}\,|\,e\,|\,Z_a^w + L_{ab}\,|\,e\,|\,Z_b^w \tag{6a}$$

$$L_{be}q_e \equiv L_{ab}\,|\,e\,|\,Z_a^w + L_{bb}\,|\,e\,|\,Z_b^w, \tag{6b}$$

the above equations can be simplified as

$$J_a = -k\left(\frac{L_{aa}}{c_a} - \frac{L_{ab}}{c_b}\right)\nabla c_a + L_{aa}Z_a{}^*eE + L_{ab}Z_b{}^*eE \tag{7a}$$

and $$J_b = -k\left(\frac{L_{bb}}{c_b}\right)\nabla c_b + L_{ab}Z_a{}^*eE + L_{bb}Z_b{}^*eE \tag{7b}$$

Expressing in this way, the atomic flux is separated into components coming from diffusion and electromigration. For dilute alloys, the kinetic analysis of Manning (23) shows that

$$L_{aa} \simeq \frac{Nc_a}{kT}D_a \text{ and } L_{bb} \simeq \frac{Nc_b}{kT}D_b f_b, \tag{8}$$

where D_a and D_b are the uncorrelated diffusion coefficients of the solvent and solute atoms, respectively, and f_b is the solute correlation factor. Using these expressions, one can write the electromigration components of the atomic flux as

$$J_a^E = \frac{Nc_aD_a}{kT}Z_a{}^{**}eE \text{ and } J_b^E = \frac{Nc_bD_bf_b}{kT}Z_b{}^{**}eE. \tag{9}$$

Here $Z_a{}^{**}$ and $Z_b{}^{**}$, being called as the apparent effective charges, are experimentally measured parameters and can be related to the actual effective charges by the following relationships:

$$Z_a{}^{**} = Z_a{}^*\left(1 + \frac{L_{ab}Z_b{}^*}{L_{aa}Z_a}\right) \text{ and } Z_b{}^{**} = Z_b{}^*\left(1 + \frac{L_{ba}Z_a{}^*}{L_{bb}Z_b{}^*}\right) \tag{10}$$

To correlate the phenomenological approach to the atomistic details of the diffusion process, the phenomenological coefficients have to be expressed in terms of the frequencies of various jumping processes involved in atomic transport. Such an analysis provides a link to examine in detail the kinetic process in electromigration as well as the effects of interactions between the moving species on the overall atomic motion. This problem has been formulated by Manning (23) and the formulation has been applied to analyze the effect of solute addition on electromigration in dilute alloys of Al, Cu, Ag, and Au (21). The results show three interesting general features about solute effects on electromigration. First, there is a universal retarding effect due to vacancy flow on solute electromigration. Second, in spite of the negative vacancy flow effect, the motion of the solute is expected to be significantly faster than the solvent atom in the pure material because of the faster electromigration rate of the solute atom. And third, most of the solute additions are expected to enhance the electromigration of the solvent except for Fe addition in Cu and Ag. All these features can be attributed to the nature of the jumping process and the interaction between the solute atom and the vacancy.

One significant result derived from this analysis is that the addition of most solutes in these metals will not reduce the overall rate of electromigration in bulk materials. Since the four metals considered are the most commonly used for interconnections in electronic devices, this raises a question regarding the mechanism of using alloy addition to improve electromigration lifetime of conductor lines. Particularly interesting is the example of Cu addition in Al, which is known to be effective in reducing electromigration damage (24); to the contrary, the kinetic analysis shows a large enhancement factor for Al electromigration due to Cu addition. This difficulty led to the suggestion that since electromigration in conductor lines is controlled by grain-boundary mass transport, the kinetic effects of solute addition at grain boundaries must be different from that in the bulk lattice, thus the results based on bulk analysis cannot apply. It is clear that a kinetic analysis of solute effects for grain boundaries is needed to understand the nature of electromigration in thin films. So far, except for some relatively simple models based on solute-vacancy bonding at grain boundaries (25), an analysis based on atomistic jumping processes at grain boundaries has not been developed.

2.2 Theory of Electromigration

The development of electromigration theory has evolved around the two components, electrostatic and electron-wind, of the driving force. The early development in theory includes the semi-classical ballistic model by Fiks in 1959 (4) and by Huntington and Grone in 1961 (5), and the quantum-mechanical model by Bosvieux and Friedel (26) in 1962. Although those early models are relatively simple in formalism, most of the key concepts of the driving force have been incorporated. The theoretical development continues to the present time with a number of new formulations as well as refinements. The course of the theoretical developments has been reviewed by Sorbello (27,28) and more recently by Verbruggen (29).

In the ballistic model, the electron wind force is formulated by calculating the momentum transfer to the jumping atom due to its collision with the charge carriers and the electrostatic force is taken to be the force exerted by the field on the bare ion. The original formulation by Fiks (4) is particularly simple where the wind force is calculated from the total momentum transfer per unit time due to collision with the electrons and expressed in terms of an effective charge parameter Z^w (defined in Eq. 3) as

$$Z^w = -n_e \lambda_e \sigma_e \tag{11a}$$

where n_e is the electron density, λ_e mean free path between collisions and σ_e the cross-section of the electron-ion collision. This expression was subsequently extended (30) to include the contribution from the positive holes as

$$Z^w = -n_e \lambda_e \sigma_e + n_h \lambda_h \sigma_h \tag{11b}$$

Huntington and Grone (5) treated the electron-ion collision by taking into account the change of the electronic states caused by the collision. In this semi-classical treatment, a probability matrix is set up to account for the transition from an initial to a final state of the electron. The state of the electron and its momentum were initially taken to be those of the free electron. Subsequently this was modified to take into account the scattering of the holes by considering the momentum transfer coming from the pseudomomentum of the electronic states, i.e. the momentum of the electron plus the lattice (31). It was also emphasized that the force is not constant during jumping, so it has to be obtained by integrating the transition matrix over the jumping path of the moving atom. The latter consideration is important for treating different types of jumping processes. For example, while jumping of the interstitial atom is not expected to change much in its interaction with electrons throughout the jumping path, the jumping of the substitutional atom is quite different since the scattering at the saddle point can be significantly higher than at the lattice sites of the starting and final positions. This effect is incorporated by means of a parameter of specific defect resistivity. Accordingly, the effective charge Z^w is expressed as

$$Z^w = -Z\left(\frac{\rho_d}{N_d}\right)\left(\frac{N}{\rho}\right)\frac{|m^*|}{m^*} \tag{12}$$

where the ratio of the specific resistivities between ρ_d/N_d of the moving defect and ρ/N of the lattice atom, provides a measure of the electron-ion scattering throughout the jumping process. The value of ρ_d/N_d varies according to the jumping mechanism. For the case of substitutional self-diffusion where the scattering power varies approximately in a sinusoidal manner from the equilibrium lattice position to the saddle point, it can be taken to be about 1/2 of that at the saddle point; for impurity diffusion, its value can be appreciably higher even for substitutional jumping.

Bosvieux and Friedel (26) formulated the first quantum mechanical treatment of the driving force. They obtained the wind force by calculating the electrostatic force exerted by the electron charge density on the bare ion in the presence of the electric field. In this formalism, the electric field has two effects on the electrons: first the wavefunction is modified because of charge polarization and second, the Fermi surface is displaced due to current flow. These two effects are assumed to be independent and were evaluated by a self-consistent, time-independent perturbation method. Interestingly, one can show that by carrying out the calculation to the first order, as in the original work, the wind force can be reduced to a form identical to that of the ballistic model.

One controversy arose from the result derived for the direct force by Bosvieux and Friedel. In their formalism, the direct force comes from the sum of the polarization effect on the charge plus the electrostatic force on a bare ion. This was calculated by a perturbation method based on free electron wavefunctions. The calculation yielded the result that an interstitial impurity would be completely screened, thus encounters no direct force, while a substitutional impurity would experience a force identical to the lattice atom. The complete screening of the interstitial impurity is surprising and contradicts the result of the ballistic model which has a direct force corresponding to a bare ion.

There is a considerable interest in developing theories for electromigration subsequent to the pioneer work discussed above. The development of formal theories includes the semiclassical formulation by Das and Peierls (32) based on the Boltzmann transport equation, the density-matrix theory developed by Sorbello and Dasgupta (33) and Schaich (34), as well as the development of a linear response theory based on quantum field theory (35). In the linear response theory, the direct force was found not corresponding to the complete screening as deduced by Bosvieux and Friedel. The nature of the difference was clarified by Sorbello (36) as due to an additional contribution which arises from the dynamic nature of the scattering process as a result of the current flow. The magnitude of this contribution depends on the extent of the interaction between the charge carrier and the impurity. For weak scatterers, the direct force corresponds to the unscreened case of the bare ionic charge, thus confirming the result of the classical ballistic model. For strong scatterers, the ionic charge is in general not completely screened and its magnitude depends on the extent of the dynamic scattering. Under certain circumstances, the dynamic effect can give rise to an effective charge of the direct force exceeding the ionic charge. Results of recent experiments in measuring the wind force of interstitials (37) seem to be in agreement with the theoretical prediction. With these results, the controversy regarding the extent of screening on the direct force appears to have resolved.

Two additional contributions to the driving force have been brought out by Landauer and co-workers. The first is called the residual resistivity dipole and is related to the wind force (38). This term arises from the voltage increase required to maintain a constant current because of the increase in the resistivity due to the impurity. The other is called the carrier-density modulation effect and is related to the direct force (39). This term originates from the decrease in the local electric field due to the increase of the electron density near the impurity in the presence of current flow. These terms are essentially second-order corrections to the driving force.

The calculation of the effective charge for real metals began initially for simple metals with a nearly full band of conduction electrons. Sorbello (40) adapted a pseudopotential approach to calculate the driving force based on the self-consistent formulation of Bosvieux and Friedel. Numerical results show an agreement within 50% of the measured effective charges for simple monovalent (e.g. Na and K) and divalent metals (e.g. Be and Mg) as well as their alloys. For metals with more complex electron structures, such as the noble metals, the agreement is less satisfactory since those metals are not as well described by pseudopotentials.

Another calculation treated the anisotropy in the driving force for divalent metals of Zn and Cd with hexagonal crystalline structures. This work was first developed (41) using Bloch wave functions and a perturbation method to treat specifically the effect on scattering due to the anisotropy of the Fermi surface at the Brillouin zone boundary. Later this was generalized by using a dielectric matrix to treat the effect on electron screening due to the band structure (42). In actual calculation, a semiempirical pseudopotential form factor was used and the result seemed to account satisfactorily for the observed anisotropy in the driving force of Zn.

2.3 Techniques of Measurement

The techniques of measurement for electromigration in bulk materials have been reviewed previously by Huntington (6). Essentially, there are two basic types of measurements: one determines the composition change induced by the current flow and the other measures the mass transport relative to a fixed reference frame. The principles of these two types of measurements apply to bulk materials as well as thin films.

The first type of measurement is similar to the usual diffusion measurement except that the sample has to be shaped to allow the flow of current with a sufficient current density to produce a measurable amount of electromigration. The principle of this type of measurement is shown in Fig. 1 (6). For bulk studies, the sample is usually made into a cylindrical form with liquid cooling of both ends. The element to be measured can be incorporated uniformly into the sample or inserted near the center of the sample. Upon passage of the current, the composition distribution will change as a result of electromigration. For the case with a uniform initial distribution, a steady-state distribution can be established after a sufficient amount of time which can be derived from Eq. 7b by setting $J_b = 0$ as

$$\nabla \ell n c_b = \frac{Z_b^{**} eE}{f_b kT} = \frac{Z_b eE}{f_b kT}\left(1 + \frac{L_{ba} Z_a^*}{L_{bb} Z_b^*}\right) \tag{13}$$

Thus a measurement of the profile of c_b can yield directly Z_b^{**}/f_b without knowing the diffusivity of the moving species. In the case of self electromigration, the vacancy-flow factor inside the bracket simplifies to unity, then the measurement gives directly Z_b^*. Since the diffusivity is usually difficult to measure precisely, this technique provides a simple means to measure the effective charge.

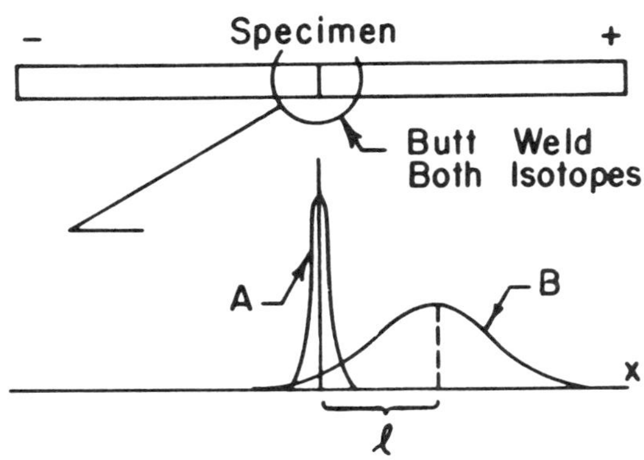

Figure 1. The isothermal isotope method: A, inert marker concentrations, B, concentration of matrix isotope. ν_a (T) $= \ell/t$ where t is duration of experiment. Temperature determined from the diffusional spreading B. (From Ref. 6)

For the case with a sharply defined source, its effective charge can be measured from the displacement of the center of the source (6) according to

$$Z_b^{**} = -\left(\frac{d}{t}\right)\left(\frac{f_b kT}{DeE}\right) \tag{14}$$

where d is the displacement of the center and t the electromigration time. In this case, the determination of Z_b^{**} requires the measurement of D. This can be done by measuring the thermal spreading of the profile c_b. Since D is difficult to measure, this method is more tedious and less accurate than the steady state. Again, for self-electromigration, one can measure Z_b^* directly.

Several methods of chemical analysis can be used to measure the composition profile with the most common being isotope-sectioning technique and ion or electron microprobe technique. Other techniques based on measurements of properties depending on the composition, such as electrical resistance and mechanical hardness, have also been used.

The second type of measurement is called the "vacancy flux" method which measures the change in the sample dimension as a result of the creation and annihilation of vacancies along the sample during the course of electromigration. The principle of this measurement is illustrated in Fig. 2. The early experiments measured simply the displacement of surface markers (scratches or indentations) separated by a uniform spacing along the length of a wire sample (5). Later in a study of Al (43), it became apparent that the change in the dimension is not confined only along the length of the sample, instead there are transverse dimensional changes. The latter was found to depend on the aspect ratio of the sample and should be included to account for the total vacancy flux. The simplicity of the measurement has made this method one of the principal techniques in electromigration studies and it has been applied to measure bulk electromigration for a large number of pure elements. Results of these measurements are in general agreement although not with the same accuracy as those obtained from chemical analysis.

The general principles of these two types of measurements have been extended to thin-film studies although the difference in the characteristics of electromigration necessitates changes in the experimental condition (44) as well as in the analysis of the composition profile. While the details of the thin-film studies can be found in a review article (11), two salient features are worth noting. First, those studies are usually carried out on films with a supporting substrate which increases significantly the dissipation of the Joule heating generated by the electrical current. As a result, the current density applied can be increased by more than an order of magnitude, i.e. from 10^3-10^4 A/cm^2 to 10^4-10^5 A/cm^2. This gives rise to a corresponding increase in the driving force, thus improving the accuracy of measuring the effective charge. Second, in thin films the temperature range where electromigration is of interest is lower than in bulk materials and lies at about half of the absolute melting point. Within this temperature range, the mass transport is dominated by grain boundaries. This requires a different analysis for electromigration at grain boundaries boundaries (45). In addition, the effect of microstructure becomes important and has to be considered.

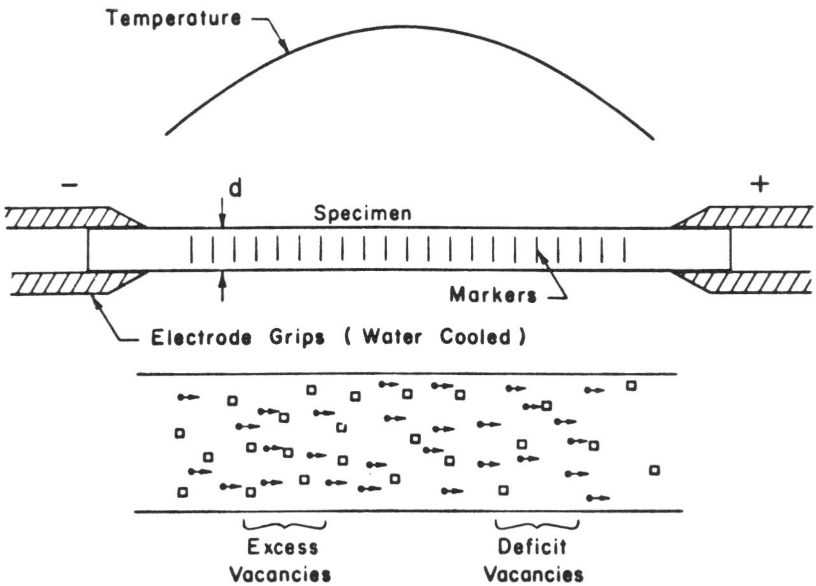

Figure 2. Vacancy flux method–schematic experiment: • electron, ¤ vacancy. (From Ref. 6)

In spite of these differences, the two basic types of measurement for bulk materials have been applied for studying thin-film electromigration. For measurement of the composition distribution, chemical analyses using electron microprobe (46) and radioactive isotopes (47) have been employed. For measurement of the mass flux, techniques have been developed to measure mass accumulation and depletion along thin-film stripes during the course of electromigration (44). Among the second type of measurements, of particular interest is the technique of drift velocity measurement (48). In this technique, one can measure the mass transport as well as the stress generated in the film due to electromigration. Details of the techniques used to measure thin-film electromigration are described in Ref. 12.

2.4 Electromigration in Substitutional Systems

The study of substitutional electromigration can be traced to the early work of Seith and Wever (2) in multiphase binary alloys in the early 1950's. The interest in such studies started to flourish in the early 1960's when the theoretical models were first formulated (see § 2.2). These models proposed the origin of the wind force as coming from electron-ion collision, thus indicating the possibility of using electromigration to probe the nature of electron-ion interaction in metals. It became clear also that it is simpler to study electromigration in pure elements instead of alloys. This led tosystematic studies for about a decade of noble metals and other monovalent and divalent metals with relatively simple electronic structures. The study subsequently extended to systems with complex electronic structures, including polyvalent, transition and refractory metals, as well as substitutional alloys.

In the early 1970's, there was a surge of interest in studying electromigration in thin films following the discovery of the cracked stripes in integrated circuits (9). It soon became clear that the problem concerned primarily with mass transport at the grain boundaries; of particular interest is the mechanism of damage formation in the conductor line. Although electromigration at grain boundaries differs in some basic aspects from its counterpart in bulk lattices, the migration of atoms in most systems of interest still seems to occur via the vacancy mechanism at grain boundaries. The type of materials studied for this problem was largely confined to a few metals and alloys with high electrical conductivity because of their usage as electrical conductor lines in semiconductor integrated circuits.

Substitutional electromigration in bulk metals has been reviewed by Huntington (6) in 1974. The corresponding study in thin films has been reviewed by d'Heurle and Ho in 1978 (12). Except for a few additional studies since then, these articles provide comprehensive summaries of the experimental results. Here only the salient features of these results will be highlighted.

In Table 1, selected values of Z^*/f for pure metals are listed from Refs. 6, 15 and 49. For the noble metals, Au, Ag, and Cu, there seems to be a general agreement on the values of Z^*/f obtained in different laboratories except for the rather high value obtained by Doan for Ag (50). Electromigration in these metals is clearly dominated by the electron wind force which has a magnitude about 5 to 10×

TABLE 1

Z*/f for Several Pure Metals

Metal	Z*/f	Temp. range (C°)	Reference
Noble metals			
Gold	−9.5 to −7.5	850-1000	Huntington and Grone (1961) from Ref. 6
	−8.0		Gilder and Lazarus (1966) from Ref. 6
Silver	−21±5	830-890	Doan (1971) from Ref. 6
	−8.3±1.8	795-900	Patil and Huntington (1970) from Ref. 6
Copper	(−5.0 to −2.0)/f	800-900	Grone (1961) from Ref. 49
	−5.5±1.5	845-1030	Sullivan (1967a) from Ref. 6
	−4.8±1.5	870-1005	Grimme (1971) from Ref. 6
Alkali metals			
Lithium	−2.5 to −1.6	90-160	Thernquist and Lodding (1968) from Ref. 6
Sodium	−3.3±0.7	45-80	Sullivan (1967b) from Ref. 6
Divalent metals			
Zinc	(−4.3 to −4.0)/f	350-370	Kuzmenko and Kharkov (1959) from Ref. 15
∥ to c axis	−2.5±0.2	366-400	Routbort (1968) from Ref. 6
⊥ to c axis	−5.5±0.6		
Cadmium	−5.0/f	250	Kuzmenko and Kharkov (1959) from Ref. 15
∥	−2.0±0.2	215-290	Alexander (1971) from Ref. 6
⊥	−4.1±0.4		
Magnesium	2.0±0.3	500-580	Wohlgemuth (1973) from Ref. 6
Trivalent metals			
Aluminum	(−28.0 to −16.0)/f	500-570	Kuzmenko (1962) from Ref. 49
	−30 to −12	480-640	Penney (1964) from Ref. 6
	Comparable	450-610	Heumann and Meiners (1966) from Ref. 6
Indium	−11.5	115-150	Lodding (1965) from Ref. 6
Gallium(liquid)	−1.3	18-312	Lodding (1967)from Ref. 6
Thallium	−4.0±0.5	233-303	Lodding et al. (1972) from Ref. 6
Transition metals			
Nickel	−3.5	1000-1400	Hering and Wever (1967b) from Ref. 6
Iron	+2±1	700-1300	Hering and Wever (1967a) from Ref. 6
Cobalt	+1.6±0.3	1260-1360	Ho (1966) from Ref. 6
Platinum	+0.28±0.04	1480-1670	Huntington and Ho(1963) from Ref. 6
Refractory metals			
Zirconium	+0.3	930-1730	Campbell and Huntington (1969) from Ref. 6
Uranium	−1.6±0.1	830-1100	D'Amico and Huntington (1969) from Ref. 6
Quadrivalent metals			
Lead	−47/f	250	Kuz'menko (1962) from Ref. 6
Tin	−80/f	190	Kharkov and Kuz menko (1960) from Ref. 6
	−18	180-213	Khosla (1973) from Ref. 6

larger than the direct force. The behavior of the alkali metals, Li and Na, is similar to the noble metals with the direction of electromigration determined by the wind force. The magnitude of the wind force is smaller, giving an effective charge of only about −2. It is interesting that the relative magnitudes of Z* for the noble metals and alkali metals seem to be consistent with their relative magnitudes of electrical conductivities. This suggests that the electrons, instead of the holes, dominate in the scattering processes underlying these two phenomena.

The studies on the divalent metals, Zn and Cd, were carried out on single-crystal specimens. The results showed clearly measurable anisotropy in the values of Z* along the a and c crystalline axes. The observed anisotropy in Z* is inconsistent with that of the electrical conductivity which is only slightly higher along the a axis. This reveals the different nature of the scattering processes responsible for electromigration and conductivity of these metals. This question has been resolved in a theoretical study by Genoni and Huntington (42). Based on the band structure of Zn, these authors showed that the anisotropies in Z* and conductivity are different primarily due to the different averages of the relaxation time of the electron-ion scattering for the two phenomena.

Results of transition and refractory metals show that the value of Z* is generally small and the wind force can be positive or negative. Thus for these metals with complex band structures, the wind force is a result of the balance of the scattering from electrons and holes and the outcome is difficult to predict. This was first demonstrated for Co (51) where the positive Z* was found to correlate with the positive Hall coefficient. However, such a correlation does not seem very strong for other metals of this category, e.g. Pt has a small positive Z* but a negative Hall coefficient. It appears the complex band structure of these metals making it difficult to establish a simple correlation. For the same reason, the effective charges of these metals have not been as satisfactorily accounted for by theory as the monovalent or divalent metals.

For the trivalent and quadrivalent metals, electromigration is in general observed to be dominated by the electron wind force. The magnitude of Z* appears to vary considerably and there seems to be no systematic patterns for these metals. It is interesting to note that Al, the most commonly used metal in microelectronics, has a high value of Z*. This high driving force presents a basic problem due to electromigration for using this metal to form conductor lines in microelectronic devices.

Electromigration of impurities in bulk metals or alloys have not been as extensively studied as self-electromigration in pure metals. Nevertheless, there are some interesting results, notably the study by Doan (52) on the valence effect on electromigration in Ag using elements in the same row of the Periodic Table as Ag (Cd, In, Sn and Sb). Results of this measurement showed a good correlation of Z* to the parameter $z(z+1)$ where z is the difference in nominal valence between the solute and Ag. This correlation was interpreted to show that Z* is an average of the wind forces for the substitutional and saddle-point configurations. However, such a correlation was found to be less clear in a subsequent study of valence effects on electromigration in Cu and Ni (53). Using another series of solute atoms: Mn, Fe, Co, and Ni, Doan (22,52) has investigated the vacancy flow effect for electromigration in dilute Ag alloy systems.

In the last several years, Huntington and coworkers have carried out electromigration studies of fast diffusers in Pb alloys. This study provides an interesting contrast to the substitutional systems discussed so far since those fast diffusants diffuse via a complex interstitial or institialcy mechanism. There is considerable technological interest as well since those alloys are commonly used to form solder contacts in microelectronics. Results of electromigration studies in these alloys have been reviewed by Hu and Huntington (54). Therefore, the results are summarized only very briefly.

In these studies of lead alloys, detailed measurements have been carried out to investigate the effects of alloy addition on diffusion and electromigration of the fast diffusants. The effect of alloy additions of Sn and In on diffusion was found to be similar for Cu, Au and Ni but different for Ag. The diffusion coefficients of Ni, Cu and Au in lead alloys are reduced compared with pure lead. The results can be accounted for using a trapping model with the binding energy decreasing in the order of Ni→Cu→Au. The effect is different for Ag which apparently is not affected by alloy addition. The alloying effect on electromigration has been measured by a steady-state technique and the values of Z^* of the fast diffusants was found not affected. This led to the conclusion that electromigration occurs primarily through the untrapped impurities and the partitioning of the trapped and the untrapped impurities appears to maintain a local equilibrium during the course of electromigration.

3. Electromigration in Metallic Thin Films

3.1 Nature of Electromigration in Metallic Thin Films

The study of electromigration in metallic thin films was initiated at the beginning of the 70's after Blech and Sello (9) identified that the crack formation in the aluminum stripe was caused by electromigration. Most of the studies were carried out in the industrial laboratories with the emphasis placed on the reliability problem of electromigration-induced failure in semiconductor integrated circuits. A number of comprehensive review articles on electromigration in metallic thin films have been published during 1970's (10-13) and recently discussed in IEEE thin-film electromigration workshop (14).

The dissipation of Joule heating in bulk metals sets a limit on the maximum current density that can be carried to 10^4 A/cm^2 before reaching the melting point of the materials. For semiconductor integrated circuits, thin-film conductors are usually deposited on insulating substrates, such as silicon, silicon oxide and nitride, which have relatively good heat conductivities and provide intimate contacts. This facilitates the dissipation of Joule heating generated by the electric current passing through the thin-film conductor. In addition, the small dimensions of the cross-section of the thin-film conductor (typically not more than 5 μm thick and 10 μm wide) compared with the thermal diffusion length (several tens of microns) allow fast heat removal in the thin-film conductor. Thus, thin-film conductors can carry a high current density in the range of 10^5-10^6 A/cm^2 and still allow the device to operate at temperatures below half of the melting point of the materials. A current density as high as 10^7 A/cm^2 has been applied in pulse mode without burning out Al thin-film stripes of 2.5 μm in linewidth and 0.4 μm in film thickness (55).

When an electric current passes through a polycrystalline thin-film conductor, the atomic flux J comes primarily from electromigration in the lattice and at the grain boundaries. For the lattice (5),

$$J_l = \frac{1}{kT} N_l D_l \bullet j\rho e Z_l^* \tag{15}$$

and for an ideal grain structure of equiaxed texture (44),

$$J_b = \frac{1}{kT} \frac{N_b \delta}{d} D_b \bullet j\rho e Z_b^* \tag{16}$$

where N is the atomic density, D the diffusivity, j the current density, ρ the resistivity, eZ* the effective charge, k the Boltzmann constant and T the absolute temperature. The quantity δ is the effective boundary width (~10Å) for mass transport and d is the average grain size. The subscript l and b denote parameters of the lattice and grain boundary respectively. For metallic thin films with less ideal or random grain structures, the grain-boundary parameters in Eq. (16) have to be replaced by an average value. An effective geometrical factor taking into account the effect of grain orientation on mass transport should also be included in Eq. (16).

The relative contributions of these two components can be estimated from the ratio of J_l and J_b,

$$\frac{J_l}{J_b} = \frac{N_l}{N_b} \frac{d}{\delta} \frac{D_l}{D_b} \frac{Z_l^*}{Z_b^*} \tag{17}$$

The measured values of Z_l^* and Z_b^* usually do not differ by more than an order of magnitude. For thin films with 1 μm grain size at 0.5 T_m,

$$N_l \cong N_b,\ \frac{d}{\delta} \cong 10^3,\ \frac{D_l}{D_b} \cong 10^{-7} \text{ and } \frac{J_l}{J_b} \cong 10^{-4} \qquad (18)$$

Thus, at moderate temperatures, the atomic flux from lattice diffusion is vanishingly small as compared with that from grain-boundary diffusion. The latter becomes the dominating mode of mass transport. The high current density, the moderate temperature and the grain-boundary controlled diffusion characteristics distinguish electromigration in thin-film conductors from bulk metals.

3.2 Damage Formation and Kinetics

In thin-film conductors, electromigration-induced damage usually appears in the form of voids and hillocks. Voids can grow and link together to cause electrical discontinuity in conductor line which lead to open-circuit failure. Hillocks can also grow and extrude out materials to cause short-circuit failure between adjacent conductor lines in the same level as well as in adjacent levels in a multilevel interconnecting structure. Alternatively, it can break through the passivation or protective coating layers and lead to subsequent corrosion induced failures. Void-open circuit failure usually appears in an earlier time frame than extrusion-short circuit failure in conductor lines using Al-based metallurgies. However, electromigration induced damage is not the only cause of failures in metallization on semiconductor integrated circuits. The metallization used in semiconductor device fabrications, due to its multilayered and multilevel structures and the use of a combination of different materials, have other reliability problems such as contact degradation, shallow junction leakage, interdiffusion, interfacial reactions, thermo and mechanical stresses, dielectric breakdown, corrosion and adhesion. The reliability concerns and metallization limitations for semiconductor devices have been reviewed previously (56-65).

Since the mass transport at grain boundaries in most metals is by the vacancy diffusion mechanism (66), so there will be a vacancy flux J_v opposite to J_b. And it is convenient to consider the role of the vacancy flux in damage formation. The vacancy flux in the grain boundary with contributions from electromigration and diffusion due to vacancy concentration gradients in the grain boundary is given by

$$J_v = - D_v \nabla C_v + J_b \qquad (19)$$

where D_v is the vacancy diffusivity in the grain boundary. The local variation of the vacancy concentration C_v can be deduced using the continuity equation:

$$\frac{dC_v}{dt} = -\nabla \bullet J_v + \frac{C_v - C_v^o}{\tau} \qquad (20)$$

where the last term expressing the local deviation of the vacancy concentration from the thermal equilibrium value C_v^o. The quantity τ is the average lifetime of a vacancy, in the presence of a sink or source. Under a steady-state condition,

$$\frac{dC_v}{dt} = 0 \qquad (21)$$

So

$$C_v - C_v^o = \tau \nabla \bullet J_v \qquad (22)$$

Hence, the extent of deviation of the vacancy concentration from equilibrium is proportional to the vacancy flux divergence.

In the ideal case of a structurally uniform conductor line with no temperature gradient, there is no flux divergence so that electromigration-induced damage will not occur. However, most conductor lines are polycrystalline thin films containing structural defects. In addition, temperature gradient cannot be completely eliminated. Whenever there is a spatial gradient in any of the parameters in Eq. (16), divergence in the atomic or vacancy flux will be present and give rise to local vacancy supersaturation or depletion. The extent of the defect supersaturation or depletion depends on the magnitude of the flux divergence as well as the effectiveness of sinks (or sources) in absorbing (or generating) the defects. Electromigration-induced damage will begin by first nucleating voids under a sufficient local vacancy supersaturation. Subsequent void growth requires sustaining vacancy flux divergence. After the void reaches a certain size, the current density and temperature can be increased significantly in a localized region around the void, thus increasing further the rate of damage formation. Damage formation in the form of hillocks occurs in a similar manner with local accumulation of atoms.

Finally, the growth and linkage of voids and hillocks would cause permanent damage in the conductor line by void-open or extrusion-short. In this last stage, the time and the location for the permanent damage to appear are basically the result of a statistical process, depending on the spatial distribution of flux divergent sites. In general, the processes of damage formation starting from the electromigration-induced atomic flux, flux divergence, vacancy supersaturation or depletion, void nucleation or hillock formation, and finally growth and linkage of voids or extrusions are very complex in nature. Computer modeling would be useful for simulating and understanding these processes.

3.3 Structural and Thermal Effects

In thin-film conductors, flux divergence can be caused by inhomogeneity in local structure and temperature. The magnitude of the atomic flux divergences will, of course, depend on the severity of the structural and thermal inhomogeneities in the metallic thin films. The structural imperfections include point defects, dislocations, grain boundaries and triple points (junction of three grain boundaries), any variations in grain morphology and non-uniformity in grain size.

In an in situ TEM study (67), voids were found to nucleate usually at grain boundaries or existing pinholes and grow into adjacent grains in the large-grained Al stripes. Compared with large-grained stripes, many more voids were seen in the fine-grained Al stripes, probably due to the existence of more nucleation sites. The void edges are relatively straight due to the straightness of the grain boundaries in large-grained samples while the voids in fine-grained samples appear more irregular due to the large numbers of grains that are involved. In another in situ TEM study (68), grain-boundary triple points were seen to be the most favorable sites for the initiation of damage because of the large divergence of the atomic flux associated with them. It should be pointed out that the divergence of a triple point can result from either the relative angles between adjacent boundaries or by a discontinuity of diffusivity at the junction. After voids had been formed at triple points in Al thin films, mass depletion or "boundary open" was observed to occur often along those boundaries having the largest component of flux in the direction of electron flow.

Since voids are present at grain boundaries in all cases, grain boundaries appear to be the preferred sites for vacancy condensation. A significant correlation was found between the damage site locations and particular grain configurations (69). Most of the void-opens occur at a point in the Al stripe where grain size changed abruptly or grain boundaries ended abruptly, i.e., where an entire grain run across the stripe, as shown in Fig. 3. Non-uniformity in the grain size of the metallic thin film can cause divergence of atomic flux at the junction across which there are different number of grain boundaries. Voids can appear at a junction between a coarse and a fine grained area while hillock formation can occur at a junction between a fine and a coarse grained area in the direction of the ion transport as shown in Fig. 4. This is because less ions are leaving the coarse grained area via grain boundaries than are entering it from the fine grained area. Flux divergences can also be caused by other structural discontinuities, e.g., at second-phase particles and at positions which have deviations in chemical composition. In addition, flux divergence can arise from variations in the atomic structure or defect concentration at the grain boundary, such as that caused by segregation of solute atoms or contaminants. This affects the atomic interactions between the migrating ion and the grain-boundary atoms, consequently altering the diffusivity parameters and the activation energy.

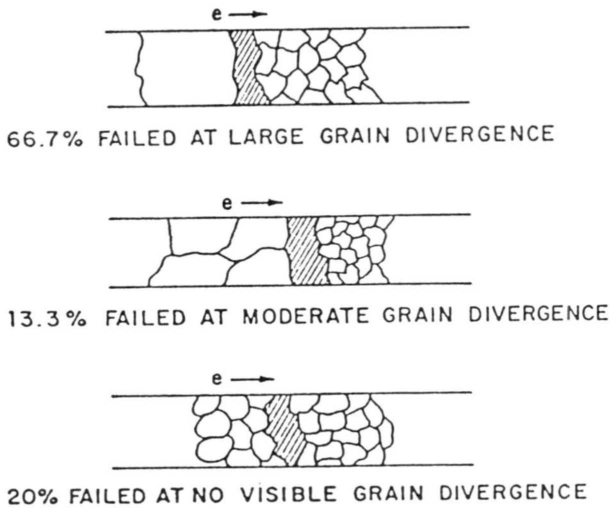

Figure 3. Configuration of sites where failures occurred, and percentage of failures. (From Ref. 69)

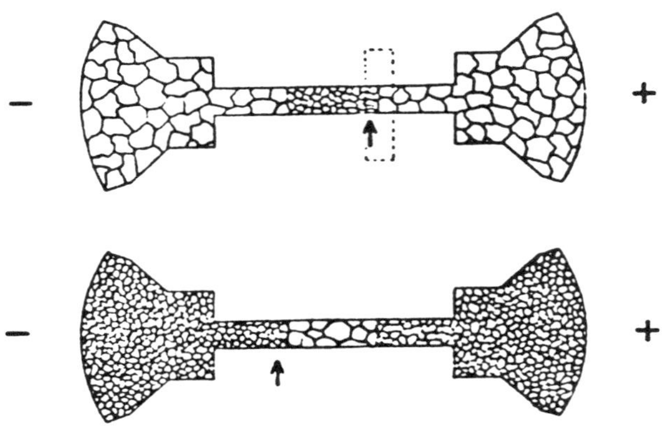

Figure 4. Thin films with grain size gradient. The arrows mark the positions where hillocks are observed in aluminum films. (From Ref. 13)

During device operation, temperature gradients exist both globally and locally in thin-film conductors since heat can be generated by active devices as well as Joule heating. However, global temperature gradient across the thin-film conductor is usually small except near the electrodes or the contact pads. Electromigration-induced damage formation due to temperature gradients in Al thin-film stripes has been observed in an in situ TEM study (70). As a result of a TEM sample preparation, very large temperature gradients are inherently produced in the Al stripe suspended over the hole in silicon substrate, on the thin oxide skin. Void formation was found in regions where the electron flow is in the direction of increasing temperature. Hillocks and whiskers were found to form where the electron flow is in the direction of decreasing temperature. Gross temperature gradient which are caused by Joule heating of the thin-film conductor in combination with cooling of the thicker electrodes has also been found to form voids and hillocks at the cathode and the anode, respectively, in a SEM study (13). These two types of damage formation due to temperature gradients are illustrated in Fig. 5. Large local temperature gradients or hot spots can be caused by poor adhesion or contaminations at the interface between metal film and substrate, or by the thickness variations of the metal film.

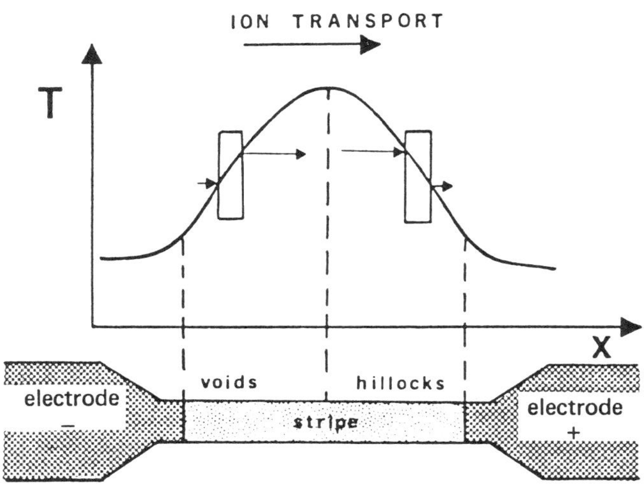

Figure 5. Schematic representation of a temperature gradient (caused by Joule heating) along a thin film stripe. The direction of ion transport due to the "electron wind effect" is indicated. (From Ref. 13)

3.4 Lifetime Measurement

The lifetime measurements of conductor lines is the most commonly used method for evaluating electromigration resistance. The results usually follow a log-normal distribution and are given in terms of a median time-to-failure (MTF), or t_{50}, which is the time to reach a failure of 50% of a group of identical conductor lines. In order to obtain the result of lifetime measurement in a reasonable time frame, the electromigration lifetime test is carried out under a set of accelerated test conditions at elevated temperatures and with high current density stressing. The data are then extrapolated to the device operating condition, which is usually at room temperature and with current density stressing below 5×10^5 A/cm^2, by an Arrhenius-like empirical equation originally formulated by Black (71)

$$\text{MTF} = A j^{-n} \exp(E_a/kT) \tag{23}$$

where A is a material constant relating to structural, electrical and diffusional properties of the thin-film conductor, E_a is the activation energy for electromigration-induced failure, j the current density, k the Boltzmann constant and T the absolute temperature.

The extrapolation assumes, of course, that the mechanisms leading to electromigration-induced failures under the accelerated lifetime test conditions are the same as those under device operating conditions. As a result, the temperatures in such accelerated lifetime tests have to be at least below a half of the melting point of the thin-film conductor, where the bulk contribution to the mass transport is not important. For polycrystalline Al and Al alloy films, values in the range of 0.4 to 0.8 eV have been reported for E_a, depending on the microstructure, metallurgy, fabrication process, heat treatment and geometry of the thin-film conductor. Experimental values of the current exponent n have been found generally in the range of 1 to 3, but large values, up to 10, have been also reported. The value of n, higher than 2, is probably due to Joule heating from a current density stressing higher than 1×10^6 A/cm^2. Under such high current density stressing, the temperature distribution of the thin-film conductor can be affected easily by local variations of adhesion and conductor geometry. Thus, the temperature of the thin-film conductor has to be measured accurately with the adjustment of any temperature distribution along the conductor line.

The reliability of the conductor line is also determined by the spread in electromigration-induced failure times, which is expressed as the standard deviation of the log-normal failure distribution, σ, and the first failure. By combining σ and t_{50}, one can predict the time for early failures. However, the values of σ and the first failure can vary considerably depending on the number of conductor lines used for the lifetime measurement and from one fabrication lot to another lot. Therefore, the extrapolation of the standard deviation and the first failure from the accelerated lifetime test conditions to the device operation condition is subject to considerable variations. In general, predictions of the reliability of the conductor line at device operating conditions must be made with extreme caution.

Other techniques of measurement for electromigration in thin films, such as measurements of mass accumulation or depletion, resistance measurements, the cross-stripe experiment and measurement of drift velocity, have been reviewed previously by d'Heurle and Ho (12). Recently, four new techniques have been proposed. The Temperature-ramp Resistance Analysis to Characterize Electromigration (TRACE) method (72) is based on resistance measurements in the early stages of electromigration-induced damage. The Standard Wafer-level Electromigration Acceleration Test (SWEAT) method (73) is to monitor electromigration susceptibility of the wafer level. The Breakdown Energy of Metal (BEM) method (74,75) is to measure the median-energy-to-fail of properly designed structures during a ramp current stress. The 1/f (76-79) and $1/f^2$ (80) noise methods are basically electrical noise measurements. At this time, all these four new techniques are in the exploratory stage. The validity of these tests and their correlation to the electromigration have yet to be established.

3.5 Methods to Improve Electromigration Resistance

One of the goals in studying electromigration in thin-film conductors is to devise methods to improve the electromigration resistance of thin-film interconnects. In the past, this practical aspect of electromigration study has been emphasized, particularly among the industrial laboratories. As seen from the discussion of the electromigration-induced damage formation process, it is clear that the basic requirement for reducing damage is to minimize the local divergence of the atomic flux. This can be accomplished, in principle, by reducing the magnitude of the atomic flux and/or the inhomogeneity of the parameters controlling the mass transport. Since the damage formation process is complex in nature, methods found to be effective often incorporate several factors for reducing electromigration-induced damage.

The magnitude of the atomic flux is determined by the electromigration driving force and the grain-boundary diffusivity. Thus to reduce the atomic flux, the choice is to reduce the driving force or the diffusivity. To reduce the driving force has some basic difficulties since it requires either a change in the scattering process responsible for the effective charge or a reduction in the current density. The scattering process is intrinsic, thus difficult to change. A reduction in the current density is not practical since the current density is dictated by device requirements and the tendency is to increase its magnitude for smaller devices. This leaves the choice to reduce the grain-boundary diffusivity. To achieve this, the most common approach is by solute addition although it should be emphasized that solute addition can contribute to the improvement by other effects, e.g. grain structure modification, and is not confined only to the reduction in diffusivity. The best known example of solute addition is the improvement of electromigration resistance in Al stripes by the addition of Cu (24). Other solute elements such as Mg, Mn and Ti have also shown to be effective in improving the electromigration lifetime in Al stripes (81). Solute elements or impurities in the thin film can cause a retardation of the electromigration-induced atomic flux, probably because of the segregation of these impurities at the grain boundaries and their interaction with the migrating ions. Since the solubility of Cu in Al is only about 0.05 at.% at 230°C (82), excess Cu atoms are likely to segregate at the Al grain boundaries. It was observed in a "marker" experiment (83) that Cu addition of an amount exceeding the solubility can reduce the electromigration flux of Al through the grain boundary. Cu also forms second compound phase particles with Al which serve as sources of Cu atoms to Al grain boundaries. The t_{50} and activation energy were found to increase with Cu concentration in Al stripes from 0 to 8 wt.% and then decrease beyond 8 wt.% (84) as shown in Fig. 6.

Efforts in reducing flux inhomogeneity have been focused on microstructure modifications which can increase the electromigration lifetime of Al fine lines by an order of magnitude. Improvements of electromigration resistance in single-crystal Al stripes (85) and in conductors composed of chains of Al single-crystal grains (86) and fabricated from large-grain Al films (87) have been reported. Microstructure modifications, such as the use of bamboo grain structure in Al (87) and Al-Cu (88,89) fine lines, have also been identified as an important factor contributing to the improvement of electromigration resistance.

Another approach for improving the electromigration resistance in thin-film conductors is to cover them with a protective layer or overcoat. The mechanism for the improvement is not well understood although it is likely that the overlayer provides stress accommodation and alters the source or the sink condition so as to enhance the defect annihilation kinetics. The increase in electromigration lifetime in Al stripes with a layer of glass was first found by Black (71). In general, the electromigration lifetime of most conductors increases by an order of magnitude or more with a complete surface coverage (90-92). The use of transition metal underlayers, such as TiN (93), Cr (94) and Ti-W (95) have been reported to increase the electromigration lifetime of thin-film conductors by an order of magnitude. This transition

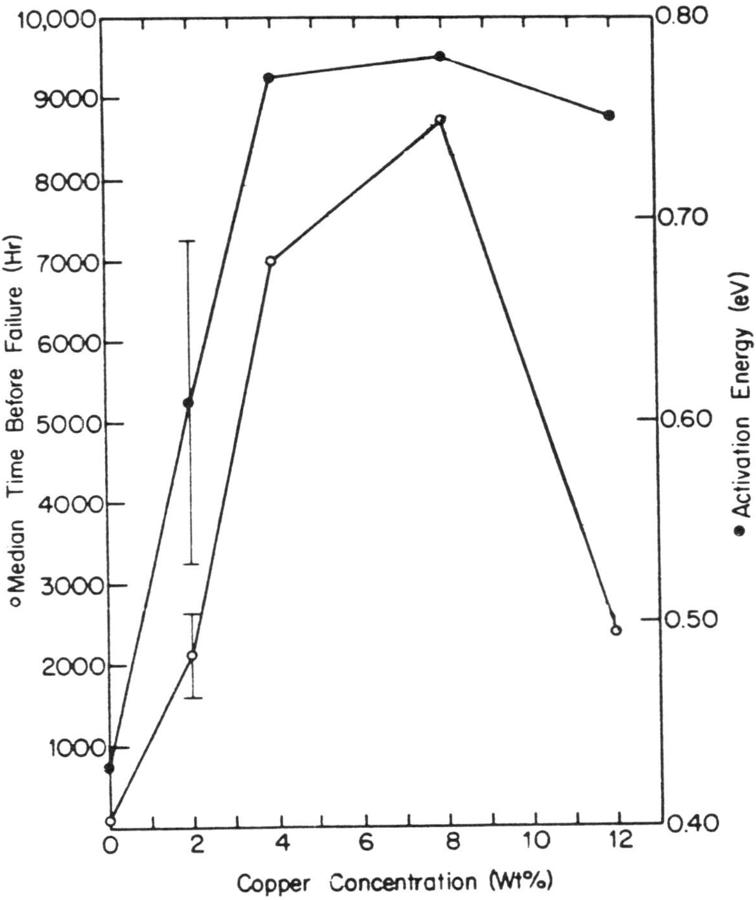

Figure 6. Activation energy for the failure process and MTBF as a function of copper content in hot-substrate aluminum-copper-1.7% silicon. The MTBF are for testing of 0.8-μ thick, 15-μ wide samples at 8×10^5 A/cm^2 and a temperature of 227°C. As discussed in the test, the bars indicate limits for as-deposited and heat-treated material. (From Ref. 84)

metal underlayer, e.g. Cr, can provide a redundant structure allowing void healing as well as a metal diffusion barrier (94).

A combination of solute addition and structure modification has been used to improve electromigration resistance of Al and Al-Cu fine lines. This is to incorporate an Al-transition metal intermetallic sandwich layer in the conductor, either in the middle or as top and bottom layers, and the electromigration lifetime was found to improve 50-100 times (96). The intermetallic layer was formed by reacting Al with a thin layer of transition metal, e.g. Cr, Hf or Ti. These transition elements have been found to serve several purposes (97). First, they reduce the electromigration rate of Al. Second, the intermetallic layer was found to serve as a barrier for linkage of voids formed in the top and bottom Al-Cu layers. And third, it causes microstructural modification, such as to induce columnar grain structure in the top and bottom Al-Cu layer as shown in Fig. 7, thus reducing structural sites for damage formation. In subsequent studies, similar approaches of using Al-Cu/Ti/Al-Cu (98), layered Al with silicide (99) and layered Al-Si with Ti or W (93,100-102) have been found to be effective in improving the electromigration resistance of Al-Cu, Al and Al-Si fine lines, respectively.

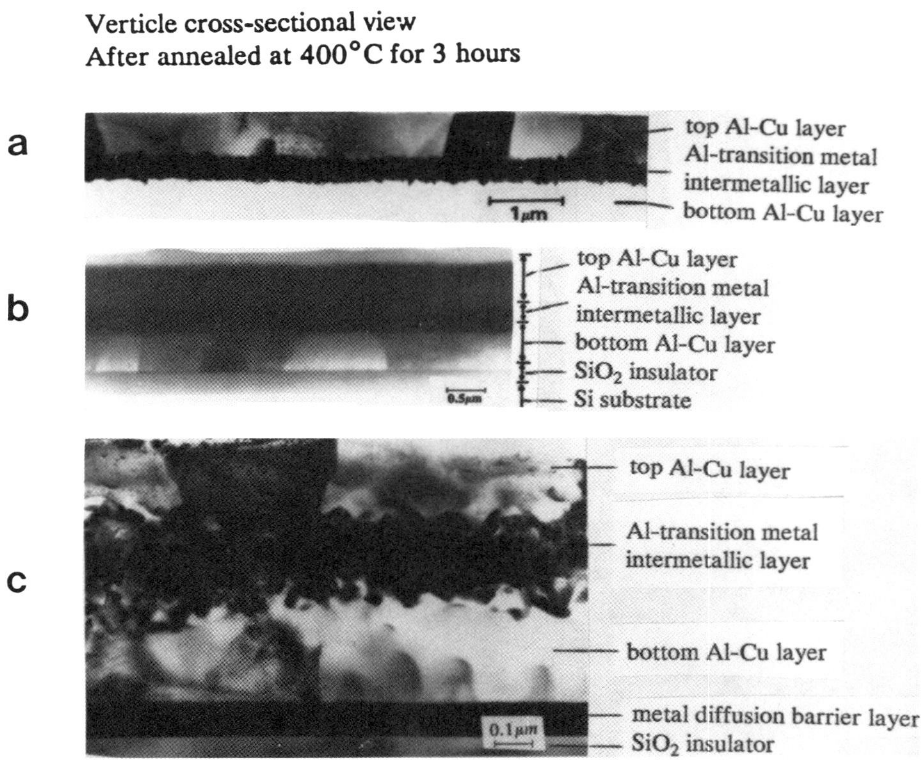

Figure 7. Cross-sectional transmission electron micrographs of Al-transition metal intermetallic layer (a) top Al-Cu layer (b) bottom Al-Cu layer (c) Al-intermetallic layer. (From Ref. 97)

4. Electromigration in Fine Lines

4.1 Scaling Trends

The purpose of device scaling is to improve the device density and performance of the integrated circuit. The effects of scaling on the functional requirements of interconnects can be assessed according to the scaling rules. Due to the different operating principles, the scaling rules differ for FET (field-effect transistors) and bipolar devices. Consider first the FET devices. Based on a constant field approximation, Dennard et al. (103) first derived a set of scaling rules by assuming that all device dimensions can be scaled linearly. They found that as the device dimension reduced by a factor of k, its performance as measured by the delay per circuit can be enhanced by the same factor if the dopant level increases by k while the voltage and current reduce by k. On this basis, one can deduce the scaling factor for other parameters pertinent to interconnect metallization, including the line resistance, capacitance, current density, and power dissipation. These factors are listed in Table 2.

TABLE 2

Scaling Results for Some Device Parameters

Device Parameter	Scaling Factor FET	Scaling Factor Bipolar
Device dimension	1/k	1/k
Voltage	1/k	~1
Current	1/k	1/k
Delay time/circuit	1/k	1/k
Power density	1	~k
Line resistance	k	k
Line current density	k	$\sim k^2$

Scaling for bipolar devices is more complicated because its vertical structure and operation principle do not allow similar linear scaling employed for FET devices. The wider spectrum of bipolar circuit design adds further complications in deducing a general set of scaling rules. Particularly relevant to the discussion of interconnect is the applied voltage. Since the switching voltage cannot be less than the voltage drop across a p-n junction, which is usually about half of the bandgap of the semiconductor (e.g. about 0.6 eV for Si), scaling of the voltage is restricted. Indeed, the voltage level used in devices is usually chosen to be a constant, as set by the industrial standard. Based on a constant-voltage scaling, Tang and Solomon (104) employed device simulation technique to analyze the effect of scaling on device parameters. They found that for optimum power-delay product, the current density increases about linearly with the reduction in device dimensions. For I^2L type circuits, Prince (105) deduced separately a set of scaling rules based on a linear scaling of the delay time. He found that the current density varies approximately with the square of the dimensional scaling. Results of these two studies lead to somewhat different scaling of device parameters and they are summarized together in Table 2.

The impact of scaling on electromigration is twofold: one arising from the increase in the current density and the other from the reduction in device dimension. From Table 2, the current density is seen to increase linearly with size reduction for FET devices and higher for bipolars. In addition to the effect on the driving force due to the increase in j, the $j^2\rho$ increase in the Joule heating can raise the conductor temperature, giving rise to higher atomic diffusivity and mass transport. This problem, together with the increase in the power density, necessitates an improvement in removing the heat generated during device operation. The combined effect of these factors will inevitably cause the electromigration flux to increase beyond that from current density alone. The values for j and $j^2\rho$ predicted for FET and bipolar devices for different periods of technology development are listed in Table 3. With the projected high current density in submicron

TABLE 3

Trends in Device Electromigration Parameters

Period	Line Width (μm)	Current Density (10^5A/cm^2) FET	Current Density (10^5A/cm^2) Bipolar	Joule Heating* $J^2\rho$ FET	Joule Heating* $J^2\rho$ Bipolar
1970's	3-6	0.2-0.5	0.5-1	1	4
1980-85	1.5-3	0.5	1	2-4	4 15
Late 1980's	0.5-1.5	1-2	8-15	15	~200

* The figures given for Joule heating are normalized with respect to the value of FET in the 70's.

metal lines, there will be increasing reliability concern of electromigration-induced failures. It has been recognized that the current-carrying capability of submicron metal lines will constitute as one of the factors limiting the scaling of VLSI devices.

4.2 Process and Material Related Aspects

To characterize the electromigration behavior in fine metal lines, it is useful to consider the three general aspects: process, material and geometry related. The process-related problems can be viewed as basically caused by any process-induced imperfection in the metal line, the substrate or the passivation layer. Since an isolated metal wire can carry only about 10^4 A/cm^2 before melting, the Joule heating generated in a metal line carrying 10^5 A/cm^2 or above must be almost completely removed through the substrate and/or the passivation layer. Processing defects can cause local current crowding because of the shrinkage of the current-carrying cross-sectional area due to the presence of defects. As a result, the local temperature will be raised.

For a spherical void in a thick film, the current density at the top of the void surface increases as much as by a factor of 2 (106). An increase in the current crowding in a perpendicular crack even exceeds the geometrical proportion (107). The effect can be significant, particularly during the last stage of damage formation where local increase in current density due to void formation and crack growth can lead to temperature increase and accelerated damage formation. Other types of defects, such as poor adhesion spot in the substrate and/or the passivation layer, can also cause local heating problems. This will increase the atomic diffusivity and lead to accelerated electromigration-induced failures or thermal runaway to destroy the metal line.

The effect of adherence of the Al thin film to the substrate on electromigration has been studied using TEM (108). Voids are found in areas of local thinning in the central region of Al stripes and in the blistered regions of Al stripes with NaCl substrates. The local thinning area and the blistered region extend over a large number of grains. The temperature and diffusivity are maximum while the temperature gradient is small in the central region. Thus, the formation of these voids is not due to either structural defects or temperature gradients. These results indicate that poor adhesion can give rise to hot spots and accelerated flux divergencies which will lead to local damage and drastically reduce electromigration lifetime.

The second aspect involves the material-related problems which comes about because other materials such as polysilicon, silicide, salicide and metal diffusion barrier, are employed together with the metallic thin film to form the multilayered structure conductors. Because of the different functional and stability requirements, these materials are often characterized by a higher resistivity and a finer grain size than those of the metal lines. Depending on the specific layered structure and device geometry, such material combinations can give rise to flux divergence and electromigration-induced damage with peculiar characteristics. The geometry-related problems, which are discussed in Section 4.4, arise basically as a result of grain morphology modification due to the reduction in metal line dimension to the submicron range.

4.3 Effect of Microstructure

It has been recognized that there is a strong correlation between microstructure and electromigration lifetime in fine metal lines. This is particularly true in VLSI technology when the linewidth and thickness of metal lines used as interconnects are reduced to the submicron range, thus becoming comparable to the grain size. For metal films, it is generally observed that the grain size is about the same as the film thickness. For a 1.0 μm thick film, there are only a few grains across a 1-3 μm wide metal line. With a small number of grains across the metal line, the contribution of each grain boundary to the overall mass transport becomes more important. Each divergent site in the grain structure also becomes potentially more damaging since a metal line can fail without requiring a statistical linkage of damage originated from several divergent sites, as would be the case for a wide metal line with many grains across. This shortens the metal line lifetime and increases the randomness of the failure statistics. Moreover, as the grain boundary area reduces, the surface and interface become increasingly important for transporting atoms in the metal line, thus playing a more important role in the formation of electromigration-induced damage.

A number of early studies on the effect of microstructure on electromigration lifetime have been reported. The median time-to-failure for 10 and 15 μm wide Al stripes fabricated from large-grain films of 8 μm grain size and a highly preferred orientation was found to be much longer than stripes fabricated from small-grain films of 2 μm grain size and a mixed orientation as shown in Fig. 8 (69). The corresponding activation energies were found to be 0.73 and 0.51 eV for large and fine-grain films, respectively, as measured by a marker velocity techniques. The dependence of activation energy on the preferred grain orientation and thus the local atomic structure of the boundaries themselves indicate that the activation energy may be related to intrinsically different transport properties of the grain boundaries. The standard deviation of the log-normal distribution for the large-grain films is less than that for fine-grain films.

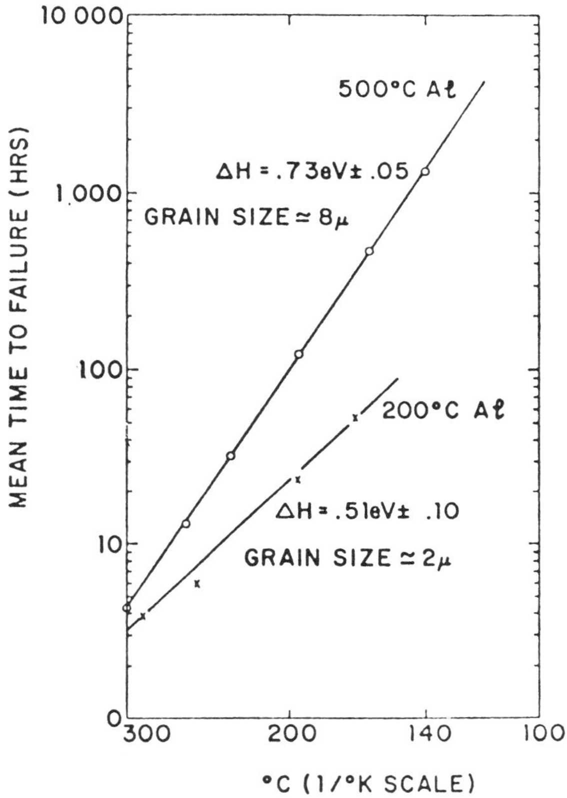

Figure 8. Activation energy determinations for fine- and large-grain conductors. Test condition was T = 150°C, J = 1×10^6 A/cm^2. (From Ref. 69)

The increase of electromigration lifetime with increasing grain size in Al conductors with width of more than several microns has been reported (109-111). The extent of improvement was found to increase approximately linearly with grain size. A subsequent study showed that such an improvement is limited by the ultimate grain size that can be reached through primary grain growth (112). By annealing a 2.0 μm wide Al-Cu line at elevated temperatures to induce grain growth, the electromigration lifetime was recently found to increase with the grain size until the grain growth was constrained by the metal line thickness (89). The microstructure in a 2.0 μm wide Al-Cu line after various heat treatment is shown in Fig. 9. The dependence of the electromigration lifetime on grain size and annealing time in the Al-Cu line is plotted in Fig. 10. The electromigration lifetime of Al-Cu stripes was also found to depend on the grain-size distribution (113) and film orientation (114). Results of these studies showed that when the standard deviation of the grain size distribution is large, the electromigration lifetime decreases because the mixed grain size gives rise to inhomogeneity of electromigration flux over the length of the stripe.

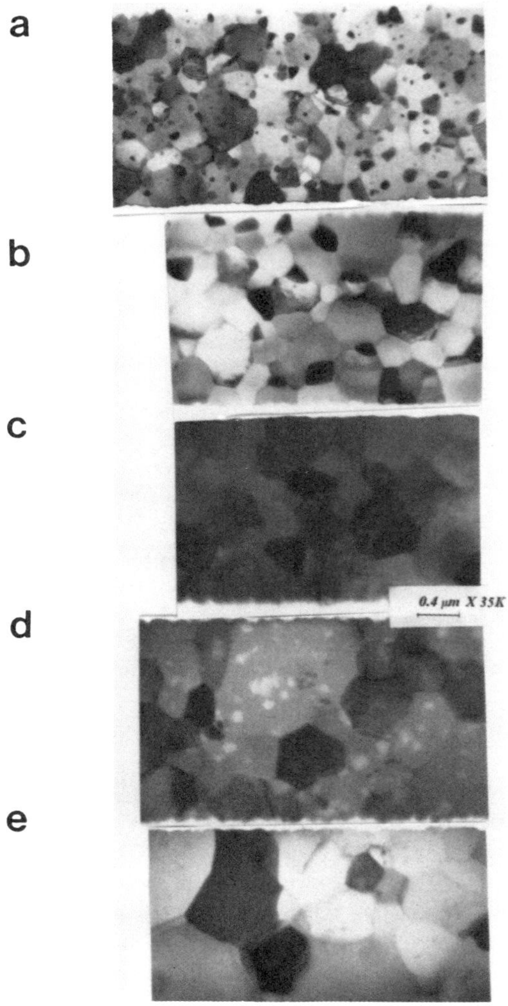

Figure 9. Transmission electron micrographs of 2.0 μm wide Al-Cu lines annealed at 400°C for various time periods. (a) As deposited, (b) 7.5 minutes, (c) 15.0 minutes, (d) 30.0 minutes, (e) 60.0 minutes. (From Ref. 89)

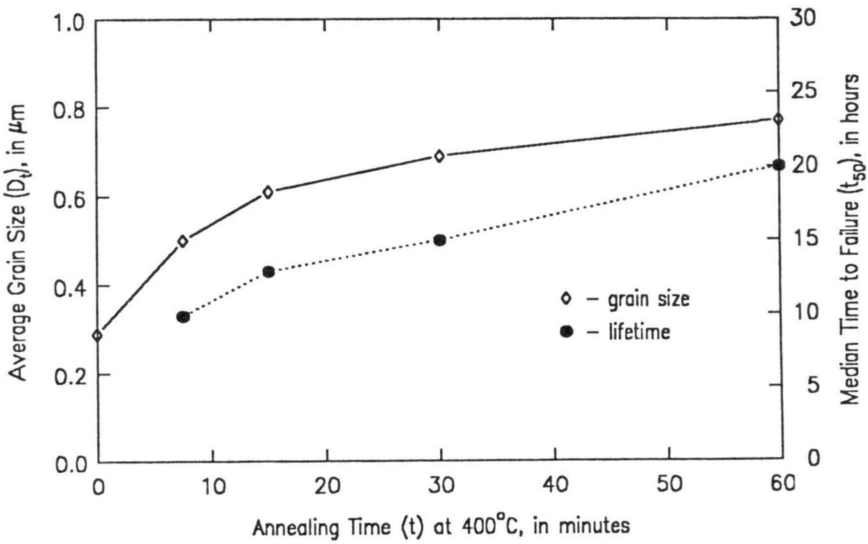

Figure 10. Grain size and electromigration lifetime of a 2.0 μm wide Al-Cu line annealed at 400°C for various time periods. Test condition was T = 127°C, J = 5×10^6 A/cm^2. (From Ref. 89)

4.4 Effect of Metal Line Geometry

4.4a Linelength Dependence. The linelength dependence of electromigration lifetime in wide Al lines was found by Agarwala et al. (115) to be a function of the linewidth. The lifetime decreases rapidly with increasing linelength until the length is greater than 62.5 and 125 μm in Al lines of 10 and 15 μm width, respectively, as shown in Fig. 11. Then, the lifetime reaches a saturation value beyond a critical length in the range of 125 to 250 μm and 250 to 500 μm, correspondingly. The standard deviation of the log-normal failure distribution decreases with increasing linelength. The effect of linelength of lifetime in Al lines of 15 μm width was also found by Learn et al. (116) to depend on the substrate temperature for metal deposition. The lifetime is longer for a hot substrate than for a cold substrate. As the linelength increases from 125 to 1250 μm and from 1250 to 5000 μm, the lifetime decreases by a factor of 10.6 and 1.5, respectively, for a 350°C substrate as shown in Fig. 12. For a room temperature substrate, the lifetime only decreases by a factor of 4.3 and 1.3, correspondingly. There is almost no effect of the substrate temperature on

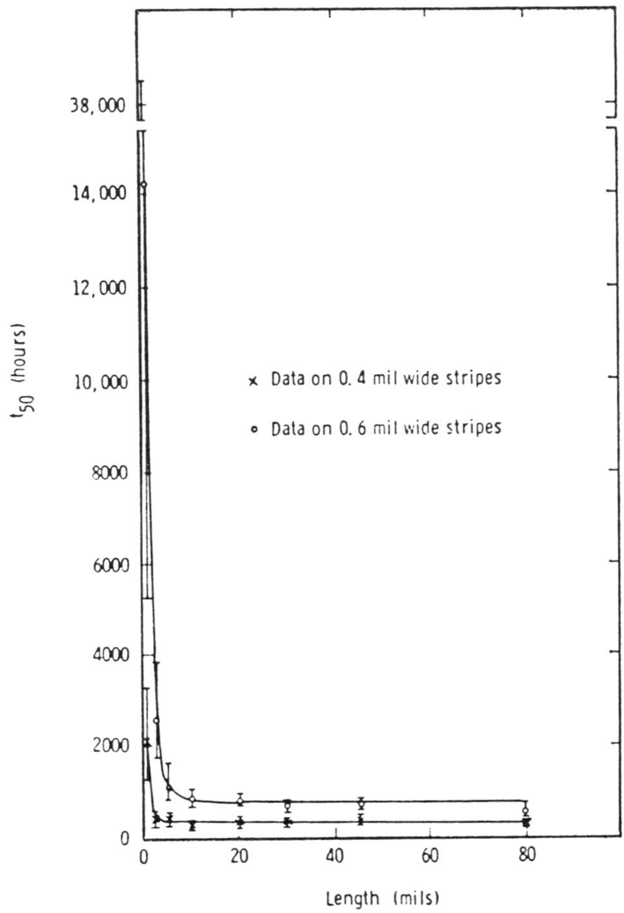

Figure 11. t_{50} vs. length of the conductors of width 0.4 and 0.6 mil. Test condition was $T = 130°C$, $J = 0.75 \times 10^6$ A/cm^2. (From Ref. 115)

lifetime for 5000 μm long Al lines. Recently, the lifetime in Al-Si lines of 3.0 μm width was found by Schafft et al. (117) to decrease by only 16% and 5% with increasing linelength from 400 to 800 μm and from 800 to 1200 μm, respectively as shown in Fig. 13. In contrast to the findings of Agarwala et al. the standard deviation of the log-normal failure distribution is not a function of linelength. The lifetime in Al-Cu lines of 0.75 μm was also found to decrease slightly with increasing linelength from 10 to 50 μm and then levels off to 500 μm as

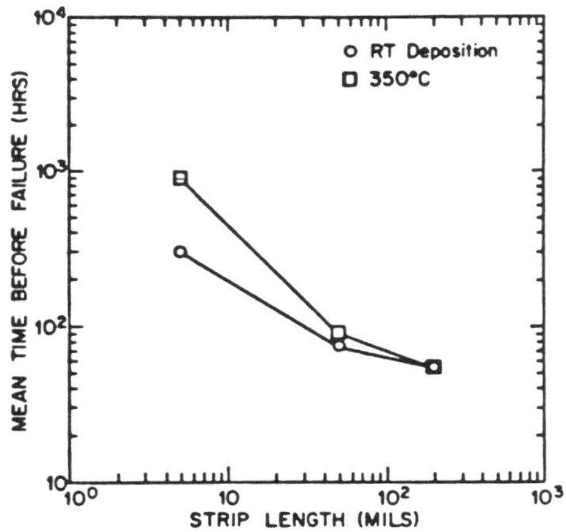

Figure 12. Mean time before failure as a function of strip length for samples of hot-substrate and cold-substrate aluminum. Test conditions were 220°C and 8×10^5 A/cm². (From Ref. 116)

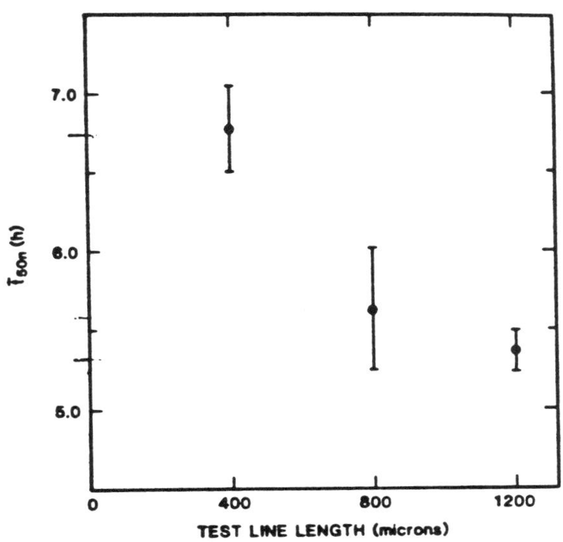

Figure 13. Mean of three replicate measurements of t_{50n} at three test-line lengths. Error bars are the standard errors of the mean. All the data were normalized to the test condition T = 175°C, J = 2.5×10^6 A/cm². (From Ref. 117)

shown in Fig. 14 (118). But the overall linelength dependence of lifetime in Al-Cu submicron lines is small. The standard deviation of the log-normal failure distribution does not vary systematically with linelength. The results of these studies indicate that the linelength dependence of electromigration lifetime in metal lines is a function of linewidth, metallurgy and metal deposition conditions. The critical length appears to decrease with decreasing linewidth.

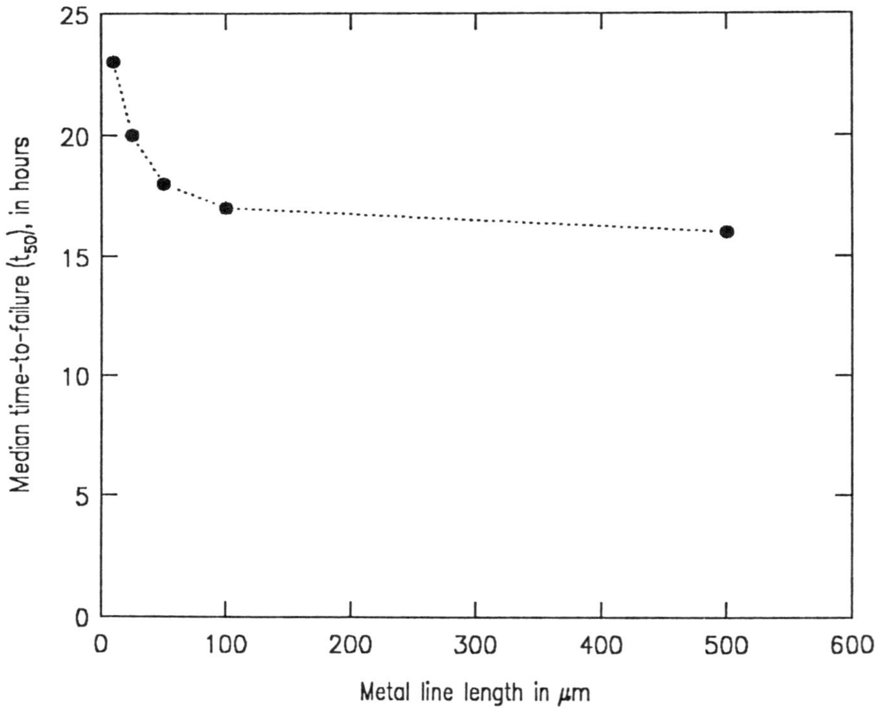

Figure 14. Effect of metal line length on the median time-to-failure in Al-Cu submicron lines. Test condition was T = 227°C, J = 2.5×10^6 A/cm^2. (From Ref. 118)

If we assume that the density of extrinsic defects, e.g. process-induced defects, and that such defects of varying severities are distributed randomly in the thin film, and further assuming that the lifetime of the metal line is also determined, not by an average value of the severities of the defects, but by the most severe defects in the metal line, then a decrease of lifetime with increasing linelength is expected.

This is because the probability of finding more severe defects is higher in a longer metal line. A decrease in the standard deviation of the lifetime distribution with increasing linelength is also expected because the probability of finding the most severe defect in a longer metal line containing a large number of defects is higher. However, if we assume that electromigration-induced failure can be caused by any intrinsic defects, e.g., point defects, dislocations and grain boundaries, of or beyond certain severity in the case where there is no extrinsic defects presented, then there would be no linelength dependence of electromigration lifetime. This is because there is an abundance of these defects in all the metal lines. This assumption, of course, depends on the nature, severity and density of the defects in the metal lines.

4.4b Linewidth Dependence. A number of studies on the linewidth dependence of electromigration lifetime in metal lines have been reported. Agarwala et al. (115) reported a linear relationship between the lifetime and linewidth of Al lines with linewidth down to 5.0 μm as shown in Fig. 15. In subsequent studies, as the linewidth reduces below a critical value, the lifetime was found either to level off or go through a minimum and then increase, reversing the trend in wider

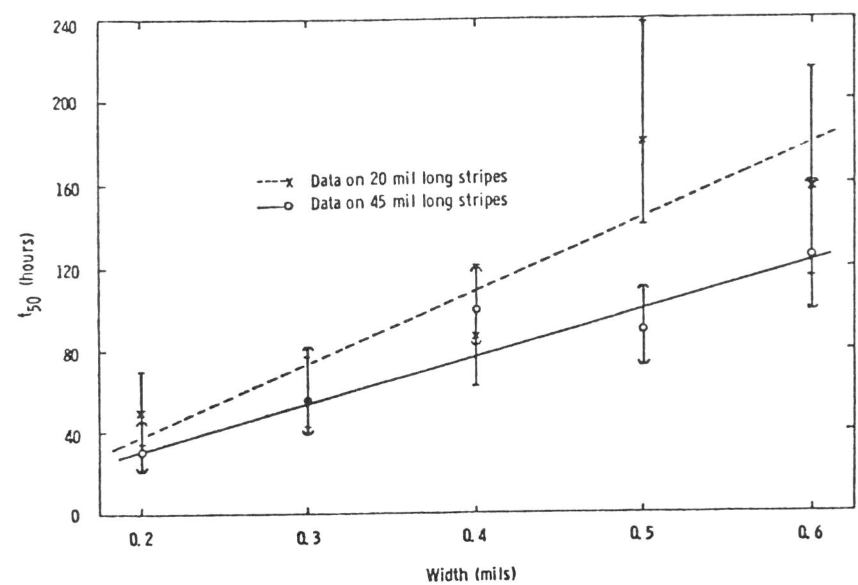

Figure 15. t_{50} vs width of the 20- and 45-mil-long conductors. Test condition was T = 130°C, J = 0.75×10^6 A/cm^2. (From Ref. 115)

lines. The lifetime in Al-Cu lines patterned by chemical etch was found by Scoggan et al. (119) to level off below a critical width of 5.5 μm, whereas the lifetime in Al-Cu lines of the same dimension but patterned by metal lift-off increases below a critical value of 3.5 μm as shown in Fig. 16. The lifetime in Al-Cu-Si and Cr-Ag-Cr lines also

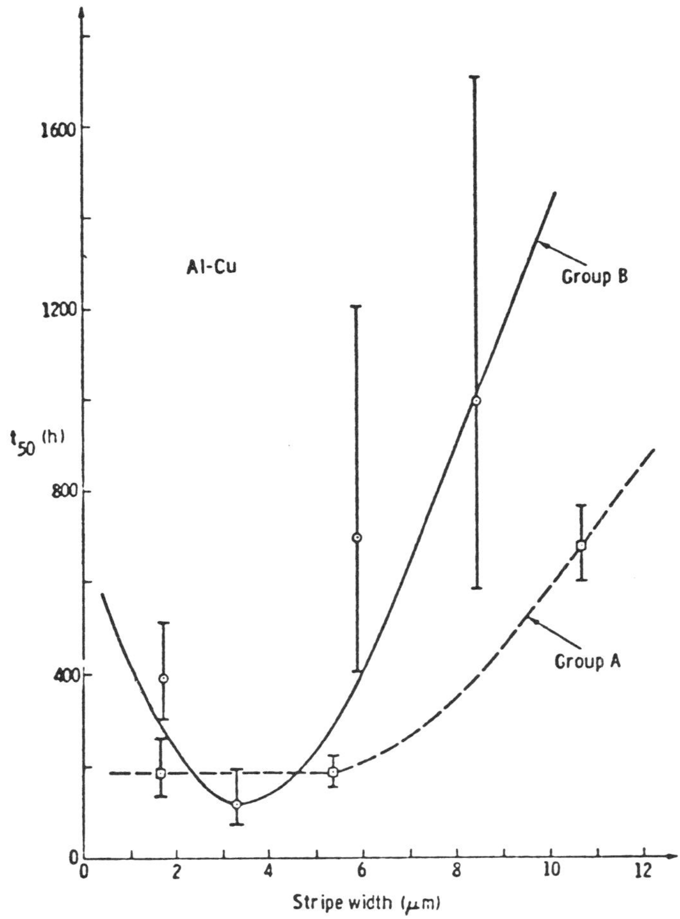

Figure 16. Plots of t_{50} vs. stripe width for Al-Cu samples of groups A and B. All the data were normalized to the test conditions T = 250°C, J = 1×10^6 A/cm^2. The error bars represent 95% confidence intervals for the t_{50} data points. Group A - Chemical etch, Group B - Metal lift-off. (From Ref. 119)

levels off below a critical widths of about 2.0 μm and 1.5 μm, respectively. Both the lifetime and the critical width increase with the grain size in Al-Cu-Si lines. For all three metallurgies, as the linewidth decreases, the standard deviation of the log-normal failure distribution first increases and then either levels off or goes through a maximum. The linewidth dependence of lifetime is much stronger for Al-Cu and Al-Cu-Si lines than for Cr-Ag-Cr lines, and the reverse is true for the linewidth dependence of the standard deviation of the log-normal failure distribution. The critical width has also been reported by several authors to be 2.0 μm (88,120) and 1.8 μm (121) and 0.75 μm (89) for Al-Cu fine lines, and 1.2 μm (122) for Al/Ti fine lines. Bamboo grain structure was observed in Al-Cu fine lines of width below the critical value (88,89). In contrast to the data of Scoggan et al., the standard deviation of the log-normal failure distribution increases as the linewidth is reduced below the critical value (120,121).

Since the electromigration-induced open-circuit failure in the metal line is across its width, an increase of lifetime with increasing linewidth is expected because the probability of linking failure-causing defects across the width is lower for a wider metal line. Moreover, when the linewidth is larger than the average grain size, the lifetime is also expected to increase with the linewidth since it becomes more difficult for a crack to propagate across a wide metal line. As the linewidth becomes comparable to or smaller than the grain size, the fine metal line assumes a more or less bamboo grain structure and the lifetime is again expected to increase. The results of all these studies indicate that the linewidth dependence of electromigration lifetime in fine metal lines is a function of the film thickness, patterning technique, metallurgy and metal deposition condition. The critical width seems to decrease with decreasing film thickness. The deposition condition (evaporation or sputtering) and the patterning technique (lift-off or subetch) are believed to have little effect on the grain morphology and electromigration lifetime, if annealing is performed after patterning.

4.4c. Effect of Film Thickness. The electromigration lifetime in Al-Cu-Si lines was found by Scoggan et al. (119) to increase with decreasing film thickness. The critical width decreases from 2.5 to 1.5 μm when the film thickness decreases from 1.1 to 0.8 μm. Recently, the change in the critical linewidth for Al and Al-Cu fine lines was investigated by varying the thickness of metal lines (123). The film thickness was found to play an important role in determining the linewidth dependence of electromigration lifetime in fine metal lines. As shown in Fig. 17, the lifetime initially decreases and then increases below a critical width as the linewidth decreases. There is also a sharp increase in the lifetime when the submicron metal line is narrower than the critical width. The lifetime increases for submicron metal lines with linewidth comparable to or smaller than film thickness. The critical linewidth, below which the electromigration lifetime starts to increase, lies between 0.625 and 2.0 μm for film thickness ranging from 0.35 to 0.9 μm. These experimental data indicate that the critical width is sensitive to the film thickness of fine metal lines.

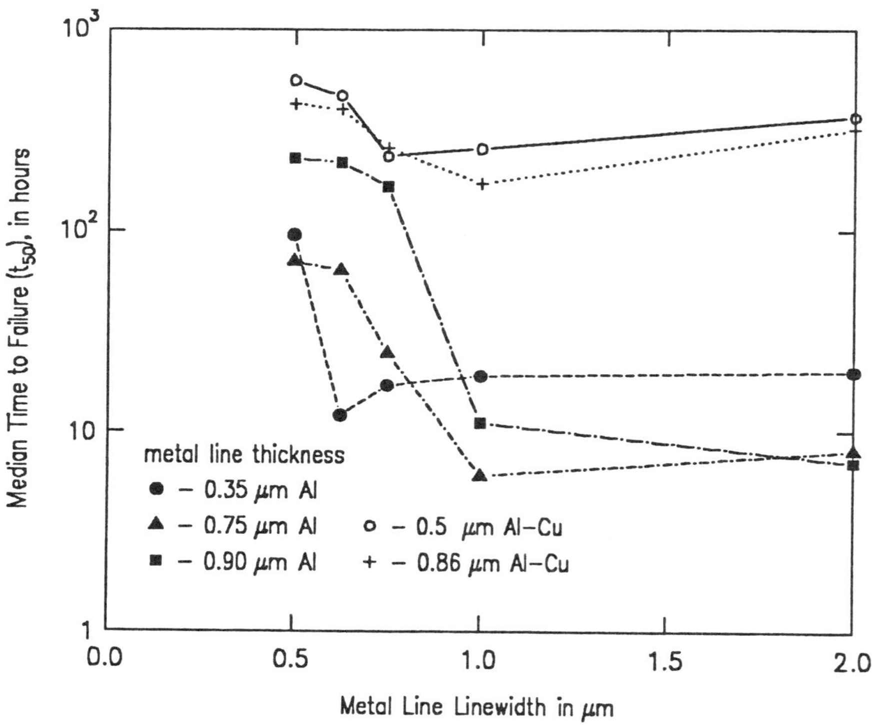

Figure 17. Effects of linewidth and film thickness on electromigration lifetime in fine metal lines. Test condition was T = 182°C, J = 1×10^6 A/cm^2. (From Ref. 123)

In Al and Al-Cu thin films of thickness less than 1.0 μm, the film thickness is comparable to the grain size in the bulk materials. So the grain size is constrained by the film thickness. As a result, columnar grain structure is generally observed. The columnar grain structure in Al-intermetallic fine lines has been identified as an important factor contributing to the improvement of electromigration resistance (97).

In Al and Al-Cu submciron lines, both the linewidth and film thickness are smaller than the grain size. So the ultimate grain size and grain structure are determined by the linewidth in addition to the film thickness. Thus, bamboo grain structure can be observed in those Al and Al-Cu submicron lines having linewidth comparable to or smaller than film thickness. Figure 18 shows schematically the formation of a columnar and bamboo grain structure in metallic thin films. The microstructure in Al-Cu fine lines of linewidth ranging from 0.5 to 2.0 μm have been studied (124) and is shown in Fig. 19. The grain structure approaches a bamboo grain structure in the 0.5 μm wide Al-Cu line. It has been demonstrated that bamboo grain structure can be produced by recrystallization in Al lines of 4.0 μm width (87). The lifetime in these bamboo-grained Al lines is substantially longer than polycrystalline Al lines.

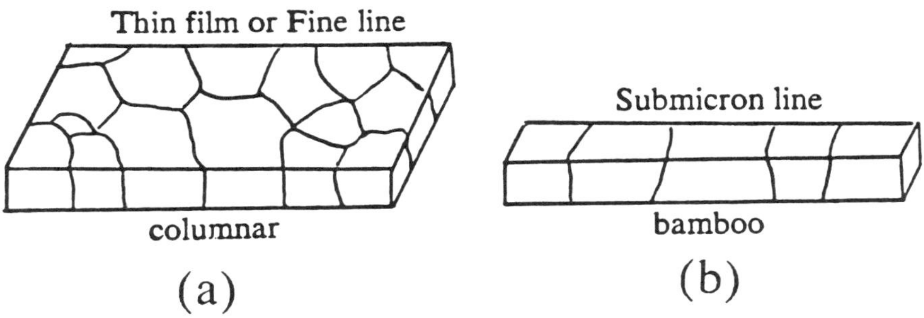

Figure 18. (a) The columnar grain structure in thin film or fine metal line, (b) The bamboo grain structure in submicron metal line.

In fine metal lines, several factors contribute to the improvement of electromigration resistance. First, when the grain size is comparable to the line width, the number of grain-boundary paths for diffusion along the metal line is small, thus the mass transport is reduced. Second, for fine metal lines with such a grain structure, the number of grain-boundary triple points per unit length is reduced, so does the number of flux divergent sites for electromigration-induced damages. And third, the electromigration-induced atomic flux is minimized in the bamboo-grained structure since most of the grain boundaries are perpendicular to the direction of current flow. In general, the effect of the line geometry on electromigration lifetime in interconnects can be a complex function of the metal line dimensions, microstructure, metallurgy, patterning technique and metal deposition conditions.

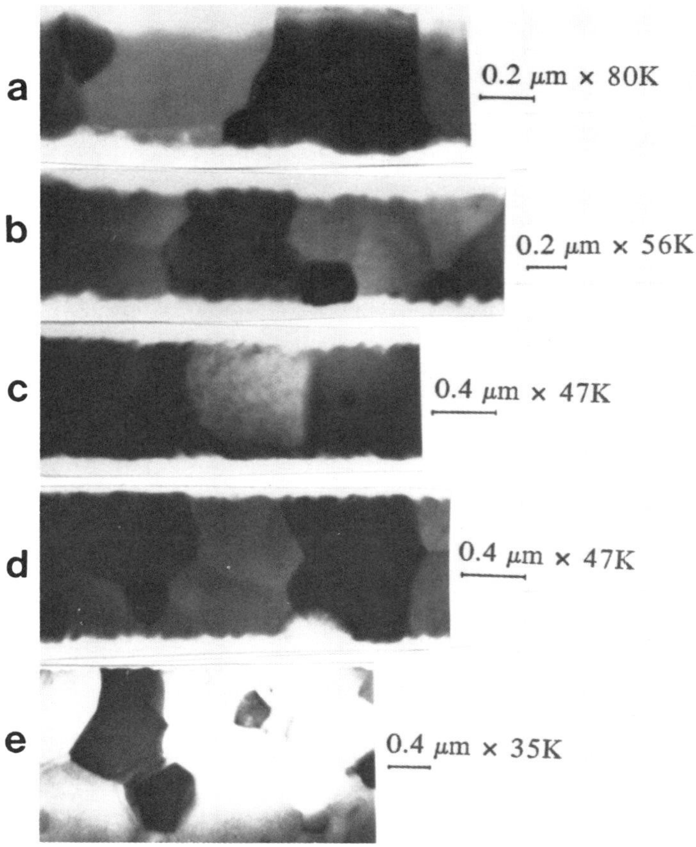

Figure 19. Transmission electron micrographs of Al-Cu fine lines with 0.5 μm film thickness after annealing at 400°C for an hour. (a) Linewidth 0.5 μm, (b) Linewidth 0.625 μm, (c) Linewidth 0.75 μm, (d) Linewidth 1.0 μm, (e) Linewidth 2.0 μm. (From Ref. 124)

Since multilevel interconnection consists of metal lines of varying geometries, when one designs the maximum current density allowable in the interconnects, one should consider the effect from the weakest line in the metal line geometry.

4.5 Pulsed Current Powering Characteristics

Almost all the electromigration lifetime measurements have been carried out under a steady direct current (dc) powering condition. In most applications, the current signals in the device are often applied under a pulsed condition as shown in Fig. 20. In FET devices, the pulsed current can be either unidirectional or bidirectional. In bipolar devices, the pulsed current is unidirectional with square waveform in addition to standby continuous dc. So the electromigration characteristics under a pulsed powering condition are of practical interest. This can be examined using a square current pulse to study the effect of on/off time on the lifetime. In practice, the actual waveform of the current pulse contains a transient part due to the charging and discharging of the circuit, so a triangular current pulse is also needed in order to investigate the effect of the pulsed current powering.

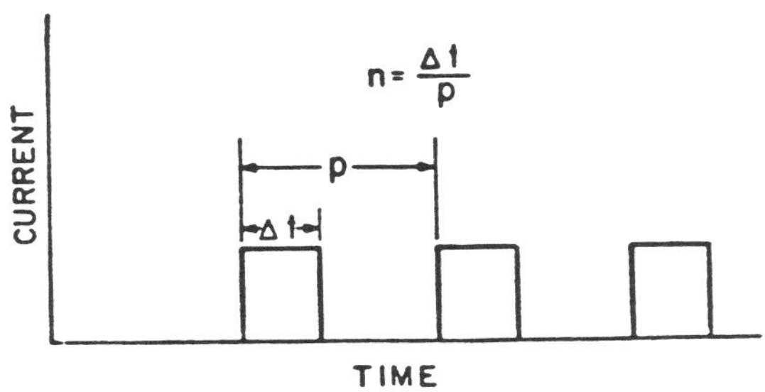

Figure 20. In ac pulsed powering, the current is applied every p seconds and remains on for a time Δt, so that the duty cycle $n = \Delta t/p$. (From Ref. 125)

Several factors distinguish the electromigration characteristics under pulse conditions. First, the current-on period when defects can accumulate to form damage and the current-off period when structural and thermal relaxation occur depend on the frequency and duty cycle of the pulsed current. The time scales of current on and off periods and their relative magnitudes can influence the nature of the structural relaxations and defect accumulation processes. Second, there will be temperature excursion and gradient in the conductor with their magnitudes depending on the current density and the thermal time constants of the conductor and its substrate. Third, the magnitudes of Joule heating and power dissipation under a pulsed condition are smaller than that under a direct current condition. A combination of these factors can change the basic process of damage formation although the mechanisms underlying such effects are not yet understood. Because of the structural and thermal relaxations during the current-off periods, an improvement in lifetime is expected under a pulsed dc powering condition.

Several models have been proposed to correlate the electromigration lifetime under a pulse condition to that under a dc condition. One assumes that damages accumulate only when the current is on and the current-off periods have no effect except to lengthen electromigration lifetime. Then, the mean time to failure (MTF or t_{50}) under a pulsed dc powering condition is inversely proportional to the duty cycle and can be expressed in terms of that under a steady dc powering condition as

$$t_{50}(\text{pulsed dc}) = \frac{1}{r}\, t_{50}(\text{steady dc}) \tag{24}$$

where r is the duty cycle defined as the percentage of time while the current is on. This on-time correction model gives the most conservative estimation of the improvement in the electromigration lifetime (126). A second model assumes that the migrating ions experience an average current density (rj) rather than the individual pulse. Then, the t_{50} under a pulsed condition can be expressed as

$$t_{50}(\text{pulsed dc}) = A_o(rj)^{-n} \exp\left(\frac{E_a}{kT}\right) \tag{25}$$

In the case when the current density exponent n equals to 2, the t_{50} under a pulsed condition is inversely proportional to the average power density and can be expressed in terms of that under a steady dc powering condition as

$$t_{50}(\text{pulsed dc}) = \frac{1}{r^2}\, t_{50}(\text{steady dc}) \tag{26}$$

This model has been found to give a good fit to some of the experimental data (127,128) as shown in Figs. 21 and 22.

A number of investigations have been reported on the effects of frequency and duty cycle on the electromigration lifetime under a pulsed current condition. Due to the large range of test conditions, there does not seem to be strong correlations between results of these measurements. Nevertheless, there exists a general trend of increasing lifetime with decreasing duty cycle, as expected. The dependence observed is not consistent; it varies from an inverse linear relationship (129) to a decrease with the square (127,128), or exponential (55) of the duty cycle. The dependence of the lifetime on the frequency is even less consistent. An increase of lifetime with increasing frequency

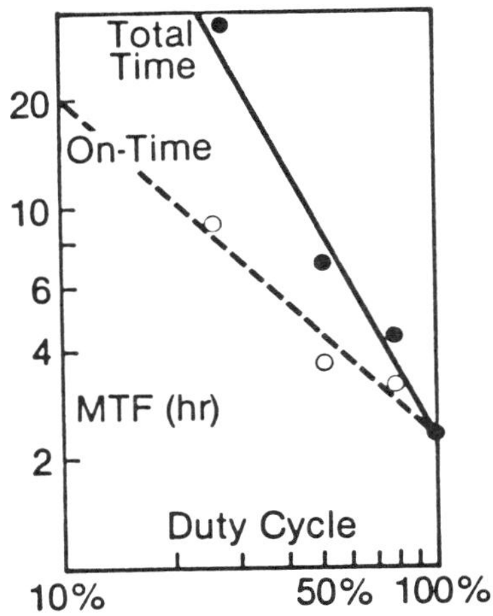

Figure 21. MTF as a function of duty cycle at a pulse frequency of 1kHz. Test condition was T = 200°C, J = 2×10^6 A/cm^2. (From Ref. 127)

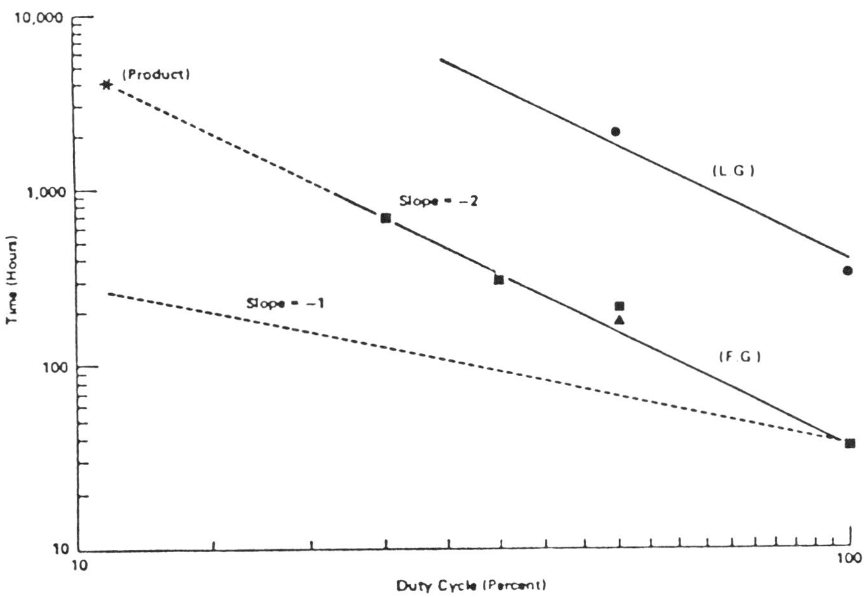

Figure 22. Median time to fail of the conductor as a function of duty cycles. Test condition was T = 150°C, J = 1.2×10&6. A/cm^2. (F.G) - Fine grain, (L.G) - Large grain. (From Ref. 128)

as shown in Fig. 23 was found in two experiments (55,130) but none was found in two others (127,128).

Interestingly, in one experiment (127) the activation energy was found not to change with duty cycle. In another experiment (128), the activation energy, the statistical distribution of the failure time and the failure mode were found to be the same under pulse tests, with square or triangular pulses, and steady dc powering. The occurrence of electromigration-induced damage in the form of voids and hillocks under pulsed dc powering conditions was observed to be a complex function of frequency and duty cycle (125). All these indicate some general difficulties in obtaining lifetest results under pulse conditions due to the high current densities used. This can introduce heating and stress relaxation which can affect lifetime in a complex manner but not yet well understood.

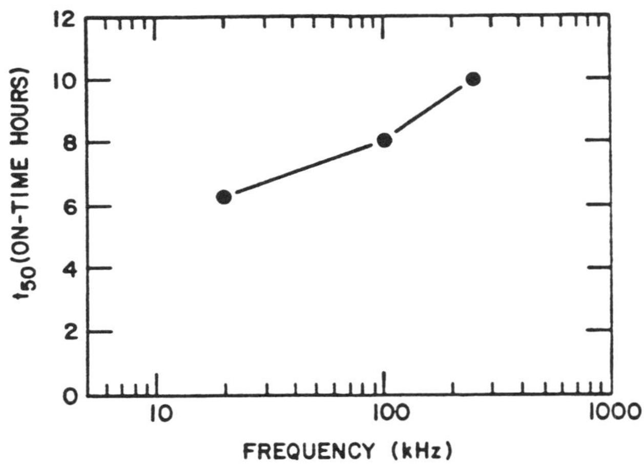

Figure 23. The dependence of conductor lifetime on pulse frequency at an ambient temperature of 100°C, with a 25% duty factor and a pulse current density of 1×10^7 A/cm^2. (From Ref. 55)

5. Electromigration in Multilevel Interconnection

5.1 Electromigration at Contacts

In VLSI circuits, the increase in the wiring density requires the use of multilevel interconnections. In such structures, there are numerous contacts connecting the metal lines from one level to an adjacent level or to a semiconductor junction. The electromigration characteristics of these contacts are usually very different from those of flat metal lines. At a contact, material reactions usually occur at the interface where two different materials join, even without the current driving force. The addition of a current will alter the temperature distribution as well as the driving force for mass transport. As a result, the nature of the flux divergence at the interface can be very different from that of a flat line. Although contact degradation has long been recognized as a reliability concern, the effect due to electromigration has attracted little attention until the late 70's. With the advent of large-scale integration, electromigration-induced contact degradation became an emerging problem at that time. In VLSI, the device scaling trend demands significant increases in the current density at the contacts (see §4.1). This, when coupled with the decrease in the contact dimension, has made contact electromigration of ever increasing concern. The problem is expected to be of particular importance to bipolar devices where the speed performance requires a high driving current.

Most of the studies on contact electromigration have been carried out on ohmic contacts in silicon with Al or Al alloy metal lines. A typical test structure used to study contact electromigration (131) is shown in Fig. 24. In this structure, two metal/silicon contacts are fabricated in an n^+ diffused resistor, so contact electromigration with opposite current flow directions can be observed simultaneously. A metal line is fabricated to link two diffused resistor contacts, a design which allows a comparative study of electromigration in metal lines and contacts.

The contact reaction at the Al/Si interface with no applied current has been investigated a long time ago (132-135). The reaction is dominated by silicon dissolution via a solid state diffusion process into Al in order to saturate the silicon solubility in Al. Due to the usual presence of a thin oxide layer on silicon, the diffusion of Si into Al occurs mainly through the pinholes in the oxide. The depleted region is refilled by Al and has the shape of a spike bounded by (111) Si planes, which can be readily observed by etching away the Al layer. The spike will grow with increasing annealing time and eventually it can punch through the diffused junction to short out the contact. A common solution to the spike problem is to saturate the Al line with about 1 at.% of Si to reduce the chemical driving force of Si dissolution. While this circumvents the junction spiking, another form of junction degradation has been observed. This is caused by the formation of Si precipitates on the Si substrate; since the Si precipitates are

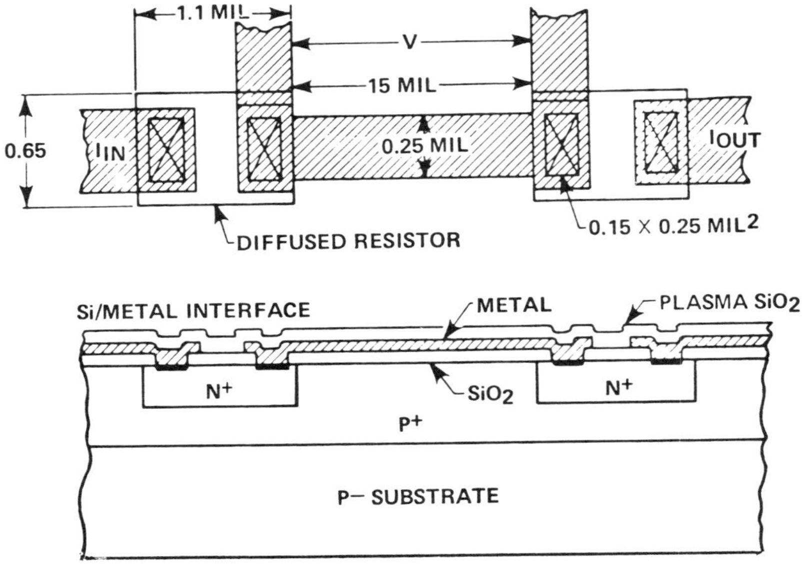

Figure 24. Test structure HCSO1 – High current stress Si/metal interface. (From Ref. 131)

saturated with Al, which is a p type dopant, this produces a highly resistive p-n junction on the contact (136).

When a direct current is applied to the contact, the electromigration driving force superimposes onto the chemical driving force and alters the nature of contact reaction. Referring to the test structure of Fig. 24, at the positive contact (source) where electrons flow out, Al and Si both migrate out through the Al matrix. Since this is added to the chemical driving force, the rate of Si dissolution into Al is increased. For Al, electromigration causes its depletion at the contact, which can eventually produce an open failure at the contact. At the negative contact (drain) where electrons flow into, electromigration of Si through the Al matrix increases the rate of accumulation of Si on the contact, thus accelerating the rate of failure due to the formation of a resistive p-n contact. Electromigration of Al to this contact is not expected to have a significant effect on the failure rate although a sufficient amount of Al accumulation can lead to an extrusion-type failure. The dominant failure mode will be determined by the relative rates of all these kinetic processes as well as by the geometry and material property of the contact, particularly the junction depth and microstructure of the metal line.

A comprehensive study has been carried out on electromigration on Si contacts with Al, Al+Cu and Al+Cu+Si metallurgies (131). With the shallow contacts (contact depth of 0.35 μm) used in this study, the primary failure mode for all contacts was junction shorting due to spike formation. Although migration of Al from one contact to the other was observed, open failure in the conductor lines has not been detected. The addition of Cu and Si improves the contact lifetime with the best results of about 100 times improvement between 150° to 215°C belonging to Al+Cu+Si. The improvement was attributed to a reduction in the rates of of Si dissolution and migration as a result of Cu and Si additions. The temperature dependence of these lifetime data shows an apparent activation energy of 0.6 ± 0.2 eV for all samples, consistent with a grain-boundary controlled kinetics.

Similar studies have been carried out on Si contacts with Al (137-140) and Al+Si (141,142) metallurgies. For pure Al, the failure mode was found to be primarily due to Al depletion at the positive contact. The depletion occurred first at the leading edge of the contact, reflecting the local current crowding. The activation energy is usually low, about 0.5 eV. With the addition of Si, the failure mode shifted to junction shorts due to spike formation or to the formation of a resistive contact. The activation energy appeared to increase to 0.8-0.9 eV. The overall results are consistent with those from Cu addition. Interestingly, in one of the studies (141), a junction of 1.35μm depth was used, and the dominant failure mode was due to Al depletion. While under similar test conditions, the failure of a 0.35 μm junction was caused mainly by short due to spike formation (131). Thus it appears that the failure mode can shift from open to short, depending on the junction depth.

When a silicide layer is imposed between Al and Si to improve contact stability, the problem becomes even more complicated because two interfaces, Al/silicide and silicide/Si, have to be considered. The material reaction at the Al/silicide/Si contact has been investigated to considerable detail (143,144). The reaction was observed to occur mainly at the thermodynamically unstable Al/silicide interface. This leads to the displacement of Si by Al in the silicide, resulting in the formation of Al-silicide ternary compound and Si precipitates. The junction degrades when the silicide layer is completely consumed or when Si particles precipitate on the substrate to form a resistive contact. The degradation process can accelerate if pinholes exist in the silicide layer, so Si precipitates can reach the substrate through the pinholes without a complete consumption of the silicide layer. With the addition of a current, the electromigration of Al and Si has to be included in considering the degradation process. While the problem is not yet completely understood, experiments have been carried out to investigate Al/$CoSi_2$ (145) and AlCuSi/$TaSi_2$ (146) contacts.

The cobalt silicide contact failed mainly due to junction leakage caused by spike formation as a result of Al and $CoSi_2$ reaction. The reaction can be inhibited by using a TiN barrier layer up to about 450°C. In the second study, the failures observed are either due to the electromigration-induced void formation on Al contact sidewalls and at the interface between Al and $TaSi_2$ above 250°C, or due to the formation of a high-resistance, non-ohmic contact as a result of Si precipitation. In this study, the high activation energies of 1.1 eV was attributed to a bulk diffusion process since the contact dimension is small relative to the average grain size in Al thin film. The large current density exponents of 6 was explained as due to the local heating caused by current crowding at the leading edge of the junction.

5.2 Current Crowding and Local Heating in Studs and Vias

Studs and vias are two basic structures which link the metal line from one conductor level to the adjacent conductor level through the insulator level in the multilevel interconnection scheme. The aspect ratio of the stud or via structure is defined as its height over its width. The width of the stud or via structure is usually 0.1 μm to 0.3 μm smaller than the linewidth to allow for lithographic overlay. The height of the stud or via structure is designed to exceed the thickness of the thin insulator film in order to improve device density and speed performance. As a result, the stud or via aspect ratio will increase, for example, from 0.4 to 4.0 as the linewidth of the metal line shrinks from 5.0 μm to 0.5 μm if the insulator film thickness remains at 2.0 μm. The high aspect ratio in combination with the projected high current density gives rise to current crowding and local heating, thus increasing the concern for electromigration-induced failures in multilevel interconnection.

Analytical expression for the current density distribution in a stud structure has been derived by the transmission line model approximation (147). Numerical method has also been used to calculate the distributions of current density and temperature in multilevel interconnection (148). In this calculation, a finite element method is used to obtain numerical solution to differential equations governing the current and heat flows subject to appropriate boundary conditions. The cross-sectional views, current density distributions and temperature contours in the 1.5 μm high and 1.5 μm wide stud and via structures with 1.5 μm metal line thickness are shown in Fig. 25 and 26 respectively. Two current density peaks located in the lower left and upper right stud corners are plotted in Fig. 25(b). However, their magnitudes are too small to change significantly the overall temperature distribution as shown in Fig. 25(c). Current density peak in studs increases with decreasing stud width and increasing aspect ratio. The

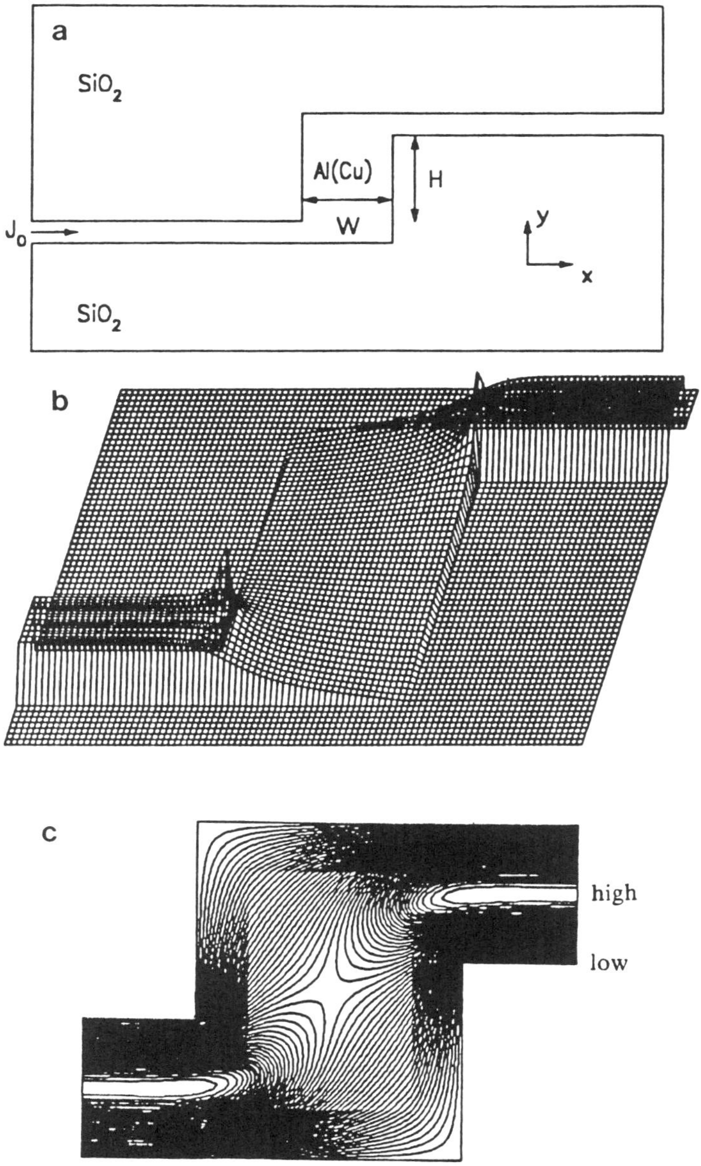

Figure 25. (a) Cross-sectional view (b) Current density distribution (c) Temperature contour in a stud structure of 2.0 μm high and wide with 0.5 μm metal line thickness. (From Ref. 148)

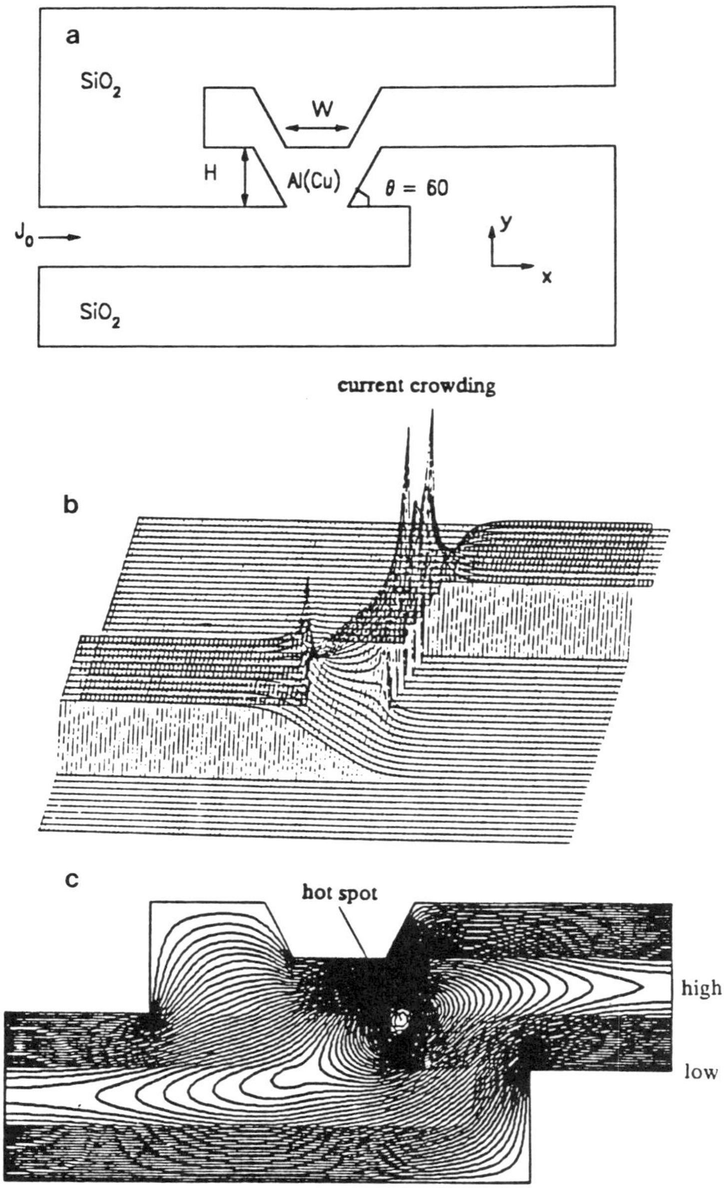

Figure 26. (a) Cross-sectional view (b) Current density distribution (c) Temperature contour in a via structure of 1.2 μm high and wide with 10 μm metal line thickness. (From Ref. 148)

rate of increase is faster for submicron lines than for wider lines but becomes approximately constant with the metal line thickness. The magnitude of the current density peak in studs can be decreased by decreasing its height if its aspect ratio is less than 1.0. The current density peaks in a 60° step angle via structure are plotted in Fig. 26(b). Their magnitudes are higher than that in a comparable stud structure. The extent of current crowding in vias increases with increasing step angle due to the reduction in the step cross-section. The location of maximum current crowding can shift into the base region from the step region when its width becomes smaller than the step thickness as the aspect ratio is increased. As shown in Fig. 26(c), current crowding in a via structure causes a hot spot in the 60° step region because of Joule heating. The current density peaks in the via can be decreased to the same magnitude of a comparable stud by increasing the metal line thickness in the upper conductor level. Thus, this problem can be circumvented. Results of this study indicate that the stud width and its ratio to the metal line thickness are two important factors in controlling the extent of current crowding and local heating in the stud structures.

Results of these finite element calculations can provide a general approach to optimize the aspect ratio and geometries of stud and via structures for minimal current crowding and local heating in multilevel interconnection. This approach can also be applied to other structures and materials with different material properties which may be employed to form multilevel interconnection. Moreover, it would be useful to extend this technique to investigate the effects of microstructure, defects and mass transport on electromigration-induced damage formation in multilevel interconnection.

6. SUMMARY

In this article, we have reviewed electromigration studies in bulk metals as well as in metallic thin films. The investigations in bulk materials have a relatively long history of more than fifty years and the objectives have been focused on the basic understanding of the electromigration phenomenon as a force-driven diffusion process. On the experimental side, electromigration has been used effectively to probe the nature of electron-ion interaction in substitutional and interstitial diffusion systems. A considerable amount of results have been accumulated, revealing a systematic trend of the effective charge for metals with electronic structures varying from the simple alkali elements to the complex refractory metals. On the theoretical side, thermodynamic formulation and kinetic analysis have provided a framework in which the macroscopic behavior of electromigration can be analyzed in terms of the atomistic jumping frequencies. An example using this approach to treat the solute effect on electromigration was discussed in §2.1. The theory on the effective charge has been relatively well developed, including the use of many-body quantum mechanical formulation to treat the scattering between ions and charge carriers. The controversy regarding the extent of screening on the electrostatic force seems to have been resolved by recent experiments and theory that takes into account the dynamical nature of the scattering process. All these efforts have provided a level of understanding of electromigration in bulk metals where the migration and the scattering processes can be examined in atomistic details.

The electromigration studies on thin films have, from the beginning, oriented toward technological applications. The efforts have been concentrated on the understanding of the electromigration-induced failure of conductor lines. The characteristics of electromigration in thin films, nevertheless, have many interesting basic aspects, including the role of grain boundaries, the nature of the divergent sites and the effects due to changes in the microstructure and geometry of the conductor line. Results obtained so far have provided mostly qualitative information on some of these problems. Only a limited amount of quantitative measurements have been carried out for understanding the basic characteristics of electromigration in thin films. This is in contrast to the studies of bulk materials and has left some important questions unanswered, including the nature of the electromigration driving force at grain boundaries and an atomistic model for explaining the improvement of electromigration lifetime by solute additions.

Looking ahead, the development of VLSI technology will continue to demand ever increasing capability of electromigration of conductor lines. This, when coupled with decreasing line dimensions, has stimulated extensive interest in studying electromigration in fine lines. The submicron line dimension and the use of multilevel interconnection structure have brought out a set of new and interesting questions, including the mechanism for damage formation in submicron lines, electromigration at device junctions and line contacts, and electromigration under fast pulse (less than 1 nsec) conditions. In spite that many of these problems originate from technological applications, the operating conditions and the material properties give rise to certain unique electromigration characteristics, for which there is little existing knowledge. With the evolution of the technology, new problems in electromigration will continuously emerge. This will provide new and challenging problems to study. The success of the future technology may well depend on how successfully some of these electromigration problems can be understood and solved.

REFERENCES

1. M. Gerardin, *Compt. Rend.* **53**, 727 (1861).
2. W. Seith and H. Wever, *Z. Elecktrochem.* **59**, 942 (1953).
3. F. Skaupy, *Verh. Deut. Phy. Ges.* **16**, 156 (1914).
4. V. B. Fiks, *Sov. Phys. – Solid State* **1**, 14 (1959).
5. H. B. Huntington and A. R. Grone, *J. Phys. Chem. Solids* **20**, 76 (1961).
6. See the review by H. B. Huntington, "Electromigration in Metals" in "*Diffusion in Solids: Recent Developments*" ed. by A. S. Nowick and J. J. Burton, Academic Press, New York (1974), pp. 303-352.
7. D. A. Rigney, in "*Electro- and Thermo-Transport in Metals and Alloys*", ed. by R. E. Hummel and H. B. Huntington, AIME, New York (1977), Chap. VIII, pp. 140-159.
8. D. T. Peterson, in "*Electro- and Thermo-Transport in Metals and Alloys*", ed. by R. E. Hummel and H. B. Huntington, AIME, New York (1977), Chap. IV, pp. 54-67.
9. I. A. Blech and H. Sello, in *Physics of Failures in Electronics* **5**, USAF-RADC Series, (1966), p. 496.
10. F. M. d'Heurle and R. Rosenberg in "*Physics of Thin Films*", Vol. 7, Academic Press, New York (1973), p. 257.
11. P. S. Ho, F. M. d'Heurle and A. Gangulee in "*Electro and Thermo-Transport in Metals and Alloys*", ed. by R. E. Hummel and H. B. Huntington, AIME, New York (1977), Chap. VII, pp. 108-139.
12. F. M. d'Heurle and P. S. Ho in "*Thin Films: Interdiffusion and Reactions*", ed. by J. M. Poate, K. N. Tu and J. W. Mayer, Wiley-Interscience, New York (1978), p. 243.
13. R. E. Hummel, in "*Electro- and Thermo-transport in Metals and Alloys,*" ed. by R. E. Hummel and H. B. Huntington, AIME, New York (1977), Chap. VI, pp. 93-107.
14. J. J. Bart, J. R. Black, P. B. Ghate, P. S. Ho and C. M. Walter, *Proc. of 20th IEEE Relia. Phys. Symp.*, (1982), p. 284.
15. J. Verhoeven, *Metallurgical Reviews* **8** (31), 311 (1963).
16. J. N. Pratt and R. G. R. Sellors, "*Electrotransport in Metals and Alloys*", Trans Tech Publications, Switzerland (1973).
17. R. S. Sorbello, in "*Electro- and Thermo-Transport in Metals and Alloys*", ed. by R. E. Hummel and H. B. Hungtington, AIME, New York (1977), Chap. I, pp. 2-19.
18. D. A. Blackburn, in "*Electro- and Thermo-Transport in Metals and Alloys*", ed. by R. E. Hummel and H. B. Hungtington, AIME, New York (1977), Chap. II, pp. 20-36.

19. H. Wever, in "*Electro- and Thermo-Transport in Metals and Alloys*", ed. by R. E. Hummel and H. B. Hungtington, AIME, New York (1977), Chap. III, pp. 37-53.
20. See, for example, Y. Adda and J. Philibert in "*La Diffusion dans les Solides*", Presses Universitaires de France, Paris (1966), Vol. 1, Chap. V.
21. P. S. Ho, *Phys. Rev.* **B8**, 4534 (1973).
22. N. V. Doan, *J. Phys. Chem. Solids* **33**, 2161 (1972).
23. J. R. Manning, "*Diffusion Kinetics for Atoms in Crystals*", Van Nostrand-Reinhold, Princeton, New Jersey (1968).
24. I. Ames, F. M. d'Heurle and R. Horstman, *IBM J. of Res. Develop.* **4**, 461 (1970).
25. R. Rosenberg, *J. Appl. Phys.,* **16**, 27 (1970); see also the review by F. M. d'Heurle and A. Gangulee, "Solute Effects on Grain Boundary Electromigration and Diffusion", in Nature and Behavior of Grain Boundaries, H. Hu, ed., Plenum Press, New York (1972), p. 339.
26. C. Bosvieux and J. Friedel, *J. Phys. Chem. Solids,* **23**, 123 (1962).
27. R. S. Sorbello, *Comments on Solid State Physics*, **6**, 117 (1975).
28. R. S. Sorbello, in "*Electro- and Thermo-Transport in Metals and Alloys*", ed. by R. E. Hummel and H. B. Huntington, AIME, New York (1977), Chap. I, pp. 2-19.
29. A. H. Verbruggen, Ph.D. Thesis, Free University in Amsterdam, Holland (1985).
30. M. D. Glichuk, *Ukr. Fiz. Zh.* **4**, 684 (1959).
31. H. B. Huntington and P. S. Ho, *J. Phys. Soc. Japan* Suppl. II, **18**, 202 (1963).
32. A. K. Das and R. Peierls, *J. Phys. C* **8**, 3348 (1975).
33. R. S. Sorbello and B. Dasgupta, *Phys. Rev.* **B16**, 5193 (1977).
34. W. L. Schaich, *Phys. Rev.* **B19**, 620 (1979).
35. P. Kumar and R. S. Sorbello, *Thin Solid Films* **25**, 25 (1975).
36. R. S. Sorbello, *Phys. Rev.* **B31**, 798 (1985).
37. A. H. Verbruggen, R. Griessen and D. G. de Groot, *J. Phys. F*: Metal Physics, **16**, 557 (1986).
38. R. Landauer and J. W. F. Woo, *Phys. Rev.* **B10**, 1266 (1974).
39. R. Landauer, *Phys. Rev.* **B14**, 1474 (1976).
40. R. S. Sorbello, *J. Phys. Chem. Solids* **34**, 937 (1973).
41. M. D. Feit and H. B. Huntington, *Phys. Rev.* **B5**, 1416 (1972); H. B. Huntington, W. B. Alexander, W. B. Feit and J. L. Routbout, in "*Atomic Transport in Solids and Liquids*", ed. A. Lodding and T. Lagerwall, Tubingen (1971), p. 91.
42. T. C. Genoni and H. B. Huntington, *Phys. Rev.* **16**, 1344 (1977).
43. R. V. Penney, *J. Phys. Chem. Solids* **25**, 335 (1964).
44. P. S. Ho and J. K. Howard, *J. Appl. Phys.* **45**, 3229 (1974).

45. K. L. Tai and M. Ohring, *J. Appl. Phys.* **48**, 28 (1977); K. L. Tai and M. Ohring, *J. Appl. Phys.* **48**, 36 (1977).
46. P. S. Ho, *J. Appl. Phys.* **49**, 2735 (1978).
47. P. S. Ho, J. E. Lewis and J. K. Howard, *Thin Solid Films* **25**, 302 (1975).
48. I. A. Blech and E. Kinsborn, *Thin Solid Films* **25**, 327 (1975).
49. S. T. Pai and J. P. Marton, *Can. J. Phys.* **55**, 103 (1977).
50. N. V. Doan, *J. Phys. Chem. Solids* **32**, 2135 (1971).
51. P. S. Ho, *J. Phys. Chem. Solids* **27**, 1331 (1966).
52. N. V. Doan, *J. Phys. Chem. Solids* **31**, 2079 (1970).
53. P. Guilmin, L. Turban and M. Gerl, *J. Phys. Chem. Solids* **34**, 951 (1973).
54. C. K. Hu and H. B. Huntington, Ch. 10 of this book.
55. R. J. Miller, *Proc. of 16th IEEE Relia. Phys. Symp.,* (1978), p. 241.
56. J. L. Vossen, G. L. Schnable and W. Kern, *J. Vac. Sci. Tech.* **11**, 60 (1974).
57. E. Philofsky and E. L. Hall, *IEEE Trans.* **PHP-11,** 281 (1975).
58. A. J. Learn, *J. Electrochem. Soc.* **123**, 894 (1976).
59. P. S. Ho, *Proc. of 20th IEEE Relia. Phys. Symp.,* (1982), p. 284; P. S. Ho, *Proc. of 3rd ECS VLSI Sci. and Tech. Symp.,* (1985), p. 146.
60. P. B. Ghate, *Proc. of 20th IEEE Relia. Phys. Symp.,* (1982), p. 292; P. B. Ghate, *Solid State Tech.,* **3**, 113 (1983); P. B. Ghate, *Physics Today,* **39**, 58 (1986).
61. A. K. Sinha, *Proc. of 1st ECS Sci. and Tech. Symp.,* (1982), p. 173.
62. D. Pramanik and A. N. Saxena, *Solid State Tech.,* **1**, 127 (1983); D. Pramanik and A. N. Saxena, *Solid State Tech.,* **3**, 131 (1983).
63. A. N. Saxena, *Proc. of 1st IEEE VLSI Multilevel Interconnection Conf.,* (1984), p. 1; A. N. Saxena, *Proc. of 3rd IEEE VLSI Multilevel Interconnection Conf.,* (1986), p. 7.
64. C. J. Dell'Oca, *Proc. of 2nd IEEE VLSI Multilevel Interconnection Conf,* (1985), p. 3.
65. G. C. Schwartz, *Proc. of 1st ULSI Sci. and Tech. Symp.,* (1987), p. 493.
66. T. Kwok, P. S. Ho, S. Yip, R. W. Balluffi, P. D. Bristowe and A. Brokman, *Phys. Rev. Lett.* **47**, 1148 (1981); T. Kwok, P. S. Ho and S. Yip, *Surf. Sci.* **144**, 44 (1984).
67. I. A. Blech and E. S. Meieran, *J. Appl. Phys.* **40**, 485 (1969).
68. L. Berenbaum, *J. Appl. Phys.* **42**, 880 (1971).
69. M. J. Attardo and R. Rosenberg, *J. Appl. Phys.* **41**, 2381 (1970).
70. I. A. Blech and E. S. Meieran, *Appl. Phys. Lett.* **11**, 263 (1967); I. A. Blech and E. S. Meieran, *J. Appl. Phys.* **40**, 485 (1969).

71. J. R. Black, *IEEE Trans.* ED-16, 338 (1969); J. R. Black, *Proc. IEEE 57*, 1587 (1969).
72. R. W. Pasco and J. A. Schwarz, *Solid State Elec.,* **26**, 445 (1983).
73. B. J. Root and T. Turner, *Proc. 23rd IEEE Relia. Phys. Symp.,* (1983), p. 100.
74. C. Hong and D. L. Crook, *Proc. 23rd IEEE Relia. Phys. Symp.,* (1983), p. 108.
75. L. Bacci, C. Caprile and G. D. Santi, *Proc. of 4th IEEE VLIS Multilevel Interconnection Conf.,* (1987), p. 463.
76. J. L. Vossen, *Appl. Phys. Lett.* **23**, 287 (1973).
77. T. M. Chen, T. P. Djen and R. D. Moore, *Proc. 23rd IEEE Relia. Phys. Symp.,* (1985), p. 87.
78. A. Diligenti, B. Neri, P. E. Bagnoli, A. Bersenti and M. Rizzo, *IEEE Electron Device Lett.* **6**, 606 (1985).
79. R. H. Koch, J. R. Lloyd and J. Cronin, *Phys. Rev. Lett.* **55**, 2487 (1985).
80. J. G. Cottle and T. M. Chen, *Proc. 4th IEEE VLSI Multilevel Interconnection Conf.,* (1987), p. 449.
81. A. Gangulee and F. M. d'Heurle, *Appl. Phys. Lett.*, **19**, 75 (1971); A. Gangulee and F. M. d'Heurle, *Thin Solid Films* **16**, 227 (1973).
82. M. Hansen, "Constitution of Binary Alloys," McGraw-Hill Book Company, Inc., New York (1958).
83. J. K. Howard and R. F. Ross, *Appl. Phys. Lett.* **18**, 344 (1971).
84. A. J. Learn, *J. Electronic Materials,* **3**, 531 (1974).
85. F. M. d'Heurle and I. Ames, *Appl. Phys. Lett.* **16**, 80 (1970).
86. C. H. Herzig and W. Wiemann, *Phys. Stat. Solidi* **A26**, 459 (1974).
87. J. M. Pierce and M. E. Thomas, *Appl. Phys. Lett.* **39**, 165 (1981).
88. S. Vaidya, T. T. Sheng and A. K. Sinha, *Appl. Phys. Lett.* **36**, 464 (1980).
89. T. Kwok, *Proc. of 4th IEEE VLSI Multilevel Interconnection Conf.* (1987), p. 456.
90. J. R. Lloyd and P. M. Smith, *J. Vac. Sci. Tech.,* A1, 455 (1983).
91. L. E. Felton, J. A. Schwarz, R. W. Pasro and D. H. Norburg, *J. Appl. Phys.* **58**, 723 (1985).
92. L. Yeu, C. Hong and D. Crook, *Proc. fo 23rd IEEE Relia. Phys. Symp.,* 81 (1985).
93. B. Grabe and H.-U. Schreiber, *Solid State Electronics,* **26**, 1023 (1983).
94. E. Levine and J. Kitcher, *Proc. of 22nd IEEE Relia. Phys. Symp.,* (1984), p. 242.
95. L. J. Fried, J. Hares, J. S. Lecdeton, J. S. Logen, G. Paal and P. A. Totta, *IBM J. Res. Develop.,* **26**, 263 (1982).

96. J. K. Howard, J. F. White and P. S. Ho, *J. Appl. Phys.* **49**, 4083 (1978).
97. T. Kwok, P. S. Ho and H.-C. W. Huang, *J. Vac. Sci. Tech.* **A2**, 241 (1984); P. S. Ho, T. Kwok and H.-C. W. Huang, *Proc. of the 16th Int'l Conf on Solid State Devices and Materials*, (1984), p. 55.
98. S. S. Iyer and C. Y. Ting, *Proc. of 22nd IEEE Relia. Phys. Symp.*, (1984), p. 273.
99. S. Shiona, T. Mosiya and M. Kashiwagi, *Proc. of 1st IEEE VLSI Multilevel Interconnection Conf.*, (1984), p. 61.
100. D. S. Gardner, T. L. Michelke, K. C. Saraswat, T. W. Barbee, Jr., J. P. McVittie and J. D. Meindl, *IEEE Trans.* ED-32, 2 (1985); D. S. Gardner, T. L. Michelka, P. A. Flinn, K. C. Saraswat and J. D. Meindl, *Proc. of 2nd IEEE VLSI Multilevel Interconnection Conf.*, (1985), p. 102.
101. R. E. Jones, Jr. and L. D. Smith, *Proc. of 2nd IEEE VLSI Multilevel Interconnection Conf.*, (1985), p. 194.
102. B. W. Shen, T. Bonifield and J. McPherson, *Proc. of 2nd IEEE VLSI Multilevel Interconnection Conf.*, (1985), p. 114.
103. R. H. Denard, F. H. Gaensslen, H. N. Yu, V. L. Rideout, E. Bassous, and A. R. LeBlanc, *IEEE J. Solid-State Circuits,* **SC-9**, 256 (1974).
104. D. D. Tang, and P. M. Solomon, *IEEE J. Solid-State Circuits,* **SC-14,** 679 (1979).
105. J. L. Prince in *Very Large Scale Integration: Fundamentals and Applications,* ed. by D. F. Barbe, Springer-Verlag (1982), Chapter 2.
106. P. S. Ho, *J. Appl. Phys.* **41**, 64 (1970).
107. Y. S. Chaung and H. L. Huang, *Jap. J. Appl. Phys.* **14**, 267 (1975); Y. S. Chaung and H. L. Huang, *J. Appl. Phys.* **47**, 1775 (1976).
108. R. Rosenberg and L. Berenbaum, *Appl. Phys. Lett.* **12**, 201 (1968).
109. J. R. Black, *Proc. of 6th IEEE Relia. Phys. Symp.*, (1968), p. 148.
110. C. B. Oliver and D. E. Bower, *Proc. of 8th IEEE Relia. Phys. Symp.*, (1970), p. 116.
111. A. J. Learn, *Appl. Phys. Lett.* **19**, 292 (1971).
112. A. Gangulee, *J. Appl. Phys.* **45**, 3749 (1974).
113. B. N. Agarwala, B. Patnaik and R. Schnitzel, *J. Appl. Phys.* **39**, 1487 (1972).
114. S. Vaidya and A. K. Sinha, *Thin Solid Films,* **75**, 253 (1981).
115. B. N. Agarwala, M. J. Attardo and A. P. Ingraham, *J. Appl. Phys.* **41**, 3954 (1970).

116. A. J. Learn and W. H. Shephard in *Proc. of 9th IEEE Relia. Phys. Symp.,* (1971), p. 129.
117. H. A. Schafft, T. C. Station, J. Mandel and J. D. Shott, *IEEE Trans.* **ED-34** (3), 673 (1987).
118. T. Kwok, J. Finnegan and D. Johnson, *Proc. of 5th IEEE VLSI Multilevel Interconnection Conf.,* (1988), in press.
119. G. A. Scoggan, B. N. Agarwala, P. P. Peressini, and A. Brouillard, *Proc. of 13th IEEE Relia. Phys. Symp.,* (1978), p. 151.
120. E. Kinsborn, *Appl. Phys. Lett.* **36**, 968, (1980).
121. S. S. Iyer and C. Y. Ting, *IEEE Trans.* **ED-31**, 1468 (1984).
122. K. Eden, W. Roth, and H. Beneking, *Microelectronic Engineering,* **1**, 263 (1983).
123. T. Kwok, *Proc. of 1st ECS ULSI Sci. and Tech. Symp.,* (1987), p. 593.
124. T. Kwok, C. Y. Ting and J. U. Han, *Proc. of 2nd IEEE VLSI Multilevel Interconnection Conf.,* (1985), p. 83.
125. A. T. English, K. L. Tai and P. A. Turner, *Appl. Phys. Lett.* **21**, 397 (1972).
126. F.M. d'Heurle, *Proc. IEEE* **59**, 1409 (1971).
127. J. M. Towner and E. P. van de Ven, *Proc. of the 21st IEEE Relia. Phys. Symp.,* (1983), p. 36.
128. R. W. Lin and D. Y. Guan, *Proc. of ECS Electromigration of Metals and 1st ECS Multilevel Metallization and Packaging Symp.,* (1985), p. 75.
129. E. Kinsbron, C. M. Melliar-Smith, A. T. English and T. Chynoweth, *Proc. of 16th IEEE Relia. Phys. Symp.,* 248 (1978); E. Kinsbron, C.M. Melliar-Smith and A.T. English, *IEEE Trans.* **ED-26**, 22 (1979).
130. J. R. Davis, *Proc. IEEE* **123**, 1209 (1976).
131. P. B. Ghate, *Proc. of 19th IEEE Relia. Phys. Symp.,* (1981), p. 243.
132. R. S. Keen, L. R. Lowenstern and G. L. Schnable, *Proc. of 6th IEEE Relia. Phys. Symp.,* (1968), p. 58.
133. P. A. Totta and R. P. Sopher, *IBM J. Res. Develop.,* **13**, 226 (1969).
134. J. R. Black, *Proc. of ECS Symp. on Ohmic Contacts to Semiconductors,* (1968), p. 311; J. R. Black, *Proc. IEEE 57*, 1587 (1969).
135. R. J. Anstead and S. R. Flyd, *IEEE Trans. ED-16,* 381 (1969).
136. T. M. Reith and J. D. Schick, *Appl. Phys. Lett.* **25**, 524 (1974).
137. G. S Prokop and R. R. Joseph, *J. Appl. Phys.* **43**, 2595 (1972); G. S. Prokop, C. Y. Ting and R. R. Joseph, *Proc. of 8th IEEE Relia. Phys. Symp.,* (1970), p. 121.

138. J. Chern, W. G. Oldham and N. Cheung, *IEEE Trans. ED-32,* 1341 (1985).
139. S. Vaidya, A. K. Sinha and J. M. Andrews, *J. Electrochem. Soc.* **130**, 496 (1983); S. Vaidya and A. K. Sinha, *Proc. of 20th IEEE Relia. Phys. Symp.,* (1982), p. 50.
140. P. A. Gargini, C. Tseng and M. H. Woods, *Proc. of 20th IEEE Relia. Phys. Symp.,* (1982), p.66.
141. J. R. Black, *Proc. of 16th IEEE Relia. Phys. Symp.,* (1978), p. 233.
142. J. Chern, W. G. Oldham and N. Cheung, *IEEE Trans. ED-33,* 1256 (1986).
143. U. Köster, P. S. Ho and J. E. Lewis, *J. Appl. Phys.* **53**, 7436 (1982); ibid **53**, 7445 (1982).
144. C. Y. Ting and M. Wittmer, *Thin Solid Films,* **96**, 327 (1982).
145. S. Vaidya, R. J. Schutz and A. K. Sinha, *J. Appl. Phys.*, **55**, 3514 (1984).
146. S. D. Steenwyk and E. F. Kankowski, *Proc. of 24th IEEE Relia. Phys. Symp.,* (1986), p. 30.
147. J. M. Pimbley and D. M. Brown, *IEEE Trans. ED-33,* 1399 (1986).
148. T. Kwok, T. Nguyen, P. Ho, and S. Yip, *Proc. of 25th IEEE Relia. Phys. Symp.,* (1987), p. 130; T. Kwok, T. Nguyen, S. Yip, and P. S. Ho, *Proc. of 4th IEEE VLSI Multilevel Interconnection Conf.,* (1987), p. 252.

8

Diffusion Barriers in Semiconductor Contact Metallization

Hannu P. Kattelus and Marc-A. Nicolet

1. INTRODUCTION

1.1 General Considerations

One consequence of the constantly shrinking dimensions of integrated circuits is that the technological problems related to contacts and interconnections have moved from a peripheral position to one of central concern. The difficulties related to contacts and interconnections have to do basically with the fact that the electrically active regions of electronic devices are made of semiconductors, while the electrically most desirable interconnections require metallic conductors.

Metallurgically, metals and semiconductors tend to react, so that when they are brought into contact with each other, the resulting configuration is unstable. For example, pure aluminum has a small but finite solubility for silicon; in direct contact with silicon such a film will tend to dissolve some silicon under thermal treatment, which precipitates out of the aluminum again upon subsequent cooling. The resulting reaction between aluminum and silicon in the contact openings can be destructive unless means are applied to prevent the process. Highly doped polycrystalline silicon instead of aluminum circumvents this problem, but the minimum achievable bulk resistivities (~300 $\mu\Omega$cm) (1) are too high for very large scale integrated circuits. One remedy is to cover the poly-silicon with a transition metal silicide that has much better electrical conductivity than poly-silicon. The advantage of this solution is that it can easily be adapted to existing poly-silicon technologies, but the resultant sheet resistances are still several times larger than those achieved with good metallic conductors (2). The engineering approach to tackling this problem of an unstable metal-semiconductor contact is to separate functions, meaning that the task of establishing a good electrical interconnection should be separated from that of forming a stable electrical contact to the semiconductors.

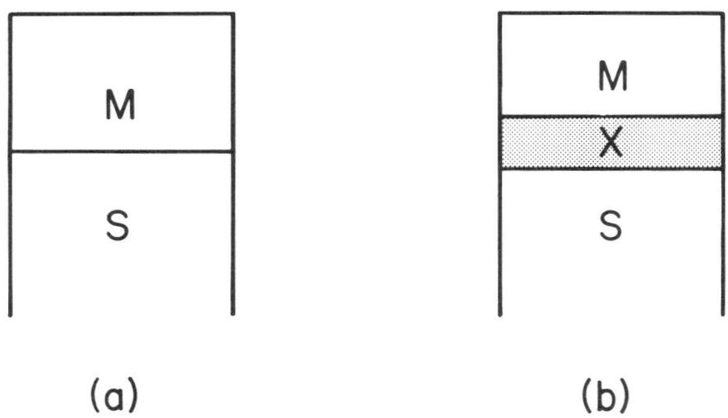

Fig. 1. Schematic diagram of a diffusion barrier, X, inserted between two materials, S and M, to minimize an existing process of degradation that would occur if M were in direct contact with S. In doing so, the interconnection problem is decoupled from that of forming a contact.

That line of thinking naturally leads to a solution that invokes a diffusion barrier, see Fig. 1. The concept is that a thin intermediate layer (X in Fig. 1) is inserted between the two components (M and S) in order to avoid a direct contact between M and S and to suppress any undesirable interaction between them (3-8). Ideally, the diffusion barrier X should be chemically inert so as not to react with M and S, and it should have insignificant diffusivities for atoms of M and S while being permeable to electron transport. In addition to these fundamental requirements, many additional conditions must be met to make a diffusion barrier practical; the barrier must have good adhesion under time-temperature stressing, it must be compatible with subsequent process requirements such as selective patterning, it must have acceptable thermal conductivity and mechanical properties, and it has to be chemically inert with respect to any ambients to which it may be exposed.

In practice, the problem posed is to find a diffusion barrier for a given selection of materials M and S. In a contact, S is a semiconductor (e.g. silicon, gallium arsenide) and M is typically an interconnection line. A contacting layer, C, whose function is to establish the desired electrical characteristics of the contact, is commonly added to this scheme when the diffusion barrier itself is inadequate for this purpose (see Figure 2). It may also serve to improve other characteristics, such as adhesion. To insert the contact layer C between the diffusion barrier X and the semiconductor S is a natural step in the engineering approach of separating functions. The diffusion barrier X promotes the stability of the contacting scheme, and the contacting layer C_S optimizes its electrical characteristics (see Figs. 6, 7). For reasons of symmetry, the general contacting scheme then also includes a contacting layer C_M between the diffusion barrier and the metal overlayer, as indicated in Figs. 2 and 12. An example of this general contact scheme was embodied in the beam-lead technology, one of the early successful contact metallizations (see Section 5.2).

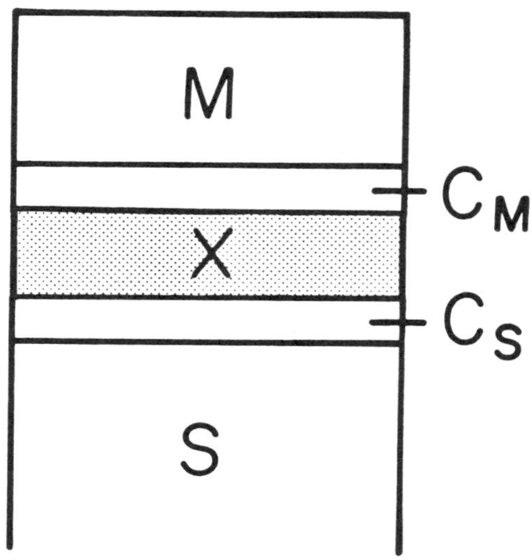

Fig. 2. A contacting layer, C, inserted between the diffusion barrier, X, and the semiconducting substrate, S, provides added flexibility in tailoring the I-V characteristic of the contact. The contacting layer C_S optimizes the electrical characteristics of the contacting scheme. In the general case, a contacting layer C_M is also inserted between the metal M and the diffusion barrier X (see Fig. 12).

1.2 Functions of a Diffusion Barrier

Two comments are appropriate at this point. The first one pertains to the direction of the electrical current flow in the barrier layer. Under the conditions prevailing in contacts, that current flows principally perpendicularly to the plane of the film, (Fig. 3(a)).

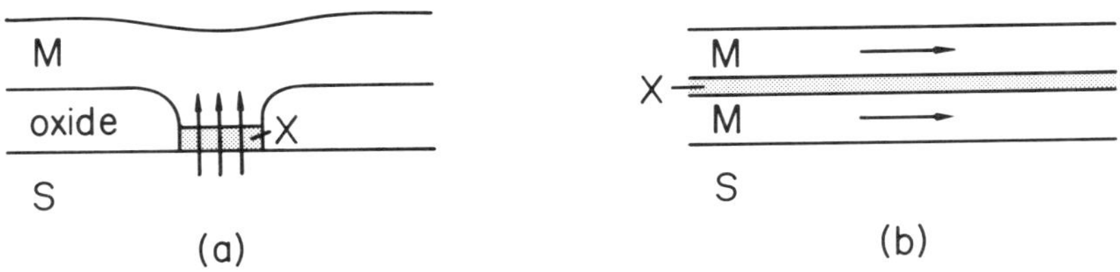

Fig. 3. The direction of the electrical current flow with respect to the barrier layer, X, in a semiconductor metallization. (a) The barrier in a contact is in a region where the current flows principally perpendicularly to the plane of the layer. (b) The barrier in an interconnection is in a region where the current flow predominantly parallel to the plane of the layer.

Another case can be visualized in which the electrical current predominantly flows parallel to the barrier. This situation prevails in interconnection lines. It has been found that the electromigration characteristics of a conductor can be vastly improved by inserting a thin layer of an intermetallic (9) or an interstitial compound (10) in it, Fig. 3(b). The effectiveness of this layer has been attributed to the fact that it is electrically conducting and maintains the continuity of the current path while hindering the propagation of voids across the whole film. This type of a barrier layer has been classified as a "driving force barrier" to distinguish it from conventional diffusion barriers (5). This review only deals with the case of electrical contacts, where the electrical current predominantly flows through the barrier layer, not parallel to it.

Secondly, one should appreciate that by adding a diffusion barrier, an existing instability between the two materials M and S is not eliminated. The barrier merely serves to slow down the degradation. Given enough time, M and S will ultimately interact with each other nevertheless, but the rate of the process is vastly reduced because the atomic diffusivities in the presence of the layer are much slower than without it. Quite evidently, it is the kinetic properties of atomic transport in the barrier that determine its value. The full meaning of this statement is discussed in the next section.

1.3 Electrical Requirements in Contacts

In a contact, a diffusion barrier must fulfill two rather different conditions depending on how it is used. Firstly, the barrier may separate two metallic conductors, such as the interconnection line above it and a contacting layer below it, (Fig. 2). In this case, the voltage drop occurs mainly as a result of the electrical current flowing uniformly through the cross section, A, of the contact. For a thickness, d, of the barrier the voltage drop, V, across it is equal to $\rho dI/A = \rho dj$, where ρ is the resistivity of the barrier material, I is the current and j the current density. Assuming that this voltage drop remains below an acceptable value, V_{max}, one notices that for a given current density, it is the product of the resistivity and the film thickness that is the relevant parameter in this application. With 25 mV for V_{max} (equals kT/q, where k is the Boltzmann constant, T is the absolute room temperature and q is the electronic charge) and $10^6 A/cm^2$ for the current density the ρd product of the barrier layer must be limited to 2.5×10^{-8} Ωcm^2. With typical barrier thicknesses of the order of 100nm, the bulk resistivity must remain below about 2500 $\mu\Omega cm$. For current densities that are less extreme than those assumed in this example (5), a bulk resistivity in the range of $10^{-1} - 10^{-2}$ Ωcm may be acceptable. In this application, relatively poor conductors are thus usable. This observation is important because it means that many materials other than simple metals qualify electrically as diffusion barriers. Among these are: amorphous metals, whose resistivities are typically above ~100 $\mu\Omega cm$ (11), and heavily doped polycrystalline semiconductors, such as silicon ($\rho \gtrsim 300$ $\mu\Omega cm$) or indium-tin oxide ($\rho \gtrsim 150$ $\mu\Omega cm$) (12).

Secondly, the diffusion barrier may be in contact with the semiconductor directly (Fig. 1b). For ohmic contacts, the contact conductivity, σ_c, is the quantity of primary concern. This quantity is defined as the differential conductance per unit area for biases near V = 0 of the I-V characteristic of the barrier material and the semiconductor, i.e. $\sigma_c = dj/dV$ ($V \simeq 0$). For an ohmic contact with the previously used values of $j = 10^6$ A/cm^2 and $V_{max} = 25$ mV, a contact resistivity $\rho_c = 1/\sigma_c = V/j = 2.5\times10^{-8}$ Ωcm^2 is required. In today's practice, values of 10^{-6} Ωcm^2 are routinely obtained for ohmic contacts. This is much larger than the theoretical estimate of 2.5×10^{-8} Ωcm^2 and shows that there is need and room for further improvements in the contact resistivities of ohmic contacts.

An optimally designed diffusion barrier will have voltage drops ρdj and $\rho_c j$ that are of comparable magnitudes. With today's contact resistivities of 10^{-6} Ωcm^2, the bulk resistivity of a diffusion barrier material begins to be a significant factor electrically only when the bulk resistivity exceeds $\rho \sim 10^{-2}$ Ωcm at 100nm thickness. This condition is easily met with many materials. In the application of Fig. 1b, where the barrier layer is in direct contact with the semiconductor, the I-V characteristic of the contact thus depends on the electrical properties of both the barrier material and the semiconductor. The electrical properties of the diffusion barrier are normally fixed when it is made of a metallic material. The resistivity of the resulting Schottky contact can then be minimized further only by manipulating the semiconductor.

The simultaneous conditions of atomic stability, bulk electrical resistivity, and interfacial contact resistivity imposed on a barrier material that is in direct contact with the semiconductor, as indicated in the scheme of Fig. 1b, may be difficult to meet. Additional flexibility is obtained by inserting a contacting layer, C_S, between the barrier and the semiconductor, as shown in Fig. 2. This configuration, in effect, is a means of tailoring the electrical I-V characteristic of a contact and its metallurgical stability in relatively independent ways. Future electrical improvements in ohmic contacts will come from reduced contact resistivities obtained by suitably matching the properties of the contacting layer C_S and the semiconductor. The pair has to be able to sustain time-temperature stresses imposed on the overall structure. It need not be compatible with the overlayer metal as well. This metallurgical stability of the contact assembly is assigned to the diffusion barrier X. The same concept is applicable to the design of rectifying Schottky contacts. In general, a contacting layer C_M can also be inserted between the diffusion barrier X and the metallic overlayer M. This contacting layer may serve to optimize specific features of the barrier/overlayer interface, such as adhesion.

1.4 Summary

Over extended periods of time at room temperature, or rapidly at elevated temperatures, the intrinsic instability of a metal-semiconductor interface leads to atomic interdiffusion. After a characteristic time-temperature cycle, the semiconductor device ceases to operate because of this interaction. An interposed diffusion barrier layer can retard this process and thus increase the lifetime of the device. Adding a diffusion barrier, however, also increases the complexity of the device and its manufacturing cost. It is an economic issue to decide when it is worthwhile to use a diffusion barrier to increase the device lifetime or to raise the operating temperature. The continuously shrinking dimensions of electronic devices steadily reduce the diffusion lengths that can be tolerated. This fact raises the issue of the thermal stability of the metallization system to a concern of primary importance.

Clearly the number of requirements that have to be met to solve a practical problem is large. In addition to the structural and electrical stability, the issues of adhesion, thermal match, corrosion, and reproducibility are all of concern. These questions will not be discussed here. They often depend on the specifics of a particular situation.

Only issues of general character are addressed here with the emphasis on the particular feature of low atomic diffusivity. The most promising material groups in this respect are described in some detail. The bulk resistivity requirements for diffusion barriers in contacts are weak and fairly poor conductors are still usable. In general, a diffusion barrier is flanked by two contacting layers whose presence allows the optimizing of specific features. The contact resistivity in ohmic contacts to a semiconductor is an example, and the adhesion of a metallic overlayer to the diffusion barrier is another.

2. IMPORTANCE OF THE CHOICE OF THE MATERIAL AND OF THE DEPOSITION PROCESS

2.1 Basic Considerations

To specify a diffusion barrier requires two choices: that of the material and that of the deposition process. The selection of the material is largely determined by the conditions enumerated in the preceding section regarding inertness, low diffusivities, and appropriate electrical, thermal, chemical and mechanical properties. This aspect of the subject of diffusion barriers has attracted most of the interest in the past. The other aspect, namely that of the choice of the deposition process, has received inadequate attention. It is as important as the choice of the material, however, as is demonstrated in the following example.

Consider a diffusion barrier material with a bulk diffusivity for some impurity atom. If that bulk diffusivity is known, it is likely to have been determined from experiments performed at high temperatures, where such experiments are generally carried out. With a known activation energy, typically several electron volts, the diffusivity can be extrapolated to the lower temperatures where thin film diffusion barriers are normally operating. Say that for such a temperature, a period of ten years is required to permeate a layer of 1 mm thickness. The time required for an atom to diffuse through a distance x changes as the square of x. With the same diffusion process, a film of 100 nm is thus permeated in 3 s. These numbers show that because of the very small dimensions of the thin film diffusion barriers, diffusion processes that appear to be slow on a macroscopic scale are quite fast on a microscopic scale. To make matters worse, estimates of the time required are often wrong by orders of magnitude. The reason is that at low temperatures, diffusion is dominated by other processes than those observed at high temperatures. [Extended defects, such as grain boundaries or dislocations, can provide diffusion paths that are much more rapid than bulk lattice diffusion (13) (see Fig. 4).]

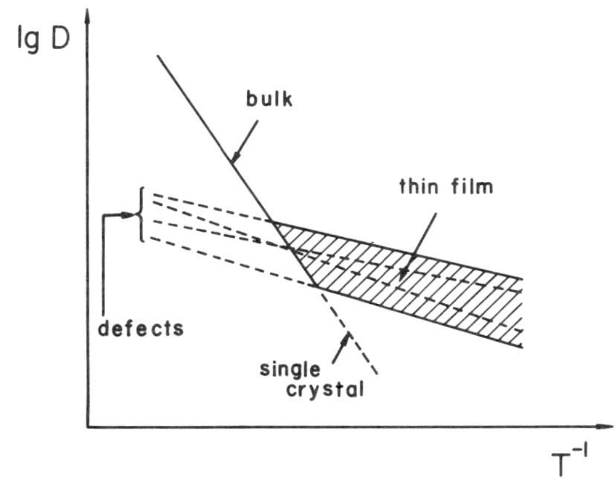

Fig. 4. Schematic representation of the diffusivity in a solid as a function of reciprocal temperature. At elevated temperatures, bulk diffusion processes dominate, which extrapolate to low temperatures only for single crystalline materials. At temperatures below about two-thirds of the melting point, extended defects dominate the diffusion process. The amount and nature of defects in thin films depends on the deposition process.

Two conclusions follow directly from this observation, namely: 1) the diffusivities applicable to thin films at low temperatures are much higher than would be expected from extrapolated high temperature bulk diffusivities, and 2) the diffusivity actually applicable to a thin film depends on the nature and the density of its extended defects. This defect structure is not an intrinsic property of the material, but primarily determined by the process of deposition. It therefore follows that the kinetic properties of a thin-film diffusion barrier are as much a function of the deposition process as they are of the material. It is therefore essential in the whole field of thin-film diffusion barriers to pay as much attention to the deposition process as is given to the choice of the material. A description of the diffusion barrier that considers only the choice of the materials and disregards the characterization of the process of deposition is nearly useless. Consistent with this, we have made an effort in the present review to always describe the method of deposition whenever a practical case is discussed.

2.2 Single Crystalline Versus Non-Single Crystalline Films

How pertinent and important these conclusions are is demonstrated by the following example (14): Two thin films of gold and lead in contact with each other react quite readily under thermal annealing at temperatures of a few hundred degrees. When a polycrystalline thin film of silver is inserted between the two layers, gold is observed to diffuse through the silver layer and to form compounds with the lead. That this process is possible because of the polycrystalline structure of the silver layer is proven convincingly by a parallel experiment where the silver film is single crystalline. Such a film is essentially impermeable to gold in comparison to the polycrystalline film. Observe that the reason why the two films have different kinetic properties is that they have different microscopic structure (polycrystalline and single crystalline), and observe further that this difference is a result of different processes that were used to form the silver film. This is self-evident, but needs to be stressed because of the fundamental importance that the deposition process has in determining the defect structure of the film. In the above example, the silver layer was electron beam evaporated epitaxially on a single crystal gold film grown on a cleaved NaCl single crystal substrate, while the polycrystalline silver film was obtained by vacuum evaporation onto collodion-coated NaCl. A number of other examples have been reported in the literature which all concur with a general conclusion, namely that grain boundaries and extended defects constitute rapid paths for atomic diffusion.

2.3 Summary

One deduces that the single crystalline structure is the preferred one for any diffusion barrier. The practical difficulties of realizing single crystalline thin films routinely have precluded the implementation of this ideal structure in practice. It is conceivable, however, that the emphasized research in growing single crystalline films on both crystalline and amorphous substrates may, in the future, open the door to this ultimate goal of structurally perfect diffusion barriers. For the present, the single crystalline diffusion barrier remains a laboratory object.

The remaining available choices for the microscopic structure of thin films are only two: polycrystalline and amorphous. We shall address both cases in the subsequent sections.

3. POLYCRYSTALLINE FILMS CONSISTING OF ONE SINGLE LAYER

It is convenient to subdivide the diffusion barriers also according to the number of sublayers they contain. The simplest case is a film consisting of one single homogeneous layer, which we shall consider first. A barrier film that reacts with the adjacent media in a laterally uniform way will be treated later, since the resulting configuration may be considered in all as a multilayered structure (see section 5).

3.1 Elemental Metallic Films

The most obvious idea for the choice of an electrically conducting diffusion barrier is to pick a pure metal. Pure metals do not exist in amorphous form at room temperature or above, so that these materials will always be polycrystalline. As a consequence, pure elemental metallic films must be ruled out as is suggested by the following example:

Sequentially deposited films of titanium and gold react upon thermal annealing by forming intermetallic compounds, like Au_2Ti, Au_4Ti, etc. (15). Titanium is frequently employed in metallization systems to assure adhesion with an underlying layer and gold is an excellent conductor for interconnection. To assure a stable combination of these two elements, a diffusion barrier is desirable. It is evident that a metal that forms solid solutions with either gold or titanium is a poor choice. Such a layer is soluble in one of the adjacent layers and is consequently not inert. A metal with very little solid solubility for both titanium and gold would seem much more promising. The solid solubility of molybdenum in bulk gold is less than 1 at. % at temperatures below 800°C, and that of gold in bulk molybdenum is exceedingly small. The two metals do not form an intermediate phase (16). Molybdenum thus satisfies the condition of inertness with gold very well. (It does not satisfy it for titanium with which molybdenum forms a solid solution.) Yet, one finds that in thin molybdenum-gold couples, molybdenum contains approximately 2-3 at. % gold (17). The explanation is that gold diffuses preferentially along the grain boundaries of molybdenum, because molybdenum, having the higher melting point (2620°C), typically forms films with smaller grains than gold, whose melting point is only 1063°C. A concentration of the order of a few at. % is typical of grain boundary decoration. (For a hypothetical polycrystalline material with cubic grains of 5 nm a side and a monolayer of foreign atoms in all grain boundaries, the apparent volume concentration of foreign atoms is of the order of 10 at. %). A film of molybdenum thus does not constitute a diffusion barrier for gold. This is also what is in fact observed (18). One concludes that the condition of low bulk solubility is not sufficient for a good barrier performance of polycrystalline films of pure metals.

There is a well-known rule according to which, other things being equal, atomic diffusivities scale with the melting point of the material (19). Conducting solids with very high melting points should therefore be particularly promising candidates for applications as diffusion barriers in thin film contacts. At first sight, this conclusion would seem to be at a variance with experimental facts in the case of elemental polycrystalline films: molybdenum has one of the highest melting points of all elements. Yet, gold diffuses fairly rapidly through a pure polycrystalline films of molybdenum. Another highly refractory metal, tungsten, has generally been considered a good diffusion barrier both in evaporated (20) and sputtered (21) form for aluminum metallization on silicon. However, experiments showing poor properties for tungsten in preventing aluminum diffusion into the silicon/silicide contact region also exist (22). Rf sputtered tungsten and molybdenum layers have also shown comparably unsatisfactory properties in GaP metallization schemes involving a Au/Zn contacting layer and a gold overlayer giving rise to zinc or gallium diffusion into the top gold (23). One possible interpretation of these differences will be discussed in the next section in connection with stuffed barriers. On the other hand, a valid comparison between molybdenum and tungsten thin films would require that the diffusivity of gold (or aluminum) be investigated in parallel experiments on molybdenum and tungsten films that are prepared under similar conditions. The problem is experimentally difficult to tackle, because refractory metals tend to be quite reactive as well, so that convincing comparisons would demand ultra-high vacuum processing. Such experiments would be very illuminating, but they have not yet been performed. It is therefore impossible at this time to state whether the melting point argument is important in this case or whether the presence of an impurity is always associated with a good barrier performance of elemental polycrystalline metal films.

The logic would imply that the statement of rapid diffusion paths should apply generally to compound polycrystalline layers as well. It is the presence of extended defects that is responsible for the rapid interdiffusion, and these defects are present in both elemental and compound polycrystalline films. The nature of the bulk, however, will affect the numerical value of the grain boundary diffusivity and its activation energy. One is thus led to select those materials for which diffusivities along extended defects are lowest.

3.2 Effect of Impurities – Stuffed Barrier

There is a body of experimental evidence that demonstrates convincingly that the presence of impurities can alter the kinetic properties of polycrystalline films dramatically. The impurities most frequently invoked are oxygen, nitrogen and carbon. But the task of demonstrating persuasively that a particular chemical species is decorating grain boundaries lies at the forefront of present instrumental capabilities, particularly when the impurity is a light element. The notion that impurities can restrict the diffusion of atoms along grain boundaries and extended defects is an appealing one, however, and it

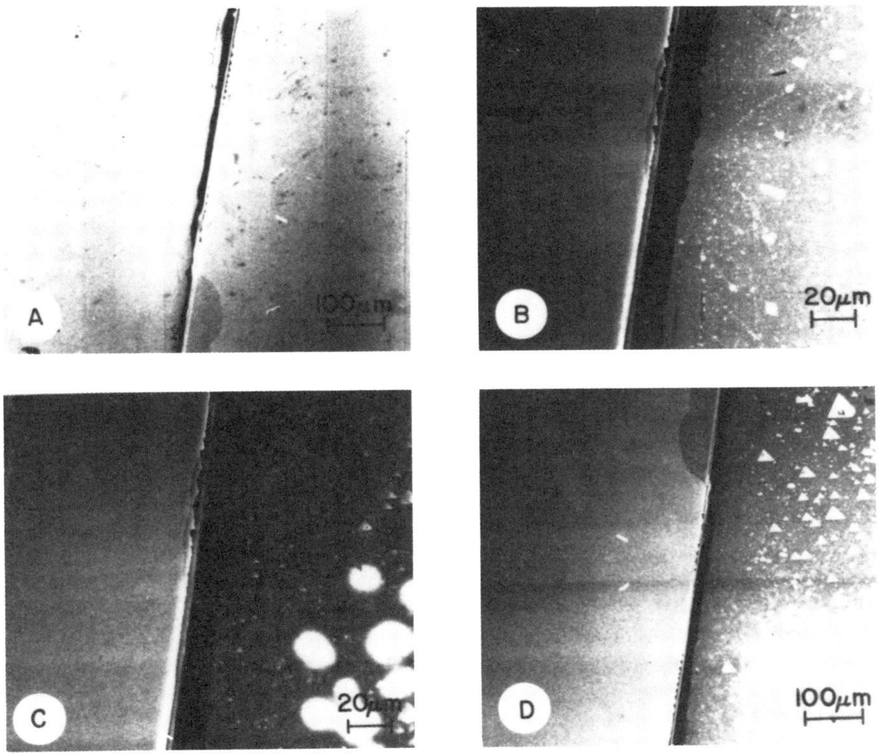

Fig. 5. Secondary electron micrographs taken at various temperatures of type 1 (left) and type 2 (right) samples mounted side-by-side on the heating fixture in the microscope (A) 23°C; (B) 290°C; (C) 300°C; (D) 310°C. The oblique line from top to bottom on each photograph corresponds to the butted edges of the two specimens. The type 1 barrier layer contains more impurities than the type 2 barrier layer, and is more effective as a diffusion barrier for gold and silicon than the purer one (24).

has been readily accepted as the explanation for the beneficial effects of impurities in diffusion barriers. It should be borne in mind, however, that in most instances where this model is invoked, it is still conjectural. The term *stuffed diffusion barrier* has been coined to refer to that situation.

A good example of what is meant by a stuffed diffusion barrier is shown in Fig. 5 (24). The figure shows SEM micrographs of two types of samples. In both cases, a ⟨111⟩-oriented single crystalline substrate of silicon is covered with a 150 nm thick film of titanium-tungsten alloy and a subsequently deposited 100 nm thick layer of gold. The films were deposited by rf diode sputtering in an argon ambient from preformed targets of titanium-tungsten and gold without breaking the vacuum between depositions. In one case, shown on the left of each figure, the sputtering chamber was evacuated to a partial pressure of mass 18 (presumably H_2O) of about 3.8×10^{-7} torr; in the other case (shown on the right of each figure) that partial pressure was 1.5×10^{-8} torr. The samples were subsequently ramp-annealed on the stage of the SEM from room temperature to 400°C over a period of one hour with continuous observation of surface morphology. Intermixing of gold and silicon is observed in the sample deposited with the lower partial pressure of mass 18, but not in the other case. The titanium-tungsten film thus works as a diffusion barrier only when it is deposited in the poorer of the two vacua. Auger depth profiles of gold, titanium, tungsten, silicon and oxygen of the two as-deposited samples are identical except for oxygen, which has a uniform oxygen content of about 6% for the "dirty" deposition as opposed to about 3% for the "clean" one. These measurements alone do not prove that oxygen decorates the grain boundaries. In fact, they do not even prove that oxygen is responsible for the beneficial effect. The concentration of oxygen, however, has the order of magnitude expected for a stuffed barrier.

As mentioned in the previous section, thin polycrystalline films of tungsten, either evaporated (20), sputtered (21), or chemical-vapor-deposited (25), can be good diffusion barriers against aluminum. One interpretation is that the success of this particular element may be based on the difficulty of producing really clean thin films of this material, meaning that the barrier is stuffed. Whether this point of view is correct or not is an open question.

The ability of impurities in the concentration level of a few atomic percent to impair beneficial attributes to polycrystalline diffusion barriers has implications. One of them is the violation of the dogma that says that in semiconductor device technology, cleaner is better. There is nothing wrong in the use of impurities to accomplish desirable results. The formation of p-n junctions is a well-know example. The treacherous aspect of such applications resides in ignoring the existence of the situation. Clearly, for a successful application of a stuffed barrier, it is necessary to carefully control the level of a beneficial impurity included in the film. This degree of control is unlikely to be present when the role of that impurity has been overlooked. The likely result is an erratic behavior of the contacting scheme whose performance fluctuates without apparent reason. This state of affairs is typical for contacting schemes that have been used on an industrial scale. Quite likely, stuffed barriers are used in practice much more often than is recognized.

Polycrystalline thin films consisting of metallic elements *are* usable as diffusion barriers in contact schemes when they are stuffed. This statement is a broad generalization of a few experimental studies, incomplete by themselves, and of general inferences on the importance of diffusion along extended defects at low temperatures. Only future detailed experimental work based on films deposited in ultra-high vacuum and the addition of controlled amounts of impurities will be able to establish how valid or not that rule in fact is.

3.3 Compound Film – Inert Barrier

The sole alternative to a polycrystalline film that is elemental is one that is a compound. In passing from elements to compounds, the number of possible materials increases enormously. The new possibility of a two- or a multi-phase film arises. It is not possible at this time to provide a systematic treatment of polycrystalline compound diffusion barriers. The number of studies available are too few. We shall therefore merely review some well-studied cases.

3.3.1 From Elemental to Compound Films. To improve the stability of the commonly used Ti/Pt/Au metallization system, tungsten was introduced to replace titanium, because tungsten has a very low solubility for gold and it does not form compounds with it. However, to promote adhesion and to improve the corrosion resistance, titanium was added into tungsten slightly in excess of the solid solubility requirements (26). A diffusion barrier was thereby created that has been used later in the semiconductor field in applications very different from those that originally gave rise to its birth.

Before proceeding, we observe that the concept of the stuffed barrier can, of course, be extended to include compound polycrystalline films. Extended defects can be decorated with impurities there, too. The distinction between a stuffed and a pure compound barrier, however, may be difficult to make. The impurity in the defects may in fact be forming a compound. The situation is exemplified by the two contacting schemes show in Figs. 6 and 7.

In both instances, the problem at hand is to stabilize an aluminum overlayer against interdiffusion with a silicon contact, which is formed

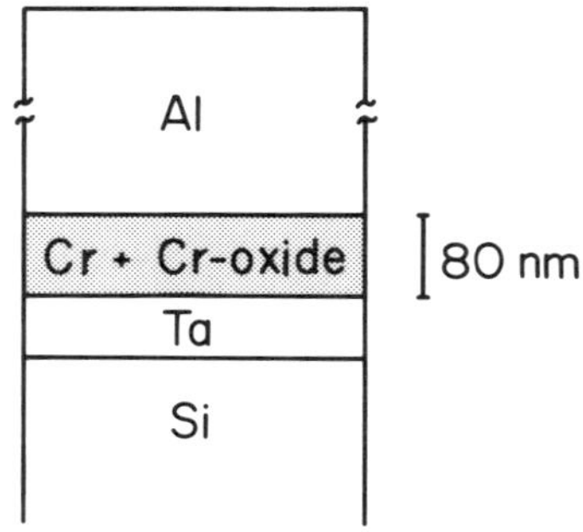

Fig. 6. An elemental metallic diffusion barrier is made effective by adding a controlled amount of impurities. When chromium is evaporated in the presence of oxidants, the resulting mixture of elemental chromium and its oxides prevents interdiffusion of aluminum with the silicon/tantalum contact region at 450°C for one hour (27).

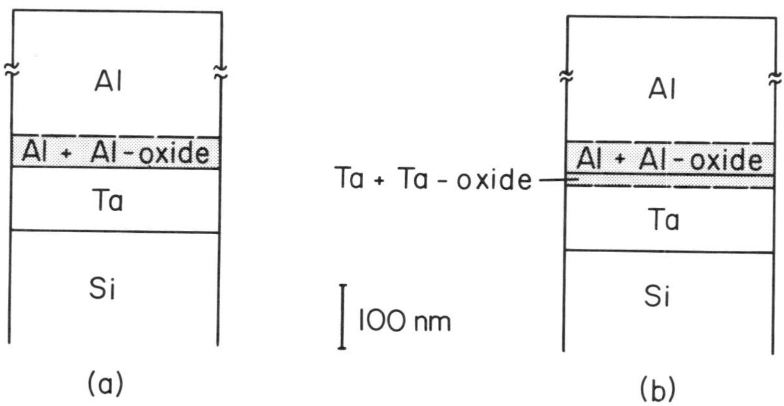

Fig. 7. The presence of aluminum oxides and tantalum oxides make aluminum and tantalum layers effective as diffusion barriers. In comparison to Fig. 6, the number of metals needed in these schemes is reduced by stuffing a part of the aluminum interconnection (a) and the tantalum contacting layer (b) (28).

using a refractory metal, like tantalum (and, in some cases, with an additional contacting layer of PtSi between the silicon and tantalum, not shown in the figures). One solution is to evaporate a chromium film of ~ 80 nm thick while taking care that a H_2O partial pressure of 10^{-5} torr is maintained during the deposition, and keeping the substrate below 160°C (27). As a result, the chromium layer contains significant amounts of oxygen, the location and chemical composition of which remains uncertain (Fig. 6). After subsequent evaporations of aluminum in a higher vacuum, this contact is sintered at 400°C for one hour, and it has been successfully tested in time-temperature cycles of two hours at 450°C.

The idea of creating a stuffed conducting layer (or a two-phased layer containing metal and oxides) can, of course, be implemented with any metal-impurity combination that is sufficiently reactive. Recognizing this, one arrives at the solution implemented in Fig. 7, where aluminum or tantalum themselves are used to form the stuffed barrier (28). Here the diffusion barrier is obtained by evaporating 50 nm of aluminum in an impure vacuum of 2×10^{-5} torr, followed by a deposition of the bulk of the aluminum in high vacuum (see Fig. 7(a)). Similarly, the tantalum evaporation may be brought to an end at increased oxygen partial pressure to doubly insure the tightness of the barrier, as shown in Fig. 7(b). The amount of aluminum-oxide in the stuffed aluminum layer is less than 10% and the amount of tantalum-oxide in the stuffed tantalum layer is less than 40%. The stability of these structures has been confirmed by electrical measurements on Schottky diodes, showing that the barrier height is unchanged in heat treatments up to 500°C for one hour. The ideality factor, however, increases from 1.0 to 1.1 during this treatment.

In another case (18), nitrogen was purposely introduced into molybdenum at levels definitely beyond those needed to merely decorate grain boundaries. Radio-frequency sputtering was used to produce molybdenum diffusion barriers and gold overlayers onto silicon substrates in a turbomolecular-pumped vacuum system. Two types of barriers were used: one was molybdenum sputtered in pure argon, the other was similarly prepared, but contained a 10-20 nm thick nitrided molybdenum layer within the molybdenum. The nitrided layer was sputtered in a gas mixture of 20% nitrogen and 80% argon for a period of time of ~1 minute, using the same elemental molybdenum target. The films were annealed at 450°C for 30 minutes in a nitrogen ambient and characterized by Auger depth profiling. Severe indiffusion of gold into silicon through the molybdenum layer can be seen in the former case whereas no interdiffusion is observed in the case of the nitrided molybdenum barrier. Transmission electron microscopic studies of the pure and nitrided molybdenum layers revealed that the film sputtered in argon has b.c.c. structure (a_o = 0.374 nm) with a grain size of 10-20 nm, while that containing nitrogen has f.c.c. structure (a_o = 0.406 nm) and grain size of 5-10 nm. Elemental molybdenum has a b.c.c. structure with a lattice constant of 0.3147 nm (29), and Mo_2N is f.c.c. with a = 0.4165 nm (30). Although no further characterization of the structural composition was made, it was concluded that a partial explanation of why nitrided molybdenum is a better diffusion barrier than the parent metal between silicon and gold may involve a phase change to a denser atomic lattice.

The limit in this direction is to provide sufficient amounts of two or more elements for the deposition, so as to end up with a uniform polycrystalline compound film. This case is considered next.

3.3.2 Single Phase Compounds. The best examples of compound diffusion barriers are provided by films of titanium nitride, TiN, and other related materials (30). These interstitial alloys are typically metallic, highly refractory, and chemically inert (31). These attributes are those that would appear to be attractive for a good thin film diffusion barrier material. That idea was recorded first by Nelson (32). The very high melting points (e.g. ~2950°C for titanium nitride) should reduce diffusivities at the testing temperature of electronic devices according to arguments of melting point scaling. The inertness should assure the chemical stability with the adjacent materials, and the electrical conductivities are well within the required range of less than ~0.01 Ω cm. The group of interstitial compounds is fairly large and incorporates nitrides, carbides, borides, and hydrides of transition metals. This diversity offers room for choice and optimization. Of all these compounds, titanium nitride has been investigated most diligently. This is the case that will therefore be reviewed in some detail here. The results published on other interstitial alloys are rather similar. Table 1 lists papers investigating properties of thin single-layer interstitial compound films when used as diffusion barriers in metallization systems of semiconductors. Cases in which a contacting layer is interposed between the substrate and the diffusion barrier are considered in section 5.2. Because some thin films of these materials have been seen to react very little with many electronic materials and to exhibit small atomic diffusivities for these materials as well, these films have been described on occasion as *inert barriers*.

We note in passing that interstitial compounds have been commonly used industrially as coatings (on the order of micrometers thick and more), where their great hardness (31) extends the lifetime of drilling tools (33). In particular, titanium nitride also finds applications in the coating of watch cases, pens, and alike, where its chemical inertness and gold-like appearance are desirable (34),(77). Titanium nitride is also used for the shading of plate glass windows of modern buildings (35). Much information has been collected on the properties of some interstitial compounds as a result of these applications. This information can be found by consulting the references just cited. In Table 1, we only list those thin-film studies that focus on the properties and applications of these materials in metallization systems of semiconductors.

Thin films of titanium nitride are typically polycrystalline with a fine-grained (~10-50 nm) f.c.c. structure (36),(37). Titanium nitride films can be deposited by a number of methods; for electronic applications, a frequently applied technique is reactive sputtering in diode, triode or magnetron mode (38-40). The electrical resistivity of the films thus obtained may be as low as 30 $\mu\Omega$cm (36),(41) and can be further decreased by thermal annealing during (37) or after (42) deposition. Care is required to maintain low partial pressures of oxidants during deposition to achieve low bulk resistivities; oxygen inclusions at a level of several atomic percent rapidly increases the resistivity to several hundred micro-ohm-cm (43). The cubic titanium nitride can exist over a large range of stoichiometries, TiN_x, going from very nitrogen deficient to slightly nitrogen rich (TiN_x, $0.6<x<1.2$). (At low nitrogen concentrations, an intermediate tetragonal Ti_2N phase can be seen (37),(39).) The actual composition obtained and therefore the properties of the layer thus depend on the deposition conditions even more so than is the case of compounds with fixed stoichiometry. Systematic studies have shown that an excess of nitrogen in films deposited by dc magnetron sputtering in an argon/nitrogen ambient is beneficial to the barrier performance between the silicon substrate and aluminum (44). The same study concludes that the presence of oxygen is detrimental to the barrier performance. The opposite has been claimed for films deposited by rf diode sputtering in an argon/nitrogen gas mixture or by e-beam evaporation of titanium in a nitrogen ambient (45). How oxygen alters the electrical properties and the barrier performance of a film, and whether the results depend on the deposition process, is not understood at this time. This subject deserves further investigations. The same thing can be said about other impurities (such as carbon, boron, chlorine, etc.).

It is commonly observed that reactively sputtered films with low oxygen content are under large compressive stress. The stress increases with applied negative bias voltage to the substrate during deposition. This bias is normally applied to reduce the oxygen content and to improve the crystalline structure by enhanced surface diffusion (46),(47). Stresses as high as 1-10 GPa (= $10^{10}-10^{11}$ dynes/cm^2) have been observed (34),(48),(49). Few quantitative measurements of stress as a function of deposition parameters and techniques are available. For films reactively sputtered in an rf magnetron system, it is known that compressive stress is reduced when the substrate bias is eliminated, but at a cost of increasing oxygen contamination. The

Table 1. List of Papers Investigating Properties of Thin Films of Interstitial Compounds as Diffusion Barriers in Complete Metallization Schemes of Semiconductors (Structure S/X/M of Figure 1).

Diffusion barrier (X)	Method of Barrier Deposition*	Barrier thickness in nm	Semi-conductor substrate (S)	Metal overlayer (M)	Metal deposition method	Device/Test	Annealing time in min	Lowest Temperature of observed instability in °C (Method), Comments	Ref.
HfN	rf magnetron sputtering in Ar and N_2,	150	Si	Pd	evap.		15	500 (SEM), 700 (RBS)	1
HfN	$p_o < 2 \times 10^{-6}$ torr,	150	Si	Cu	evap.		15	500 (SEM), 750 (RBS)	
HfN	$r = 0.67$ nm/s, $V_b = -50$ V	150	Si	Au	evap.		15	500 (SEM), 500 (RBS)	
LaB_6	dc magnetron sputtering in Ar from LaB_6 target, $p_o < 2.7 \times 10^{-6}$ torr, $r = 0.12$ nm/s, $T_a = 1000$°C (N_2)	370	Si	Al	evap.		30	600 (RBS)	2
$Ti_{0.37}C$	rf magnetron sputtering in Ar and CH_4,	35	Si	Al	evap.		30	550 (AES)	4,5
TiC	p_o = low 10^{-7} torr,	20	Si	Al	evap.		30	550 (AES)	
$Ti_{1.5}C$	$r = 0.58 - 1.1$ nm/s,	73	Si	Al	evap.		30	500 (AES)	
$Ti_{3.1}C$	$V_b = 30$ V (peak-to-peak) (3)	100	Si	Al	evap.		30	500 (AES)	
TiN	rf magnetron sputtering in N_2, $p_o = 10^{-6}$ torr, $r = 0.17 - 1.0$ nm/s (6)	200	Si	Ag	evap.		40	600 (RBS)	7

Table 1. List of Papers Investigating Properties of Thin Films of Interstitial Compounds as Diffusion Barriers in Complete Metallization Schemes of Semiconductors (Structure S/X/M of Figure 1).

Diffusion barrier (X)	Method of Barrier Deposition*	Barrier thickness in nm	Semi-conductor substrate (S)	Metal overlayer (M)	Metal deposition method	Device/Test	Annealing time in min	Lowest Temperature of observed instability in °C (Method), Comments	Ref.
TiN	rf magnetron sputtering in N_2	70	Si	Ag	evap.		80	>600 (RBS)	8
TiN	or in Ar from TiN target ,	100	Si	Au	evap.		60	500 (SEM), 450 (RBS)	
TiN	$p_o = 10^{-6}$ torr ,	~100	Si	Al	evap.		40	550 (SEM), 500 (RBS)	
TiN	r = 1.0 nm/s	100	Si	Cu	evap.		30	700 (RBS)	
TiN	(6)	~100	Si	Pd	evap.		80	>600 (RBS)	
TiN		100	Si	/Ti/Pd/Ag		solar cell/ I-V curve	20	575 (electrical)	9
TiN	rf magnetron sputtering in N_2, $p_o = 10^{-6}$ torr, r = 0.17 nm/s (6)	170	Si	/Ti/Ag	evap.	solar cell/ I-V curve	10	>600 (electrical)	10
TiN	rf magnetron sputtering in Ar and N_2	40	GaAs	Au(or Pt)	evap.		15	600 (SEM), 600 (RBS)	11
TiN	$p_o < 2 \times 10^{-6}$ torr, r = 0.17 - 0.33 nm/s, V_b = -50 - 0 V		Si	Pd(or Au or Cu)	evap.			(see Ref. 1)	

Table 1. List of Papers Investigating Properties of Thin Films of Interstitial Compounds as Diffusion Barriers in Complete Metallization Schemes of Semiconductors (Structure S/X/M of Figure 1).

Diffusion barrier (X)	Method of Barrier Deposition*	Barrier thickness in nm	Semi-conductor substrate (S)	Metal overlayer (M)	Metal deposition method	Device/Test	Annealing time in min	Lowest Temperature of observed instability in °C (Method), Comments	Ref.
TiN	evaporation in N_2, $p_o = 1.5 \times 10^{-7}$ torr, $r = 0.5 - 1.5$ nm/s, $T_a = 900°C$ (Ar or N_2)	100	Si	Al	evap.		30	600 (RBS)	12
TiN	evaporation in N_2, $p_o = 1.5 \times 10^{-7}$ torr, $r = 0.5 - 1.5$ nm/s, $T_a = 900°C$ (Ar or N_2)		Si	Al	evap.	shallow junction leakage, contact resistivity to n-Si	30	400 to 550: increasing leakage, $\rho_c > 10^{-4}\ \Omega cm^2$	13
TiN	magnetron sputtering in Ar and N_2, (T_a = 800°C?)	100	Si	/Ti/Al ↓ $TiAl_3$	sputt.		120	>600(RBS,optical)	14
TiN	rf magnetron sputtering in Ar and N_2,	50, 100	Si	Pd	evap.		15	750 (SEM), 750 (RBS)	1
TiN	$p_o < 2 \times 10^{-6}$ torr,	50, 100	Si	Cu	evap.		15	800 (SEM), 800 (RBS)	
TiN	$r = 0.05 - 0.25$ nm/s, $V_b = -50 - 0$ V	50, 100	Si	Au	evap.		15	>800(SEM),>800 (RBS) (for the best barrier; i.e. 50 nm thick, bias sputtered TiN)	

Table 1. List of Papers Investigating Properties of Thin Films of Interstitial Compounds as Diffusion Barriers in Complete Metallization Schemes of Semiconductors (Structure S/X/M of Figure 1).

Diffusion barrier (X)	Method of Barrier Deposition*	Barrier thickness in nm	Semi-conductor substrate (S)	Metal overlayer (M)	Metal deposition method	Device/Test	Annealing time in min	Lowest Temperature of observed instability in °C (Method), Comments	Ref.
TiN	dc magnetron sputtering in N_2, $p_0 = 2 \times 10^{-7}$ torr, $r \sim 0.2$ nm/s (15)	50	InP	Au	evap.**	In diffusion to Au	30	600 (SEM/EDX), activation energy for In diffusion in TiN:2.5 eV	16
TiN	dc magnetron sputtering in Ar and N_2, $r = 0.1 - 0.2$ nm/s	100	Si	Al	evap.		30	>600 (AES,optical)	17
TiN	evaporation in N_2,	100	Si	poly-Si: As	CVD	As and B diffusion to Si		$T \geq 900$°C:TiN is a better diffusion barrier for B than for As (SIMS)	18
TiN	$p_0 < 2 \times 10^{-7}$ torr, $r = 0.5$ nm/s, $T_a = 900$°C (Ar) (12)	100	Si	poly-Si: B	LPCVD				

Table 1. List of Papers Investigating Properties of Thin Films of Interstitial Compounds as Diffusion Barriers in Complete Metallization Schemes of Semiconductors (Structure S/X/M of Figure 1).

Diffusion barrier (X)	Method of Barrier Deposition*	Barrier thickness in nm	Semi-conductor substrate (S)	Metal overlayer (M)	Metal deposition method	Device/Test	Annealing time in min	Lowest Temperature of observed instability in °C (Method), Comments	Ref.
TiN	evaporation in N_2, $p_o = 2 \times 10^{-8}$ torr, $r = 0.1$ nm/s, $T_s = 200°C$, $T_a = 800°C$ (vacuum)	50	Si	Al	evap.	shallow junction leakage	30	540(RBS,optical, electrical)	19
TiN	rf diode sputtering in Ar and N_2, $p_o = 3\text{-}4 \times 10^{-7}$ torr, $r = 0.07$ (dynamic) -0.24(static) nm/s, $T_a = 800°C$ (vacuum)	37, 41	Si	Al	evap.	shallow junction leakage	30	520-540 (RBS, optical, electrical)	19
W_2N	rf magnetron sputtering in Ar and N_2,	100	Si	Ag	sputt.**		30	800 (RBS, SEM)	21
	$p_o < 3 \times 10^{-6}$ torr,	100	Si	Al	sputt.**		30	600 (RBS), 550 (SEM)	
	$r = 0.38 - 0.83$nm/s (20)	100	Si	Au	sputt.**		30	600 (RBS, SEM)	
ZrB_2	rf diode sputtering in Ar from ZrB_2 target, $p_o \sim 10^{-6}$ torr, $r = 0.17$ -0.25 nm/s, $V_b \sim -80$ V, $T_a > 1000°C$ (rapid) (22)	~200	Si	Al	evap.		120	625 (RBS,AES)	23

Table 1. List of Papers Investigating Properties of Thin Films of Interstitial Compounds as Diffusion Barriers in Complete Metallization Schemes of Semiconductors Structure S/X/M of Figure 1).

Diffusion barrier (X)	Method of Barrier Deposition*	Barrier thickness in nm	Semi-conductor substrate (S)	Metal overlayer (M)	Metal deposition method	Device/Test	Annealing time in min	Lowest Temperature of observed instability in °C (Method), Comments	Ref.
ZrN	rf diode sputtering in Ar and N_2 (or Kr and N_2), $p_o = 2.7 \times 10^{-7}$ torr, T_s = 100-300°C, T_a = 900°C (He)	80-150	Si	Al	evap.		30	550 (RBS)	24
ZrN	rf diode sputtering in Ar and N_2, $p_o < 3 \times 10^{-7}$ torr, r = 0.23 nm/s, $V_b \sim -100$ V	120	Si	Al	evap.		30	>600 (RBS)	25

*Deposition using an elemental metal source, unless noted.

p_o = base pressure before deposition.

r = deposition rate.

T_s = substrate temperature during deposition.

T_a = preannealing temperature before metal overlayer deposition.

V_b = substrate bias voltage.

**All depositions in the same vacuum.

TABLE 1 – REFERENCES

1. Suni, I., Mäenpää, Nicolet, M-A., and Luomajärvi, M., Thermal Stability of Hafnium and Titanium Nitride Diffusion Barriers in Multilayer Contacts to Silicon, *J. Electrochem. Soc.*, **103**, 1215 (1983).
2. Ryan, J. G., and Roberts, S. R., The Formation and Characterization of Rare Earth Boride Films", presented at the International Conference of Metallurgical Coatings, San Diego (1984).
3. Eizenberg, M., and Murarka, S. P., Reactively Sputtered Titanium Carbide Thin Films: Preparation and Properties , *J. Appl. Phys.*, **54**, 3190 (1983).
4. Eizenberg, M., Murarka, S. P., and Heimann, P. A., Interaction of Reactively Sputtered Titanium Carbind Thin Films with Si, SiO_2, Ti, $TiSi_2$, and Al, *J. Appl. Phys.*, **54**, 3195 (1983).
5. Eizenberg, M., Brener, R., and Murarka, S. P., Thermal Stability of the Aluminum/Titanium Carbide/Silicon Contact System", *J. Appl. Phys.*, **55**, 3799 (1984).
6. Mäenpää, M., von Seefeld, H., Chéung, N., Nicolet, M.-A., and Cullis, A. G., Characterization of Titanium Nitride Films Prepared by rf Sputtering, in: *Proc. Symp. Thin Film Interfaces and Interactions,* (J. E. E. Baglin and J. M. Poate, eds.), vol. 80-2, p. 316, The Electrochemical Society, Princeton (1980).
7. von Seefled, H., Cheung, N. W., Mäenpää, M., and Nicolet, M.-A., Investigation of Titanium-Nitride Layers for Solar-Cell Contacts, *IEEE Trans. Electron Devices,* **27**, 873 (1980).
8. Cheung, N., von Seefeld, H., and Nicolet, M.-A., Titanium Nitride as a Diffusion Barrier in Metallization Schemes, in: *Proc. Symp. Thin Film Interfaces and Interactions,* (J. E. E. Baglin and J. M. Poate, eds.), vol. 80-2, p. 323, The Electrochemical Society, Princeton (1980).
9. Weizer, V. G., and Broder, J. D., On the Cause of the Flat-Spot Phenomenon Observed in Silicon Solar Cells at Low Temperatures and Low Intensities, *J. App. Phys.*, **53**, 5926 (1982).
10. Cheung, N. W., von Seefeld, H., Nicolet, M-A., Ho, F., and Iles, P., Thermal Stability of Titanium Nitride for Shallow Junction Solar Cell Contacts, *J. Appl. Phys.* **52**, 4297 (1981).
11. Mäenpää, M., Suni, I., Sigurd, D., Finetti, M., and Nicolet, M.-A., Stable Metallization Systems for Solar Cells, *Phys. Stat. Sol.* (a), **72**, 73 (1982).
12. Ting, C. Y., TiN Formed by Evaporation as a Diffusion Barrier Between Al and Si, *J. Vac. Sci. Technol.* **21**, 14 (1982).
13. Ting, C. Y., and Wittmer, M., The Use of Titanium-Based Contact Barrier Layers in Silicon Technology, *Thin Solid Films* **96**, 327 (1982).

14. Norström, H., Donchev, T., Östling, M., and Petersson, C. S., $TiSi_2$/TiN – A Stable Multilayered Contact Structure for Shallow Implanted Junctions in VLSI Technology, *Phys. Scr.* **28**, 633 (1983).
15. Noël, J.-P., Houghton, D. C., Este, G., Sheperd, F. R., and Plattner, H., Characteristics of dc Magnetron, Reactively Sputtered TiN_x Films for Diffusion Barriers in III-V Semiconductor Metallization, *J. Vac. Sci. Technol.* **A2**, 284 (1984).
16. Houghton, D. C., TiN as a Diffusion Barrier for III-V Metallization, in: *Thin Films and Interfaces II,* Mat. Res. Soc. Symp. Proc. Vol. 25 (J. E. E. Baglin, D. R. Campbell, and W. K. Chu, eds.), p. 149, Elsevier Science, New York (1984).
17. Pan, A. I., Morgan, A. E., and Ritz, K., Reactive Sputter Deposition and Barrier Properties of Titanium Nitride, presented at the International Conference on Metallurgical Coatings, San Diego (1984).
18. Ting, C. Y., TiN as a High Temperature Diffusion Barrier for Arsenic and Boron, *Thin Solid Films* **119**, 11 (1984).
19. Sinke, W., Stout, P. K., and Saris, F. W., Pre-Annealing of TiN Barriers in Al Metallization of Silicon, presented at the Fall Meeting of the Materials Research Society, Boston (1984), (to be published in the Proceedings).
20. Affolter, K., Kattelus, H. P., and Nicolet, M.-A., Characterization of W-N Alloys Formed by Sputter Deposition, presented at the Spring Meeting of the Materials Research Society, San Francisco (1985).
21. Kattelus, H. P., Kolawa, E., Affolter, K., and Nicolet, M.-A., Sputtered W-N Diffusion Barriers, presented at the International Conference of Metallurgical Coatings, Los Angeles, (to be published in *J. Vac. Sci. Technol.*)
22. Shappirio, J. R., and Finnegan, J. J., Synthesis and Properties of Some Refractory Transition Metal Diboride Thin Films, *Thin Solid Films* **107**, 81 (1983).
23. Shappirio, J. R., Finnegan, J. J., Lux, R. A., and Fox, D. C., Resistivity, Oxidation Kinetics and Diffusion Barrier Properties of Thin Film ZrB_2, *Thin Solid Films* **119**, 23 (1984).
24. Krusin-Elbaum, L., Wittmer, M., Ting, C. Y., and Cuomo, J. J., ZrN Diffusion Barrier in Aluminum Metallization Schemes, *Thin Solid Films* **104**, 81 (1983).
25. Östling, M. Nygren, S., Petersson, C. S., Norström, H., Wiklund, P., Buchta, R., Blom, H.-O., and Berg, S., Reactively Sputtered ZrN Used as an Al/Si Diffusion Barrier in a Zr Contact to Silicon, *J. Vac. Sci. Technol.* **A2**, 281 (1984).

large compressive stress involved is clearly one cause of failure of titanium nitride diffusion barriers. Delamination, blistering and fracturing initiate local breakdown (50). It is not evident, though, that unstressed films will also function as effective diffusion barriers, as information on this topic is unavailable. The remedy commonly applied is to use as thin a titanium nitride layer as possible to minimize the total tension in the film (50).

Another known cause of failure is related to reactions. Aluminum films in direct contact with titanium nitride react above about 550°C. Compounds such as AlN and $TiAl_3$ have been identified (51), and a simplified ternary Ti-Al-N phase diagram has been sketched to show that a reaction $1TiN+4Al \rightarrow 1AlN+1TiAl_3$ can indeed take place (52). The details of the reaction still remain open. Lack of adhesion is a reoccurring source of irreproducibility and of failures. It is difficult to say whether this is a consequence of stresses or of the inertness of titanium nitride films – in many cases probably of a combination of both. Adhesion can be improved by interposing thin interfacial titanium layers. This may be accomplished simply by shutting off the flow of nitrogen to the titanium nitride deposition system (32). All the interfaces in the metallization system must fulfill increased adhesion requirements when a highly stressed layer is involved in the contacting scheme.

Although chemically inert, titanium nitride will oxidize fairly readily above 500°C (53-55). The inertness causes a paucity of acceptable etches for titanium nitride. Selective etching is possible, but the etch rate depends on the details of the deposition process. Two HF-containing chemical etches have been proposed (39) as well as an organic EDTA (ethylenediaminetetraacetic acid) solution (56-57). Titanium nitride can also be etched in a CF_4 plasma (39) or a chlorine-based reactive ion etch (49,44).

In spite of the many open questions about titanium nitride thin films, the consensus is that these films can indeed form useful diffusion barriers (8,58-59). Early applications (6) were those of the improved beam-lead process with a configuration of Si-substrate/PtSi/Ti/TiN/Pt/Au (60), and of Nelson's metallization scheme Si-substrate/PtSi/TiN/Ti/Au (32). Figure 8 shows an example involving a silicon solar cell of a flat array type (61,4). In that application the diffusion barrier was reactively sputter-deposited in an rf magnetron system at a deposition rate of ~0.17 nm/s onto silicon solar cells with a shallow junction (~200 nm), and patterned using mechanical masking during deposition. The performance of the cell after annealing at 600°C for 10 min. is basically unaltered (Fig. 8(b)), while a test cell with a conventional metallization system is destroyed, see Fig. 8(a). In this application, the current density through the Si/TiN interface is modest, ~1 A/cm^2, so that the measured resistivity of 500 $\mu\Omega$cm for the titanium nitride (no bias voltage was used during deposition) and a contact resistivity of $<10^{-3}$ Ωcm^2 with silicon suffice. In fact, no major difference is seen in the I-V characteristics of the sintered (400°C) samples of Figs. 8(a) and (b) without and with the barrier, indicating that the titanium nitride film used has acceptable electrical properties for this application. Detailed investigations of the contact properties of these titanium nitride films on silicon have been carried out (57,62-64), It is found that on n- and p-type silicon, titanium nitride forms Schottky-barriers with a barrier height of about 0.45 and 0.69 V, respectively. These values converge to a half-gap height value of about 0.55 V after vacuum annealing at 600°C. This behavior has been attributed to the out-annealing of donor-like traps created during the deposition of the film. On heavily doped silicon, the contact resistivity is about 2×10^{-5} Ωcm^2 on n^+ and 3×10^{-4} Ωcm^2 on p^+ layers, which convert to 6×10^{-5} Ωcm^2 after annealing for the same reason (65). Similar results are quoted by Wittmer et al. (66). Other electrical measurements of the TiN – semiconductor interface include Schottky-barrier heights on n-type GaAs (ϕ_{bn}~0.83V) (67) and characterization of TiN gate metal-oxide-semiconductor field-effect transistors on silicon (68).

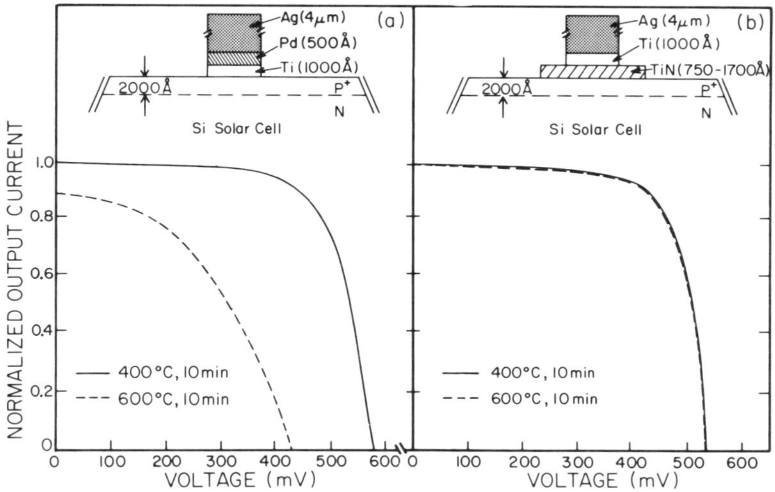

Fig. 8. I-V characteristic of a shallow-junction (200 nm) silicon solar cell under AMO illumination with the conventional Ti-Pd-Ag metallization system (a), and with a metallization scheme containing a reactively sputtered titanium nitride diffusion barrier (b). The solid lines were measured after the usual 400°C, 10 min sintering treatment applied to the cells to minimize the contact resistance. The dashed lines were measured after thermal annealing at 600°C for 10 min in dry nitrogen (4).

Another example demonstrates the effectiveness of a titanium nitride film as a diffusion barrier for indium in a structure InP substrate/TiN/Au (69). The titanium nitride layer was reactively sputtered from a dc magnetron source in pure nitrogen prior to gold e-beam evaporation in the same vacuum cycle. A detailed investigation of the diffusion of indium atoms carried out on these films has led to a determination of the activation energy, Q, for the rate-limiting diffusion process and an estimate of the diffusivity, D_{In}. The value $D_{In} = 0.3\ cm^2/s \times exp(-244\ kJ\ mol^{-1}/kT)$ is quoted. This corresponds to a diffusivity of $7.5 \times 10^{-19}\ cm^2/s$ at 450°C and an activation energy of 2.5 eV. Auger analysis and visual inspection of gold-metallized samples proved that titanium nitride is superior to platinum or chromium as a diffusion barrier for indium from an InP substrate. It is the mark of a reliable diffusion barrier that it should be applicable with comparable success on various substrates and for various metal overlayers. The two previous examples make that point for TiN. A further example is discussed in Section 5.2.

Most diffusion barrier studies with thin films of titanium nitride and other interstitial compounds have so far been executed with reactively sputtered layers. The films can also be obtained by other deposition techniques, such as: a) evaporation through a nitrogen or ammonia gas onto a substrate which may be at ambient or at elevated temperature (70-71,67), b) high-dose implantation of nitrogen into titanium (72), c) thermal nitridation of titanium at ~1000°C (73), d) evaporation through a nitrogen containing gas that is electrically ionized ("activated reactive evaporation") (74), e) ion plating, which is similar to activated reactive evaporation, but where the sample is additionally biased to a high negative potential (75), or f) chemical vapor deposition (CVD) (76). The latter techniques are used commercially for the application of thick coatings to improve the mechanical characteristics of working tools (77). Only a few investigations have been carried out with thin film diffusion barriers of interstitial alloys produced by these alternate processes (22,45,70). It would be desirable to extend these investigations, in particular with regards to CVD, which emerges as a dominant contender among very large scale integration technologies.

The success of these diffusion barriers has also created a need for an improved understanding of their basic physical and chemical characteristics. In particular, self-diffusion in these films needs to be studied, as well as the diffusion of foreign species (78), and the way the films react with adjacent metallic layers such as aluminum. The role of stress in the films also needs to be clarified.

The term "interstitial compound" traditionally refers to combinations of transition metals with hydrogen, boron, carbon, or nitrogen. The term might be extended to encompass the oxides, silicides, sulfides, and phosphides as well (30-31). Of these, oxides have been considered before as candidates for diffusion barriers (3,8), and silicides have, in fact, been incorporated in contacts in an attempt to improve the stability (8,58-59).

Metallic conductivities in the required range ($<10^{-2}$ Ωcm) for applications as ~100 nm thick diffusion barriers in semiconductor contact metallizations have been reported for dioxides of chromium, iridium, molybdenum, osmium, rhodium, ruthenium, and tungsten as well as for titanium monoxide (8). Another group of low resistivity oxides is wide band-gap (~3 eV) semiconductors, like In_2O_3, SnO_2, or ZnO (12), in which oxygen vacancies or degenerate doping can induce resistivities as low as 100 μΩcm (79). Reported melting points for all these compounds are below 2000°C. Most oxides are insulators. They can be used as diffusion barriers in contacts only if they are thin enough (typically $\lesssim 5$ nm) to permit the transmission of electrons by tunneling. It is uncertain that films of such small thickness can offer practically acceptable solutions (3).

It is a very common practice in silicon integrated circuits to use silicides as the primary metallic layer to the silicon substrate. The near-noble metal (nickel, palladium, platinum, cobalt) silicides have been used since the beginning of integrated circuit technology, because of the ease with which uniform layers can be produced by thermal reaction of the silicon in a SiO_2 window and a film. A silicide/silicon interface is thereby obtained which reproducibly yields low contact resistivities on highly-doped silicon and Schottky barriers on lightly-doped silicon. It is found, however, that these contacting silicide layers react with aluminum at temperatures in the range of 200-400°C. Typically, these silicides form compounds of the near-noble metals with aluminum. The reaction can fully consume the silicide layer (8,58-59). The refractory metal silicides, $TiSi_2$ and $MoSi_2$, behave differently with aluminum, in that they fail predominantly by diffusion of aluminum through the film in a localized fashion (21,80-81). At least with aluminum, these silicides have not performed as good diffusion barriers.

Logically, one would be led to consider for diffusion barriers first those refractory metal disilicides with the highest congruent melting point (i.e. $TaSi_2$,WSi_2, $MoSi_2$, $ReSi_2$, or $NbSi_2$ in order of decreasing melting point (82)). The performance as diffusion barriers for only $TaSi_2$(2,83) and $MoSi_2$ (21,80) have been studied in contact with an aluminum overlayer. The results for $TaSi_2$ look promising at least up to 500°C as far as the general structural integrity is concerned (2), but electrical interface measurements of sputter-deposited $TaSi_2$ on silicon show that the contact is stable at 475°C only if this silicide is tantalum-rich (83). In the $MoSi_2$ case the failure is not attributed to intrinsic material properties, but to pinholes. These silicides all tend to have electron barrier heights of approximately 0.6 V, which makes them suitable for ohmic contacts on complementary integrated circuits, but less so as Schottky barriers. Another fact which may possibly act in favor of these refractory silicides is their high affinity for oxygen. It is difficult to deposit metal films of these silicides that are free of impurities such as carbon, nitrogen, or oxygen. But it is known that oxygen, for example, tends to accumulate in the unreacted metal as the silicide layer forms under heat treatment of such a film on silicon. This oxygen-enriched layer could be advantageous for diffusion barrier application if it were stable in the presence of an aluminum overlayer. It might promote the formation of a thin layer containing aluminum, transition metal, and oxygen that would be transparent to electrons but impermeable to atoms (3).

3.3.3 Multiphase Compounds. Conceptually, a possibility of multiphase compound films also exists. For example, molybdenum and nitrogen are known to form both Mo_2N and MoN phases that can be mixed in a thin film (84-85). That situation arises whenever more than one equilibrium compound forms in a binary system. When amorphous phases are included, possible choices become even more numerous. An example of the latter type is discussed in section 4.

3.4 Summary

In this section, polycrystalline diffusion barriers have been reviewed. These layers contain grain boundaries that constitute more rapid diffusion paths than the bulk material at typical operating temperatures of electronic devices. Whether polycrystalline materials can be used as effective diffusion barriers depends on the details of the grain boundary structure and on the corresponding diffusivities along them. These properties are controlled directly by the method that is used to deposit the film. At a fixed temperature, diffusivities tend to decrease with increasing melting point of the material.

The experimental evidence indicates that elemental metallic polycrystalline films do not act as effective diffusion barriers in a pure state. Polycrystalline alloys containing only metallic elements have shown similar behavior. The atomic transport through the metal layer is, however, reduced when impurities like oxygen or nitrogen are used to stuff the barrier.

Polycrystalline compound layers also have grain boundaries. These may naturally contain excess amounts of one of the constituent elements. One of the most studied compound diffusion barriers, titanium nitride, seems to be most effective when the titanium to nitrogen ratio is not exactly 1:1. Films produced by rf sputtering are normally under high compressive stress.

Other refractory interstitial nitrides, borides, and carbides have shown equally promising characteristics as diffusion barriers as titanium nitride. Work is still needed to determine how structural details of these layers affect the diffusivities.

4. AMORPHOUS FILMS CONSISTING OF ONE SINGLE LAYER

The fast diffusion paths that extended defects provide in polycrystalline films can be eliminated by forming amorphous materials. This possibility has been ignored until fairly recently. Results are now beginning to emerge that are promising and that suggest that films of amorphous metallic alloys offer valid options for stable electrical contacts (7). For our purpose, a film is considered to be amorphous if a Debye-Scherrer x-ray diffraction pattern produces only halos and no sharp lines. This practical classification does not distinguish between polycrystalline films with extremely small grains (a few nanometers) and a truly random distribution of atoms. The property that distinguishes these films from those that are really polycrystalline is the absence of crystalline defects that extend over distances that are comparable to the thickness of the film.

4.1 General Observations

For applications, some of the important parameters that characterize an amorphous diffusion barrier are the crystallization temperature, T_c, the diffusivity of impurities, D_i, the resistivity, ρ, and the chemical stability of the layer with its adjacent media. When the amorphous film is in direct contact with the semiconductor, the barrier height and the resulting contact resistivity are of primary concern, too.

Measurements of T_c for a number of amorphous metallic alloys show that values in excess of 1000°C are possible (86). There is thus no reason to reject amorphous layers on the grounds that they are metastable. Elemental metals instead cannot be made amorphous above cryogenic temperatures. The use of elemental nonmetals may be restricted by their high electrical resistivity. Amorphous silicon, for example, does not meet the electrical requirements even in a doped form. Substances like silicon, carbon, phosphorus, or boron can, however, be used in combination with metallic elements to form stable amorphous alloys with low resistivity.

The low atomic diffusivities resulting from the amorphous structure is the attractive feature of these layers. Typically, the atomic diffusivities in the amorphous phase are orders of magnitude below those of a corresponding polycrystalline phase at a temperature where grain boundary diffusion dominates in the latter. This is so, roughly, below two-thirds of the melting point of the material (87). The diffusivity in the amorphous phase may still much exceed that of a crystalline bulk, however. For example, the diffusivity of implanted gold in Pd-Cu-Si glass at 300°C is about 5×10^{-17} cm^2/s in the as-quenched specimen, and $\sim 5 \times 10^{-19}$ cm^2/s in material that was preannealed at 350°C for 5 min before gold implantation. In contrast, the grain boundary diffusion constant in polycrystalline palladium at the same temperature is about 10^{-15} cm^2/s (88). This example also demonstrates that the diffusivity depends on the thermal history of the amorphous material. The available data on diffusivities in amorphous alloys are quite limited when measured against the large number of elements (metals, intentional dopants, and accidental impurities) that are of relevance in semiconductor technology. It will be essential to build up a reliable and pertinent data-base on these atomic diffusivities in amorphous alloys for an orderly progress in the understanding of these types of diffusion barriers.

Amorphous metals have resistivities that are typically a few hundred micro-ohm-cm, which is quite acceptable for thin-film diffusion barrier applications in contacts.

Amorphous metallic films can react with adjacent metallic layers or semiconducting substrates. However, the rate at which a particular element in the amorphous alloy reacts can be much less than would be the case for the corresponding polycrystalline film of that element under otherwise similar conditions. This by itself suggests that amorphous metallic alloys may be useful for controllably reducing the rate of a solid-state reaction, for instance in the formation of ohmic contacts to shallow p-n junctions. This idea seems worth pursuing.

What determines the barrier height, ϕ_b, of an amorphous metallic alloy in contact with the semiconductor remains to be established. One expects that the first few atomic layers at the barrier-semiconductor interface play a decisive role in determining that barrier height. This implies that how, exactly, the layer is deposited should be important, and that small interfacial modifications could lead to major changes. Table 2 lists the electrical contact properties of amorphous alloy films reported so far when sputter-deposited on silicon or gallium arsenide. For comparison, barrier heights of the constituent elements are listed as well. On silicon, the sum $\phi_{bn}+\phi_{bp}$ of the barrier heights on n-type and p-type substrates closely equals the band-gap of silicon in all measured cases. Thermal annealing changes the barrier heights little, and that is true for $Ni_{36}W_{64}$ even when localized formation of nickel silicides is known to occur (at 550°C annealing for 30 min). This can be explained by observing that the metals involved in these

Table 2. Schottky-Barrier Heights and Contact Resistivities of Some Amorphous Metallic Alloys on Silicon or Gallium Arsenide Substrates Before and After Annealing Steps.

	ϕ_{bn}(V)		ϕ_{bp}(V)		ρ_{cn}($\mu\Omega cm^2$)		ρ_{cp}($\mu\Omega cm^2$)		Ref.
	as-sputtered	annealed	as-sputtered	annealed	as-sputtered	annealed	as-sputtered	annealed	
Si	on n-Si		on p-Si		on n^+-Si		on p^+-Si		
$Fe_{45}W_{55}$/Ag	0.61	300°C 30 min: 0.65			0.10	400°C 30 min: 0.20 500°C 30 min: 0.10	2.8	400°C 30 min: 1.1 500°C 30 min: 1.1	1
(Fe)	0.62-0.68								
(W)	0.64-0.67								
$Mo_{30}Ni_{70}$	0.63	500°C 30 min: 0.63	0.47	500°C 30 min: 0.44	2.4	500°C 30 min: 3.3	3.1	500°C 30 min: 1.8	2
$Mo_{54}Ni_{46}$	0.68	0.63	0.42	0.42	4.1	3.5	3.0	1.4	
$Mo_{58}Ni_{42}$	0.68	0.64	0.42	0.42	1.1	1.0	1.3	0.8	
(Mo)	0.55-0.59								
(Ni)	0.59-0.66								
$Ni_{36}W_{64}$,	0.65	400°C 30 min: 0.65	0.45	400°C 30 min: 0.43	1.4	400°C 30 min: 1.1	0.89	400°C 30 min: 0.18	3,4
$Ni_{36}W_{64}$/Al		550°C 30 min: 0.66		550°C 30 min: 0.39					

Table 2 (Continued)

$Ir_{58}Ta_{42}/Au$	0.83	350°C 1 h 0.84 400°C 77 h 0.59				5,6
(Ir) (Ta)	0.88-0.91 ∿ 0.6					
$Nb_{47}Ni_{53}/Au$	0.51	350°C 16 h 0.54				5,6
$N_{21}Ni_{30}W_{49}$, $N_{21}Ni_{30}W_{49}/Al$	0.59	450°C 30 min 0.65				d8
$W_{35}Zr_{65}$ $W_{70}Zr_{30}$ (Zr)	0.54 0.56 ∿ 0.55	650°C 30 min 0.56 0.58				7
GaAs	on n-GaAs					
$Ir_{56}Ta_{44}/Au$	0.79	350°C 48 h 0.76				6
$Ir_{58}Ta_{42}/Au$	0.64	350°C 2 h 0.71				5
$Nb_{47}Ni_{53}/Au$	0.72	350°C 3 h 0.73				5,6
Si[illegible]Ti[illegible]W[illegible]/Au	0.79	300°C 944 h 0.79				8

TABLE 2 – REFERENCES

1. Finetti, M., Pan, E., T.-S., Suni, I., and Nicolet, M.-A., Electrical Characteristics of Amorphous Iron-Tungsten Contacts on Silicon, *Appl. Phys. Lett.* **42**, 987 (1983).
2. Kung, K. T.-Y., Suni, I., and Nicolet, M.-A., Electrical Characteristics of Amorphous Molybdenum-Nickel Contact to Silicon, *J. Appl. Phys.* **55**, 3882 (1984).
3. Zhu, M. F., Suni, I., and Nicolet, M.-A., Electrical Characteristics of Amorphous $Ni_{36}W_{64}$ Contacts on Silicon, in: *SPIE Vol. 463 – Advanced Semiconductor Processing and Characterization of Electronic and Optical Materials* (D. K. Sadana and C. M. Lampert, eds.), p. 19, SPIE, Bellingham, Washington (1984).
4. Zhu, M. F., So, F. C. T., and Nicolet, M.-A., Amorphous Ni-N-W Films as a Diffusion Barrier Between Al and Si, *Thin Solid Films* (to be published).
5. Todd, A. G., Harris, P. G., Scobey, I. H., and Kelly, M. J., Amorphous Metal-Semiconductor Contacts on Silicon, in: *SPIE Vol. 463 – Advanced Semiconductor Processing and Characterization of Electronic and Optical Materials* (D. K. Sadana and C. M. Lampert, eds.), p. 19, SPIE, Bellingham, Washington (1984).
6. Kelly, M. J., Todd, A. G., Sisson, M. J., and Wickenden, D. K., Contacts Between Amorphous Metals and Semiconductors, *Electron. Lett.* **19**, 474 (1983).
7. So, F. C. T., and Nicolet, M.-A., unpublished.
8. Anderson, W. T., Jr., Christou, A., and Davey, J. E., Amorphous Thin Film Diffusion Barriers on GaAs and InP, *Thin Solid Films* **104**, 57 (1983).

layers (as well as their silicides) have quite similar barrier heights. Such is not the case for an amorphous $Ir_{58}Ta_{42}$ film. The high value of 0.83 V for the as-deposited sample suggests that Ir may be dominantly present at the interface, because it reacts more readily with silicon than tantalum. After a long annealing at 400°C, the barrier height decreases substantially, presumably as a result of a reaction involving tantalum, which has a barrier height that must be in the vicinity of 0.6 V (82). An interesting question that arises here is whether the barrier height of as-deposited samples can be adjusted by varying the alloy composition at the interface.

The few measured contact resistivities on silicon agree with theoretical estimates that were obtained from the measured barrier heights; they are in the $\mu\Omega cm^2$ range for doping concentrations on the order of 10^{20} cm^{-3} (89).

4.2 Examples of Amorphous Diffusion Barriers

There are only relatively few experimental studies of amorphous diffusion barriers at this date. Table 3 lists papers investigating properties of thin amorphous metallic films as diffusion barriers in semiconductor metallization systems. In the table, the lowest temperature of observed instability is reported. It should be noted that in the case of amorphous diffusion barriers, not involving rapid localized grain boundary motion, this temperature does not necessarily mean a lost integrity of the whole metallization system if the interaction is laterally uniform and slow (90). The layer can thus still act as a sacrificial barrier, which is described in Section 5.1. In this section, the case of nickel-tungsten films is briefly discussed to demonstrate the relevant features of thin amorphous metallic alloys described above.

Amorphous films of a composition of $Ni_{36}W_{64}$ have been obtained by co-sputtering nickel and tungsten on a room temperature substrate. The sputtering was done in an rf planar magnetron system with a bare tungsten cathode partially covered by a thin nickel mask, using pure argone gas (89). The films remain amorphous upon annealing in vacuum for 30 min up to 600°C on both silicon and SiO_2 substrates. The room temperature value of the resistivity is about 170 $\mu\Omega$ cm.

Upon crystallization, the resistivity drops precipitously (91). On silicon, the crystallization is accompanied by the formation of silicides. Much below that temperature, however, the film reacts locally with the silicon substrate and forms pyramidal intrusions into the silicon that consist of orthorombic NiSi and cubic $NiSi_2$, of which some are epitaxially registered (see Figure 9). The localized nature of the process may be caused by non-uniformities in the native oxide at the interface or by some lateral instability of the reaction. The growth of the

Table 3. List of Papers Investigating Properties of Thin Films of Amorphous Metallic Alloys as Diffusion Barriers in Complete Metallization Schemes of Semiconductors (S/X/M Structure of Figure 1).

Diffusion barrier (X)	Method of Barrier Deposition*	Barrier thickness in nm	Semi-conductor substrate (S)	Metal overlayer (M)	Metal deposition method	Device/Test	Annealing time in min	Lowest temperature of observed instability in °C (Method), comments	Ref.
$Fe_{45}W_{55}$	rf magn. sputtering of Fe-W in Ar, $p_o \sim 10^{-6}$ torr, $r \sim 0.33$ nm/s	100	Si	Ag	sputt.**	Schottky barrier height, contact resistivity		see Table 2	1
$Fe_{37}W_{63}$	rf magn. sputtering of Fe-W in Ar, $p_o \sim 10^{-6}$ torr, $r \sim 0.08$ (dynamic) – 0.8 (static) nm/s	100	Si	Al	evap.		30	550 (XRD)	2
"		100	Si	Au	evap.		30	500 (RBS)	
"		100	Si	Cu	evap.		30	650 (XRD)	
"		100	Si	Ni	evap.		30	600 (XRD)	
"		100	Si	Pd	evap.		30	600 (XRD, RBS)	
"		100	Si	Pt	evap.		30	600 (RBS)	
$Ir_{58}Ta_{42}$	dc triode sputtering of Ir-Ta in Ar	200	GaAs	Au	evap.	Schottky diode/ Barrier height	1440	600 (AES)	3-5
"		200	Si	Au	evap.		1440	600 (AES), for electrical measurements see Table 2.	
$N_{21}Ni_{30}W_{49}$	rf magn. sputtering of Ni-W in Ar & N_2, $p_o < 2 \times 10^{-6}$ torr, $r \sim 0.08$ nm/s (dynamic)	45	Si	Al	evap.	Schottky diode/ Barrier height	150	500 (XRD), for electrical measurements, see Table 2.	6
$N_{23}W_{77}$	rf magn. sputtering of W in Ar & N_2, $p_o < 3 \times 10^{-6}$ torr, $r \sim 1.0$ nm/s (12)	100	Si	Ag	sputt.**		30	800 (RBS), 600 (optical)	13
"		100	Si	Al	"		30	600 (RBS), 500 (optical)	
"		100	Si	Au	"		30	900 (RBS), 500 (optical)	

Table 3. List of Papers ... (continued)

Diffusion barrier (X)	Method of Barrier Deposition*	Barrier thickness in nm	Semi-conductor substrate (S)	Metal overlayer (M)	Metal deposition method	Device/Test	Annealing time in min	Lowest temperature of observed instability in °C (Method), comments	Ref.
$N_{33}W_{67}$	rf magn. sputtering of W in Ar & N_2, $p_o < 3 \times 10^{-6}$ torr, $r \sim 1.0$ nm/s (12)	100	Si	Ag	sputt.**		30	700 (RBS), 600 (optical)	13
"		100	Si	Al	"		30	600 (RBS), 500 (optical)	
"		100	Si	Au	"		30	900 (RBS), 500 (optical)	
$Nb_{43}Ni_{57}$	rf sputtering of Nb-Ni in Ar, $r \sim 0.5$ nm/s	~ 1000	Si	Cu			600	~ 500 (AES)	7-8
$Nb_{45}Ni_{55}$	rf sputtering	~ 1000	Si	Au	evap.		15	550 (RBS)	9
$Nb_{45}Ni_{55}$	rf sputtering	40	Si	Au	evap.		20	600 (RBS)	10
"		40	GaAs	Au	evap.		20	600 (RBS)	
"		40	GaP	Au	evap.		20	600 (RBS)	
$Nb_{47}Ni_{53}$	dc triode sputtering of Nb-Ni in Ar, $r \sim 0.17$ nm/s	50-500	GaAs	Au	evap.	Schottky diode/ Barrier height	1440	400 (AES), for electrical measurements, see Table 2.	3-5
"		50-500	Si	Au	evap.				
$Ni_{36}W_{64}$	rf magn. sputtering of Ni-W in Ar, $p_o < 2 \times 10^{-6}$ torr, $r \sim 0.07$ nm/s (dynamic)	45	Si	Al	evap.	Schottky diode/ Barrier height	30	500 (RBS), for electrical measurements, see Table 2.	6
$Si_{10}W_{90}$	rf sputtering of Si-W in Ar, $r \sim 0.5$ nm/s	~ 1000	Si	Au	evap.		600	> 500 (AES)	7

Table 3. List of Papers (continued)

Diffusion barrier (X)	Method of Barrier Deposition*	Barrier thickness in nm	Semi-conductor substrate (S)	Metal overlayer (M)	Metal deposition method	Device/Test	Annealing time in min	Lowest temperature of observed instability in °C (Method), comments	Ref.
$Si_{10}W_{90}$	Sputtering of Si/W or Si/Ti-W multilayers, T_a ~ 500 °C (rapid cooling)	55-130	GaAs	Au		FET/Barrier height, noise figure, gain	960	> 400 (AES)	11
$Si_{10}W_{90}$		"	InP	Au			480	> 450 (AES)	
$Si_{10}Ti_{10}$-W_{80}		"	GaAs	Au			56640	> 300 (electrical) barrier heights in Table 2.	14
$Si_{10}Ti_{10}$-W_{80}		"	InP	Au					

* p_o = Base pressure before deposition.

r = Deposition rate.

T_a = Preannealing temperature before metal deposition.

** All depositions in the same vacuum.

TABLE 3 – REFERENCES

1. Finetti, M., Pan, E., T.-S., Suni, I., and Nicolet, M.-A., Electrical Characteristics of Amorphous Iron-Tungsten Contacts on Silicon, *Appl. Phys. Lett.* **42**, 987 (1983).
2. Suni, I., Nicolet, M.-A., Pai, C. S., and Lau, S. S., Stability of Amorphous Fe-W Alloys in Multilayer Metallizations on Silicon, *Thin Solid Films* **107**, 73 (1983).
3. Kelly, M. J., Todd, A. G., Sisson, M. J., and Wickenden, D. K., Contacts Between Amorphous Metals and Semiconductors, *Electron. Lett.* **19**, 474 (1983).
4. Todd, A. G., Harris, P. G., Scobey, I. H., and Kelly, M. J., Amorphous Metal-Semiconductor Contacts for High Temperature Electronics – I. Materials and Characterization, *Solid-St. Electron.* **27**, 507 (1984).
5. Wickenden, D. K., Sisson, M. J., Todd, A. G., and Kelly, M. J., Amorphous Metal-Semiconductor Contacts for High Temperature Electronics – II. Thermal Stability of Schottky Barrier Characteristics, *Solid-St. Electron.* **27**, 515 (1984).
6. Zhu, M. F., So, F. C. T., and Nicolet, M.-A., Amorphous Ni-N-W Films as a Diffusion Barrier Between Al and Si, *Thin Solid Films* (to be published).
7. Wiley, J. D., Perepezko, J. H., and Nordman, J. E., High Temperature Metallization System for Solar Cells and Geothermal Probes, *Sandia Laboratories Report* No. SAND80-7167 (1980).
8. Wiley, J. D., Perepezko, J. H., Nordman, J. E., and Guo, K.-J., Amorphous Metallizations for High-Temperature Semiconductor Device Applications, *IEEE Trans. Ind. Electron.* **29**, 154 (1982).
9. Doyle, B. L., Peercy, P. S., Wiley, J. D., Perepezko, J. H., and Nordman, J. E., Au Diffusion in Amorphous and Polycrystalline $Ni_{0.55}Nb_{0.45}$, *J. Appl. Phys.* **53**, 6186 (1982).
10. Doyle, B. L., Peercy, P. S., Thomas, R. E., Perepezko, J. H., and Wiley, J. D., Atomic Interdiffusion in Au/Amorphous Ni-Nb/Semiconductor Systems, *Thin Solid Films* **104**, 69 (1983).
11. Anderson, W. T., Jr., Christou, A., and Davey, J. E., Amorphous Thin Film Diffusion Barriers on GaAs and InP, *Thin Solid Films* **104**, 57 (1983).
12. Affolter, K., Kattelus, H. P., and Nicolet, M.-A., Characterization of W-N Alloys Formed by Sputter Deposition, presented at the Spring Meeting of the Materials Research Society, San Francisco (1985).
13. Kattelus, H. P., Kolawa, E., Affolter, K., and Nicolet, M.-A., Sputter W-N Diffusion Barriers, presented at the International Conference of Metallurgical Coatings, Los Angeles, (to be published in *J. Vac. Sci. Technol.*)
14. Papanicolaou, N. A., Anderson, W. T., Jr., and Christou, A., Low Noise MESFET with Sputtered Amorphous Metal Gate Defined by Lift-Off, Inst. Phys. Cong. Ser. No. 65, Chapter 5, 407 (1983).

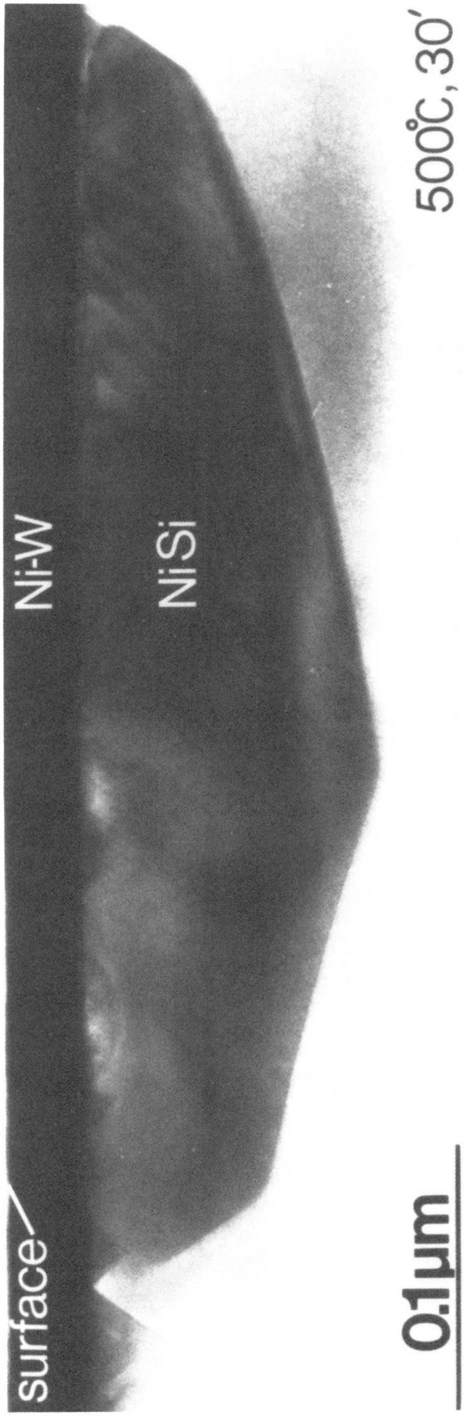

Fig. 9a. Cross-sectional TEM micrograph of an amorphous $Ni_{36}W_{64}$ layer on ⟨100⟩ oriented Si substrate after vacuum annealing at 500°C for 30 min. A large orthorhombic NiSi grain projects into the silicon; the metallic layer remains amorphous.

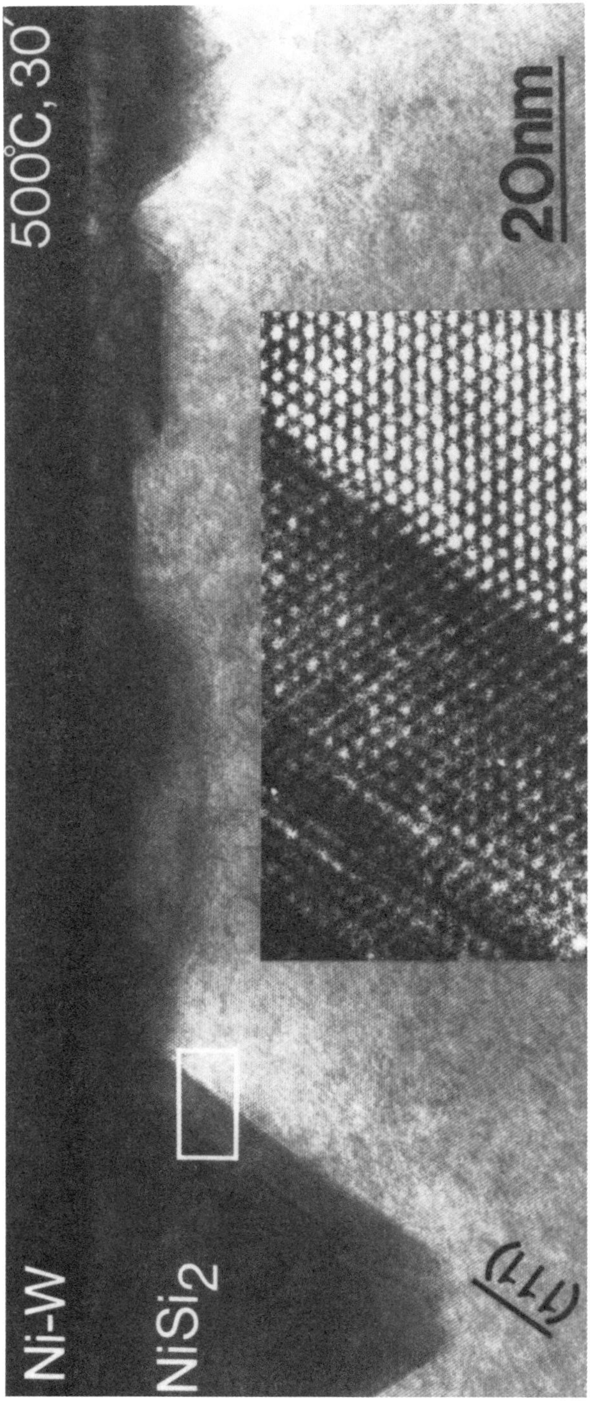

Fig. 9b. Cross-sectional HREM micrograph of another interfacial region of the sample shown in Fig. 9a. Note the epitaxially registered pyramids of $NiSi_2$ in the silicon substrate (91).

pyramids increases linearly with time, which shows that the limiting process most probably is not transport, but reaction.

In contact with an aluminum overlay, both nickel and tungsten will react with the metal and form aluminides at 500°C when the film is still amorphous (92). A similar reaction of tungsten with aluminum has been reported for amorphous iron-tungsten and has been associated with predominant grain boundary migration and reaction of tungsten in polycrystalline aluminum (90). This mode of interaction that is dominated by grain boundary diffusion is also documented for pure tungsten films in contact with aluminum (93-94).

The barrier height and the contact resistivity of amorphous $Ni_{36}W_{64}$ on silicon have also been investigated (see Table 2). Both parameters are quite insensitive to the onset of the local nickel silicide formation for reasons suggested in the preceding section. The ideality factor of the I-V characteristic, however, does rise from an initial 1.04 to a maximum of 1.35.

4.3 Summary

Amorphous metallic alloys form an emerging group of promising diffusion barrier materials. In addition to the absence of grain boundaries, the good corrosion resistance and the flexible structure can be attractive features in hindering chemical or mechanical failure modes of devices. The material selection is wide, especially when passing from binaries to ternaries, quarternaries, etc. Thus, an optimal structure for each application can be sought when more information about the interaction of various types of amorphous metallic alloys with interesting electronic materials is available. Such optimization is exemplified by two cases: 1) increasing the silicide forming temperature of Ni-W by replacing the Ni by Zr (95) or 2) increasing the aluminide forming temperature by adding nitrogen to form Ni-N-W (92).

5. MULTILAYER CONFIGURATIONS

The conceptual solution described in Fig. 1 attempts to stabilize an unstable contact structure by inserting one additional layer, the diffusion barrier, into the structure between the reacting pair of materials. Configurations with more than one additional layer can, of course, be conceived. Contacting schemes of this type are considered in this section. We subdivide them into three categories. The first deals with arrangements that exploit the tendency of thin films to react with each other, so that although a single layer is interposed initially, a multilayer arrangement forms subsequently as a result of thin-film reactions. The second category considers schemes in which the electrical characteristics of the contact to the semiconductor are optimized by inserting an additional layer (the "contacting layer") between the semiconductor and the diffusion barrier. Finally, contacting schemes can be constructed which incorporate more than one type of diffusion barrier in series ("tandem barriers").

5.1 Sacrificial Barrier

Figure 10 depicts the idea of a sacrificial diffusion barrier. It exploits the fact that thin adjacent films often react in a laterally uniform fashion. If a barrier layer, X, reacts uniformly with a material S and M on either side of it, then that layer effectively maintains a separation of S and M as long as it is not fully consumed in the reactions. That point in time is predictable when the rates for both reactions are known as a function of temperature. Under those conditions, the minimum barrier thickness can be quoted for any temperature cycle that the contact should survive. This *sacrificial diffusion barrier* concept was first implemented and recognized as such with a titanium film between a silicon single crystal and an aluminum metallization layer (96). Another scheme uses a sputtered bilayer of tantalum/$TaAl_3$ instead of titanium (97). A tantalum film reacts at a lower temperature with a superimposed aluminum film forming $TaAl_3$, than with the silicon substrate below. The tantalum/$TaAl_3$ metallization system thus constitutes a sacrificial tantalum layer whose consumption by the aluminum is anticipated through the sputter deposition of $TaAl_3$.

The unique feature of the sacrificial barrier is that the point of potential failure can be predicted, i.e. the time when the layer is fully

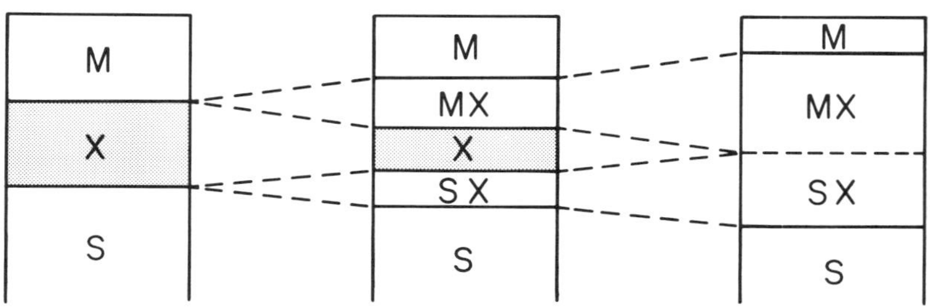

Fig. 10. A sacrificial barrier maintains the separation between M and S only for a while, but that duration is predictable when the reactions with M and with S are laterally uniform and their rates are known.

consumed. The accuracy of these predictions has recently been put to test with an evaporated chromium film as a sacrificial barrier between various silicide layers on a silicon single crystal substrate and an aluminum overlayer by comparing failure points calculated on the basis of published reaction rates (98) with actual failures observed on Schottky diodes (99). The first series of tests, including NiSi, Pd_2Si and PtSi for the contacting layer, yielded good agreement with the predicted behavior, and appeared to validate the model. When the chromium sacrificial barrier layer and the aluminum overlayer were deposited in another vacuum evaporator at rates that were two to five times faster than in the previous case, it was observed, however, that the chromium reaction rate with aluminum was increased up to 100 times, though the reaction still maintained its uniform nature. The difference in the two cases was attributed to a reduced impurity level in the metal layers of the latter case. It was indeed observed that the intentional exposure to impurities during the evaporation phase of the metals or of the annealing phase of the complete metallization system reduced that reaction rate much. One is thus forced to conclude that the rate of the chromium/aluminum reaction depends on the deposition and annealing conditions of the films, and that the agreement between the failure points calculated and observed in the first test was fortuitous.

The identification of the impurities that affect thin-film reactions and their quantitative characterization emerges as one of the important issues raised by that study. The notion of the sacrificial barrier itself may have to be reconsidered, because it is conceivable that in reality impurities may always have to be present to assure a laterally uniform controllable reaction between polycrystalline metallic films (4).

In light of these comments, one would expect that inconsistencies in the holding time reported for sacrificial barriers should be the rule. Indeed, conflicting results as to the usefulness of titanium as a sacrificial barrier exist. Two papers state that titanium works successfully as a sacrificial barrier between aluminum and platinum or palladium silicides on silicon (100-101), but another questions its usefulness (102). The evidence thus seems to support the idea that the sacrificial diffusion barrier may merely be a special case of a stuffed barrier, in the sense that impurities may be essential in assuring the beneficial performance of the system. For the practitioner, this means that the reaction rates involved in a sacrificial diffusion barrier must be considered as system-dependent quantities, whose values and reproducibility must be established and monitored in each particular case.

5.2 Contacting Layers

One can maintain the advantages inherent to the silicides, either near-noble or refractory, by using them merely as a primary contacting layer, not as a diffusion barrier. To accomplish stability, a diffusion barrier is then introduced additionally to separate this contacting layer from the metallization overlay. This approach has the advantage that it assigns the tasks of establishing a desirable electrical contact property and a desirable contact stability to separate functional entities, thereby implementing the proven engineering principle of separating functions to meet distinct goals. The cost is an added complexity that ultimately must be justified by economical considerations.

An example of a scheme embodying a silicide contacting layer (PtSi) and a diffusion barrier (TiN) is that of the beam-lead technology (60). The complete contact configuration can be qualitatively described as Si/PtSi/Ti/TiN/Pt/Au (6). Another scheme of the same type uses a sputtered titanium-tungsten layer to separate aluminum from PtSi (103-104). An application involving a gold overlay, titanium nitride as the diffusion barrier, and a contacting layer of chromium on GaInAs on InP substrate is shown in Fig. 11 (105). That contact scheme is part of an InP injection laser.

The titanium nitride layer was deposited by dc magnetron sputtering in nitrogen gas. The cross-sectional SEM pictures show that a planar surface morphology is maintained only when the TiN diffusion barrier is used. The structures of the cleaved surfaces and the energy-dispersive x-ray analyses of the top gold layers also show a restricted interdiffusion when the TiN barrier is involved. This example confirms the effectiveness of sputtered titanium nitride films as diffusion barriers. In concept, a good diffusion barrier should be successful regardless of the nature of the substrate or the function of the device. Reactively sputtered titanium nitride evidently meets that test. This is demonstrated again by yet another application of a reactively sputtered titanium nitride film in an ohmic contact to the n-type source and drain of a GaAs MESFET (106). The contacting layer is an approximately 200 nm thick layer of a conventional Au-Ge-Pt type alloy. The metal overlay is Ti/Pt/Au. Contacts with the diffusion barrier have a contact resistivity of 10^{-6} Ωcm^2 and survive lifetime tests at 250°C for 100 h practically unchanged, while those without the barrier fail ($\rho_c \sim 10^{-4}$ Ωcm^2). The small signal gains of the two types of transistors have similar values after deposition. However, those with the barrier layer remain unaffected by a thermal treatment at 250°C for 300 h while the three measured devices without the barrier totally fail.

A contacting layer that is formed by reaction with the substrate can be combined with an inert diffusion barrier using the same metal. For example, thermal annealing of a Si/Ti/TiN system at a high temperature (700-900°C) converts the underlying titanium to $TiSi_2$ without affecting the titanium nitride barrier, when evaporated (70) or

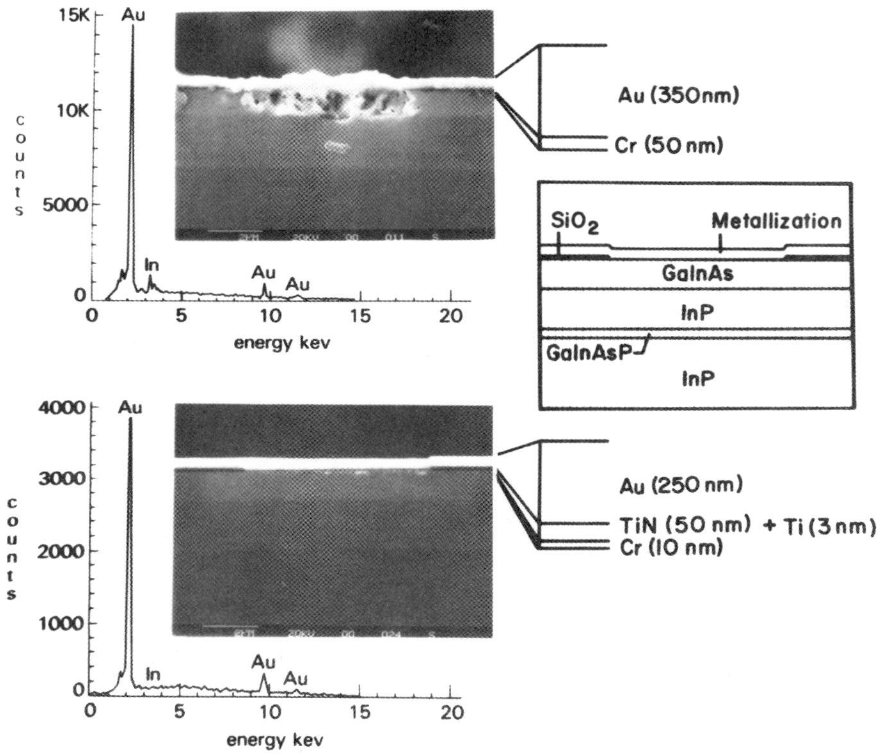

Fig. 11. SEM micrographs of a cleaved cross-section of a laser metallized with Cr/Au (upper part) or Cr/TiN/Au (lower part) after an annealing treatment. Kirkendall void formation extends deep into the semiconductor in the case without TiN, but is restricted almost by an order of magnitude in depth, when the barrier is used. The EDX spectra are acquired from the metallization surface, with the beam energy adjusted so that the full analysis depth is within the gold layer. Indium diffusion into gold can only be seen in the case without TiN (105).

essential not only in forming a reliable electrical contact, but also in promoting the stability against an aluminum overlay in the final $Si/TiSi_2/TiN/Al$ scheme. The averaged large-area integrity measured by backscattering spectrometry (107) can then be maintained up to 550-575°C for 30 min, but electrical measurements show a localized degradation at temperatures around 500°C (22,108). The use of the same element in both the contacting layer and the diffusion barrier is technologically attractive. As shown in these examples, the transition from the titanium to the titanium nitride deposition can simply be performed by opening the nitrogen leak valve to the deposition chamber. This type of contact can even be implemented in a single annealing step by depositing also the low resistivity overlay metal before sintering. This idea has been realized by replacing titanium with zirconium in the just described silicon/aluminum metallurgy (109) as well as in other systems (106,110-111).

It has also been shown that the instability between an aluminum overlay and a titanium nitride diffusion barrier can be circumvented by employing a less reactive compound overlay film, such as $TiAl_3$ (112). The resistivity of this material is several times higher than that of aluminum (113), but the compound is thermally stable in contact with titanium nitride.

Table 4 lists papers studying the stability of complete metallization systems of semiconductors involving a contacting layer together with an inert diffusion barrier. Investigations of other high temperature diffusion barrier types in such structures are few (7,103).

5.3 Tandem Barriers

It is, of course, possible to connect several diffusion barriers in series, to try to further enhance the stability of a contact (Figure 12). The system last discussed can serve as an example. If the compound $TiAl_3$ is not admissible as a metallic overlay, but aluminum is mandatory, the titanium nitride diffusion barrier and the aluminum metal layer can be separated by a titanium film. Upon annealing, the titanium film will act as a sacrificial barrier and prevent the direct contact of aluminum with titanium nitride. Such a scheme may be readily implemented without introducing distinct deposition steps, but by changing from a reactive deposition to an inert deposition in a single operation.

We are unaware of studies or applications in which the concept of tandem barriers has been pursued explicitly. There are at least two reasons why tandem barriers should be considered. Firstly, the existence of two unequal sides in the tandem barrier offers the possibility to optimize the performance of each side with respect to chemical stability against the adjacent material. Secondly, when a single barrier fails predominantly by a localized breakdown, the probability for this

Table 4. List of Papers Investigating Properties of Thin Films of Interstitial Compounds as Diffusion Barriers in Complete Metallization Schemes of Semiconductors using a Contacting Layer (Structure S/C/X/M of Figure 2).

Diffusion barrier (X)	Method of Barrier Deposition*	Barrier thickness in nm	Semi-conductor substrate (S)	Contacting layer (C)	Metal overlayer (M)	Metal deposition method	Device/Test	Annealing time in min	Lowest temperature of observed instability in °C (Method), comments	Ref.
MoN_x	rf sputtering in Ar & N_2, $p_o < 7 \times 10^{-3}$ torr, $r \sim 0.2$ nm/s	10-20	Si	Mo	/Mo/Au	sputt.**		30	> 450	1
TaC	triode sputtering in Ar & CH_4	~ 80	Si	Pt ↓ PtSi	Al	sputt.**			~ 600	2
TaN	or in Ar & N_2, r = 0.15-0.2 nm/s	~ 100	Si		Al	sputt.**		90	600(RBS, optical)	
TaN	triode sputtering in Ar & N_2	100	Si	Cr ↓ $CrSi_2$	Ni	sputt.**		90	> 600(RBS)	3
TaN	triode sputtering in Ar & N_2	81		Ni ↓ NiSi	/Ta/Al ↓ $TaAl_3$	sputt.**		1200	> 600(RBS)	4
TiC	triode sputtering in Ar & CH_4, r = 0.1 nm/s	95	Si	Pt ↓ PtSi	Al	sputt.**		50	600(RBS, optical)	2
Ti_yC	rf magn. sputtering in Ar & CH_4 to form $Ti_{3.1}C$, P_o = low 10^{-7} torr, r = 1.1 nm/s, V_b = 30V (peak-to-peak), Ta = 750 °C (5)	100	Si	Ti_yC ↓ $TiSi_x$	Al	evap.		30	550(AES)	6,7

Table 4. List of Papers Investigating Properties of Thin Films of Interstitial Compounds as Diffusion Barriers in Complete Metallization Schemes of Semiconductors using a Contacting Layer (Structure S/C/X/M of Figure 2).

Diffusion barrier (X)	Method of Barrier Deposition*	Barrier thickness in nm	Semi-conductor substrate (S)	Contacting layer (C)	Metal overlayer (M)	Metal deposition method	Device/Test	Annealing time in min	Lowest temperature of observed instability in °C (Method), comments	Ref.
TiN	rf diode sputtering in Ar & N_2 or dc triode sputtering in Ar & N_2	15	Si	Ti (or PtSi/Ti)	Pt (or/Pt/Au)	sputt.		30	500; 325°C 16h: no PtTi formation (SIMS, electron microprobe electrical)	8-10
TiN	rf magn. sputtering in N_2, $p_o = 10^{-6}$ torr, r = 0.17-1 nm/s (11)	75	Si	Ti	Ag	evap.		10	600 (RBS)	12
TiN	triode sputtering in Ar & N_2	75	Si	Pt → PtSi	Ni	sputt.**		90	> 600(RBS)	3
TiN	rf magn. sputtering in N_2,	70	Si	Ti	Ag	evap.		80	600 (RBS)	13
TiN	or in Ar from TiN target, $p_o = 10^{-6}$ torr,	85	Si	Ti	Au	evap.		{10 60	400 (SEM) 450 (RBS)	
TiN	r = 1.0 nm/s (11)	85	Si	Ti	Al	evap.		{10 40	550 (SEM) 500 (RBS)	
TiN		100	Si	Ti→$TiSi_2$	Cu	evap.		30	700 (RBS)	
TiN		~ 100	Si	Ti	Pd	evap.		80	> 600(RBS)	

Table 4. List of Papers Investigating Properties of Thin Films of Interstitial Compounds as Diffusion Barriers in Complete Metallization Schemes of Semiconductors using a Contacting Layer (Structure S/C/X/M of Figure 2).

Diffusion barrier (X)	Method of Barrier Deposition*	Barrier thickness in nm	Semi-conductor substrate (S)	Contacting layer (C)	Metal overlayer (M)	Metal deposition method	Device/Test	Annealing time in min	Lowest temperature of observed instability in °C (Method), comments	Ref.
TiN	triode sputtering in Ar & N_2	95	Si	Pt	/Ti/Al → $TiAl_3$	sputt.**		1200	> 600(RBS)	4
TiN		60	Si	↓	→ $TiAl_3$/Al	sputt.**		120	600(RBS)	
TiN		95	Si	PtSi	→ Ti/$TiAl_3$	sputt.**		300	> 600(RBS)	
TiN	triode sputtering in Ar & N_2	103	Si	Pt ↓	Al	sputt.**		20	600(RBS)	2
TiN	r = 0.1 nm/s	78	Si	PtSi	Ni	sputt.**		1200	> 700(RBS)	
TiN	evaporation in N_2, $p_o = 1.5 \times 10^{-7}$ torr, r = 0.5-1.5 nm/s, T_a = 900 °C (Ar or N_2)	70	Si	Ti ↓ $TiSi_2$	Al	evap.		30	600(RBS)	14
TiN	evaporation in N_2, $p_o = 1.5 \times 10^{-7}$ torr, r = 0.5-1.5 nm/s, T_a = 900 °C (Ar or N_2)	70	Si	Ti ↓ $TiSi_2$	Al	evap.	shallow junction leakage, contact resistivity to n-Si	30	400 to 550: increasing leakage and ρ_c(2 to 5 x 10^{-6} Ω cm^2)	15
TiN	rf sputtering in Ar & N_2	50	Si	Ti ↓	Al	evap.	diffused resistor/ electro-migration in contacts		TiN effectively prevents Si out-diffusion into Al	16
TiN		50	Si	$TiSi_2$	Al:Cu	sputt.				

Table 4. List of Papers Investigating Properties of Thin Films of Interstitial Compounds as Diffusion Barriers in Complete Metallization Schemes of Semiconductors using a Contacting Layer (Structure S/C/X/M of Figure 2).

Diffusion barrier (X)	Method of Barrier Deposition*	Barrier thickness in nm	Semi-conductor substrate (S)	Contacting layer (C)	Metal overlayer (M)	Metal deposition method	Device/Test	Annealing time in min	Lowest temperature of observed instability in °C (Method), comments	Ref.
TiN	rf magn. sputtering in Ar & N_2, $p_o = 1 \times 10^{-7}$ torr, $r = 0.01$ (dynamic)-0.2 (static) nm/s $V_b = 70V$ (peak-to-peak)	36	Si	PtSi	Al:Cu	evap.	Schottky barrier diode	30	~ 500(RBS, electrical)	17
TiN		36	Si	$CoSi_2$	Al:Cu	evap.		30	~ 500(RBS, electrical)	
TiN	rf diode sputtering in Ar & N_2, $r = 0.11$ nm/s, $V_b = -100V$	50	Si, poly-Si	PtSi/Ti	/Ti/Pt/Au	evap.	transistor/ junction leakage		TiN increases lifetime (by a factor of 10^4 at 125 °C) and activation energy	18
TiN	magn. sputtering in Ar & N_2 $p_o = 7.5 \times 10^{-8}$ torr, $T_a = 800$ °C (vacuum)	160	Si	Ti → $TiSi_2$	Al	evap.		30	> 600(RBS) 600 (optical)	19
TiN	rf magn. sputtering in Ar & N_2, $p_o \sim 2 \times 10^{-6}$ torr, $r \sim 0.06$ nm/s (dynamic), $V_b = -40V$	60	Si	NiSi	Al	evap.	Schottky barrier diode	30	> 600(RBS) 550 (electrical)	20

Table 4. List of Papers Investigating Properties of Thin Films of Interstitial Compounds as Diffusion Barriers in Complete Metallization Schemes of Semiconductors using a Contacting Layer (Structure S/C/X/M of Figure 2).

Diffusion barrier (X)	Method of Barrier Deposition*	Barrier thickness in nm	Semi-conductor substrate (S)	Contacting layer (C)	Metal overlayer (M)	Metal deposition method	Device/Test	Annealing time in min	Lowest temperature of observed instability in °C (Method), comments	Ref.
TiN	rf sputtering in Ar & N_2 (17)	35	Si	$CoSi_2$	Al:Cu	evap.	shallow junction/ leakage, electro-migration		TiN increases lifetime by orders of magnitude	21
TiN	rf magn. sputtering in Ar & N_2, $p_o \sim 2 \times 10^{-6}$ torr, $r \sim 0.06$ nm/s (dynamic), $V_b = -40V$	75	GaAs	Ti + compounds	Ag	sputt.**	contact resistivity	120	> 500(RBS, electrical) $\rho_c \sim 7 \times 10^{-3}$ Ω cm^2	22
TiN	evaporation in N_2, $p_o < 2 \times 10^{-7}$ torr,	80	Si	Ti +	poly-Si: As	CVD	I-V measurement	30	~ 950 (electrical)	23
TiN	$r \sim 0.5$ nm/s, T_a = 900 °C (Ar) (14)	80	Si	$TiSi_2$	poly-Si: B	LPCVD		30	~ 950 (electrical)	
TiN	rf magn. sputtering in Ar & N_2, $p_o < 10^{-7}$ torr, (T_a = 800 °C)	150	Si	Ti + ($TiSi_2$)	Al	sputt.**	shallow junction leakage, contact resistivity	30	600(RBS) 525(optical) 525(ρ_c) 475(leakage)	24
TiN	rf magn. sputtering in Ar & N_2, $p_o \sim 2 \times 10^{-6}$ torr,	77	GaAs	Pt	Ag	sputt.	contact resistivity	30	> 550(RBS, electrical) $\rho_c \sim 10^{-4}$ Ω cm^2	25
TiN	$r \sim 0.07$ nm/s (dynamic) $V_b = -60V$	77	GaAs	Mg/Pt + compounds	Ag	sputt.		30		

Table 4. List of Papers Investigating Properties of Thin Films of Interstitial Compounds as Diffusion Barriers in Complete Metallization Schemes of Semiconductors using a Contacting Layer (Structure S/C/X/M of Figure 2).

Diffusion barrier (X)	Method of Barrier Deposition*	Barrier thickness in nm	Semi-conductor substrate (S)	Contacting layer (C)	Metal overlayer (M)	Metal deposition method	Device/Test	Annealing time in min	Lowest temperature of observed instability in °C (Method), comments	Ref.
TiN	rf magn. sputtering in Ar & N_2, $p_o \sim 2 \times 10^{-6}$ torr, $r \sim 0.05$ nm/s (dynamic) $V_b = -40V$	50	GaAs	/Ni/Ge/Au/ Pt ↓ compounds	/Ti/Pt/Au		MESFET, contact resistivity		TiN increases lifetime and the stability of ρ_c	26
ZrN	rf diode sputtering in Ar & N_2, $p_o < 3 \times 10^{-7}$ torr, $r = 0.23$ nm/s, $V_b \simeq -100V$	~ 120	Si	Zr ↓ ZrSi	Al	sputt.**		30	> 600(RBS)	27

*Deposition using an elemental metal source, unless noted

p_o = base pressure before deposition.

r = deposition rate.

T_s = substrate temperature during deposition.

T_a = preannealing temperature before metal overlayer deposition.

V_b = substrate bias voltage.

**All depositions in the same vacuum.

TABLE 4 – REFERENCES

1. Nowicki, R. S., and Wang, I., Improvement of the Diffusion Barrier Properties of Rf-Sputtered Molybdenum, *J. Vac. Sci. Technol.* **15**, 235 (1978).
2. Wittmer, M., Interfacial Reactions Between Aluminum and Transition-Metal Nitride and Carbide Films, *J. Appl. Phys.* **53**, 1007 (1982).
3. Wittmer, M., TiN and TaN as Diffusion Barriers in Metallizations to Silicon Semiconductor Devices, *Appl. Phys. Lett.* **36**, 456 (1980).
4. Wittmer, M., High-Temperature Contact Structures for Silicon Semiconductor Devices, *Appl. PHys. Lett.* **37**, 540 (1980).
5. Eizenberg, M., and Murarka, S. P., Reactively Sputtered Titanium Carbide Thin Films: Preparation and Properties, *J. Appl. Phys.* **54**, 3190 (1983).
6. Eizenberg, M., Murarka, S. P., and Heimann, P. A., Interaction of Reactively Sputtered Titanium Carbide Thin Films with Si, SiO_2, Ti, $TiSi_2$, and Al, *J. Appl. Phys.* **54**, 3195 (1983).
7. Eizenberg, M., Brener, R., and Murarka, S. P., Thermal Stability of the Aluminum/Titanium Carbide/Silicon Contact System, *J. Appl. Phys.* **55**, 3799 (1984).
8. P. R. Fournier, U.S. Patent 3,879,746; April 22, 1975 and P. R. Fournier, U.S. Patent 3,798,145, March 19, 1974.
9. Garceau, W. J., and Herb, G. K., TiN as a Diffusion Barrier in the Ti-Pt-Au Beam-Lead Metal System, *Thin Solid Films* **53**, 193 (1978).
10. Garceau, W. J., Fournier, P. R., and Herb, G. K., TiN as a Diffusion Barrier in the Ti-Pt-Au Beam-Lead Metal System, *Thin Solid Films* **60**, 237 (1979).
11. Mäenpää, M., von Seefeld, H., Cheung, N., Nicolet, M.-A., and Cullis, A. G., Characterization of Titanium Nitride Films Prepared by rf Sputtering, in *Proc. Symp. Thin Film Interfaces and Interactions* (J. E. E. Baglin and J. M. Poate, eds.), Vol. 80-2, p. 316, The Electrochemical Society, Princeton (1980).
12. von Seefeld, H., Cheung, N. W., Mäenpää, M., and Nicolet, M.-A., Investigation of Titanium-Nitride Layers for Solar-Cell Contacts, *IEEE Trans. Electron Devices* **27**, 873 (1980).
13. Cheung, N., von Seefeld, H., and Nicolet, M.-A., Titanium Nitride as a Diffusion Barrier in Metallization Schemes, in: *Proc. Symp. Thin Film Interfaces and Interactions* (J. E. E. Baglin and J. M. Poate, eds.), Vol. 80-2, p. 323, The Electrochemical Society, Princeton (1980).
14. Ting, C. Y., TiN Formed by Evaporation as a Diffusion Barrier Between Al and Si, *J. Vac. Sci. Technol.* **21**, 14 (1982).

15. Ting, C. Y., and Wittmer, M., The Use of Titanium-Based Contact Barrier Layers in Silicon Technology, *Thin Solid Films* **96**, 327 (1982).
16. Mori, M., Kanamori, S., and Ueki, T., Degradation Mechanism in Si-Doped Al/Si Contacts and an Extremely Stable Metallization System, *IEEE Trans. Comp., Hybrids, Man. Tech.* **6**, 159 (1983).
17. Schutz, R. J., TiN as a Diffusion Barrier Between $CoSi_2$ or PtSi and Aluminum, *Thin Solid Films* **104**, 89 (1983).
18. Kanamori, S., and Matzumoto, T., Suppression of Platinum Penetration Failure in Ti/Pt/Au Beam Lead Metal Systems Using a TiN Diffusion Barrier, *Thin Solid Films,* **110**, 205 (1983).
19. Norström, H., Donchev, T., Östling, M., and Petersson, C. S., $TiSi_2$/TiN – A Stable Multilayered Contact Structure for Shallow Implanted Junctions in VLSI Technology, *Phys. Scr.* **28**, 633 (1983).
20. Finetti, M., Suni, I., and Nicolet, M.-A., Titanium Nitride as a Diffusion Barrier between Nickel Silicide and Aluminum, *J. Electronic Mat.* **13**, 327 (19840.
21. Vaidya, S., Schutz, R. J., and Sinha, A. K., Shallow Junction Cobalt Silicide Contacts With Enhanced Electromigration Resistance, *J. Appl. Phys.* **55**, 3514 (1984).
22. Zhu, M. F., Hamdi, A. H., Nicolet, M.-A., and Tandon, J. L., Stable Ohmic Contact to GaAs With TiN Diffusion Barrier, *Thin Solid Films* **119**, 5 (1984).
23. Ting, C. Y., TiN as a High Temperature Diffusion Barrier for Arsenic and Boron, *Thin Solid Films* **119**, 11 (1984).
24. Nygren, S., Buchta, R., Gustafsson, V., Norström, H., Wiklund, P., Östling, M., and Petersson, C. S., The Ti/TiN/Al Metallization System – A Study of Stability, Contact Resistivity and Integrity to Shallow Junctions, in: *VTT Symposium 45: 11th Nordic Semiconductor Meeting* p. 271, VTT, Espoo, Finland (1984).
25. Kattelus, H. P., Tandon, J. L., Hamdi, A. H., and Nicolet, M.-A., Solid-Phase Ohmic Contacts to GaAs with TiN Diffusion Barriers, presented at the Fall Meeting of the Materials Research Society, Boston (1984).
26. Remba, R. D., Suni, I., and Nicolet, M.-A., Use of TiN Barrier to Improve GaAs FET Ohmic Contact Reliability, presented at the International Conference on Metallurgical Coatings, Lost Angeles (April 1985); to be published in *J. Vac. Sci. Technol.*
27. Östling, M., Nygren, S., Petersson, C. S., Norström, H., Wiklund, P., Buchta, R., Blom, H.-O., and Berg, S., Reactively Sputtered ZrN Used as an Al/Si Diffusion Barrier in a Zr Contact to Silicon, *J. Vac. Sci. Technol.* **A2**, 281 (1984).

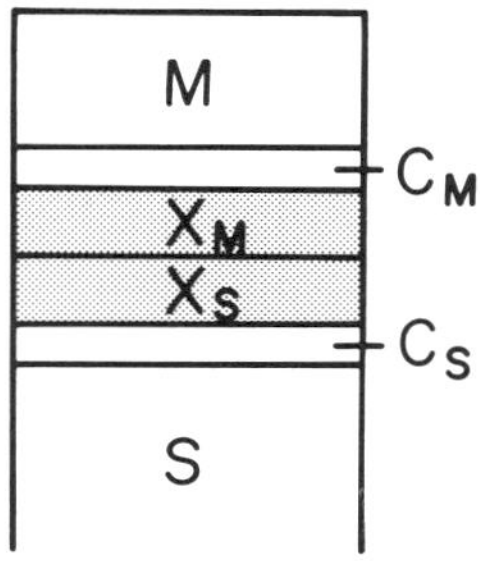

Fig. 12. Schematic diagram of a metallization scheme with a tandem diffusion barrier. The lower part of it, X_S, is optimized for the semiconductor, and the upper part, X_M, for the metal. Two contacting layers are included to complete the scheme.

type of failure may be significantly reduced in a case of two distinct layers. This concept is used commonly in the fabrication of thin film capacitors (114). If we identify four distinct types of diffusion barriers (stuffed, inert, amorphous and sacrificial) there are ten different ways of combining them in pairs. The number of possible approaches to solve any given particular problem is thus large, which gives hope that good solutions should normally be available with a proper insight and the means to study the subject.

The idea of a tandem barrier is only appealing if it can be assumed that two diffusion barriers in contact with each other are stable. In that case, the general problem can be broken down into two separate, individual problems. When and how two different diffusion barriers interact with each other thus is an important question. It has received no attention in current studies relating to diffusion barriers. Anticipating in a long extrapolation that different diffusion barriers can be found which interact little with each other, one is then led to view the tandem barrier with two contacting layers as a general conceptual scheme to enhance the stability of contacts (Fig. 12).

5.4 Summary

The concept of sacrificial diffusion barrier may have to be reconsidered in the light of some recent experiments. A uniform reaction between simple metals having a rate slow enough to make it controllable in practical systems may depend on the presence of impurities. Some stuffed or amorphous layers still serve as good examples of sacrificial barriers, but the reaction rates may have to be considered as system-dependent quantities.

It is not clear whether silicides (or arsenic and gallium compounds in the case of GaAs) can act as effective diffusion barriers by themselves. Presently, they are mainly used as contacting layers to Si (GaAs) to produce desirable electrical interface properties. An additional diffusion barrier layer is necessary to insure stability against the metallic overlay.

For future needs, a metallization scheme with a tandem diffusion barrier appears as a logical next step. In a tandem diffusion barrier, one side of the diffusion barrier is optimized for the semiconductor, the other for the overlayer. Two contacting layers are added for completeness to meet interfacial requirements, such as Schottky-barrier height or adhesion.

ACKNOWLEDGEMENTS

J. E. Baker (University of Illinois, Urbana, Illinois), R. S. Nowicki (Versatec Co., Santa Clara, California), and F. R. Shepherd (Bell-Northern Research, Ottawa, Canada) provided figures, some not yet published, for this review; we thank them for having made them available to us. We are also indebted to J. D. Wiley and J. H. Perepezko (University of Wisconsin, Madison, Wisconsin), and to C. Y. Ting and M. Wittmer (IBM, T. J. Watson Research Center, Yorktown Heights, New York) for providing us re- and pre-prints of their work. In our own work on diffusion barriers, we benefitted from active collaborations with J. L. Tandon (Applied Solar Energy Corporation, City of Industry, California) and D. R. Burger (Jet Propulsion Laboratory, California Institute of Technology, Pasadena, California). We take this opportunity to express our special gratitude to D. B. Bickler (Jet Propulsion Laboratory, California Institute of Technology, Pasadena, California), to J. R. Shappirio (U.S. Army Electronics Technology and Devices Laboratory, Fort Monmouth, New Jersey), and to I. Suni (Technical Research Centre of Finland, Espoo, Finland) for their personal interest in this subject and their constant encouragement. This work was financially supported in parts by Sandia National Laboratories under Contract 47-3967 at the California Institute of Technology, by the U.S. Department of Energy through an Agreement with the National Aeronautics and Space Administration and monitored by the Jet Propulsion Laboratory, California Institute of Technology, and by the U.S. Army, Electronics Technology and Devices Labroatory (ERADCOM), Ft. Monmouth, New Jersey. Our thanks also go to the

Technical Research Centre of Finland, Academy of Finland, and the Finnish Cultural Foundation, which provided financial support through a fellowship to H. P. Kattelus and granted him an extension of his sojourn at Caltech to complete this review, and to the Applied Solar Energy Corporation for a financial contribution. We also thank M. Parks for the excellent typing of the manuscript.

6. REFERENCES

1. Ghandhi, S. K., "*VLSI Fabrication Principles*", p. 435 (1983).
2. Murarka, S. P., "*Silicides for VLSI Applications*", New York: Academic Press (1983).
3. Nicolet, M-A., Diffusion Barriers in Thin Films, *Thin Solid Films*, **52**, 415 (1978).
4. Nicolet, M-A., and Bartur, M., Diffusion Barriers in Layered Contact Structures, *J. Vac. Sci. Technol.*, **19**, 786 (1981).
5. Ho, P. S., General Aspects of Barrier Layers for Very-Large-Scale Integration Applications I: Concepts, *Thin Solid Films,* **96**, 301 (1982).
6. Nowicki, R. S., and Nicolet, M-A., General Aspects of Barrier Layers for Very-Large-Scale Integration Applications II. Practice, *Thin Solid Films,* **96**, 317 (1982).
7. Nicolet, M-A., Suni, I., and Finetti, M., Amorphous Metallic Alloys in Semiconductor Contact Metallizations, *Solid-St. Technol.*, **26**, 129 (1983).
8. Wittmer, M., Barrier Layers: Principles and Applications in Microelectronics, *J. Vac. Sci. Technol.*, **A2**, 273 (1984).
9. Howard, J. K., White, J. F., and Ho, P. S., Intermetallic Compounds of Al and Transition Metals: Effect of Electromigration in 1-2- m-Wide Lines, *J. Appl. Phys.*, **49**, 4083 (1978).
10. Grabe, B., and Schreiber, H.-U., Lifetime and Drift Velocity Analysis for Electromigration in Sputtered Al Films, Multilayers, and Alloys, *Solid-St. Electron.*, **26**, 1023 (1983).
11. Duwez, P., Structure and Properties of Glassy Metals, *Annual Review of Materials Science,* Vol. 6, p. 105, Annual Reviews Inc., Palo Alto, CA (1976).
12. Lampert, C. M., Heat Mirror Coatings for Energy Conserving Windows, *Solar Energy Materials*, **6**, 1 (1981).
13. Gupta, D., Campbell, D. R., and Ho, P. S., in *Thin Films – Interdiffusion and Reactions* (J. M. Poate, K. N. Tu, and J. W. Mayer, eds.), p. 161, Wiley-Interscience, New York (1978).

14. Tu, K. N., and Rosenberg, R., Mass Transport in Layered Polycrystalline Thin Films, *Thin Solid Films*, **13**, 163 (1972).
15. Baglin, J. E. E., and Poate, J. M., in : *Thin Films – Interdiffusion and Reactions* (J. M. Poate, K. N. Tu, and J. W. Mayer, eds.), p. 352, Wiley-Interscience, New York (1978).
16. Hansen, M., *Constitution of Binary Alloys*, p. 217, New York: McGraw-Hill (1958).
17. Harris, J. M., Lugujjo, E., Capisano, S. U., Nicolet, M-A., and Shima, R., Studies of the Al_2O_3-Ti-Mo-Au Metallization System, *J. Vac. Sci. Technol.*, **12**, 524 (1975).
18. Nowicki, R. S., and Wang, I., Improvements of the Diffusion Barrier Properties of Rf-Sputtered Molybdenum, *J. Vac. Sci. Technol.*, **15**, 235 (1978).
19. Gjostein, N. A., in: *Diffusion*, American Society of Metals, Metals Park, Ohio (1973).
20. Bartur, M., and Nicolet, M-A., Electrical Characteristics of Al Contact to NiSi Using Thin W Layer as a Barrier, *Appl. Phys. Lett.*, **39**, 822 (1981).
21. van Gurp, G. J., and Reukers, W. M., Aluminum-Silicide Reactions. II. Schottky-Barrier Height, *J. Appl. Phys.*, **50**, 6923 (1979).
22. Ting, C. Y., and Wittmer, M., The Use of Titanium-Based Contact Barrier Layers in Silicon Technology, *Thin Solid Films*, **96**, 327 (1982).
23. Yamashita, M., and Oana, Y., Barrier Metal Against Ga and Zn Out-Diffusion in p-GaP/Au:Zn Contact System, *J. Appl. Phys.*, **52**, 7304 (1981).
24. Baker, J. E., Blattner, R. J., Nadel, S., Evans, C. A., Jr., and Nowicki, R. S., Thermal Annealing Study of Au/Ti-W Metallization on Silicon, *Thin Solid Films,* **69**, 53 (1980).
25. Morgan, A. E., private communication.
26. Cunningham, J. A., Fuller, C. R., and Haywood, C. T., Corrosion Resistance of Several Integrated-Circuit Metallization Systems, *IEEE Trans. Reliab.*, **19**, 182 (1970).
27. Dalal, H. M., Ghafghaichi, M., Kasprzak, L. A., and Wimpfheimer, H., U.S. Patent 4,214,256; July 22, 1980.
28. Chu, W. K., Howard, J. K., and White, J. F., U.S. Patent 4,206,472; June 3, 1980.
29. Hansen, M., *Constitution of Binary Alloys,* p. 1267, McGraw-Hill, New York (1958).
30. Goldschmidt, H. J., *Interstitial Alloys*, Plenum Press, New York (1967).

31. Bennett, L. H., McAlister, A. J., and Watson, R. E., Interstitial Compounds, *Physics Today*, **30**, 34 (1977).
32. Nelson, C. W., Metallization and Glassing of Silicon Integrated Cricuits, in: *Proc. Intl. Symp. Hybrid Microelectronics, Dallas (1969)*, p. 413, International Society of Hybrid Microelectronics, Montgomery (1969).
33. Hintermann, H. E., Thin Solid Films to Combat Friction, Wear, and Corrosion, *J. Vac. Sci. Technol.* **B2**, 816 (1984) – see also *Thin Solid Films,* **118** (No. 3) and references therein (1984).
34. Ahn, K. Y., Wittmer, M., and Ting, C. Y., Investigation of TiN Films Reactively Sputtered Using a Sputter Gun, *Thin Solid Films,* **107**, 45 (1983).
35. Karlsson, B., and Ribbing, C. G., Optical Properties of Transparent Heat Mirrors Based on Thin Films of TiN, ZrN, and HfN, *Proc. SPIE,* Vol. 324, p. 52 (1982).
36. Igasaki, Y., and Mitsuhashi, H., The Effects of Substrate Bias on the Structural and Electrical Properties of TiN Films Prepared by Reactive R.F. Sputtering, *Thin Solid Films*, **70**, 17 (1980).
37. Sundgren, J.-E., Johansson, B.-O., Karlsson, S.-E., and Hentzell, H. T. G., Mechanisms of Reactive Sputtering of Titanium Nitride and Titanium Carbide II: Morphology and Structure, *Thin Solid Films,* **105**, 367 (1983).
38. Lemperiere, G., and Poitevin, J. M., Influence of the Nitrogen Partial Pressure on the Properties of D.C. Sputtered Titanium and Titanium Nitride Films, *Thin Solid Films,* **111**, 339 (1984).
39. Wittmer, M., and Melchior, H., Applications of TiN Thin Films in Silicon Device Technology, *Thin Solid Films*, **93**, 397 (1982).
40. Noël, J.-P., Houghton, D. C., Este, G., Shepherd, F. R., and Plattner, H., Characteristics of dc Magnetron, Reactively Sputtered TiN_xFilms for Diffusion Barriers in III-V Semiconductor Metallization, *J. Vac. Sci. Technol.* **A2**, 284 (1984).
41. Schiller, S., Beister, G., and Sieber, W., Reactive High Rate D.C. Sputtering: Deposition Rate, Stoichiometry and Features of TiO_x and TiN_xFilms with Respect to the Target Mode, *Thin Solid Films,* **111**, 259 (1984).
42. Norström, H., Donchev, T., Ostling, M., and Petersson, C. S., $TiSi_2$/TiN – A Stable Multilayered Contact Structure for Shallow Implanted Junctions in VLSI Technology, *Phys. Scr.,* **28**, 633 (1983).
43. Mäenpää, M., von Seefeld, H., Cheung, N., Nicolet, M-A., and Cullis, A. G., Characterization of Titanium Nitride Films Prepared by rf Sputtering, in *Proc. Symp. Thin Film Interfaces and Interactions,* (J. E. E. Baglin and J. M. Poate, eds.), Vol. 8--2, p. 316, The Electrochemical Society, Princeton (1980).

44. Pan, A. I., Morgan, A. E., and Ritz, K., Reactive Sputter Deposition and Barrier Properties of Titanium Nitride, presented at the International Conference on Metallurgical Coatings, San Diego (1984).
45. Sinke, W., Stout, P. K., and Saris, F. E., Pre-Annealing of TiN Barriers in Al Metallization of Silicon, presented at the Fall Meeting of the Materials Research Society, Boston (1984).
46. Poitevin, J. M., Lemperiere, G., and Tardy, J., Influence of Substrate Bias on the Compositon, Structure, and Electrical Properties of Reactively D.C.-Sputtered TiN Films, *Thin Solid Films*, **97**, 69 (1982).
47. Sundgren, J.-E., Johansson, B.-O., Hentzell, H. T. G., and Karlsson, S.-E., Mechanisms of Reactive Sputtering of Titanium Nitride and Titanium Carbide III: Influence of Substrate Bias on Composition and Structure, *Thin Solid Films,* **105**, 385 (1983).
48. Kanamori, S., and Matzumoto, T., Suppression of Platinum Penetration Failure in Ti/Pt/Au Beam Lead Metal Systems Using a TiN Diffusion Barrier, *Thin Solid Films,* **110**, 205 (1983).
49. Schutz, R. J., TiN as a Diffusion Barrier Between $CoSi_2$ or PtSi and Aluminum, *Thin Solid Films,* **104**, 89 (1983).
50. Suni, I., Mäenpää, Nicolet, M-A., and Luomajärvi, M., Thermal Stability of Hafnium and Titanium Nitride Diffusion Barriers in Multilayer Contacts to Silicon, *J. Electrochem. Soc.*, **103**, 1215 (1983).
51. Wittmer, M., Interfacial Reactions Between Aluminum and Transition-Metal Nitride and Carbide Films, *J. Appl. Phys.*, **53**, 1007 (1982).
52. Beyers, R., Sinclair, R., and Thomas, M. E., Phase Equilibria in Thin-Film Metallizations, *J. Vac. Sci. Technol.* **B2**, 781 (1984).
53. Sigurd, D., Suni, I., Wielunski, L., Nicolet, M-A., and von Seefeld, H., Thermal Oxidation of Sputtered TiN Diffusion Barriers, *Solar Cells*, **5**, 81 (1981-1982).
54. Wittmer, M., Noser, J., and Melchior, H., Oxidation Kinetics of TiN Thin Films, *J. Appl. Phys.*, **52**, 6659 (1981).
55. Suni, I., Sigurd, D., Ho, K. T., and Nicolet, M-A., Thermal Oxidation of Reactively Sputtered Titanium Nitride and Hafnium Nitride Films, *J. Electrochem. Soc.*, **130**, 1211 (1983).
56. Garceau, W. J., Fournier, P. R., and Herb, G. K., TiN as a Diffusion Barrier in the Ti-Pt-Au Beam-Lead Metal System, *Thin Solid Films*, **60**, 237 (1979).
57. Finetti, M., Suni, I., and Nicolet, M-A., Contact Resistivity of TiN on p^+-Si and n^+-Si, *Solar Cells,* **9**, 179 (1983).
58. Murarka, S. P., Interactions in Metallization Systems for Integrated Circuits, *J. Vac. Sci. Technol.* **B2**, 693 (1984).

59. Ting, C. Y., Silicide for Contacts and Interconnects, presented at the Int. Electron Devices Mtg., San Francisco (Dec. 1984).
60. Fournier, P. R., U.S. Patent 3,879,746; April 22, 1975 and Fournier, P. R., U.S. Patent 3,798,145; March 19, 1974.
61. Cheung, N. W., von Seefeld, H., Nicolet, M-A., Ho, F., and Iles, P., Thermal Stability of Titanium Nitride for Shallow Junction Solar Cell Contacts, *J. Appl. Phys.* **52**, 4297 (1981).
62. Mäenpää, M., Nicolet, M-A., Suni, I., and Colgan, E. G., Contact Resistivities of Sputtered TiN and Ti-TiN Metallizations on Solar-Cell-Type Silicon, *Solar Energy*, **27**, 283 (1981).
63. Mäenpää, M., Suni, I., Nicolet, M-A., Ho, F., and Iles, P., Contact Resistivities Between Solar-Cell-Type Si and Transition Metal Nitrides, in: *Proc. 15th IEEE Photovoltaic Specialists Conference,* p. 518 (1981).
64. Finetti, M., Suni, I., Bartur, M., Banwell, T., and Nicolet, M-A., Schottky Barrier Height of Sputtered TiN Contacts on Silicon, *Solid-St. Electron.,* **27**, 617 (1984).
65. Finetti, M., Suni, I., and Nicolet, M-A., Finite Metal-Sheet-Resistance in Contact Resistivity Measurements: Application to Si/TiN Contacts, *Solid-St. Electron.,* **26**, 1065 (1983).
66. Wittmer, M., Studer, B., and Melchior, H., Electrical Characteristics of TiN Contacts to N Silicon, *J. Appl. Phys.,* **52**, 5722 (1981).
67. Waldrop, J. R., Titanium Nitride Schottky-Barrier Contacts to GaAs, *Appl. Phys. Lett.,* **43**, 87 (1983).
68. Wittmer, M., Noser, J. R., and Melchior, H., Characteristics of TiN Gate Metal-Oxide-Semiconductor Field Effect Transistors, *J. Appl. Phys.,* **54**, 1423 (1983).
69. Houghton, D. C., TiN as a Diffusion Barrier for III-V Metallization, in: *Thin Films and Interfaces II*, Mat. Res. Soc. Symp. Proc. Vol. 25, (J. E. E. Baglin, D. R. Campbell, and W. K. Chu, eds.), p. 149, Elsevier Science, New York (1984).
70. Ting, C. Y., TiN Formed by Evaporation as a Diffusion Barrier Between Al and Si, *J. Vac. Sci. Technol.,* **21**, 14 (1982).
71. Igasaki, Y., Mitsuhashi, H., Azuma, K., and Muto, T., Structure and Electrical Properties of Titanium Nitride Films, *Jpn. J. Appl. Phys.,* **17**, 85 (1978).
72. Armigliato, A., Celotti, G., Garulli, A., Guerri, S., Lotti, R., Ostoja, P., and Summonte, C., Properties of TiN Obtained by N_2^+-Implantation on Ti-Coated Si Wafers, *Appl. Phys. Lett.,* **41**, 446 (1982).
73. Dawson, P. T., and Stazyk, S. A. J., The Preparation and Characterization of Transition Metal Nitride Films, *J. Vac. Sci. Technol.*, **20**, 966 (1982).

74. Bunshah, R. F., The Activated Reactive Evaporation Process: Developments and Applications, *Thin Solid Films,* **80**, 255 (1981).
75. Kobayashi, M., and Doi, Y., TiN and TiC Coating on Cemented Carbides by Ion Plating, *Thin Solid Films,* **54**, 67 (1978).
76. Karlsson, B., Shimshock, R. P., Seraphin, B. O., and Haygarth, J. C., Optical Properties of CVD-Coated TiN, ZrN and HfN, *Solar Energy Materials,* **7**, 401 (1983).
77. Bunshah, R. F., Overview of Coating Technologies for Large Scale Metallurgical, Optical, and Electronic Applications, *J. Vac. Sci. Technol.*, **B2**, 789 (1984).
78. Chamberlain, M. B., Diffusion of Copper in Thin TiN Films, *Thin Solid Films*, **91**, 155 (1982).
79. Randhawa, H. S., Matthews, M. D., and Bunshah, R. F., SnO_2 Films Prepared by Activated Reactive Evaporation, *Thin Solid Films,* **83**, 267 (1981).
80. van Gurp, G. J., Daams, J. L. C., van Oostrom, A., Augustus, L. J. M., and Tamminga, Y., Aluminum-Silicide Reactions. I. Diffusion, Compound Formation, and Microstructure, *J. Appl. Phys.,* **50**, 6915 (1979).
81. Ting, C. Y., and Wittmer, M., Investigation of the Al/$TiSi_2$/Si Contact System, *J. Appl. Phys.*, **54**, 937 (1983).
82. Nicolet, M-A., and Lau, S. S. Formation and Characterization of Transition-Metal Silicides, in: *VLSI Electronics: Microstructure Science*, Vol. 6, (N. G. Einspruch and G. B. Larrabee, eds.), p. 329, Academic Press, New York (1983).
83. Neppl. F., and Schwabe, U., Barrier Properties of $TaSi_2$ in Contact with Al-Metallization, in *Thin Films and Interfaces II*, Mat. Res. Soc. Symp. Proc. Vol. 25, (J. E. E. Baglin, D. R. Campbell, and W. K. Chu, eds.), p. 587, Elsevier Science, New York (1984).
84. Shoji, F., and Nagata, S., Structure and Deposition Mechanism of Molybdenum Nitride Films Prepared by Reactive Sputtering, *Jpn. J. Appl. Phys.*, **13**, 1072 (1974).
85. Kim, M. J., Brown, D. M., and Katz, W., Molybdenum Nitride Film Formation, *J. Electrochem. Soc.,* **130**, 1196 (1983).
86. Donald, I. W., and Davies, H. A., Predictions of Glass-Forming Ability for Metallic Systems, *J. Non-Crystalline Solids,* **30**, 77 (1978).
87. Balluffi, R. W., and Blakely, J. M., Special Aspects of Diffusion in Thin Films, *Thin Solid Films*, **25**, 363 (1975).
88. Chen. H. S., Kimerling, L. C., Poate, J. M., and Brown, W. L., Diffusion in a Pd-Cu-Si Metallic Glass, *Appl. Phys. Lett.* **32**, 461 (1978).

89. Zhu, M. F., Suni, I., and Nicolet, M-A., Electrical Characteristics of Amorphous $Ni_{36}W_{64}$ Contacts on Silicon, in *SPIE Vol. 463 – Advanced Semiconductor Processing and Characterization of Electronic and Optical Materials,* (D. K. Sadana and C. M. Lampert, eds.), p. 19, SPIE, Bellingham, Washington (1984).
90. Suni, I., Nicolet, M-A., Pai, C. S., and Lau, S. S., Stability of Amorphous Fe-W Alloys in Multilayer Metallizations on Silicon, *Thin Solid Films,* **107**, 73 (1983).
91. Zhu, M. F., Suni, I., and Nicolet, M-A., Reaction of Amorphous Ni-W and Ni-N-W Films with Substrate Silicon, *J. Appl. Phys.,* **56**, 2740 (1984).
92. Zhu, M. F., So, F. C. T., and Nicolet, M-A., Amorphous Ni-N-W Films as a Diffusion Barrier Between Al and Si, *Thin Solid Films* (to be published).
93. Krafcsik, I., Gyulai, J., Palmstrom, C. J., and Mayer, J. W., Influence of Cu as an Impurity in Al/Ti and Al/W Thin-Film Reactions, *Appl. Phys. Lett.*, **43**, 1015 (1983).
94. Bartur, M., and Nicolet, M-A., Aluminum Contact to Nickel Silicide Using a Thin Tungsten Barrier, *Thin Solid Films,* **91**, 89 (1982).
95. Zhu, M. F., So, F. C. T., Pan, E. T-S., and Nicolet, M-A., Investigation of Amorphous $W_{60}Zr_{40}$ Film as a Diffusion Barrier in Metallization Schemes, *Phys. Stat. Solidi* (to be published).
96. Bower, R. W., Characteristics of Aluminum-Titanium Electrical Contacts on Silicon, *Appl. Phys. Lett.*, **23**, 99 (1973); Thesis, California Institute of Technology (1973).
97. J. K. Howard, F. E. Turene, and J. F. White; U. S. Patent 4,201,-999; May 6, 1980.
98. Howard, J. K., Lever, R. F., Smith, P. J., and Ho, P. S., Kinetics of Compound Formation in Thin Film Couples of Al and Transition Metals, *J. Vac. Sci. Technol.* **13**, 68 (1976).
99. Bartur, M., and Nicolet, M-A., Chromium as a Diffusion Barrier Between NiSi, Pd_2Si, or PtSi and Al, *J. Electrochem. Soc.,* **131**, 1118 (1984).
100. Salomonson, G., Holm, K. E., and Finstad, T. G., A Study of Ti as a Diffusion Barrier Between PtSi or Pd_2Si and Al, *Phys. Scr.,* **24**, 401 (1981).
101. Merchang, P., and Amano, J., Thermal Stability of Diffusion Barriers for Aluminum Alloy-Platinum Silicide Contacts, *J. Vac. Sci. Technol.*, **A1**, 459 (1983).
102. Ting, C. Y., and Crowder, B. L., Electrical Properties of Al/Ti Contact Metallurgy for VLSI Application, *J. Electrochem. Soc.*, **129**, 2590 (1982).

103. Ghate, P. B., Blair, J. C., Fuller, C. R., and McGuire, G., E., Application of Ti:W Barrier Metallization for Integrated Circuits, *Thin Solid Films,* **53**, 117 (1978).
104. Ghate, P. B., Blair, J. C., and Fuller, C. R., Metallization in Microelectronics, *Thin Solid Films*, **45**, 69 (1977).
105. Sheperd, F. R., unpublished.
106. Remba, R. D., Suni, I., and Nicolet, M-A., Use of TiN Barrier to Improve GaAs FET Ohmic Contact Reliability, presented at the International Conference on Metallurgical Coatings, Los Angeles (April 1985); to be published in *J. Vac. Sci. Technol.*
107. Chu, W. K., Mayer, J. W., and Nicolet, M-A., *Backscattering Spectrometry*, Academic Press, New York (1978).
108. Nygren, S., Buchta, R., Gustafsson, U., Norström, H., Wiklund, P., Ostling, M., and Petersson, C. S., The Ti/TiN/Al Metallization System – A Study of Stability, Contact Resistivity and Integrity to Shallow Junctions, in: *VTT Symposium 45: 11th Nordic Semiconductor Meeting,* p. 271, VTT, Espoo, Finland (1984).
109. Ostling, M., Nygren, S., Petersson, C. S., Norström, H., Wiklund, P., Buchta, R., Blom, H-O., and Berg, S., Reactively Sputtered ZrN Used as an Al/Si Diffusion Barrier in a Zr Contact to Silicon, *J. Vac. Sci. Technol.*, **A2**, 281 (1984).
110. Wittmer, M., TiN and TaN as Diffusion Barriers in Metallizations to Silicon Semiconductor Devices, *Appl. Phys. Lett.,* **36**, 456 (1980).
111. Zhu, M. F., Hamdi, A. H., Nicolet, M-A., and Tandon, J. L., Stable Ohmic Contact to GaAs with TiN Diffusion Barrier, *Thin Solid Films,* **119**, 5 (1984).
112. Wittmer, M., High-Temperature Contact Structures for Silicon Semiconductor Devices, *Appl. Phys. Lett.,* **37**, 540 (1980).
113. Samsonov, G. V., and Vinitskii, I. M., *Handbook of Refractory Compounds,* IFI/Plenum, New York (1980).
114. Gerstenberg, D., Thin Film Capacitors, in: *Handbook of Thin Film Technology* (L. I. Maissel and R. Glang, eds.), p. 19-29, McGraw-Hill, New York (1970).

9

Thermal Stability of Pb-Alloy Josephson Junctions

Masanori Murakami

I. INTRODUCTION

Computer performance has significantly improved in the last few decades: computing speed has increased by a factor of ten every thirteen years and the density of devices on a chip has increased by a factor of two every two years. The present computer technology is primarily based on the Si technology. A big concern of the Si technology is large power dissipation to operate the devices. Each device dissipates power of the order of several microwatts. Thus, total power dissipation increases significantly with increasing device density. Development of new devices which would overcome the limitations of current Si technology has started actively since the last decade.

One of the candidates for the new devices is the Josephson superconductive device which operates at liquid helium temperature (4.2K). Excellent reviews on this device are given in Ref. (1). The device possesses two key properties that are essential to building such computers: fast switching speeds and low power dissipation. Compared with the current computers made with the Si technology, the speed is expected by about an order of magnitude faster and the power dissipation three orders of magnitude smaller. The low power dissipation of the devices should allow them to be packed densely without incurring heat removal penalties.

The type of Josephson device most suitable for high speed computer applications is a SQUID (Superconducting Quantum Interference Device) comprised of several tunneling junctions which share common electrodes (2,3). A schematic cross-section through the center of one such currently used device is shown in Fig. 1 (3). The junction portions of the device consist of two superconductive electrodes separated by an ultra-thin (6nm thick) tunnel barrier. The junction regions are defined by windows in a thicker insulating layer.

Selection of electrode materials used in the devices is one of the key factors which influence the success in building Josephson computers in the near future (4). We wish to mention in passing that currently there is excitement and optimism in the superconductivity field following the discovery of a new class of materials which exhibit transition temperatures in excess of 80K. An experimental SQUID based on $Y_1Ba_2Cu_3O_7$ has recently been reported (5). It is too early to say if these materials would someday be a serious contender for a SQUID operating at liquid nitrogen temperature. Since these materials are complex ceramics, their fabrication would require considerable developmental effort. Our discussion is therefore confined to the existing technology. Refractory materials such as Nb, NbN or A15 compounds are attractive from the view point of thermal stability. However, it is extremely difficult to obtain reproducible electrical properties with these materials. Research on these refractory materials was focussed on the goal of establishing a reproducible process and significant progress has been made (6-8). Low melting point Pb-alloy materials are attractive because of the excellent electrical characteristics obtained for Josephson devices: Pb and Pb alloys have a relatively high superconducting transition temperature, oxide tunnel barrier thicknesses can be controlled within one atomic layer in average thickness by growing them on Pb-alloy films (9,10), and the oxides have small dielectric constants which are required for fast switching and resetting times of the junctions. To operate the devices, they are immersed at liquid helium. Also, the present Pb-alloy fabrication process requires heating the devices up to 350K.

A principal materials-related concern with Pb-alloy Josephson junctions is the reliability of the devices during thermal cycling between room and liquid He temperature (4.2K) and between room temperature and 350K. The failure occurring in some Pb-alloy junctions during thermal cycling is believed to be caused by the rupture of the ultra-thin tunnel barrier oxide when the microstructure change of the base or counter electrode is larger than the thickness of the tunnel barrier (11). The microstructure changes are caused by relaxation of the strain in the electrode films induced by thermal expansion coefficient mismatch between the Pb-alloy films and the Si substrate upon cooling to 4.2K or heating to 350K. (Silicon is chosen as the substrate material, because the surface polishing and via-hole formation techniques used to package the chips are well established.) Thus, to reduce the device failure, the strain behavior and corresponding microstructure changes upon thermal cycling have been studied extensively.

The present chapter will review studies carried out for Pb-alloy electrode materials in order to improve reliability of the Josephson devices. Before we describe details on thermal stability of Pb-alloy electrode materials, thermal strain of thin films bonded to oxidized Si substrates expected upon thermal cycling will be defined in Section II. Strain behavior upon cooling to 4.2K and correlation between strain relaxation mechanisms and microstructure changes observed after repeated thermal cycling will be described for pure Pb thin films in Section III.1. The goal is to obtain a clear understanding for strain behavior. Then, thermal stabilities of Pb-In-Au and Pb-Bi films used for base and counter electrode materials will be described in Sections III.2 and 3, respectively. Since the base electrode material must be heated to 350K, strain behavior of Pb-In-Au alloy films above room temperature will be described in Section IV. Finally, it is shown that diffusion in the lattice and along the grain-boundaries plays an important role in the strain relaxation mechanisms.

II. THERMAL STRAIN OF FILMS ON RIGID SUBSTRATES

As shown in Fig. 1, the Josephson junctions are made by depositing several Pb-alloy layers onto the oxidized Si substrates. Since the melting points of the Pb-alloy films are relatively low, the films have negligibly small intrinsic stress in the as-deposited state. The strain (or stress) is introduced into the films when the devices are cooled to 4.2K or heated to 350K due to thermal mismatch between the Pb-alloy films and the Si substrates.

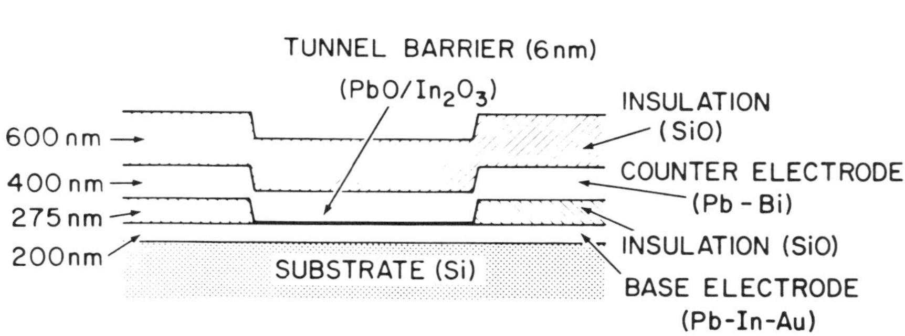

Fig. 1. Schematic cross-section of Pb-alloy Josephson junction device.

When the films are stressed by an amount smaller than the elastic limit, they deform elastically and return to their original shape after the stress is relieved. However, when the stress is applied to the film beyond the elastic limit, the film deforms plastically and does not return to the original shape after the stress is relieved. The plastic deformation (i.e. strain relaxation) causes device failure as explained later. The strain relaxation mechanisms, including diffusion, are strongly dependent on film microstructure, ambient temperature and applied stress levels.

Here, thermal strain in a Pb film on a Si substrate is defined. Cross-sections of the film/substrate composite before and after temperature change are shown schematically in Fig. 2. Figures 2(a)-(c) show the cooling process down to temperature T_1. Suppose a Pb film is deposited at ambient temperature (T_0) onto the Si substrate and, then, subsequently annealed at T_0 for sufficient time so that the film is essentially strain-free at this temperature. The cross-section of this film/substrate composite is shown in Fig. 2(a). Then, the film/substrate composite is cooled to the temperature T_1. If there is no adhesion between the film and the substrate (i.e., free-standing film), the film shrinks isotropically as shown by a solid line in Fig. 2(b) and the film is strain-free at the temperature T_1. (The dashed line shows the cross-section of the film at T_0 for reference). However, if the film has strong adhesion to the Si substrate, the film is strained in the planar direction, because the thermal expansion coefficients of Pb are much larger than those of Si. The strained film is illustrated in Fig. 2(c). The thickness (h_e) of the elastically strained film is smaller than that of the strain-free film (h_o). The thermal strain ε_{therm} (in the planar direction) is given by,

$$\varepsilon_{therm} = \int_{T_o}^{T_1} (\alpha_{film} - \alpha_{sub})dT \qquad (1)$$

where α_{film} and α_{sub} are thermal expansion coefficients of the film and the substrate, respectively. Thermal strain introduced upon heating the film/substrate composite is also calculated by Eq. (1). (The sign of ε_{therm} is "+" for tension and "−" for compression). The tensile and compressive strain is expected in the film upon cooling and heating, respectively.

When the ε_{therm} value is within the elastic limit of the film, it can support the total strain elastically, i.e., no strain relaxation occurs in this case. Since the strain is introduced biaxially into the film from the film/substrate interface, non-uniform distribution of the elastic strain in such a film could cause the microstructure to change. When the ε_{therm} value is beyond the elastic limit, the total strain cannot be supported elastically by the film and a part of the total strain is relaxed upon cooling from T_0 to T_1 or heating from T_0 to T_2. The microstructure changes upon thermal cycling are dependent on the amounts of strain relaxed and the relaxation mechanisms. This strain behavior is dependent on the film microstructure, the applied stress level, the temperature and the cooling or heating rate. It should be kept in mind that upon cooling (or heating) the film/substrate composite, the strain

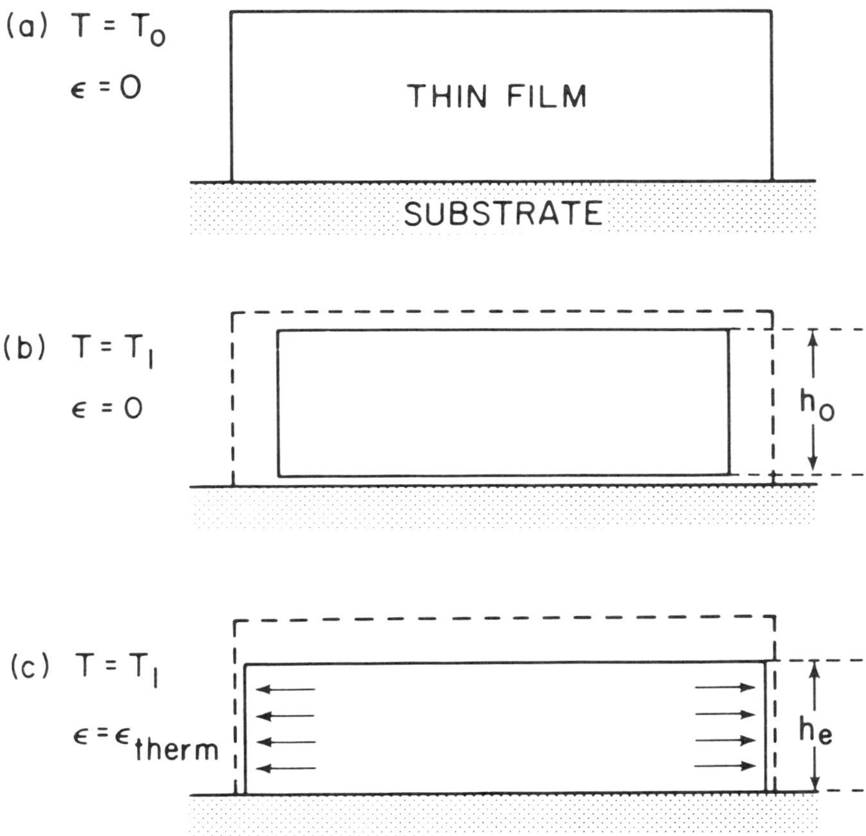

Fig. 2. Cross-section of thin film bonded to the rigid substrate: (a) strain-free film at $T = T_o$, (b) strain-free (or free-standing) film at $T = T_1$, and (c) strained film at $T = T_1$.

(or stress) level and the ambient temperature change simultaneously. There are typically various stress-temperature regions in which different strain relaxation mechanisms are dominant. Therefore, it is important to study the relaxation mechanisms of the Pb thin film bonded to the Si substrate at temperatures at which the devices are cooled or heated. The strain behavior and microstructure changes upon cooling to 4.2K or heating to 350K are explained in the next two sections.

III. THERMAL STABILITY BELOW ROOM TEMPERATURE

1. Pb Thin Films

(a) Strain Behavior. Strain behavior of Pb thin films which are deposited onto the Si substrates and then cooled from room temperature to 4.2K is described in this section. As explained in Fig. 2, the film thickness h is indicative of strain behavior in the film. When the measured h value is close to the h_o value of the strain-free film, the film is almost strain-free and when the h value is close to the h_e value of the elastically strained film, the film is elastically strained. Since it is extremely difficult to measure accurately the film thickness at a given temperature, the interplanar spacing (d) between crystal planes parallel to the substrate surface was measured by using the Bragg-Brentano x-ray diffraction method. (For films with column-type grains like Pb films, the mean d value is proportional to the film thickness h; $d = h/n$ where n is number of crystal planes with a particular orientation parallel to the substrate surface).

The calculated and measured interplanar spacings between (111) planes parallel to the film surfaces are shown as a function of temperature in Fig. 3. The d_o curve describes the strain free d-values of bulk Pb which were calculated using the reported thermal expansion coefficients of Pb (12). The d_{max} curve describes the d-values for the maximum strain reached in films upon cooling from 298 to 4.2K. These maximum values were calculated using a biaxial strain model (13-15) and assuming no strain relaxation occurred upon cooling.

The experimental d values for (111) planes measured by using the x-ray diffraction technique during the thermal cycling are indicated by various symbols (16). The open marks show the cooling process and the closed marks show the heating process. From the d-values the strain ε'_{33} normal to the film surface is calculated by,

$$\varepsilon'_{33} = (d-d_o)/d_o \qquad (2)$$

where d_o is the strain-free value at a given temperature. When $\varepsilon'_{33}<0$ the film is under tensile strain and when $\varepsilon'_{33}>0$ the film is under

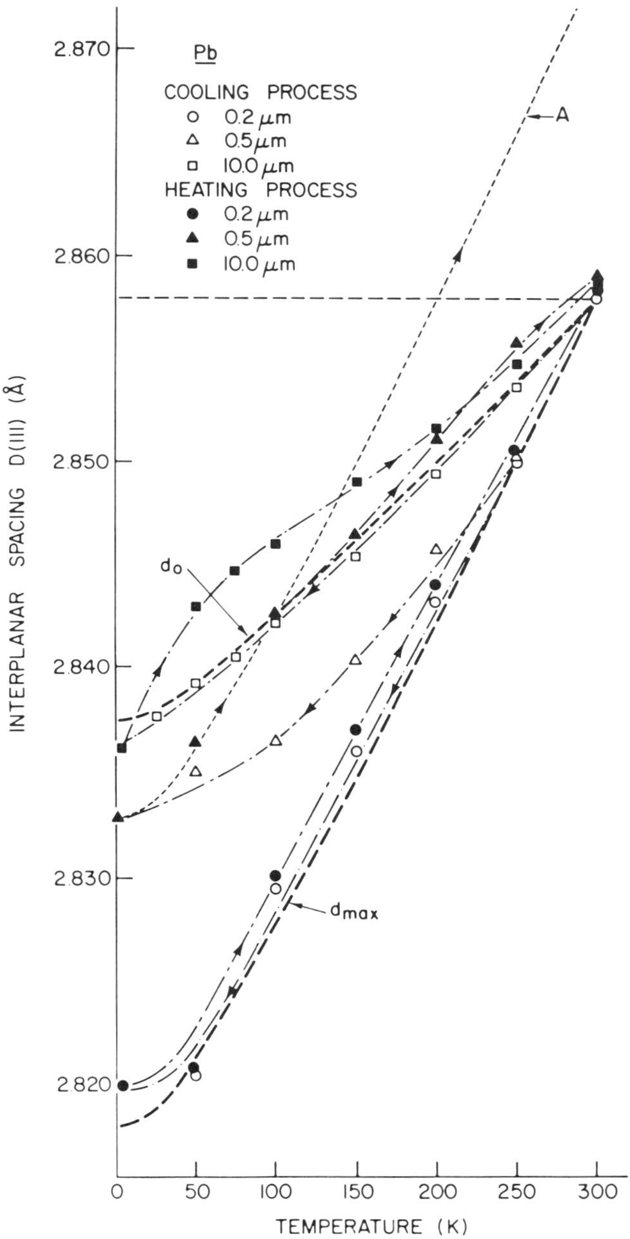

Fig. 3. Interplanar spacing changes of Pb films during thermal cycling between 298 and 4.2K. (After M. Murakami, Thin Solid Films **55**, 101, 1978)

compressive strain along a direction parallel to the film surface.

The d-values of the 0.2μm thick film were found to be close to the d_{max} values, indicating that the film is elastically strained in tension and almost no strain relaxation was observed in the film during the cycling. On the other hand, the d-values of the film with 10μm thickness were observed to be close to the d_o values during the cooling process. This means that the tensile strains which were produced due to the thermal contraction difference between the film and the substrate were completely relaxed upon cooling. However, compressive strain was observed in the 10μm film upon heating from 4.2 to 298K. For the 0.5μm thick film a tensile strain is introduced upon cooling. The film is elastically strained above 250K and the strains start to be relaxed plastically upon further cooling. During the heating process the d value becomes close to the d_o value at approximately 100K. At this temperature the average elastic strain is essentially zero. Upon subsequent heating compressive strain is introduced into the film. The dotted curve (indicated by A) shows the d values expected for the 0.5μm thick film upon heating from 4.2K which were calculated based on an assumption that there was no strain relaxation. Comparing the measured and calculated dotted curves it is noted that most of the compressive strain is relaxed during heating. The d values of a 1μm thick film are almost the same as those of the 0.5μm thick film.

From the strain measurement, it is noted that the relatively thick films do not support the strain elastically upon cooling to 4.2K. The amounts of strain relaxed upon cooling are dependent on the film thickness: more strain is relaxed by increasing the thickness (15,16). The strain relaxation mechanism at low temperature was studied by scanning electron (17,18) and transmission electron microscopy (19) and the study carried out by TEM is reviewed below.

Microstructure change of Pb thin films with 0.2μm thickness upon cooling was observed in-situ by TEM for the films which were deposited onto Si_3N_4 substrates at room temperature (19). The films were observed to have column-type grains with various sizes and no dislocation was observed in these grains before cooling. Upon cooling the film/substrate composite to 100K, the dislocations were observed. A typical TEM micrograph (19) taken at 100K is shown in Fig. 4. The dislocations were observed to emanate from grain boundaries and glide across the grains on (111) planes that are either parallel or slightly inclined to the substrate surface. All dislocation motions are confined in each grain by the native surface oxide, the substrate, grain boundaries and twin boundaries.

These dislocations were observed only in grains larger than 0.6μm in size (19). The dislocation density was measured to be $\sim 10^{10}$ cm^{-2} and was not affected by the grain sizes. The observed dislocation

Fig. 4. Transmission electron micrograph of a Pb alloy thin film deposited onto the Si_3N_4 substrate at room temperature and then cooled at 100K. (After J. H. Basson, M. Murakami and H. Booyens, J. Appl. Phys. **53**, 337, 1982).

density can account for the value of strain relaxation which was measured by the x-ray diffraction technique (15).

The dislocation glide was observed in Pb, Pb-In(20), Pb-Bi(21), and Pb-In-Au films with face-centered-cubic (fcc) structure and it was concluded that the dominant strain relaxation mechanism at low temperature is dislocation glide for these films.

(b) Microstructure Changes After Repeated Thermal Cycling. The microstructure change after repeated thermal cycling between 298 and 4.2K is discussed in this section for films in which strain is relaxed upon thermal cycling or for those which support the total strain elastically. As discussed in the previous section, the amounts of strain relaxed upon cooling depend on the film thickness. The microstructure changes for thick or thin films (corresponding to films with or without strain relaxation, respectively) are discussed below.

For films which do not support the total strain elastically at low temperatures, the strain was observed by TEM to be relaxed by dislocation glide upon cooling from 298K to 4.2K. Thus, a high density of dislocation slip steps were observed inside grains after repeated thermal cycling (17,18,22,23). A scanning electron micrograph of a 1.0μm thick Pb film taken after 100 cycles between 298 and 4.2K is shown in Fig. 5, where a high density of dislocation slip lines are observed inside grains (22). At grain boundaries formation of hillocks (protrusion from the film surface) is observed. These hillocks are formed to relieve the compressive strain introduced at room temperature after thermal cycling (22-28); atoms in the film preferentially diffuse to the film surface through the grain boundaries, particularly through triple points of the boundaries (28).

Another major microstructure change induced by strain relaxation upon repeated thermal cycling between 298 and 4.2K is grain rotation (26). Pb films prepared at ambient temperature have a strong (111) fiber structure (i.e., (111) crystal planes of grains are close to parallel to the film surface), but their (111) planes are not perfectly parallel to the film surface. A cross-section of a Pb film bonded to a Si substrate before thermal cycling is shown schematically in Fig. 6(a) where the (111) planes are illustrated by fine lines (29). An angle ω indicates

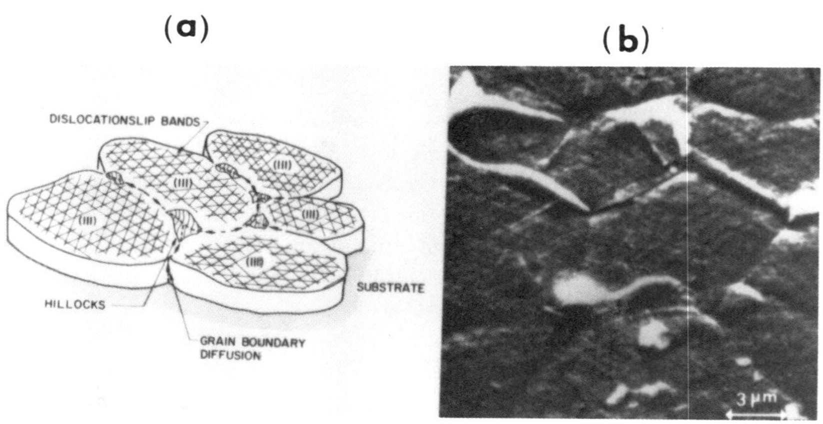

Fig. 5. Scanning electron micrograph of a Pb thin film after repeated cycling between 298 and 4.2K. (After M. Murakami, Thin Solid Films **55**, 101, 1978).

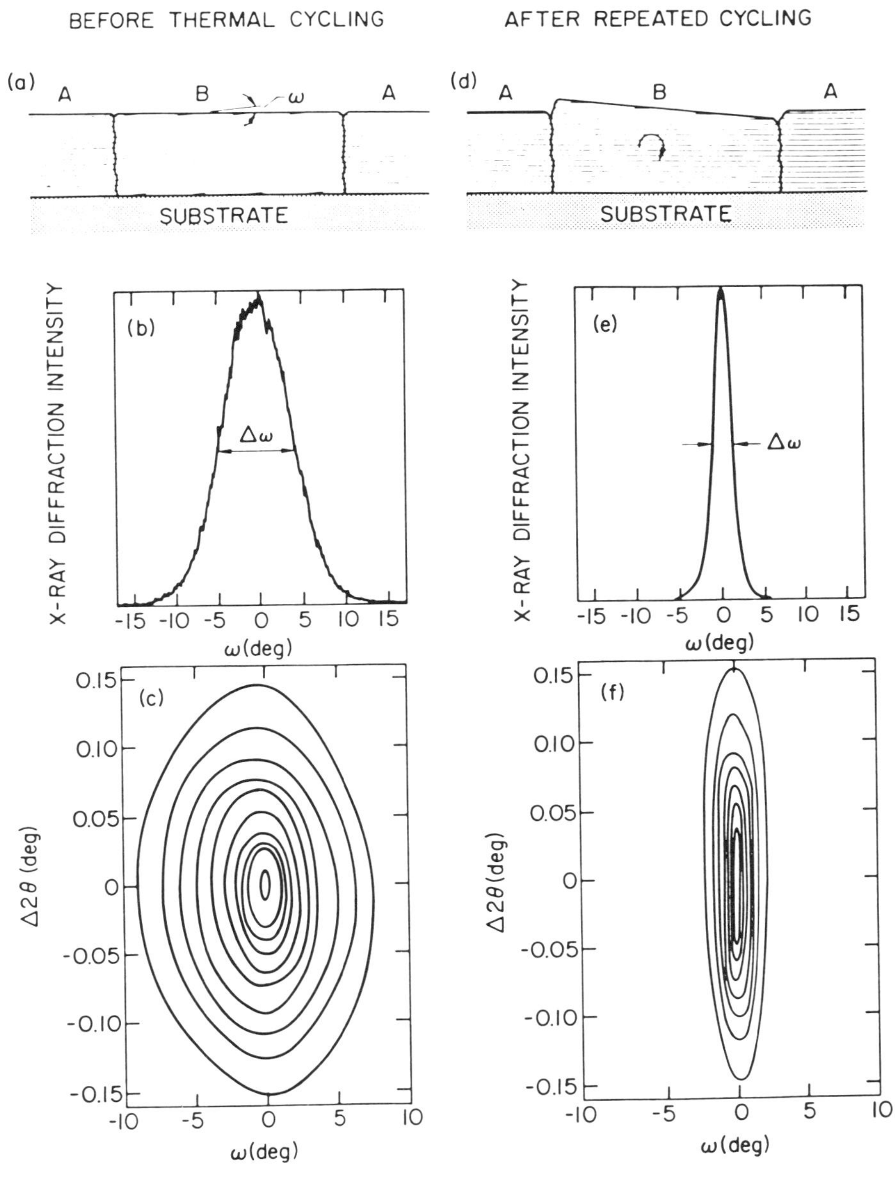

Fig. 6. Cross-sections of film/substrate composite, ω-rocking curve and $\Delta 2\theta - \omega$ curves for a Pb thin film before and after repeated thermal cycling between 298 and 4.2K. (After M. Murakami, CRC Cri. Rev. Sol. Stat. Mat. Sci. **11**, 317, 1984).

the angle between the (111) planes and the film surface, and can be measured by an x-ray ω-rocking technique. The x-ray ω-rocking curve of a $0.5\mu m$ thick Pb film before thermal cycling is shown in Fig. 6(b). The profile is broad, the width ($\Delta\omega$) at the half-maximum-intensity is about $10°$, indicating that typically (111) oriented grains are $\sim 5°$ away from being parallel to the film surface. The x-ray intensity contour map of the Pb film is shown in Fig 6(c), where $\Delta 2\theta$-ω are the angle deviations of the Pb (222) diffraction peak. (In this case $\Delta 2\theta$ corresponds to the instrumental broadening of the x-ray diffractometer). From this intensity contour map, the (111) planes of the grains that tilted away from being close parallel to the film surface do not show a preference for tilting in a particular direction.

After repeated thermal cycling such a film between 298 and 4.2K, its grains were rotated such that the (111) planes became coplanar and parallel to the film surface (26,29). A schematic cross-section of the Pb film after repeated cycling is shown in Fig. 6(d) and the corresponding ω-rocking curve and $\Delta 2\theta$-ω curve of the Pb film cycled 100 times between 298 and 4.2K are shown in Figs. 6(e) and 6(f), respectively. The mean spread in the $\Delta\omega$ decreased to $\pm 1°$. This indicates that (111) planes in each grain are parallel to the film surface and (111) oriented grains are coplanar as shown in Fig. 6(d). Such a large grain rotation causes film surface roughness as seen, for example, in Fig. 5(b). The amount of change in the $\Delta\omega$ angle was found to be proportional to the accumulated amount of strain relaxed by dislocation glide upon thermal cycling (26).

The crystal direction toward which (111) oriented grains of Pb films rotated after repeated thermal cycling between 298 and 4.2K was found to be influenced by alloying elements of Pb-alloy films as described later (30,31).

Microstructure change is also expected for films in which no strain relaxation occurs (15). Most Pb and Pb-alloy films with fcc structure are polycrystalline with strong (111) fiber structure, i.e., surface crystal planes of almost all grains are parallel to (111) planes. However, the crystal orientations of some grains are not parallel to (111) as schematically illustrated in Fig. 7. Upon cooling the film/substrate composite, the same amount of the isotropic strain ε'_{11} (in the film plane) is introduced into each grain. However, the ε'_{33} value (normal to the film surface) of each grain is different and dependent on the grain orientation (hkl). The strain ε'_{33} is related to the strain ε'_{11} by

$$\varepsilon'_{33} = \nu(\mathrm{hkl})\varepsilon'_{11,} \qquad (3)$$

where ν(hkl) is the Poisson's ratio of a biaxially strained film. The ν(hkl) value is a function of the elastic constants and the crystal orien-

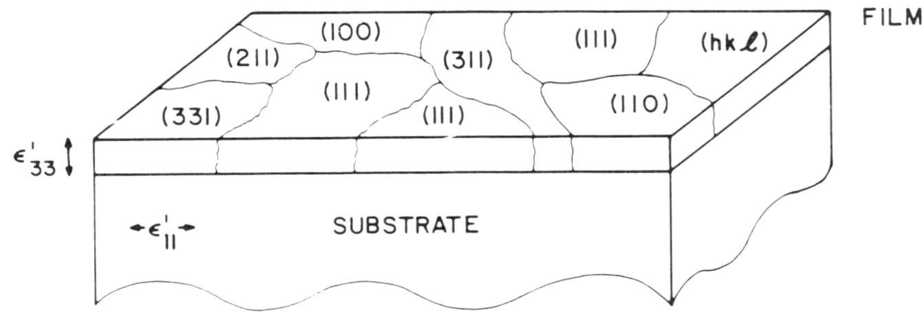

Fig. 7. Schematic illustration of a polycrystalline film on the substrate.

tation (32). Strong crystal orientation dependence of the ν(hkl) value is expected for films with strong anisotropy in the elastic constants. The ν(hkl) values were calculated for Pb films at 4.2K (16) using the elastic constants measured at 4.2K by Waldorf and Alers (33). In order to read off easily the ν(hkl) values of any oriented grains, the values were plotted on the upper half-sphere of (111) stereographic projection as shown in Fig. 8, where 111 direction is normal to the figure and the $\bar{1}10$ direction is the vertical axis. Usually, the film surfaces of most grains are not parallel to specific crystal orientations such as (111), (100), etc. and always several degrees off from these orientations. Thus, to read the ν(hkl) values of grains with any crystal orientation, it is convenient to plot the ν(hkl) values on this stereographic projection. From Fig. 8, it is seen that the ν value has the minimum value of 0.96 for (111) orientation and the maximum value of 1.635 for (100) orientation.

The validity of using the biaxial strain model to analyze the strain or stress components of thermally strained films was evaluated by measuring the ε'_{33} values of grains with various crystal orientations (15). Pb films were cooled from 298 to 4.2K and the strains ε'_{33} were measured at 4.2K by the Bragg-Brentano x-ray diffraction technique. The strains ε'_{33} measured for grains with various (hkl) orientations (shown in the upper horizontal axis) of a film are shown by open circles in Fig. 9. As expected, the ε'_{33} values have a strong dependence on the crystal orientations. A large strain of ~1.2% was obtained for (311) oriented grains and a small strain of ~0.6% was obtained for (111) oriented grains. This indicates that shrinkage of

(311) oriented grains is about a factor of two times larger than that of (111) oriented grains.

Strains ε'_{33} of various crystal orientations were calculated by using $\varepsilon'_{11} = 0.706\%$ (this amount of strain is expected to be introduced into a Pb film upon cooling from 298 to 4.2K and is calculated using Eq.(1)) and ν(hkl) values that were read from Fig. 8. These calculated ε'_{33} values are shown by a dashed line in Fig. 9. It is seen that the calculated and measured values agree very well. The result indicates that the film is strained biaxially and the biaxial strain model should be used to analyze the strain and stress components of thermally strained films. (The similar crystal orientation dependence of ε'_{33} was measured in Pb-In-Au film (31)).

The maximum shrinkage difference between grains normal to the film surface would occur for the grains which have the minimum and maximum ν(hkl) values, respectively. It can be seen from Fig. 8 that ν(hkl) has the minimum value of 0.96 for (111) orientation and the maximum value of 1.635 for (100) orientation. If (111) and (100)

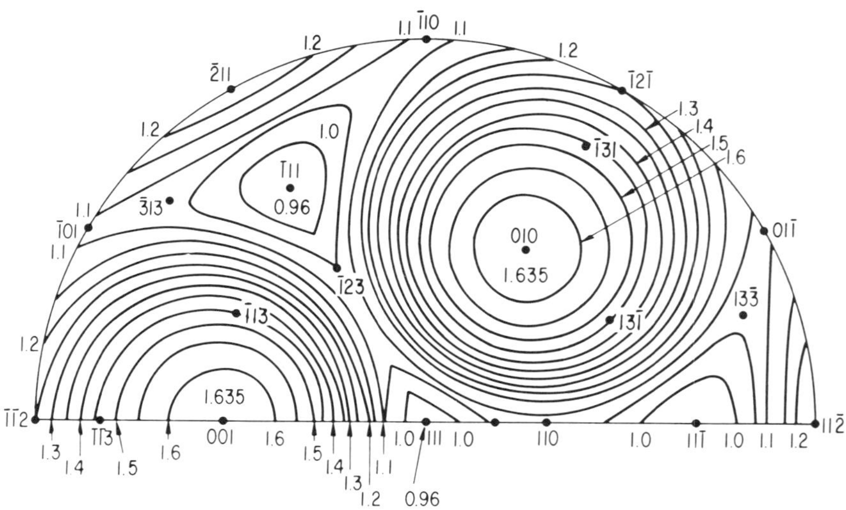

Fig. 8. A stereographic projection of equicontour curves of the strain ratio $\nu(hkl) = \varepsilon'_{33}/\varepsilon'_{11}$ for a Pb film at 4.2K. (After M. Murakami and P. Chaudhari, Thin Solid Films **46**, 109, 1977).

oriented grains are adjacent, the shrinkage difference (Δh) is calculated by, $\Delta h = [\varepsilon'_{33}(100) - \varepsilon'_{33}(111)] \times h$, where h is the film thickness. For a 0.2 μm thick film, a surface roughness $\Delta h \simeq 0.9$nm would be expected in the elastically strained polycrystalline Pb film due to the anisotropy in the elastic constants.

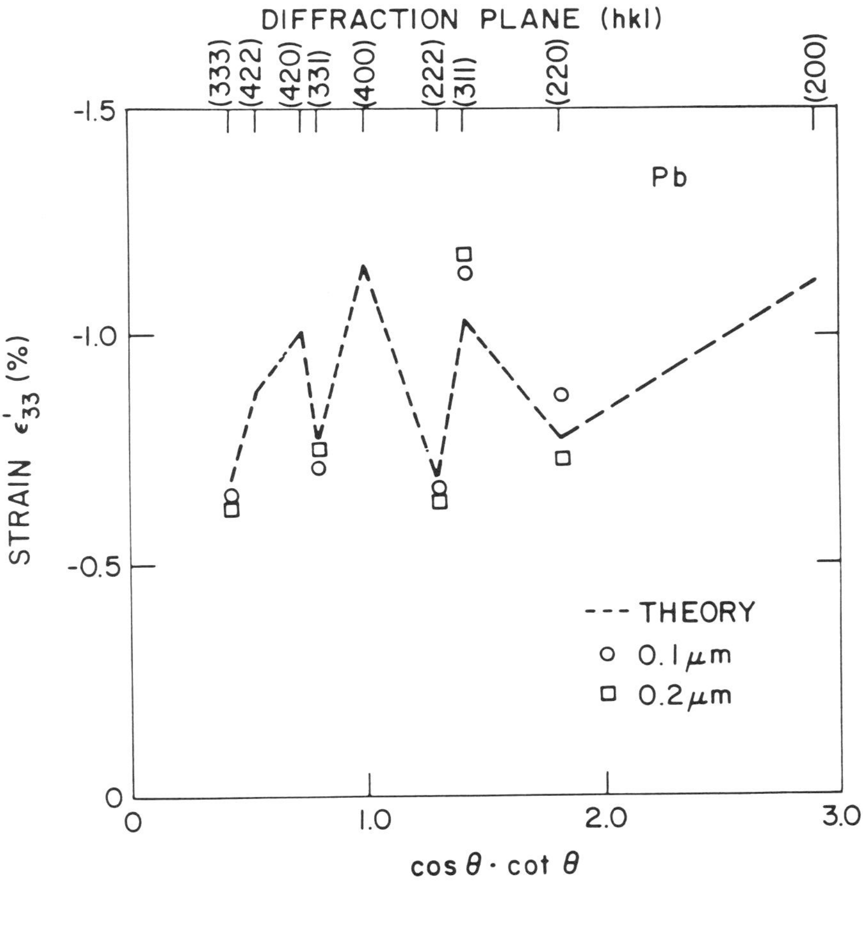

Fig. 9. Strains in grains with (hkl) planes parallel to the film surface measured at 4.2K for a Pb film. (After M. Murakami, Acta Metall. **26**, 175, 1978).

(c) Reduction of Strain Relaxation. The maximum microstructure change expected in Pb films which are strained elastically is smaller than the tunnel barrier thickness and negligibly small compared with those observed for films in which the strain was relaxed upon thermal cycling. Thus, to improve the device reliability after repeated thermal cycling, the strain relaxation should be reduced to near zero. This may be accomplished if dislocation glide can be prevented upon cooling from 298 to 4.2K, so that the films can support the maximum strain elastically and return to 298K in their original zero strain condition.

Several techniques to prevent the dislocation motion have been developed for Pb-alloy bulk materials: precipitation hardening, solid solution hardening and dispersion hardening. These techniques are not very effective for thin films. In thin films prepared by a vacuum deposition technique, it is difficult to form precipitates with sizes smaller than 10 nm inside grains, because excess vacancy concentrations are small due to the presence of near-by grain boundaries and surfaces. Besides, the choice of additional elements are limited for Josephson junction electrode materials: elements which deteriorate the electrical properties are not allowed. Therefore, the best technique is believed to be control of the strain relaxation by changing average grain size (g) and/or film thickness (h). In order to explore the possibility of eliminating the strain relaxation at 4.2K by changing g and/or h alone, their dependence on elastic strain supported in the film was studied theoretically (19, 23, 34) and experimentally (19, 35). Here, the theory (19) developed using an energy criterion model (36) is reviewed.

In order to calculate theoretically an elastic strain (ε^*) supported at 4.2K in Pb thin films with various combinations of g and h, a new thin film strength model has been developed based on the dislocation behavior observed by TEM (19). In this model the film was considered to be elastically strained until the minimum value was obtained in the sum of the elastic strain energy and the energy associated with dislocation glide.

The ε^* was expressed as a function of g and h as:

$$\varepsilon^* = 2\frac{\sqrt{3}}{bY}\left(\frac{E_1}{h} + \frac{E_2}{g}\right) \tag{4}$$

where b is the Burgers vector, Y is Young's modulus and E_1 and E_2 are the energies per unit length associated with dislocation glide along directions normal and parallel to the film surface, respectively. These two glide directions were observed by TEM (19). From Eq. (4) it is predicted that elastic strain increases by reducing g and/or h, or by increasing the energy associated with dislocation glide.

The values of the strain ε^* expected in Pb thin films upon cooling from 298 to 4.2K were calculated using Eq. (4) and compared with those measured by an x-ray diffraction technique (35). In this calculation interaction energies of dislocations with the Si substrate, the native lead oxide and grain boundaries were taken into account and calculated using a method developed by Head (37). These energies were about four times larger than the dislocation self-energy. The TEM result in which the dislocation glide was observed to be prevented by the substrate, the native oxide and grain boundaries justifies qualitatively such large calculated interaction energies. The calculated strains at 4.2K in Pb thin films are plotted by a solid curve in Fig. 10 as a function of inverse film thickness (l/h) for films with various thicknesses and with constant average grain sizes of ~4.5μm. Circles shown in this figure are the strains which were measured for films that

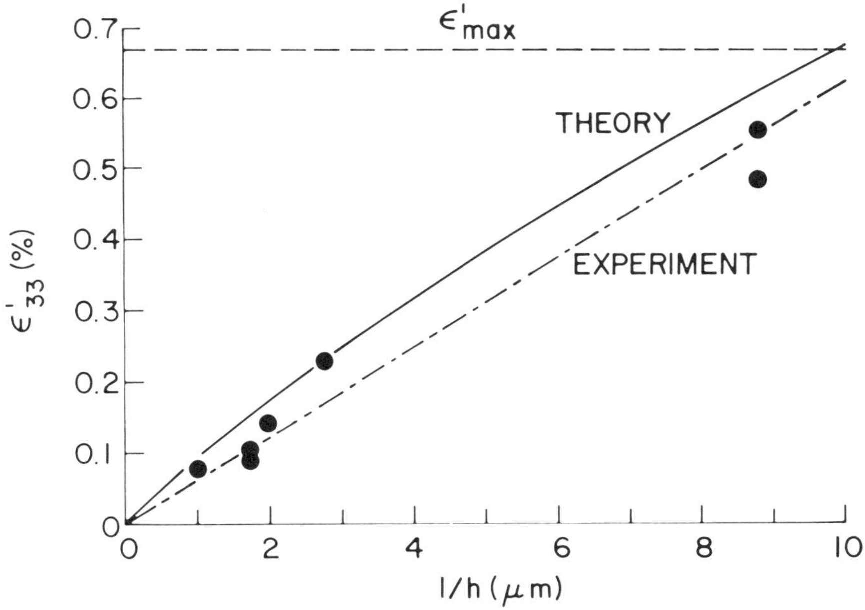

Fig. 10. Elastic strains supported at 4.2K in Pb films with various thicknesses with constant mean grain size. (After T. S. Kuan and M. Murakami, Trans AIME **13A**, 383, 1982).

were prepared by using a sputter-thinning technique (35). These films had constant average grain sizes of ~4.5μm. An excellent agreement between the experimental and theoretical strain values is seen in Fig. 10 without the use of adjustable parameters in the calculation. The ε^*_{max} shown in Fig. 10 is the strain (normal to the film surface) expected in (111) oriented Pb thin films when no strain relaxation was assumed upon cooling from 298 to 4.2K. It is seen that the strain reaches the ε'_{max} level by reducing h below ~0.1μm.

From the present result, it is indicated that the strain relaxation in Pb thin films upon cooling from 298 to 4.2K could be eliminated by

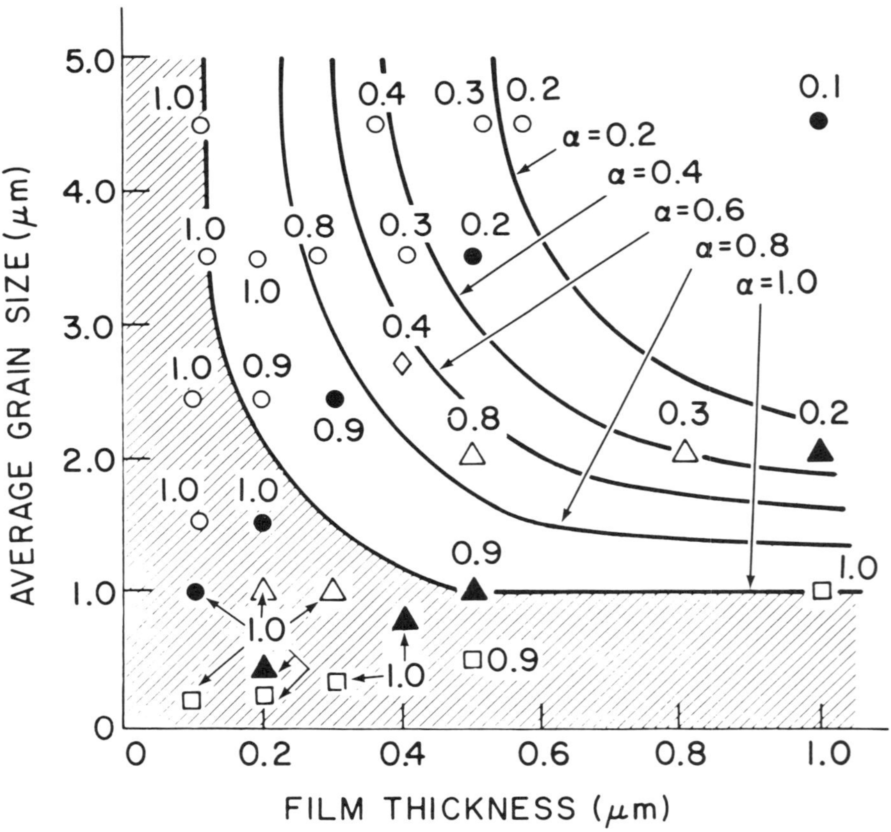

Fig. 11. The strain ratios α plotted on a film thickness versus grain size map measured at 4.2K. (After M. Murakami, Thin Solid Films **59**, 105, 1979).

reducing h and/or g. Therefore, other strengthening techniques which require additional elements are not necessary to eliminate the relaxation. Also, since E_2 was calculated to be larger than E_1, the strain relaxation is expected from Eq. (4) to be limited more effectively by reducing g than by reducing h.

In order to systematically study the effects of h and g on the strain at 4.2K by an experiment, Pb films with various combinations of h and g were prepared on oxidized substrates by using sputter-thinning, low temperature deposition or seeding techniques (35). The strains at 4.2K were measured for these films. In Fig. 11 the strain ratios α of all samples are indicated on an h versus g plot. The α values which represent the measured strains for each sample type normalized by the maximum strain values are indicative of the amount of strain elastically supported in the films. Large α values show that the film supports the strain elastically. The approximate constant strain contours are shown by the curves. It is obvious from this plot that for $h<g/5$ the film thickness limits the strain relaxation and that for $h>g/5$ the grain size limits the relaxation. The region in which the films can support elastically the thermal strain at 4.2K is shaded in Fig. 11. No significant strain relaxation was observed in the films when $g<1$ μm or when $h<0.15$ μm, if the strain relaxation at grain boundaries is negligibly small. It should be noted that the critical sizes for inhibiting strain relaxation are about six times smaller for the film thickness than for the average grain size. This indicates that the strain relaxation was limited more efficiently by reducing the grain size than by reducing the film thickness, as predicted by the theory (19).

2. Pb-In-Au Base Electrode Material.

The choice of Pb-In-Au as a base electrode material was an outgrowth of attempts to use Pb as electrode materials (38). Junctions fabricated through the use of Pb films developed shorts upon one cycling to 4.2K. Initial attempts at strengthening of Pb base electrodes involved the use of a relatively thin layer of In (39). Junctions formed through the use of Pb-rich Pb-In base electrode films showed considerably improved resistance to short upon cycling to 4.2K. This was believed due to the formation of thermally stable indium oxides at the barrier by adding In to Pb. Gold was added to strengthen grain boundaries and to retard grain boundary migration (40). Significant improvements in junction stability was achieved by adding both In and Au to Pb (41-43). The Au In_2 intermetallic compounds formed in the ternary alloy films were believed to strengthen the films and to retard further grain boundary migration at temperatures around room temperature (42). In addition, these compounds improved uniformity of the strain relaxation observed upon repeated thermal cycling (44). Analysis of strain behavior of Pb films (Sect. III. 1) indicated that the strain relaxation upon cooling to 4.2K was reduced by reducing the film thickness and/or more effectively by reducing the average grain size.

Here, effect of grain size of Pb-In-Au base electrode materials on thermal stability over repeated cycling between 298 and 4.2K is described. For this study, Pb-alloy samples with large or fine grains were prepared. A deposition sequence of Au+Pb+In (the sign "+" will be used to denote the sequence of evaporation) was used to obtain "large-grained" films (4). To obtain "fine-grained" films, the deposition sequence of Pb+Au+Pb+In was used at a substrate temperature of 77K (31, 45). The thickness of each layer was measured using a calibrated quartz thickness monitor placed near the substrate. Average grain sizes of these films with 0.2 μm thickness were measured by TEM to be 0.18 and 0.08 μm for large and fine-grained films, respectively (31). The strain at 4.2K, mosaic spread angles, fiber structure and surface morphology changes were compared between large and fine-grained films below.

(a) Strain Behavior. The average elastic strain ε'_{33} normal to the film surface was obtained from measured interplanar spacings (15) using Eq. (2). The strains at 4.2K are shown in Fig. 12 where the ε'_{max} is the strain expected when no strain relaxation occurs upon cooling from 298 to 4.2K. The ε'_{max} value of 0.67% was calculated using a biaxial strain model for (111) oriented grains. The experimental ε'_{33} values are shown by symbols (31). For large-grained films the strain values increase with decreasing film thickness, i.e. the strain relaxation was reduced by decreasing the film thickness. This strain behavior is similar to the one of pure Pb (15). For the fine-grained films the strain values were almost constant, independent of the film thickness in a range of 0.2~1.0μm. The strain values are close to the maximum value of ε'_{max}. It should be noted that the strain values for the fine-grained films are much larger than those of the large-grained films when the comparison is made at the same thickness of $h \geq 0.2\mu$m.

The increase in the elastically supported strain observed in the fine- grained films is believed to be due to the effectiveness with which fine grains in these films inhibit strain relaxation by dislocation glide. The stress supported in the fine-grained films was calculated to be about 4.5×10^9 dyn/cm^2 from the strain measured by the x-ray diffraction technique. It should be noted that this stress level is very close to the theoretical shear stress of Pb (33).

(b) Microstructure Change After Repeated Thermal Cycling. The mosaic spread angle ($\Delta\omega$) of (111) oriented grains was measured using an x-ray ω-rocking technique for films before and after repeated thermal cycling between 298 and 4.2K (31). The angle $\Delta\omega$ is close to zero when the (111) planes in each grain are perfectly parallel to each

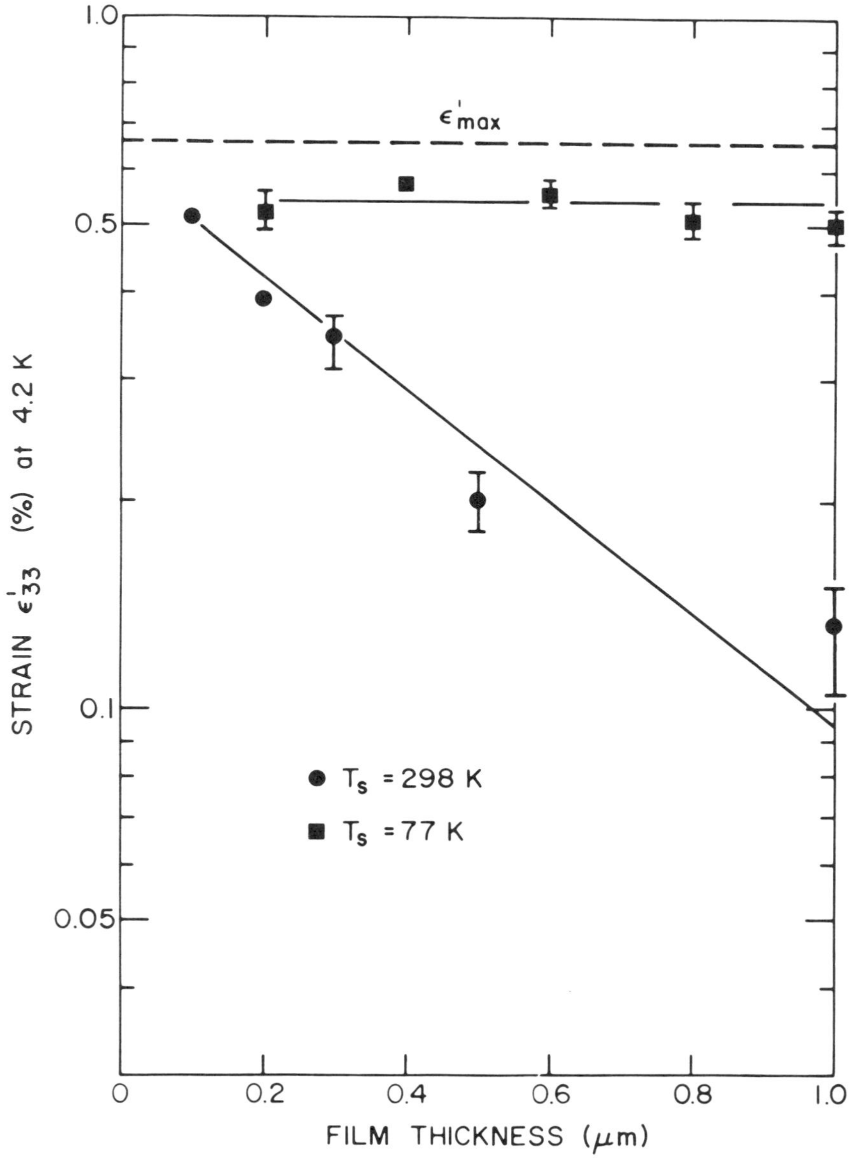

Fig. 12. Average elastic strain at 4.2K supported by Pb-In-Au films deposited at 298 or 77K. (After M. Murakami, J. Appl. Phys. **52**, 1309, 1981).

other like a single crystal, and the $\Delta\omega$ is infinitely large when the grains have a random crystal orientation.

The differences of $\Delta\omega - \Delta\omega_o$ are shown in Fig. 13 as a function of the number of thermal cycles for large-grained films prepared at 298K (a) and for fine-grained films prepared at 77K (b) with various film thicknesses, where $\Delta\omega_o$ is the spread angle of these films before the cycling. For the large-grained films, $\Delta\omega$ decreased for the 0.2 μm thick film and the $\Delta\omega$ increased for 0.5 or 1.0 μm thick film with increasing number of thermal cycles. For the fine-grained films, the $\Delta\omega$ values did not change even after 725 thermal cycles as seen in Fig. 13(b). The changes in $\Delta\omega$ are believed to be related to the amount of the strain relaxed upon thermal cycles.

Surface morphology of Pb-In-Au films with large or fine grains after repeated cycling between 298 and 4.2K was observed by SEM (5,31). Before cycling, the surface of large-grained films was relatively rough and the roughness increased with increasing film thickness. The very smooth surface was observed in fine-grained film and the smoothness was not significantly influenced by the film thickness. In Fig. 14(a) a micrograph is shown for a 0.2 μm thick film with large grains that was cycled 109 times. Hillocks are evident that developed during the thermal cycling. The typical size of the larger hillocks is 1~1.5 μm and the density of the hillocks is $\sim 1 \times 10^6/\text{cm}^2$. The regions of film surface between hillocks appeared by SEM to be the same as before cycling. For the fine-grained films, hillocks were not observed after repeated cycling. A typical micrograph for a 0.2 μm thick film cycled 725 times is shown in Fig. 14(b).

As discussed before, compressive strain is introduced into a film upon warming from 4.2 to 298K when the tensile strain is relaxed during the cooling process. More tensile strain was relaxed during cooling to 4.2K in the large-grained films than in the fine-grained films. Due to the inhomogeneous nature of polycrystalline thin film material, the relaxation of the compressive strain is believed to occur non-uniformly at room temperature. The hillocks observed by SEM are the largest manifestation of non-uniform strain relaxation in Pb-In-Au films and the only one that was large enough to be observed by SEM.

(c) Correlation Between Device Reliability and Microstructure Change. Microstructure changes in Pb-In-Au base electrode films are believed to be the primary cause of Pb-alloy Josephson device failure during repeated thermal cycling between 298 and 4.2K. From the present results, it was shown that refinement of the grains reduced the strain relaxation upon cooling, which resulted in minimizing the microstructure changes during the repeated thermal cycling. The average grain size of Pb-In-Au alloy film prepared at 298K was reduced by a factor of ~two by preparing the films at 77K (31). An additional

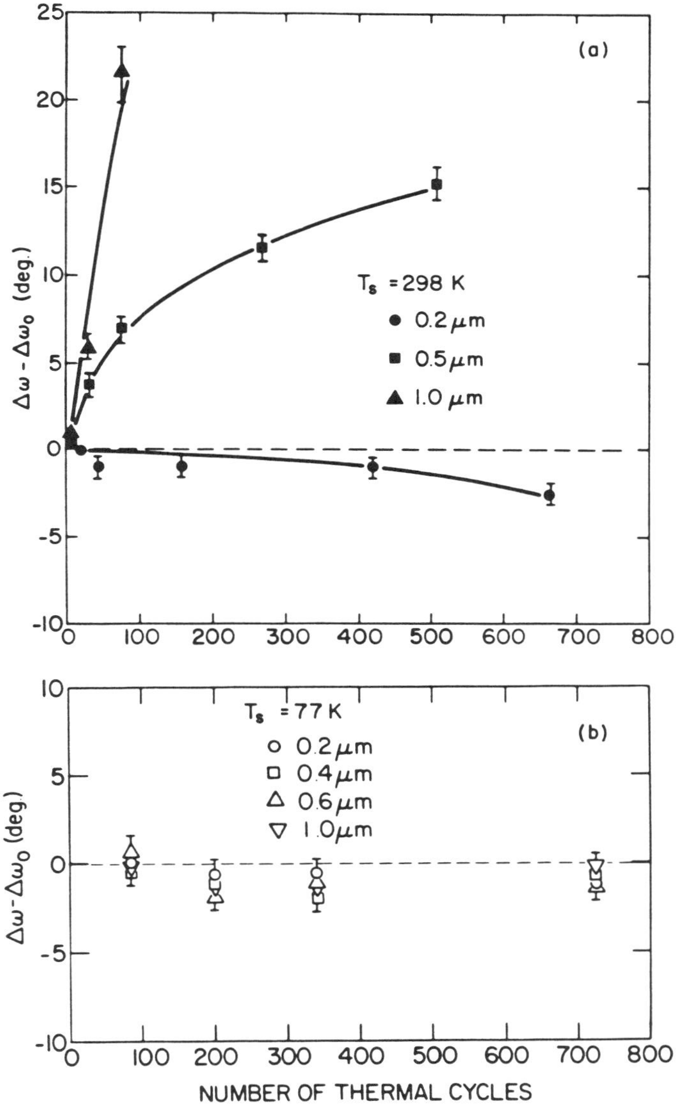

Fig. 13. Mosaic spread angle change ($\Delta\omega - \Delta\omega_0$) for Pb-In-Au films which were deposited at (a) 298K or (b) 77K, and then repeatedly thermally cycled between 298 and 4.2K. (After M. Murakami, J. Appl. Phys. **52**, 1309,1981).

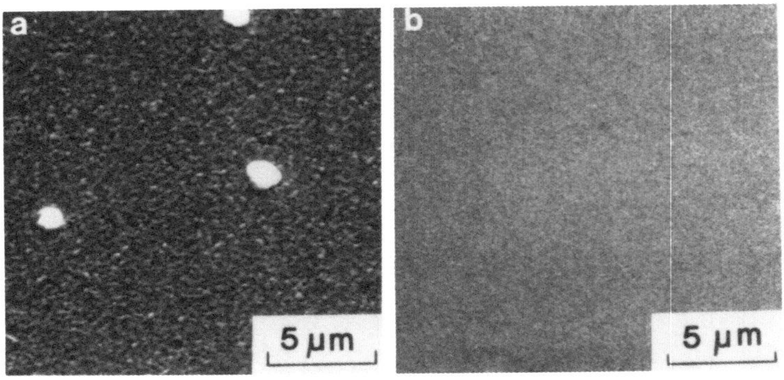

Fig. 14. Scanning electron micrographs for Pb-In-Au films with 0.2 μm thickness after repeated thermal cycling between 298 and 4.2K: (a) large-grained film and (b) fine-grained film. (After C. J. Kircher and M. Murakami, Science **208**, 944, 1980).

factor of ~two in the grain size reduction was obtained by doping 5×10^{-5} Torr oxygen at 77K (46). These results are similar to those obtained in pure Pb films (47) and would be expected for other Pb-rich Pb alloys. Junctions fabricated with reduced-grained base electrode films should thus exhibit improved cyclability. In order to test this hypothesis, an experiment was carried out (45, 46) to compare the thermal cycling stability of Pb-alloy junctions made with three different base electrode metallurgies: the recently used large-grained base electrode prepared at 298K (4), the fine-grained base electrode prepared at 77K at high vacuum (31), and ultra-fine-grained base electrode that was prepared at 77K by doping 5×10^{-7} Torr O_2 during deposition (46). Only the base electrode metallurgies were varied. A cycling experiment was carried out for the large-area (101.6 μm$\times$76.2 μm), overlap junction structure (45). The test pattern for large junctions consists of 1321 junctions on a 25.4 mm diameter Si wafer.

Figure 15 shows the best results obtained for each case in several controlled experiments (45, 46). The junctions with the fine-grained and ultra-fine-grained base electrodes appeared to follow a log-normal statistical distribution with the first failures occurring only after 400 cycles between 298 and 4.2K. The junctions with large-grained base electrodes showed ~100 failures at this point. The number of thermal cycles to reach 5% cumulative failure increased from 10 to 1000 cycles by reducing the average grain size (g) of the base electrode by a factor of ~two and further increased from 1000 to 5000 cycles by reducing g by an additional factor of ~two (48).

The numbers of thermal cycles to reach 50% failure level are plotted in Fig. 16 as a function of average grain sizes of Pb-In-Au base electrodes. It is noted that the number increases exponentially with reduction of grain size. Several techniqures, such as a low temperature deposition, seeding small amounts of nuclei of grains (44,49)

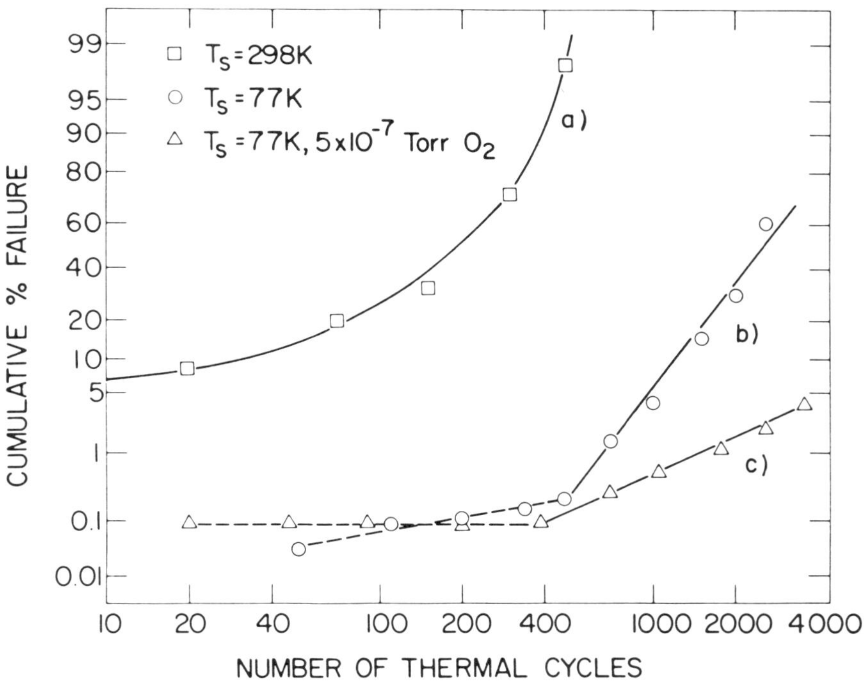

Fig. 15. Cumulative percentage failure upon cycling to 4.2K for Josephson junctions made with (a) large-grained, (b) fine-grained and (c) ultra-fine-grained base electrodes. (After H.-C. W. Huang and M. Murakami, J. Appl. Phys. **53**, 4826, 1982).

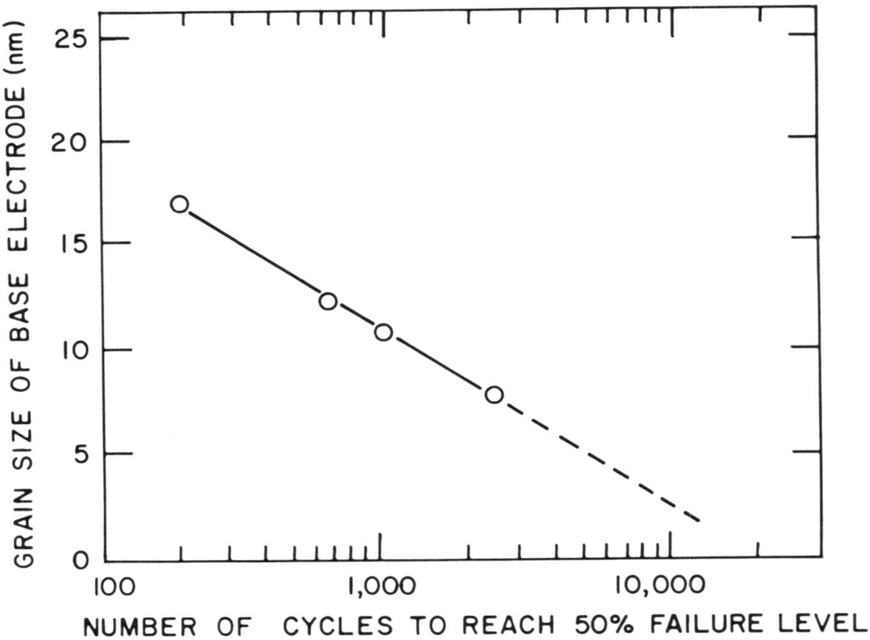

Fig. 16. Number of thermal cycles to reach 50% failure level for junctions made of base electrode materials with various grain sizes. (After H.-C. W. Huang, unpublised work).

and oxygen doping, were developed to prepare fine-grained Pb-In-Au base electrode materials.

The result indicates that there is a strong correlation between the microstructure changes induced by strain relaxation in the base electrode and the device cyclability, supporting the hypothesis that the microstructure changes in the base electrode causes the rupture of the ultra-thin tunnel barrier oxide, resulting in a device short. These experiments were focused on the microstructure of the base electrode. It is also expected that microstructure changes of counterelectrode materials could cause the device failure as described in the next section.

3. Pb-Bi Counterelectrode Material

Since the counterelectrode material is deposited after the ultra-thin tunnel oxide barrier is formed, the deposition techniques which were developed for the base electrode material, such as a low temperature deposition or seeding technique, cannot be used to prepare the counterelectrode material due to deterioration of the oxide barrier. In addition, to avoid undesirable penetration of magnetic field, the counterelectrode material is about twice thicker (0.4 μm) than that of the base electrode material (5). Thus, the amount of strain relaxed upon cooling to 4.2K is larger than that of the base electrode material if the same material is used in the counterelectrode material.

Pb-rich Pb-Bi alloy films have been used for the counterelectrode material which has a high superconductive transition temperature (50,51) and is more resistant to strain relaxation by dislocation glide compared with Pb-In-Au alloy film (52). This Pb rich alloy system has two different structures of α-phase (fcc structure) and ε-phase (hcp structure) (53,54). Here, first, the effects of crystal structure on the strain relaxation and the microstructure change after repeated thermal cycling between 298 and 4.2K are discussed. Then, a method to reduce the microstructure change expected after repeated thermal cycling between 298 and 4.2K is described.

(a) Strain Behavior. To study the effect of crystal structure on the strain relaxation, strains supported elastically at 4.2K by α or ε-phase Pb-Bi films were measured by an x-ray diffraction technique (21). These values are plotted in Fig. 17 as a function of film thickness. The maximum strain value $\varepsilon'_{max}(\alpha)$ of the α-phase Pb-Bi films is assumed to be close to that of a Pb film which was calculated to be ~0.67%. Even for films as thin as 0.2 μm, strain was observed to relax by ~0.2% upon cooling to 4.2K. The amount of strain relaxation in the α-phase Pb-Bi films at 4.2K is larger than that of Pb films, where for films thinner than 0.2 μm, less than 0.05% of strain was observed to be relaxed. In the case of ε-phase Pb-Bi films thinner than 0.8 μm, the strains were observed to vary little with film thickness. Little strain was relaxed until the film thickness exceeded ~0.8 μm. Hence, although the maximum expected strain cannot be calculated for the ε-phase Pb-Bi films, the strain of the 0.2 μm thick film is believed to be very close to the maximum value.

From the strain measurement, the ε-phase Pb-Bi film was found to be more resistant to the strain relaxation than the α-phase film when compared at the film thickness of 0.4 μm which is desirable for the counterelectrode. However, it is noted that the ε-phase film (with 0.4 μm thickness) does not support the total strain elastically at 4.2K and, thus, the microstructure change after repeated thermal cycling is expected.

(b) Microstructure Change After Repeated Thermal Cycling. As described in the previous Section, strain relaxation was observed at 4.2K in the ε-phase films and the amounts of the strain relaxed were dependent on the film thickness. Microstructure changes induced by strain relaxation after repeated thermal cycling were studied for the films with different thickness. Since the actual devices are covered by the SiO layer, the effect of the overlayer on the microstructure changes was also studied (55).

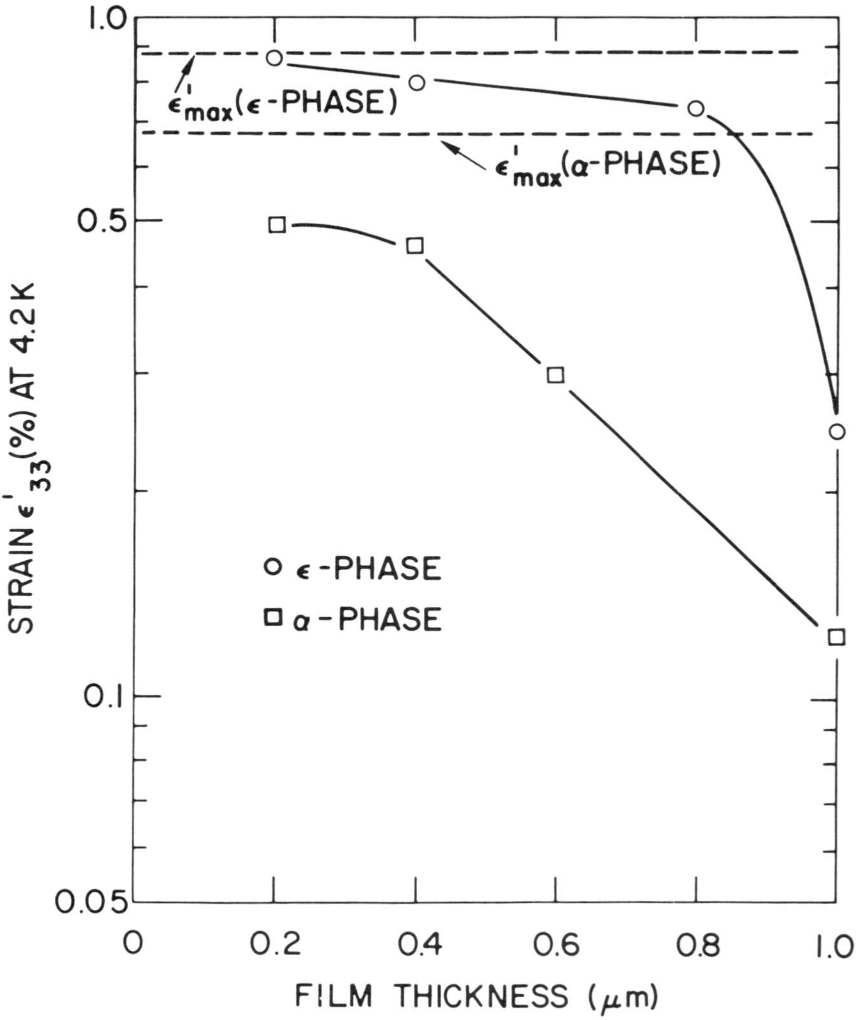

Fig. 17. Strain ε'_{33} at 4.2K for α or ε-phase Pb-Bi films. (After J. H. Basson, M. Murakami and H. Booyens, J. Appl. Phys. **53**, 337, 1982).

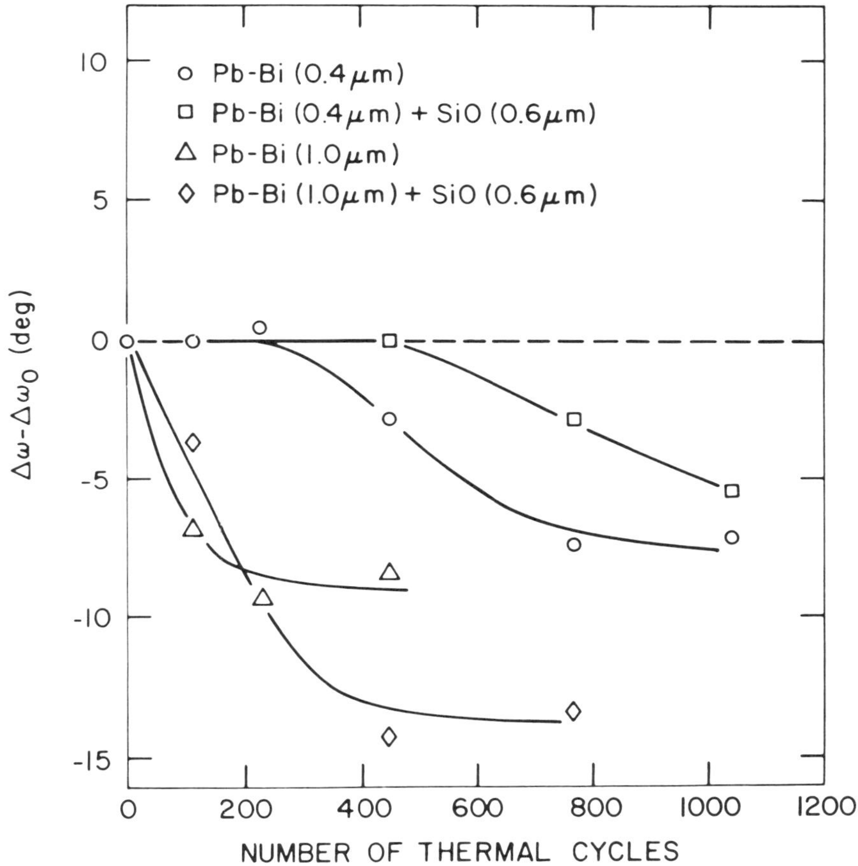

Fig. 18. Changes in mosaic spread angles ($\Delta\omega - \Delta\omega_0$) during repeated thermal cycling between 298 and 4.2K for Pb-Bi films. (After M. Murakami and J. H. Basson, J. Appl. Phys. **53**, 346, 1982).

Mosaic spread angles between (110) oriented grains were measured (55) using an x-ray ω-rocking technique which is more sensitive to microstructure change than the conventional diffraction technique. The measured $\Delta\omega$ values, given by the width at half height of the ω-rocking curve, indicated that the average variations in orientation of the (110) grains were about 17° for as-deposited films within the thickness range of 0.2~1.0μm. The variation in the $\Delta\omega$ values with cycling is shown in Fig. 18 for films with or without a 0.6 μm thick SiO layer. For a 0.4 μm thick film without the SiO layer (circles), the

$\Delta\omega$ value started decreasing at 200 cycles. At this number of cycles, the change in the fiber structure was also first observed. The first change in the $\Delta\omega$ was observed at ~400 cycles for the film with a SiO layer (squares), indicating that the grain rotation was reduced by the SiO layer. For 1.0 μm thick films with or without an SiO layer, the grains were observed to rotate immediately upon cycling.

Observation of the surface morphology of the films without SiO layers after repeated thermal cycling was carried out by SEM (55). The scanning electron micrographs of a 1.0 μm thick film with N = 0(a), N = 223(b), N = 450(c), or N = 763(d) are shown in Fig. 19. The surface morphology of the as-deposited film was observed to be relatively smooth. After repeated cycling, humps and groovings were observed at grain boundaries. After 450 cycles, surface steps (marked S), perhaps dislocation slip lines, were observed inside large grains. After further cycling, the surface became extremely rough so that grain boundaries could not be distinguished. This rough surface was also observed for films with 0.4 μm thickness and was caused by grain rotation after repeated thermal cycling.

For ε-phase Pb-Bi film with 0.4 μm thickness, the strain relaxation observed upon cooling to 4.2K was found to be due to grain

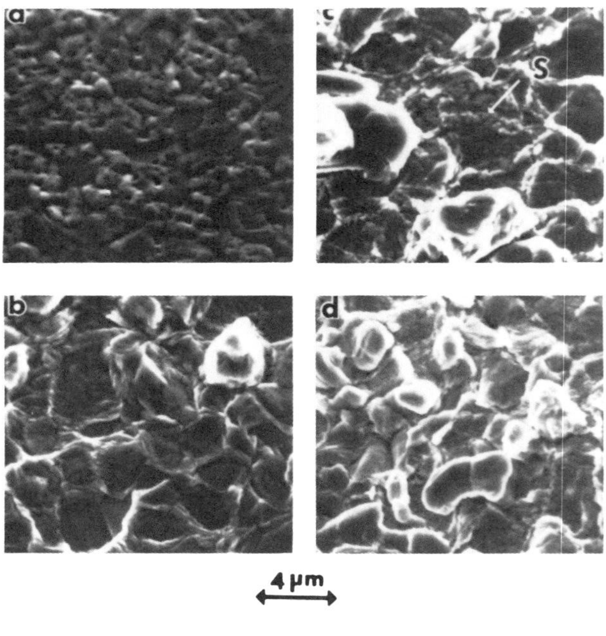

Fig. 19. Scanning electron micrographs of 1 μm thick Pb-Bi film with (a) N=0, (b) N=223, (c) N=450 and (d) N=763. (After M. Murakami and J. H. Basson, J. Appl. Phys. **53**, 346, 1982).

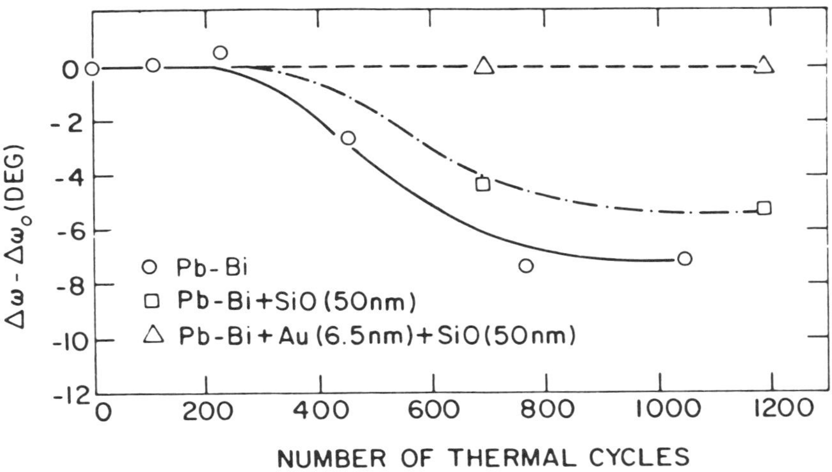

Fig. 20. Changes in mosaic spread angle ($\Delta\omega-\Delta\omega_0$) during repeated thermal cycling between 298 and 4.2K for Pb-Bi and Pb-Bi-Au films. (After M. Murakami, H.-C. W. Huang and C. J. Kircher, J. Appl. Phys. **54**, 743, 1983).

rotation. The grain rotation led to local deformation at grain boundaries. In a 0.2 μm thick film that supported most of the strain elastically at 4.2K, the grain boundaries were observed by transmission electron microscopy to tilt after ~400 cycles (55). Thus, if the grain boundary strength was improved, the tilt of the grain boundary could be reduced, resulting in the reduction of the changes in the film microstructure, such as grain rotation and hillock formation upon repeated thermal cycling.

Since Au was found to strengthen the grain boundaries of Pb by forming the intermetallic compounds at the grain boundaries (40), Au was chosen as an additive element to Pb-Bi film to reduce microstructure changes (56). In order to study the effect of the Au addition on film microstructure change after repeated thermal cycling, mosaic spread angles $\Delta\omega$ for (110) oriented grains were measured by an x-ray ω-rocking curve technique for Pb-Bi and Pb-Bi-Au films that had been thermally cycled many times between 298 and 4.2K in an automated cycling machine (57). Changes in $\Delta\omega$ are plotted in Fig. 20 as a function of the number of thermal cycles. As explained in Fig. 18, the $\Delta\omega$ value of the Pb-Bi film before cycling was $\Delta\omega_0$~17°, the $\Delta\omega$ value of the Pb-Bi film without an SiO layer (shown by circles in Fig. 20) started to decrease after ~200 cycles, and the changes in $\Delta\omega$ were reduced by a 50 nm thick SiO coating (squares). Further reduction of the change in $\Delta\omega$ was observed for samples containing Au (triangles in Fig. 20). No change in $\Delta\omega$ was observed even after 1200 cycles for a Pb-Bi-Au film coated with a 50 nm thick SiO layer. It was observed by using an x-ray diffraction technique that small Pb_3Au compounds were formed in the Pb-Bi-Au films. These compounds could strengthen the grain boundaries and, thus, the grain rotation was reduced after repeated thermal cycling between 298 and 4.2 K.

The effect of the Au addition on changes in the film surface morphology of Pb-Bi films after repeated thermal cycling was studied by optical microscopy using a hillock test structure in which 100×100 μm-square regions of Pb-alloy film were covered by a 100 nm thick layer of evaporated SiO (56). Optical micrographs of Pb-Bi and Pb-Bi-Au films with 4 nm thick Au layer after 207 cycles between 298 and 4.2K are shown in Figs. 21(a) and 21(b), respectively. Hillocks (protrusions from the film surface) are observed in the Pb-Bi film. However, no surface morphology change was observed in the Pb-Bi-Au film. After subsequent 700 cycles, no change in the surface morphology was observed in this Pb-Bi-Au film. Thus hillock formation was significantly reduced by the 4 nm thick Au addition.

(c) Correlation Between Device Reliability and Microstructure Changes. Addition of small amounts of Au to Pb-Bi films was found to significantly improve the stability of Pb-Bi films during thermal cycling, as described in the previous section. Thus junction cyclability is

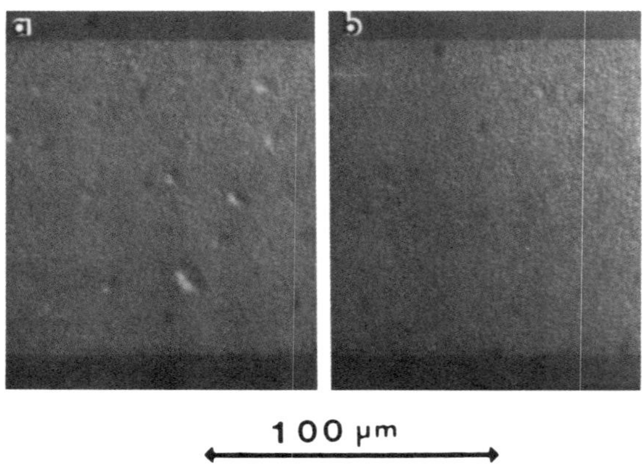

Fig. 21. Optical micrographs of (a) Pb-Bi film and (b) Pb-Bi-Au film after 207 cycles between 298 and 4.2K. (After M. Murakami, H.-C. W. Huang and C. J. Kircher, J. Appl. Phys. **54**, 743, 1983).

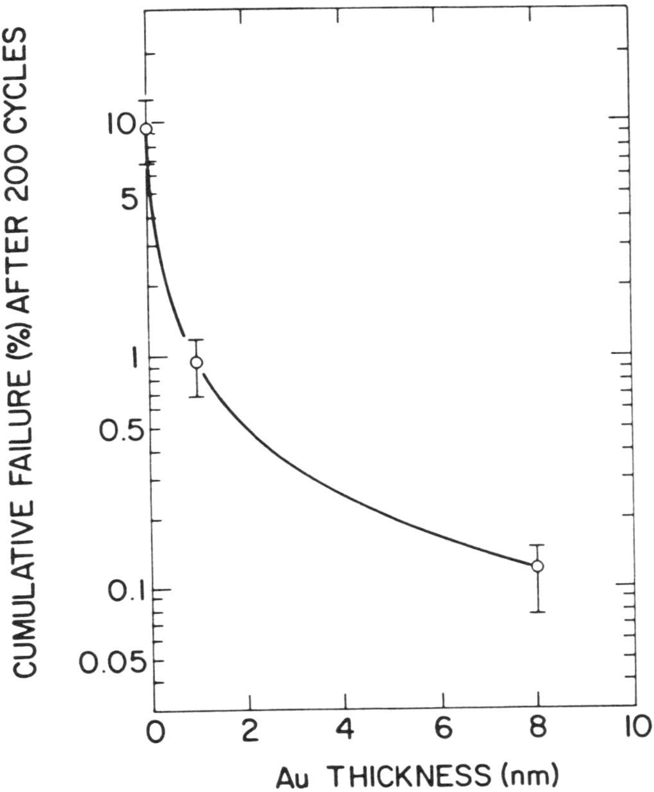

Fig. 22. Cumulative junction failure percentage after 200 cycles between 298 and 4.2K for junctions having Pb-Bi-Au counterelectrodes of various Au content. (After M. Murakami, H.-C. W. Huang and C. J. Kircher, J. Appl. Phys. **54**, 743, 1983).

expected to improve when Pb-Bi-Au films are used as the counterelectrode material. To study the junction cyclability, junctions were fabricated using Pb-Bi or Pb-Bi-Au with 1 or 8 nm thick Au layers as the counterelectrodes. Six wafers, each containing 1321 large junctions, were thermally cycled 200 times. The percentages of failure are plotted in Fig. 22 as a function of the Au layer thickness. It is observed that the junction failure level was reduced significantly by the addition of Au. Two orders of magnitude lower level of failure was observed for the junctions having 8 nm of Au added to the Pb-Bi film. The result indicates that there is a strong correlation between the microstructure changes of the counterelectrode materials and the junction cyclability in agreement with the previous Pb-In-Au base electrode experience.

However, the failure level of the control junctions having no Au in the Pb-Bi counterelectrodes shown in Fig. 22 is higher than that shown in Fig. 15. The principle difference between the two experiments is the material used for a protective coating on the junctions. The data in Fig. 22 were obtained from junctions that were coated with photoresist instead of the more rigid SiO used in the previous work to obtain the data of Fig. 15. Nevertheless, the controlled nature of these experiments (the junction samples were processed together through all fabrication steps; the Au was added to the Pb-Bi-Au samples after the Pb-Bi alloy was simultaneously deposited on all samples) make it clear that adding Au to the Pb-Bi alloy can significantly improve junction cyclability.

Addition of Au to Pb-Bi deteriorated a device design parameter V_m which is a measure of normalized tunnel current at low biases and frequently sensitive to alloy composition (4). The deterioration is believed to be due to the presence of Pb_3Au compounds (non-superconductive materials at 4.2K) near the tunnel barrier.

To form the compounds at the top far away from the barrier, thin layer of In was deposited onto Pb-Bi-Au film at 200K (58). In this film $AuIn_2$ compounds remained near the top away from the barrier and no deterioration of the device design parameter was observed. In addition, the Pb-Bi-Au-In had smooth surface morphology after repeated thermal cycling between 298 and 4.2K and stronger corrosion resistance to ambient atmosphere and moisture than Pb-Bi or Pb-Bi-Au. This corrosion resistance may be due to formation of thin In_2O_3 layer at the surface of the Pb-Bi-Au-In films.

IV. THERMAL STABILITY ABOVE ROOM TEMPERATURE

After the junctions are prepared, the devices must be heated to 350K to assemble the chips. This high temperature cycle occurs only a few times. Upon heating the junctions, the electrode materials are strained due to thermal expansion mismatch between the electrode materials and the Si substrate. The strain could cause the microstructure to change. Since the junction area is confined by the hole in the SiO layer as seen in Fig. 1, the microstructure change is concentrated inside the hole. If the change is larger than the thickness of the barrier, the junction would short. In this section strain behavior and microstructure change upon thermal cycling between 298 and 350K are described.

1. Strain Behavior

The (111) interplanar spacings d normal to the film surfaces were measured from x-ray (333) diffraction peaks for films which had been thermally cycled between 298 and 350K in steps of 10K (59). All samples were annealed at 298K for 1 day before further thermal treatments. The d values measured at each temperature for films with different film thickness are shown in Fig. 23. The D_o line shown in

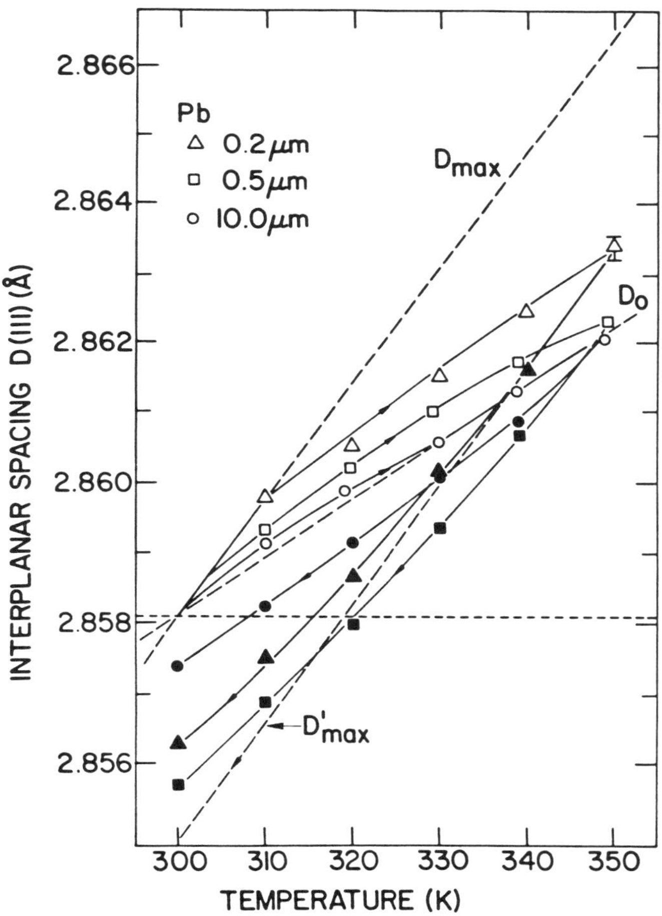

Fig. 23. Interplanar spacings for Pb films which were deposited at 298K and then thermally cycled between 298 and 350K. (After M. Murakami and T. S. Kuan, Thin Solid Films **66**, 381, 1980).

Fig. 23 denotes the calculated strain-free values of bulk lead which were calculated using the reported thermal expansion coefficients of lead (12). The D_{max} line denotes the calculated d values of the maximum strain for lead on a silicon substrate. The values were calculated for (111)-oriented lead grains using the biaxial strain model and assuming no strain relaxation upon heating. The details of the calculation procedure are given in Refs. (15,59). In the calculation, the temperature dependencies of elastic constants C_{ij} were obtained from the literature (33): $\delta C_{11}/\delta T = -2.128$, $\delta C_{44}/\delta T = -1.571$ and $\delta C_{12}/\delta T = -1.114$ in units of 10^8 dyn/cm^{2}K.

The typical experimental values of d are shown by different symbols for films of various thicknesses. The d values of a film with 10 μm thickness are very close to the D_o values upon heating. This means that the compressive strain at the film surface within the x-ray penetration distance of about 1 μm is relaxed almost completely upon heating. However, a tensile strain was observed upon cooling from 350 to 298K. For a film with 0.2 μm thickness no strain relaxation was detected after 15 min at 310K; however, relaxation was observed at temperatures above 310K. The d value at 350K is higher than that of a film with 0.5 μm thickness. The broken curve labelled as D'_{max} indicates the calculated d values of the film with 0.2 μm thickness when no strain relaxation was assumed upon cooling from 350 to 298K. The deviation of the experimental data from the curve indicates that part of the strain is also relaxed upon cooling.

When a Pb thin film was isothermally annealed at temperatures above room temperature, the thermal strain induced into the film was found to be relaxed at two distinguishable rates. A typical strain relaxation process is shown in Fig. 24 for a Pb thin film which was deposited at room temperature and then isothermally annealed at 350K. The strains denoted by closed circles were measured by an x-ray diffraction technique. The maximum strain value ε'_{max} was calculated to be 0.146% when no strain relaxation was assumed upon heating. In Fig. 24 it is found that the strain was relaxed very quickly from ε'_{max} to 0.06% upon heating (primary relaxation process) and then relaxed at a much slower rate (secondary relaxation process) to the near-zero strain level.

If the relaxation process is diffusion-controlled, then we could determine the diffusion coefficient D from the measured strain relaxation rate. Before doing that, we must first identify the mechanisms behind the two relaxation processes.

(a) Primary Strain Relaxation Process. The mechanisms responsible for the primary relaxation are believed to be a combination of grain growth, dislocation glide, whisker and hillock growth, and grain boundary diffusion creep. Since the relaxation rates by these mechansims are extremely fast as seen in Fig. 24, it is difficult to deduce a specific mechanism for the primary relaxation process from the relaxation rate measurement.

When the temperature is raised high enough to allow grain boundary migration, grain growth can occur which will reduce the grain boundary and surface energies. In the present lead films, grains were observed to grow concurrently with heating to temperatures above

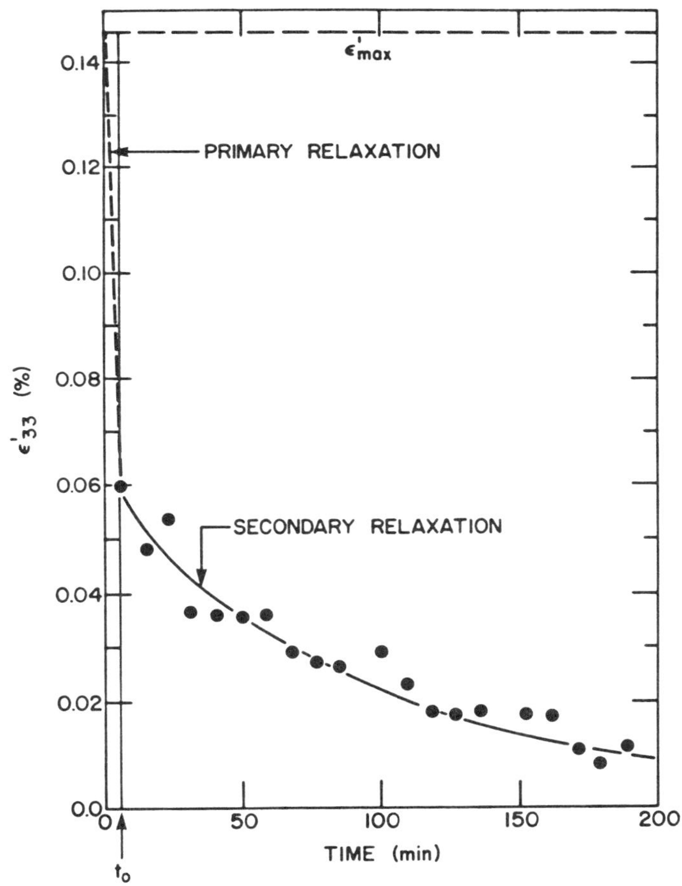

Fig. 24. Strain changes during isothermal annealing for a 0.2 μm thick Pb film which was deposited at 298K and then annealed at 350K. (After M. Murakami and T. S. Kuan, Thin Solid Films **66**, 381, 1980).

340K (59). It has been suggested by Chaudhari (60) that in a thin film deposited on its substrate the compressive strain would be relaxed upon grain growth. He has estimated the critical grain size at which the changes in total energies of the elastic strain (induced by the thermal mismatch) and of the grain boundary are at a minimum. He has shown that if the stress produced by the coalescing of two boundaries is tensile, the grain growth can favorably relax the compressive stress in the films on the substrates.

The grain growth upon heating to 350K was observed by TEM (59). It was seen that the as-deposited lead film has a very wide grain size distribution ranging from about 0.1 to 4 μm. Twins and a very low density of grown-in dislocations were observed inside grains. Large grains were observed to grow at the expense of small grains and the smallest grain size observed at 350K was about 0.6 μm. Inside the grown grains, grain boundary grooving was observed, some of which replicated the grain boundaries observed in the film before heating. Upon heating, most of the twins observed at room temperature disappeared. Slight movement of the grown-in dislocations was observed. This dislocation motion would possibly relax part of the compressive strain. However, no additional dislocations were detected inside the grains at 350K. During subsequent isothermal annealing at 350K no significant change in the microstructure was observed except that a very small number of grains were still growing.

Whiskers and hillocks were found to grow during the heating process even at a temperature at which the grain growth was not observed. The compressive stress induced in the film on heating was considered to be relaxed in the vicinity of the whiskers and hillocks (18, 61, 62). Lahiri and Wells (62) have observed using scanning electron microscopy that the hillocks in lead thin films which formed upon heating shrank upon cooling. This indicates that hillocks locally relaxed the compressive strain in the film.

Another mechanism which may have contributed to the primary relaxation of the compressive strain is considered to be grain boundary diffusion creep, although no direct experimental evidence for it was observed in the present experiment. The compressive strain is considered to be extremely high in the vicinity of grain boundaries. Thus, atom diffusion through grain boundaries would be enhanced by the induced strain. The atoms at grain boundaries would flow to the film surface, film-substrate interface, or laterally to a hillock that is growing.

The primary strain relaxation at 350K was observed to be reduced by the SiO layers coated on the Pb films (59). The strain measurement during isothermal annealing was carried out for 0.5 μm thick Pb films with or without SiO layers and changes in strain are shown in

Fig. 25. It is noted that the primary relaxation is slightly reduced by coating 0.01 μm thick SiO layer and significant reduction is observed when the layer thickness is 0.4 μm. This indicates that microstructure changes such as grain growth and hillock formation which were observed during the primary strain relaxation process are inhibited by the SiO layer. The same SiO effect was observed in Pb-In-Au base electrode material at 350K (63). However, the strain relaxation rates during the secondary relaxation process were not influenced by the SiO layers.

Concurrently with cooling from 350 to 298K, significant amounts of strain was observed to be relaxed (Fig. 23). Using TEM dislocation slip bands were observed within large grains, and the dislocation glide was believed to be the primary strain relaxation mechanism during cooling process. Since the Josephson devices are required to cycle between 298 and 350K only few times, microstructure change large enough to rupture the tunnel barrier is not expected by the dislocation glide.

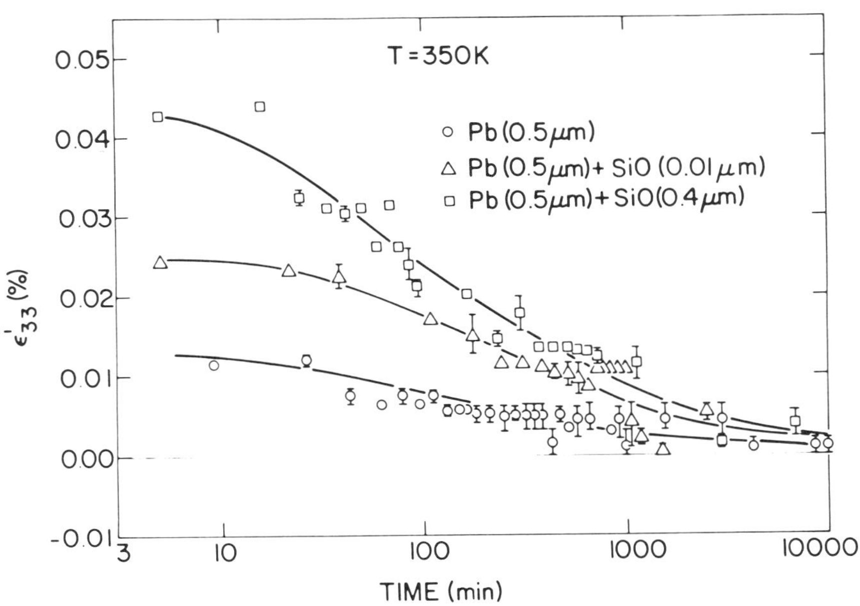

Fig. 25. Strain changes during isothermal annealing at 350K for Pb films with or without SiO coatings which were deposited at 298K and then heated to 350K. (After M. Murakami and T. S. Kuan, Thin Solid Films **66**, 381, 1980).

(b) Secondary Relaxation Process. After the primary strain relaxation process was completed, the rest of the strain was observed to relax very slowly during subsequent isothermal annealing at 350K as seen in Figs. 24 and 25. This relaxation is called the secondary relaxation. During this process, the microstructure change was not detected by in-situ TEM observation. By applying a diffusion creep model (64,65) developed in thin films (66), diffusion coefficients (D) during the secondary strain relaxation process were determined (59). The D values were very close to those obtained in bulk diffusion (67). Thus, the relaxation was concluded to be by the volume diffusion creep which has minimum effect on the film microstructure change.

The major microstructure change which ruptures the tunnel barrier is hillock formation and the details of the mechansim are described below.

2. Hillock Formation of Pb-In-Au Base Electrode

(a) Mechanism. The mechanism of hillock formation for Pb-In-Au base electrode materials was studied by measuring hillock growth rates during isothermal annealings using SEM (28). Since the hillock formation rates were very fast, special samples in which atoms diffuse long distances to form hillocks inside small areas were prepared so that the rates can be measured accurately. The line-shaped Pb-In-Au films with various line lengths were covered by SiO layers except small holes in the center of the lines as shown in Fig. 26. When the sample is heated, the volume of the film covered by the SiO layer expands and the film is under compressive stress. The excess atoms in the film diffuse toward the stress-free window and form hillocks inside the window as shown by arrows in Fig. 26. The driving force for hillock

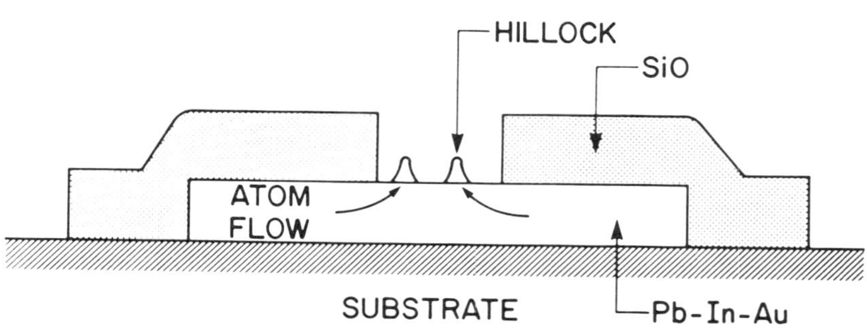

Fig. 26. Atom flows toward an opening in a film coated by a SiO layer.

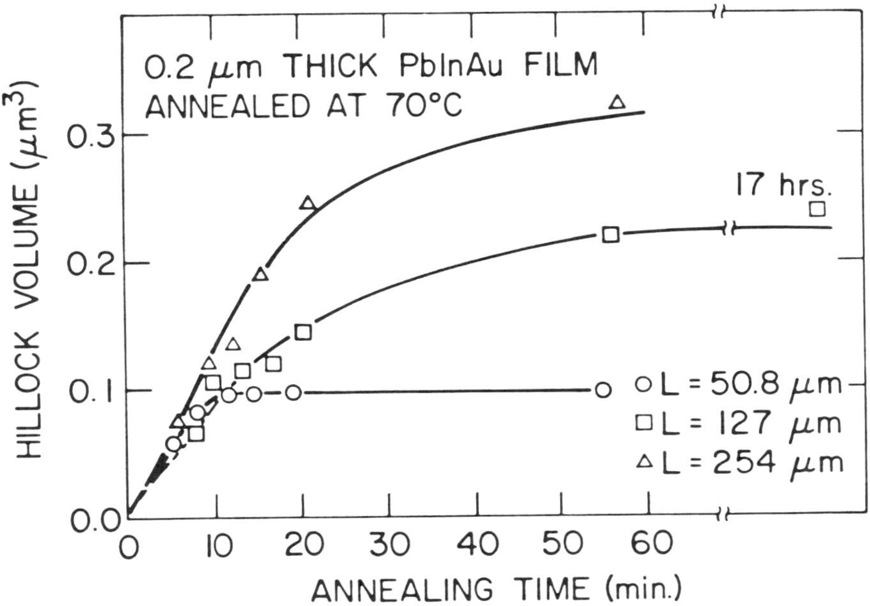

Fig. 27. Changes in hillock volumes for Pb-In-Au films with various line lengths.

formation is the strain gradient in the film with or without SiO coverage.

To understand the diffusion mechanism of the excess atoms for hillock formation, the total volume of the hillocks formed inside the window was measured after annealing for predetermined times (28). As an example, changes in the hillock volumes during isothermal annealing at 343K are shown in Fig. 27 for samples with line length of 50.8, 127 or 254 μm.. The volumes of the hillocks increase with time at the initial stages of annealing and saturate at the later stages. The time to reach the plateau is observed to be shorter for samples with shorter line length and the final total volume of hillocks is found to be proportional to the volume enclosed by the SiO layer.

The mechanism of hillock formation was determined by analyzing growth rates of hillocks. The diffusion coefficients D were determined by applying a "thin-slab" diffusion model (68) to the hillock formation (28). The experimental D values are plotted in Fig. 28 as a function of annealing temperatures. It is noted that the experimental D values are close to those reported for the grain boundary diffusion of Pb and Pb alloy (69). Thus, the hillocks are believed to be formed by extrud-

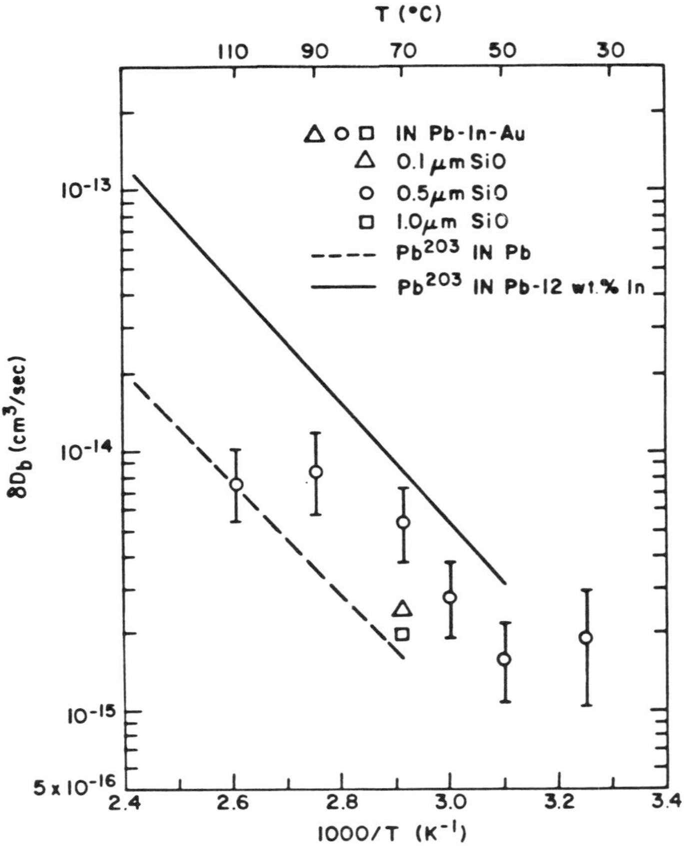

Fig. 28. Diffusivity of Pb-In-Au films during hillock formation. (After H.-C. Huang, P. Chaudhari, C. J. Kircher and M. Murakami, Philos. Mag. **54**, 583, 1986).

ing excess atoms in the Pb alloy films covered by the SiO layer through grain boundaries into the window. To reduce the device failure due to hillocks, the hillock formation should be reduced.

(b) Reduction of Hillock Formation. From analysis of hillock formation mechanism, it was found that a driving force of hillock formation inside the SiO window is the gradient of the strain in the Pb alloy film between inside and outside the window. The excess atoms under the SiO layer were extruded through grain boundaries into the window to form hillocks. Thus, reduction in the hillock formation could be achieved by reducing the strain gradient and/or reducing the grain boundary diffusivity. Since the diffusivity cannot be changed if the annealing temperature and alloy composition are fixed, it is not expected that hillock formation would be reduced by reducing the atom mobility. Also, the gradient of the strain inside and outside the SiO window cannot be completely eliminated. Thus, the best way to reduce the formation of hillocks which are large enough to cause the device failure is diversify the nucleation sites of hillocks, which results in reduction of hillock sizes. Since hillocks are formed at grain boundaries, the number of nucleation sites is increased by increasing the grain boundary density. This is achieved by reducing the grain sizes.

Hillock observation inside the SiO window after heating to 350K was carried out by SEM for large or fine-grained Pb-In-Au films which are prepared by the method described in Sect. II.2 (45). The SEM micrographs of the large and fine grained films are shown in Figs. 29 (a) and (b), respectively. It is seen that no hillocks are observed for the fine-grained films within the SEM resolution limit. Thus, the compressive strain is believed to be relaxed "more uniformly" by increasing grain boundaries.

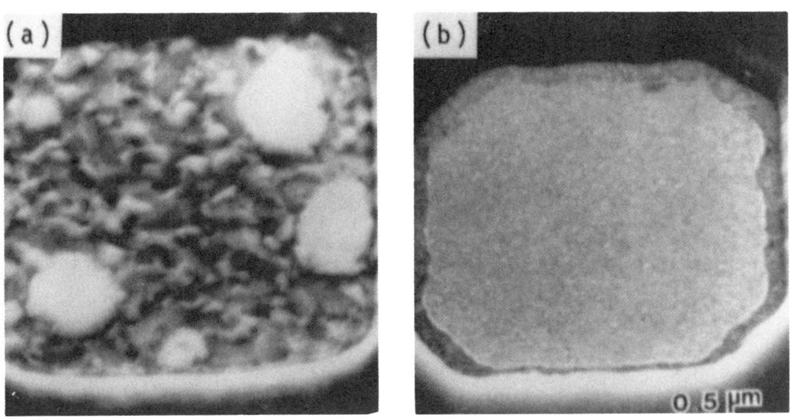

Fig. 29. Scanning electron micrographs of (a) large or (b) fine-grained Pb-In-Au films after heating to 350K. (After H.-C. W. Huang et al, IEEE Trans. Elect. Decice ED-27, 1979, 1980).

TABLE I. Development of Pb-alloy Josephson Electrode Materials

Pb/Pb-ALLOY JUNCTION

BASE ELECTRODE	COUNTER ELECTRODE
Pb	Pb
Pb-In	Pb-Au
Pb-In-Au	Bi+Pb(Layered)
Pb-In-Au (Fine grains) 77K	Pb-Bi (Coevap.) 273K
Pb-In-Au (Ultra-fine grains) 77K, oxygen doped	Pb-Bi-Au 273K
	Pb-Bi-Au-In 200K

V. SUMMARY

Pb-alloy films are attractive as electrode materials of superconductive Josephson junction devices from an electrical viewpoint. A major concern with the use of Pb-alloys is the junction stability after fabrication: the device failure is observed after temperature cycling between room and liquid helium temperature and also after heating to 350K.

In this chapter studies carried out to improve the thermal stability of the electrode materials are reviewed and progress in development of Pb-alloy electrode materials is summarized in Table I. Significant improvement in the thermal stability has been obtained by changing metallization and film microstructure. Diffusion played an important role for the Pb-alloy junction stability above room temperature. The primary relaxation mechanisms above room temperature were grain growth, grain boundary diffusion creep, hillock formation, etc., which were directly related to the diffusion along grain boundaries. The secondary relaxation mechanism is attributed to bulk diffusion within grains. Since the strain relaxation by these mechanisms cannot be eliminated when low melting point Pb alloys are used as electrode materials, the junction stability was successfully improved by distributing the deformation uniformly over the junction areas. Use of the fine-grained Pb-alloys electrodes was a unique solution to disperse the strain relaxation above room temperature. However, it is noted that tracing back Pb-alloy material development new metallization has always resulted in additional process complexity for junction fabrication. Thus, for Pb-alloy electrode materials, the junction stability and the process complexity are always trade-offs. To avoid the further process complexity, it is essential to develop refractory electrode materials which are extremely stable to heat treatment after device fabrication.

ACKNOWLEDGMENT

The author would like to acknowledge T. N. Theis for reviewing the manuscript and to Mrs. S. Manos for the typing.

REFERENCES

1. J. Matisoo, IBM J. Res. Develop. **24**, 113 (1980).
2. T. R. Gheewala, IEEE J. Solid-State Circuits, SC-14, **787**, (1979).
3. J. H. Greiner, C. J. Kircher, S. P. Klepner, S. K. Lahiri, A. Warneche, S. Basavaiah, E. T. Yen, J. M. Baker, P. R. Brosious, H-C. W. Huang, M. Murakami, and I. Ames, IBM J. Res. Develop. **24**, 195 (1980).
4. C. J. Kircher and M. Murakami, Science, **208**, 944 (1980).
5. R. H. Kock, C. P. Umbach, G. J. Clark, P. Chaudhari, and R. B. Laibowitz, Appl. Phys. Lett. **50**, 200 (1987).
6. R. F. Broom, R. B. Laibowitz, T. O. Mohr. and W. Walter, IBM J. Res. Develop. **24**, 212 (1980).
7. R. F. Broom, S. I. Raider, A. Oosenberg, R. E. Drake and W. Walter, IEEE Trans. Elect. Devices, ED-27, 1998 (1980).
8. M. Hikita, K. Takei and M. Igarashi, J. Appl. Phys. **54**, 7066 (1983).
9. J. H. Greiner, J. Appl. Phys. **42**, 5151 (1971).
10. A. T. Fromhold and J. M. Baker, J. Appl. Phys. **58**, 6377 (1980).
11. S. Basavaih, M. Murakami and C. J. Kircher, J. de Physique **39**, 1247 (1978).

12. D. E. Gray, American Institute of Physics Handbook, McGraw-Hill New York (1972).
13. R. W. Vook and F. Witt, J. Appl. Phys. **36**, 2169 (1965).
14. F. Witt and R. W. Vook, J. Appl. Phys. **39**, 2773 (1968).
15. M. Murakami, Acta Metall. **26**, 175 (1978).
16. M. Murakami, and P. Chaudhari, Thin Solid Films **46**, 109 (1977).
17. H. L. Caswell, J. R. Priest and Y. Bado, J. Appl. Phys. **34**, 3261 (1963).
18. S. K. Lahiri, J. Appl. Phys. **46**, 2791 (1975).
19. T. S. Kuan and M. Murakami, Trans. AIME **13A**, 383 (1982).
20. T. S. Kuan, private communication (1984)
21. J. H. Basson, M. Murakami and H. Booyens, J. Appl. Phys. **53**, 337 (1982).
22. M. Murakami, Thin Solid Films **55**, 101 (1978).
23. M. Ronay, Philos, Mag. A **40**, 145 (1979).
24. P. Chaudhari, J. Appl. Phys. **45**, 4339 (1974).
25. M. Ronay and C. F. Aliotta, Philos. Mag. A **42**, 161 (1980).
26. M. Murakami, J. Angilello, H.-C. W. Huang, A. Segmuller and C. J. Kircher, Thin Solid Films **59**, 105 (1979).
27. M. Murakami, T. S. Kuan and I. A. Blech, Treatise on Materials Science and Technology **24**, 162 (1982).
28. H.-C. W. Huang, P. Chaudhari, C. J. Kircher and M. Murakami, Philos. Mag. **54**, 583 (1986).
29. M. Murakami, CRC Critical Rev. in Sol. Stat. and Mat. Sci. **11**, 317 (1984).
30. M. Murakami, Trans. Japan Inst. Metals **23**, 488 (1982).
31. M. Murakami, J. Appl. Phys. **52**, 1309 (1981).
32. J. I. Nye, "Physical Properties of Crystals," Oxford Univ. Press, London and New York (1972).
33. D.L.Waldorf and G. A. Alers, J. Appl. Phys. **33**, 3266 (1962).
34. P. Chaudhari, Philos. Mag. A **39**, 507 (1979).
35. M. Murakami, Thin Solid Films **59**, 105 (1979).
36. J. W. Matthews, "Epitaxial Growth," J. W. Matthews, ed. Academic Press, New York, New York (1975) p. 560.
37. A. K. Head, Philos. Mag. **44**, 92 (1953).
38. See for example, I. Ames, IBM J. Res. Dev. **24**, 188 (1980).
39. W. Anacker, K. R. Grebe, J. H. Greiner, S. K. Lahiri, K. C. Park and H. Zappe, U. S. Patent 3773526 (1973).
40. J. H. Westbrook and K. T. Aust, The Relation Between the Structure and Mechanical Properties of Metals, Proceedings of the National Physical Laboratory (Teddington, England) Symposium, No. **15** (1963).
41. S. K. Lahiri, J. Vac. Sci Technol. **13**, 148 (1976).
42. S. K. Lahiri and S. Basavaiah, J. Appl. Phys. **49**, 2880 (1978).

43. S. Basavaiah and J. H. Greiner, J. Appl. Phys. **48**, 4630 (1977).
44. M. Murakami and C. M. Serrano, J. Appl. Phys. **53**, 326 (1982).
45. H.-C. W. Huang, S. Basavaiah, C. J. Kircher, E. P. Harris, M. Murakami, S. P. Klepner and J. H. Greiner, IEEE Trans. on Electron Devices **ED-27**, 1979 (1980).
46. H.-C. W. Huang and M. Murakami, J. Appl. Phys. **53**, 4826 (1982).
47. M. Murakami, Thin Solid Films **69**, 253 (1980).
48. H.-C. W. Huang, unpublised work.
49. M. Ronay and C. M. Serrano, J. Appl. Phys. **54**, 652 (1983).
50. L. Kopf. Rev. Sci. Instrum. **38**, 734 (1967).
51. W. H. Henkels and C. J. Kircher, IEEE. Trans. Magn. **MAG-13**, 637 (1977).
52. S. K. Lahiri, S. Basavaiah and C. J. Kircher, Appl. Phys. Lett. **36**, 334 (1980).
53. M. Hansen and K. Anderko, Constituion of Binary Alloys (McGraw-Hill, New York, 1958) p. 324.
54. M. Murakami, H.-C. W. Huang, J. Angilello and B. L. Gilbert, J. Appl. Phys. **54**, 738 (1983).
55. M. Murakami and J. H. Basson, J. Appl. Phys. **53**, 346 (1982).
56. M. Murakami, H.-C. W. Huang and C. J. Kircher, J. Appl. Phys. **54**, 743 (1983).
57. S. Basavaiah and S. J. Singer, Cryogenics **18**, 11 (1978).
58. M. Murakami, Shingakugiho, SCE **83-35**, 49 (1983).
59. M. Murakami and T. S. Kuan, Thin Solid Films **66**, 381 (1980).
60. P. Chaudhari, J. Vac. Sci. Technol. **9**, 520 (1972).
61. W. B. Pennebaker, IBM Res. Rep. RC2015, (1968).
62. S. L. Lahiri and O. C. Wells, Appl. Phys. Lett. **15**, 234 (1969).
63. M. Murakami, J. Appl. Phys. **52**, 1320 (1981).
64. F. R. N. Nabarro, Report of a Conference on the Strength of Solids, Physical Society, London (1948).
65. C. Herring, J. Appl. Phys. **21**, 437 (1950).
66. G. B. Gibbs, Philos. Mag. **13**, 589 (1966).
67. H. Mehrer and S. Seeger, Cryst. Lattice Defects **3**, 1 (1972).
68. P. Shewman, "Diffusion in Solids" McGraw-Hill, New York (1963).
69. D. Gupta and K. K. Kim, J. Appl. Phys. **51**, 2066 (1980).

10

Diffusion and Electromigration of Impurities in Lead Solders

Chao-Kun Hu and Hillard B. Huntington

1. INTRODUCTION

The progress toward greater miniaturization in the field of integrated circuitry leads to increased concern for packaging as well as for interconnections reliability. The use of solder balls for joining semiconductor devices (chip) to the carrier substrates has been of great interest to the microelectronic industry since it was introduced in the 1960s (1,2). The surface-mount technology where components are soldered directly on the surface of the printed circuit board (PCB) is also frequently used in the various electronic products (3,4). In addition to the advantages of increasing packaging density and shortening the signal path by this technique, it has been shown that the solder reflow connection is a very reliable and effective method (5). Various metallic layers are required on the electrical contact pad areas of chip, ceramic substrate and PCB for this technique. The terminal pad metallurgy often used is a multilayer film of Cr, Cu and Au on the bond pads of the chips and Ni, Cu and Au on those of the ceramic substrate. The thin Cr film serves as an adhesion layer to the glass and to the Al conductor line. The role of Ni film is the same as that of Cr for the adhesion to ceramic and to the Mo conductor lines in the bottom of the ceramic hole. The Cr and Ni films are also diffusion barrier layers for molten lead-tin solder. The Cu film is a solderable layer and the Au film is for prevention of Cu oxidation. The main solder which has been extensively used is lead-tin. Metal reactions and interdiffusion will occur during the reflow, joining and rework of the chip-substrate assembly. The changes in the distribution of the impurities in lead-tin solders will be significant during the passage of electrical current through interconnections owing to the ultrafast mobilities of the impurities and the appreciable electromigration driving forces to which they are subjected (6).

In addition to technological interest the rather remarkable behavior of atom motions of these impurities (noble metals or transition metals), the so-called "fast diffusers" in lead, tin, thallium or indium, has been of considerable scientific interest for several years (7). The dilute impurities have the mobilities 10^2 to 10^4 times that of self-diffusion. In spite of a great amount of work devoted to their study, the picture of how atom motions take place in these materials is still not clear. The observed values of the self-diffusion enhancement indicate that vacancy diffusion is not possible (8). High mobilities of the impurities suggest that they move by the interstitial mechanism but the picture is complicated by the low values for the isotope effect in diffusion coefficient E which have been measured (9,10) (for its definition, Chapter I of this may be referred to).

Before we describe atom motions (diffusion and electromigration) of metallic impurities in the solders, it would be helpful to estimate the drift velocity of these metallic elements. The diffusivity of these fast diffusers has values which are comparable to the solvent grain boundary diffusion (11), for example in the range of 10^{-5} to 10^{-8} cm^2/sec at 180°C. The values of drift velocity would be in the range of 10^{-1} to 10^{-4} $\mu m/sec$ at 180°C for a current density of $10^4 A/cm^2$. From these estimated values, one would expect these impurities to cross the solder joints within a few days. A formation of intermetallic compounds at solder ball joints interfaces would occur. Quite a lot of intermetallic compounds between Cu, Au, Ag, Ni, Pb and Sn can be found in Hansen (12) and these should have significant influence on the mechanical properties of solder joints (13), since any change of phase at interfaces might change the mechanical properties of solder joint. An experimental investigation of the atom motions of these impurities in Pb(Sn) and Pb(In) solders should offer an additional understanding of physical properties of solders. Here we describe an investigation of atom movements in lead-tin and lead-indium solders. The addition of the second component to lead to form the solder can have a marked effect on the atom motions of dilute impurities. We studied these for several cases in some detail. In Sec. 2 there will be a brief description of the properties of the "fast diffuser" in pure lead. The measurements of atom motion in lead-tin solders are presented in the Sec. 3. In Sec. 4 the studies of the high mobile impurities in lead-indium solders are described.

2. FAST DIFFUSERS IN PURE LEAD

2.1. Diffusion

The noble metals and transition metals which show fast diffusion in lead are known as part of "fast diffusers". The diffusion of gold in lead was firstly investigated by Roberts-Austen (14) around the 1900s, further studied by Seith and co-worker (15) and later intensively investigated by Turnbull's group at Harvard University (7). Figure 1 shows the diffusivities of metal impurities in lead (7). Cu, Pd, Au, Ni, and Ag are classified as "fast diffusers". Cd (16) and Hg (17) which

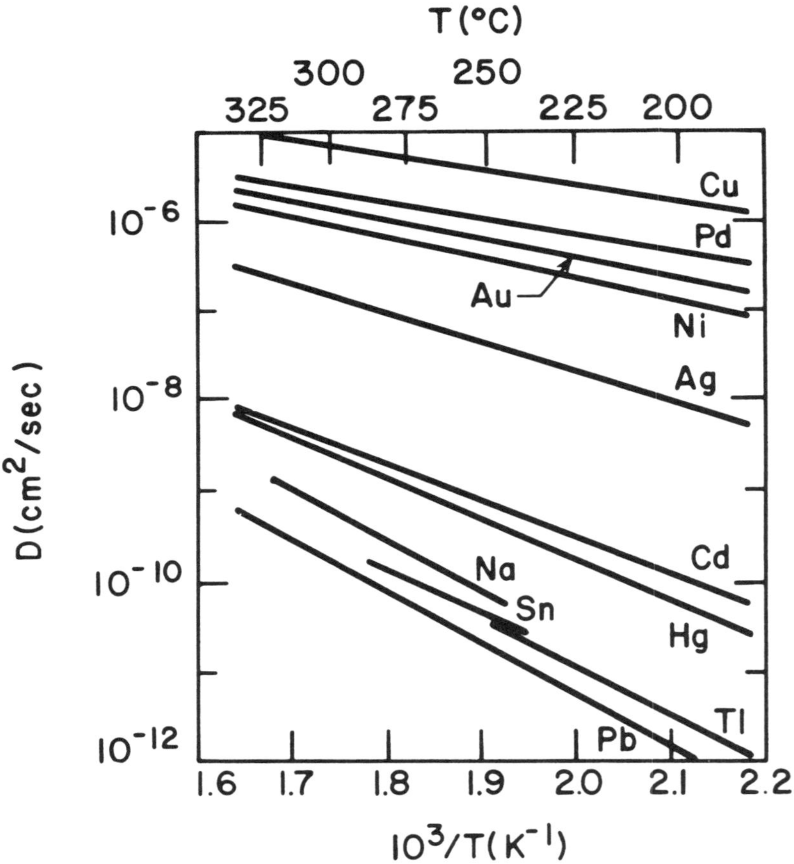

Fig. 1. Diffusivity vs 1/T for various impurities in lead (Ref. 7).

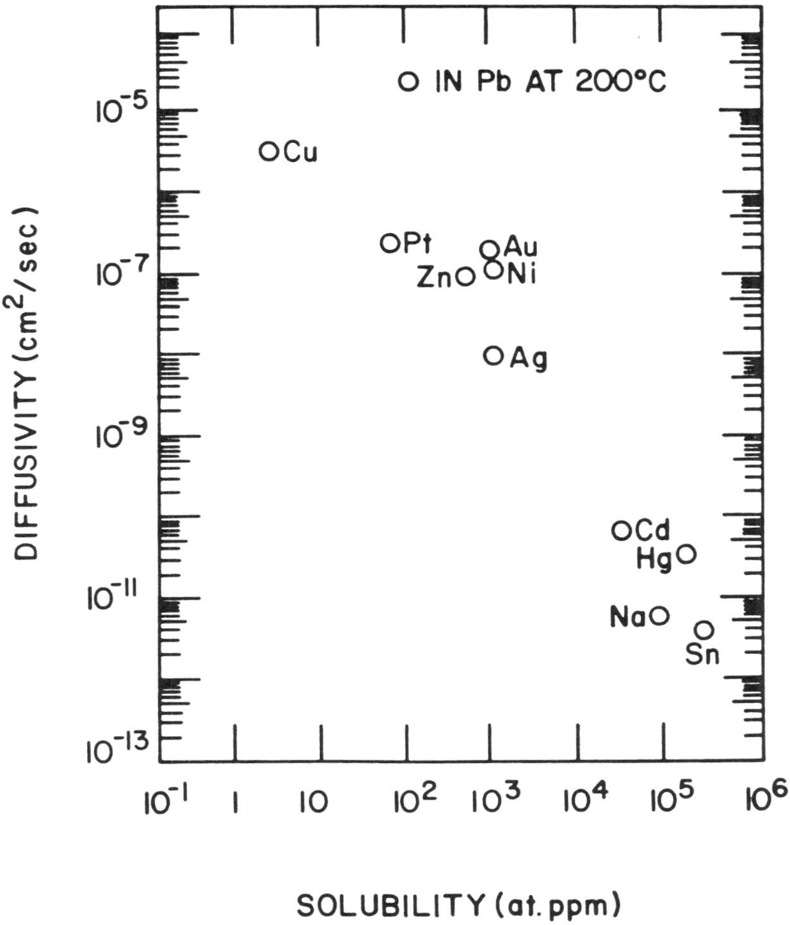

Fig. 2. Solubility of impurities in lead vs diffusivity.

diffuse in lead by the vacancy-interstitial pair mechanism lie in a intermediate group. The slowest diffusion group of Sn, Na and Tl which move in Pb by the vacancy mechanism (7) are the substitutional impurities in Pb. The strong correlation between the diffusivity and the solubility limit of the impurity in lead is shown in Fig. 2. The inverse correlation between diffusivity and solubility has not currently been explained on a quantitive basis. It is reasonable to assume that the low solubility impurities are weakly bound at all points which result in the small activation energies on moving about in the solvent. The high diffusion, however, does not imply low solubility since an impurity could be strongly bound but nearly uniformly so along a suitable diffusion path.

These fast diffusers are found to be diffusion through the crystal since the diffusion along the high-diffusivity paths (free surfaces, grain boundaries or dislocations) was ruled out by the experimental results from Kidson (18) and Ascoli (19).

The general behavior of these fast diffusers indicates some form of an interstitial or interstitialcy-type mechanism. In fact, the enhancements of lead self-diffusion by gold and silver additions (8,20) show that the diffusion mechanism is not vacancy, since the observed values of self-diffusion enhancements are well below the minima predicted (8,20). Also the pressure dependence of diffusion which has been explored by the group at Brigham Young University (21) seems to support an interstitial-type diffusion. The data from these measurements show the small activation volumes, of order of one third or less the lead atomic cell size(V_0). For instance, the activation volume of lead self-diffusion is about 0.64 V_0 (22), while the activation volume of nickel, gold and silver in lead are 0.13, 0.28 and 0.34 V_0 (21), respectively. In general a small activation volume tends to accord with the prevalence of interstitial-type mechanism.

Yet the rapid impurity motion does not appear to fit a simple interstitial mechanism, apparently, judging from the isotope effect coefficient E in these fast diffusion systems has been measured (9,10). The theoretical calculations indicate that the values of E should be close to 1 for a simple interstitial diffusion, but instead this quantity frequently turns out $\sim 1/4$ or smaller. If not a simple interstitial mechanism, then there is the possibility for an interstitialcy-type motion. For impurity this would involve "mixed dumbbell" motion, a typical "two-step mechanism" (23), for which one step would be the usual translation-rotation motion of the inhomogeneous dipole and the second step might be pure rotation for the dumbell, although computer simulation indicates that considerable energy would be required for this motion (24). An alternate possibility for this second step could be a process whereby the dumbell breaks up and recombines in another orientation. In such complex processes quite low values for E can be expected. Recently an interesting approach to the problem of the isotope effect for fast diffusers has been put forward by Ishioka et al. (25). They have suggested that, under certain conditions of linkage to surrounding atoms the E values for such interstitial migration might lie in range <0.4.

Another interesting aspect of these system is the frequent tendency to show diffusion de-hancement at high impurity concentration or with the appearance of a second mobile impurity. The effect was first studied intensely by Warburton (26) looking at the diffusion of gold as a function of gold concentration. He postulated pairs of gold atoms, or "diplon", which formed at the higher concentrations and lower temperatures and were relatively immobile. The silver impurity (27) showed no such tendency about several other examples have been found (28) where the introduction of second mobile impurity gave rise to reduced diffusion, presumably due to "trapping". We shall have occasion to refer later to this marked propensity on the part of these fast diffuser atoms to interact with other impurities.

2.2. Electromigration

Because several aspects of the fast diffuser behavior are still not clearly understood, it appeared reasonable to apply another rather different type of measurement to their motion, namely electromigration. The high mobility of the moving element made feasible steady state distribution measurements (6) on specimens that had an initially uniform concentration of the mobile impurity. The steady state technique yields results for electromigration drive of higher precision and reproducibility than other methods (6). This follows because the diffusivity effectively cancels out in the evaluation of the electromigration driving force F. It is customary to express the strength of the electromigration driving force as $F = |e|EZ^*$, where $|e|$ is the absolute value of the electronic charge, E stands for electric field, and Z^* is the effective charge number, a quantity which measures the strength of the electromigration driving force.

In the rest system of the solvent matrix the mass flow equation becomes

$$\frac{\partial c}{\partial t} = \nabla \bullet \left[D_{ch} \nabla c - \frac{|e|EZ^*}{kT} \frac{Dc}{f} \right] \qquad (1)$$

where D_{ch} is the chemical diffusion coefficient, D is the tracer diffusion coefficient, c is the impurity concentration and f is the correlation coefficient appropriate to the mechanism responsible for the motion.

For a steady state situation in one-dimension geometry, $\partial c/\partial t = 0$, one has

$$\frac{\partial c}{\partial x} = \frac{Z^*|e|E}{fkT} \left(\frac{D}{D_{ch}} \right) c \qquad (2)$$

For the case where the concentration of impurity is uniform distribution and has an initial level c_o in a sample of length L. The solution of this equation is an exponential distribution which has the form:

$$c(x) = \frac{c_o e^{+bx}}{1 - e^{-bL}} \qquad (3)$$

where b = $|e|EZ^*/(\phi fkT)$, x is the distance along the sample. $Z_a^* = Z^*/(\phi f)$ is the effective charge number for electromigration and is the parameter we seek to evaluate, $\phi = D_{ch}/D$ is the thermodynamic factor, and f is the correlation factor for diffusion. The concentration of tracer impurity is about a few ppm or less in the most of cases that are sufficiently dilute solution to have $\phi = 1$. For a simple interstitial mechanism f is also unity but that is probably not applicable in these systems.

The experimental procedure was to prepare the specimen in a cylindrical disk. Initially a uniform distribution of radioactive isotope of mobile impurity was prepared by mixing in the molten state. The sample was solidified by quenching in ice water. Aquadag was painted on the side of specimen to provide a well-known and reproducible emisivity for a temperature reading by an optical pyrometer. The temperature calibration was obtained by melting a specimen of pure lead. The lateral surface of the specimen was encased in a precision bored quartz ring to prevent plastic deformation and was placed between two watercooled stainless-steel electrodes which pressed the sample ends with a cantilever action. In spite of considerable care in preparing and aligning the specimen surfaces the initial contact resistance was usually high and the current had to be increased slowly while the specimen deformed to make improved contact with the hard metal electrodes. Once the resistance came close to its ideal value, the current and the sample temperature were maintained constant for a

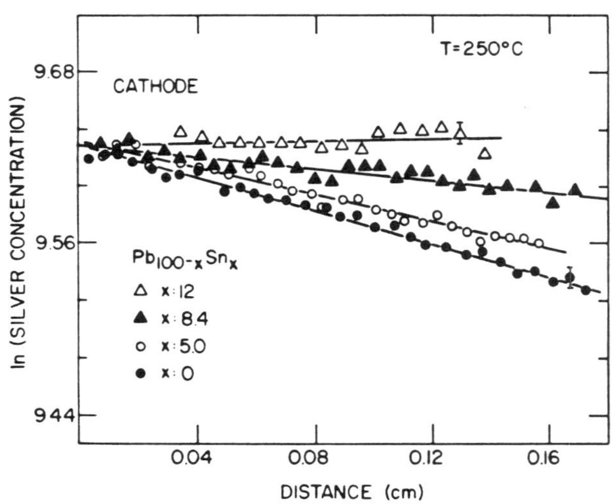

Fig. 3. Electromigration profiles of silver in lead and lead-tin solders.

time $t > L^2/D$ long enough to ensure the establishment of the steady state condition. At the end of the electromigration anneal the specimen was sectioned sequentially, usually by a microtome. Prior to counting, all the sections were dissolved in an acid solution. A NaI(Tl) well type scintillation counter was used to measure the activity of the γ ray from the radioactive isotope and a liquid scintillation spectrometer to determine the amount of tracer which decayed by a β ray emission. Specific activities of each section were calculated from the measured total activity, the background and the weight of each section. The concentration of impurity in each section can be estimated from specific activity of the radioactive isotope as determined from a fixed time in the past and the relative efficiency of counting technique. The thickness of each section was determined from its weight, its area and the density. The typical steady state concentration profiles of silver in pure lead and Pb(Sn) solders are shown in Fig. 3. From the plot of the log of the impurity concentration vs. x one could determine the apparent Z_a^*.

It appears that Z_a^* can be decomposed into two parts,

$$Z_a^* = Z_d - A/\rho(T), \tag{4}$$

where Z_d arises from the direct force of purely electrostatic nature and $A/\rho(T)$ is the contribution to Z_a^* from the so-called "wind force", which arises from momentum exchange with the charge carriers. Here ρ is the resistivity of the matrix.

The concept of the direct force has been subject to considerable controversy. Usually Z_d is taken to be the nominal valence of the moving entity but recently Sorbello (29) has shown that a correction should be expected if the electromigration driving force is strong. The values of Z_d and A are resolved by fitting Z_a^* to Eq. (4).

A summary of the results for Z_a^* on the fast diffusers in lead is shown in Table I. The vales of Z_a^* are in a good agreement between

TABLE I Z_a^*'s for Fast Diffusers in Lead

Impurity	Z_a^*	Z_d	A(Ω-cm)	Laboratory & Ref.	
Gold	−1.47	1.5± .45	120±20	Münster	30
Gold	−1.5	2.3±1.2	150±42	Troy	23
Silver	+1.0	1	0	Münster	30
Silver	+0.9	2.6±1.7	75±72	Troy	23
Copper	+1.1	1.8±1.4	29±62	Troy	31
Copper	+1.07	1.7±0.4	27±13	Münster	30
Zinc	0.2±0.2			Troy	32
Nickel	−5.8	1.4±0.2	288±13	Troy	33

different laboratory. The values for Z_d are around +1 to +3, the numbers consistent with the electrostatic charges on the respective ions.

3. DILUTE IMPURITIES IN LEAD-TIN SOLDERS

As a natural outgrowth of the work of the preceding sections we have become involved in studying the diffusion and electromigration of metallic impurities in lead solders, namely, lead-tin and lead-indium solders. This investigation had considerable technical importance since one procedure in making the back connections to the integrated circuitry chips is through solder balls coated with metallic films of nickel, gold, copper and chromium. Because the tin atom has the same nominal valence as the lead and enters the lattice substitutionally, we expected that its introduction would not greatly alter the mobility of metallic impurities. To the contrary, however, several very interesting results were revealed. Experiments were performed with dilute impurities of radioactive tracers gold, copper, nickel and silver into the lead solders in the single phase region. The results differed markedly for gold, copper, and nickel from those for silver.

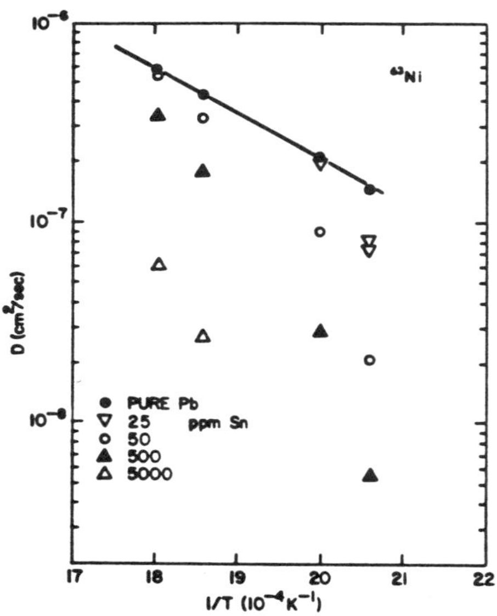

Fig. 4. Dehancement of the diffusivity of nickel in lead-tin alloys. The straight line is the Arrhenius plot for nickel in pure lead and agrees well with the literature.

3.1. Impurities Subjected to Trapping by the Added Tin

A. Diffusion

Gold, copper, and nickel all manifested a strong trapping proclivity with the addition of tin. The effect was made apparent by simple measurements of bulk diffusivity of the mobile metal impurity. As described originally (34) the standard sequential sectioning of cylindrical specimens by a microtome was employed to determine the penetration profile of the appropriate radioisotope after the diffusion anneal. By fitting the profile to the appropriate one dimensional solution of the diffusion equation the corresponding value of the diffusivity was determined. The diffusivity in each case decreased markedly as the concentration of tin was increased.

The effect was most apparent in the case of nickel (34). It turned out that the addition of a 1/20th% of tin induced a decrease of a factor of about 100 in the nickel diffusivity at 212°C (see Fig. 4). Presumably this effect must be considered as evidence for impurity trapping by the tin atoms. The data were analyzed according to a simple trapping model, wherein it was assumed that the mobile impurity (nickel) travels rapidly through the lattice, presumably in some form of interstitialcy, until it meets with a substitutional tin atom where it is trapped, initially at least as an interstitial.

Let the tin concentration be denoted by x and that of the mobile impurity by n, part of which are free, n_f, and part trapped, n_t. Then by a direct application of the mass action law one has

$$\frac{n_t}{n_f(x - n_t)} = z \exp(-F/kT) \tag{5}$$

Here z is the number of degenerate positions at a trapping tin atom for an adjacent nickel (presumably 6) and F is the free energy of trapping. On the basis that the trapped nickel makes negligible contribution to the diffusion one obtains the ratio of diffusion with x concentration of tin to that in the pure lead, $D_x/D = n_f/n$. Since n occurs only in tracer amounts, it is much less than x.

There results then,

$$((D/D_x) - 1)/x = z \exp(B/kT + S) \tag{6}$$

where B is the binding energy of trapping and S is the associated entropy. The left hand side of Eq. (6) is plotted against 1/T in Fig. 5 to give an evaluation of F. The data show some systematic curvature. From the slope of the lower part one obtains a value for B of 1.0 eV or, fitting to the whole curve, a value of 1.25 eV. In both cases the entropy gives a surprisingly large negative value perhaps reflecting the high frequency of the trapping bond.

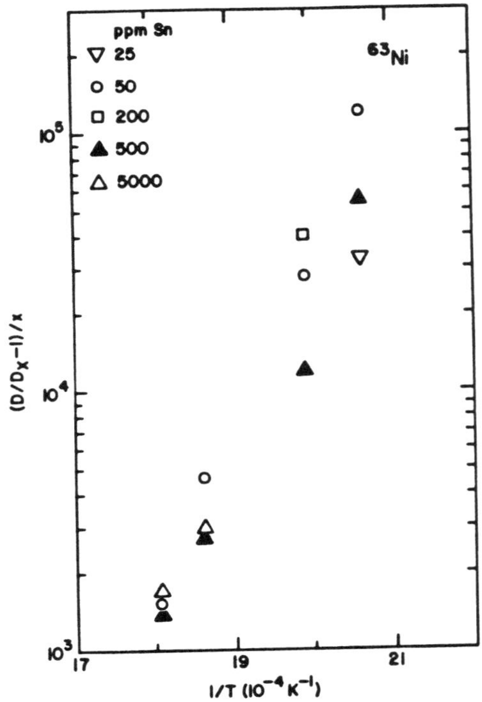

Fig. 5. Plot of $(D/D_x-1)/x$ vs. $1/T$.

Something of the same sort of a situation, although with much weaker binding, became evident in the study (35) of the diffusion of copper in these same lead-tin alloys. There was one complication in the analysis of Cu diffusion run which did not occur earlier. Because the radioisotope used in these measurements, ^{64}Cu, has a short half-life (12.8 hrs.), there was a lower temperature limit to which such studies could be pushed. Shipments of this isotope from the Oak Ridge National Laboratory at concentrations higher than those commercially available helped to extend the temperature range some degrees below that reported earlier in the literature (20). While the higher-temperature studies were ordinarily performed shortly after the isotope shipment arrived, and the studies used material of high specific activity, those measurement performed at lower temperatures required the longer anneal times and hence used the isotope at lower levels of specific activity. To maintain the counting rates at a suitable level, these lower-temperature runs required a higher concentration of copper so that the solubility limit at the surface was often exceeded. In these cases the Gaussian thin-film solution which worked very well for the high-temperature runs, was no longer satisfactory. Instead, one fitted the data appropriately to the complimentary error-function, $c(x,t) = C \text{ erfc } (x/(4Dt)^{1/2})$ where the fitting parameters were C (the saturation composition) and D. In this way the solubility limit C was determined as a function of temperature. In Fig. 6, we show examples of both Gaussian and complimentary error-function fitting for copper diffusion at temperature T = 282°C. An added dividend in this procedure was that one could obtain the solubility of copper in lead (35) from fitting parameters. A compilation of the diffusion data plotted against 1/T is shown in Fig. 7. The rapid decrease of diffusivity with added tin is again evident, although by no means as drastic as in the case for diffusing nickel. For the pure lead at temperature above 227°C straight lines have been drawn through the data. Even with no added tin it is apparent that the purity of the material has an

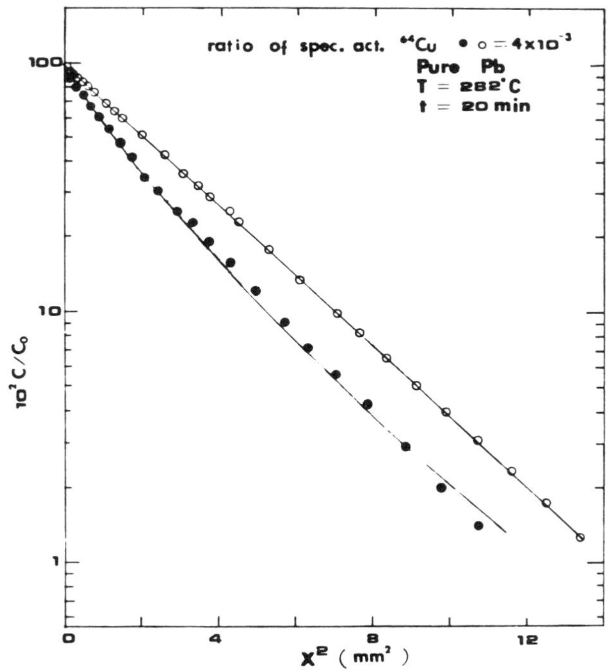

Fig. 6. Diffusion penetration profiles of copper in pure lead. The specific activity of the copper corresponding to the solid circles is 4×10^{-3} of that corresponding to the open circles. The upper and lower curves, respectively, are a Gaussian and an error-function fit to the appropriate data.

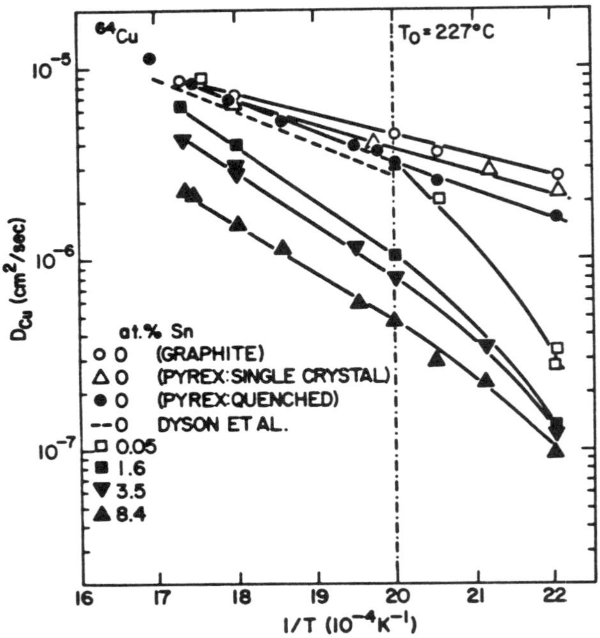

Fig. 7. Diffusivity of copper in lead and lead-tin solders as a function of 1/T above 180°C. The straight lines above 227°C are least square Arrhenius fits.

observable effect. In contrast to the usual situation for substitutional diffusion where increasing the purity tends to lower the diffusivity, here the purer material shows higher mobility, presumably because of decreased trapping. Measurements made on 99.9999% lead were higher than those for 99.999% stock, and specimens shaped in graphite molds gave diffusivities were appreciably higher than those formed in pyrex tubes.

Downward deviations of the Arrhenius line (not shown in Fig. 7) then became evident for even the specimens of highest purity showing that these also possessed some trapping impurities at concentrations well below what can be ordinarily detected.

A careful inspection of Fig. 7 shows that the data exhibiting marked trapping appear irregular. The exact nature of the irregularity become evident when one plots D_x/D versus x for various temperatures as shown in Fig. 8. It is evident that, at the lower temperatures, D_x does not decrease monotonically with x but goes through a mini-

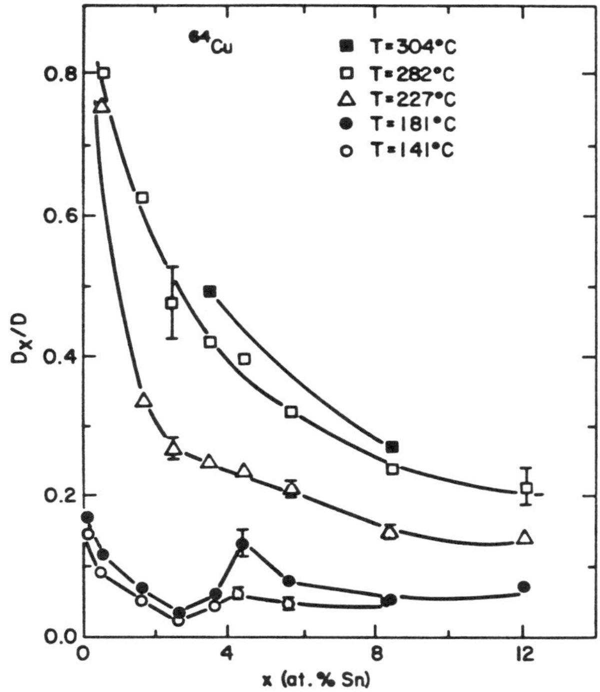

Fig. 8. Ratio of D_x/D as a function of the tin composition. Lines drawn to guide the eye.

mum and then increases to maximum or perhaps a constant value. More comment is made below on this quite surprising trend.

The plot of ln $((D/D_x)-1)/x$ versus $1/T$ in Fig. 9 again reveals a reasonably straight line, particularly at the higher temperatures, from whose slope one can deduce a binding energy of 0.32 eV and a corresponding entropy of $-(4.8 \pm 1.4)k$. As might be expected the anomaly mentioned in the preceding paragraph was responsible for considerable scatter at the lowest temperature.

Turning now to gold as the mobile impurity we come to a system which has been extensively explored, namely the gold in pure lead (19,20). Particular attention has been paid to effect of gold concentration on the mobility of gold (36) ($B \sim 0.36$ eV in terms of the trapping model) and the possibility of various gold complexes in the lead have been discussed. More recently there have been careful x-ray measurements (37) of lattice contents and Huang scattering (38) of

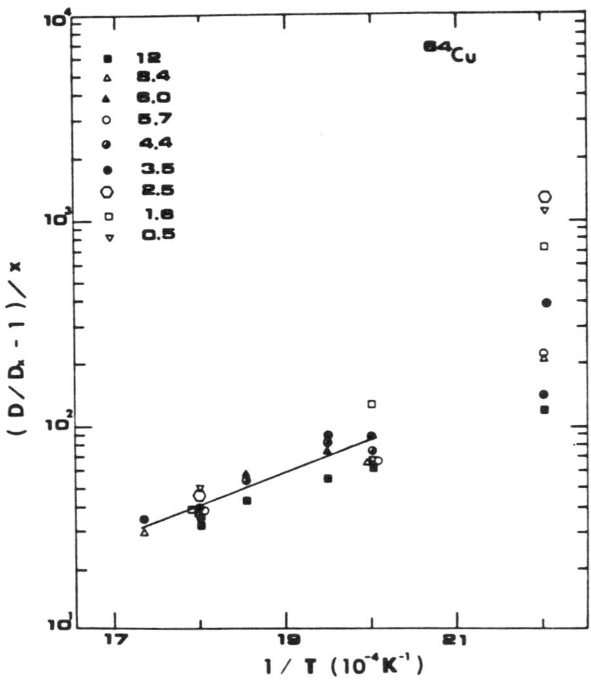

Fig. 9. Plot of $(D/D_x-1)/x$ vs. $1/T$ for copper in Pb/Sn alloys. The line is a least-square fit for $T \geq 240°C$.

these alloys which have indicated that, in spite of the conclusive evidence that the gold does not move by a vacancy mechanism, its equilibrium configuration is substitutional. Whether this situation applies to the other fast diffusers in lead is not known at this time.

The diffusion of gold in these same lead-tin alloys (39) follows much the same pattern as the copper. The data for ln D versus 1/T at various concentrations of tin (39) are shown in Fig. 10. Again the plots of D_x/D versus x for various temperatures (Fig. 11) show, at the lower temperatures, the same anomalous behavior that was described above for the case of the mobile copper impurity. An attempt to deduce a binding energy from these data gave $B = (0.22 \pm 0.12$ eV) with rather wide error limits.

Because of the longer lifetime of the isotope (^{195}Au) in this case, it was decided to explore this system in more detail, rather than that of the mobile Cu. Up to this point the lead-tin alloys have been prepared by a slow quench from the molten state on the premise that this would ensure that the tin could in this way be retained in solution. Almost by accident it was found that specimens that had been annealed at

190°C or above no longer exhibited the anomalies in their diffusivities described above. Instead the diffusivities at low tin concentration increased so that the anomalous minimum disappeared and the diffusivity decreased monotonically with tin concentration. The typical change is illustrated in Fig. 12 for diffusivity versus concentration at two different temperatures. Consequently plots in Fig. 11 are then replaced by those shown in Fig. 13. From these data the plots of ln $((D/D_x)-1)/x$ showed greatly reduced scatter as seen in Fig. 14. From the slope of the best line through this data one obtained B = (0.21±0.03) eV and a corresponding S = (−3.3±2.6)k.

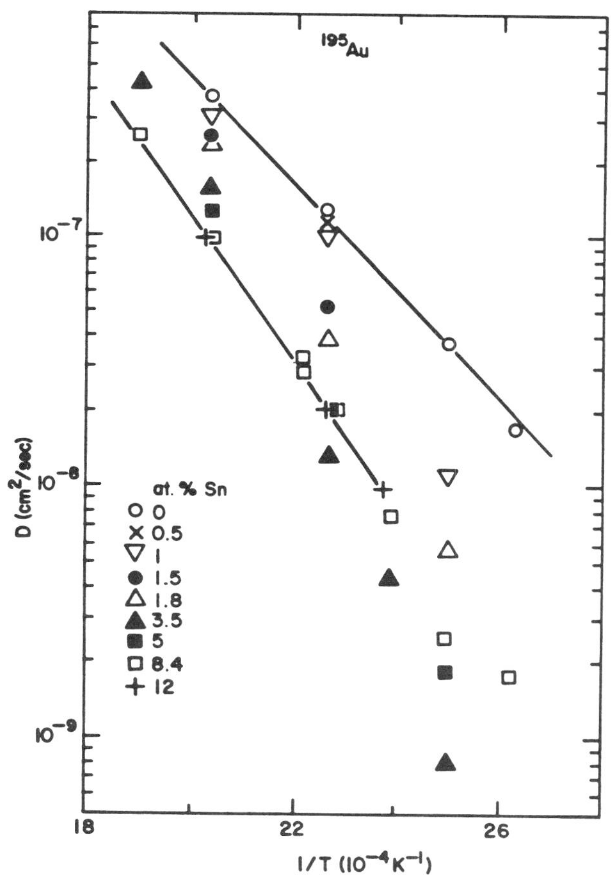

Fig. 10. Data for D_{Au} as a function of T^{-1} for several lead-tin alloys.

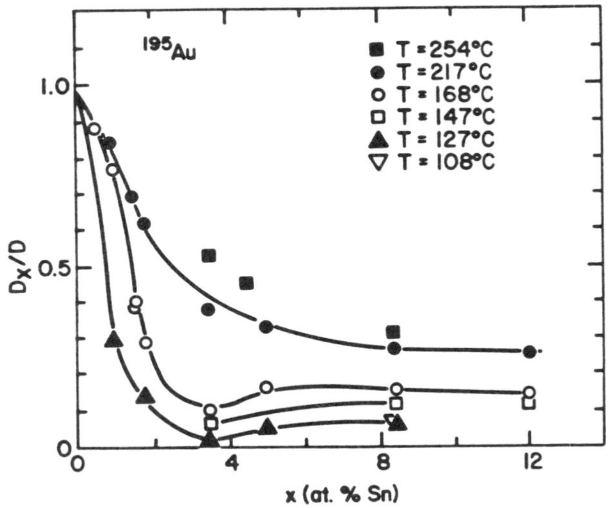

Fig. 11. Ratio of D_x/D for ^{195}Au in quenched lead-tin alloys.

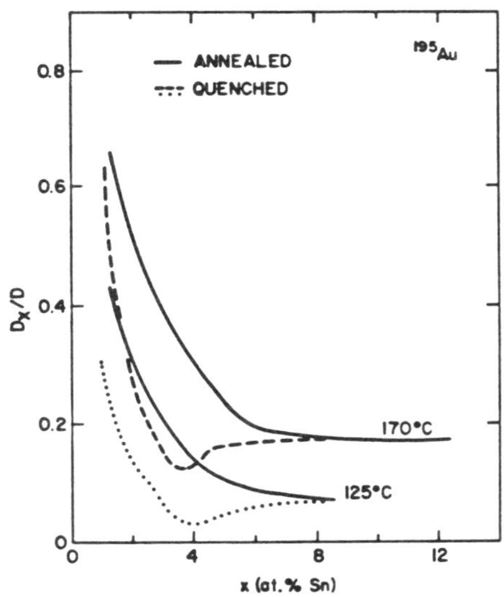

Fig. 12. The effect of annealing on the anomaly in Fig. 11 —— annealed specimens, ----- quenched specimens at 170°C, and ······ quenched specimens at 125°C.

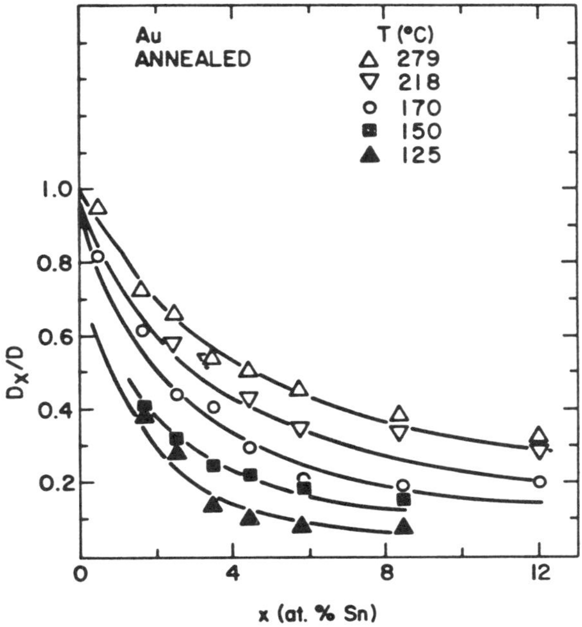

Fig. 13. D_x/D vs. x in annealed solder specimens.

Apparently the annealing process dispensed with whatever artifact had been reducing the diffusivity at temperatures below 170°C for specimens containing 1 to 4% tin. It is not certain exactly what is occurring but the most revealing procedure consisted in repeated diffusion runs at varying times on specimens of the same compositions and at the same temperature – in the critical range from 170° to 190°C where the diffusivity was changing. For each such run the concentration profile was fitted to the thin film solution of the diffusion equation and the corresponding values for Dt were obtained and plotted against annealing time (t). Of course the slope of such a plot should give D were D constant. Instead the curves were of the form $D(x,T,t) = D_f(x,T) - \Delta e^{-t/\tau(T)}$ where D_f is the diffusivity of the well annealed specimens, x gives the tin content, $\tau(T)$ is the time constant for annealing process at T, and Δ is a constant. Initially it was presumed that $\tau(T)$ would have the form $\tau \sim e^{Q/kT}$. Results from runs at various temperatures are given in Table II. The very rapid variation of $\tau(T)$ seemed to rule out an activated process with reasonably low activation energy.

Additional efforts were made to understand the nature of the annealing process. All specimens were examined metallographically but the annealed specimens appeared to have no noticeably different characteristics from those that had been quenched. Likewise the specimens in the critical composition range of 1 to 4 at.% tin did not show any marked difference, except perhaps for the appearance of some dendritic structure in this range. Hardness testing was also employed and showed the expected increase of hardness with alloying. There was otherwise no observations relevant to the puzzling diffusion behavior.

B. Electromigration

Electromigration studies have also been carried out for these three impurities that show such marked trapping. The procedure is to use the steady state method which is described in Sec. 2.

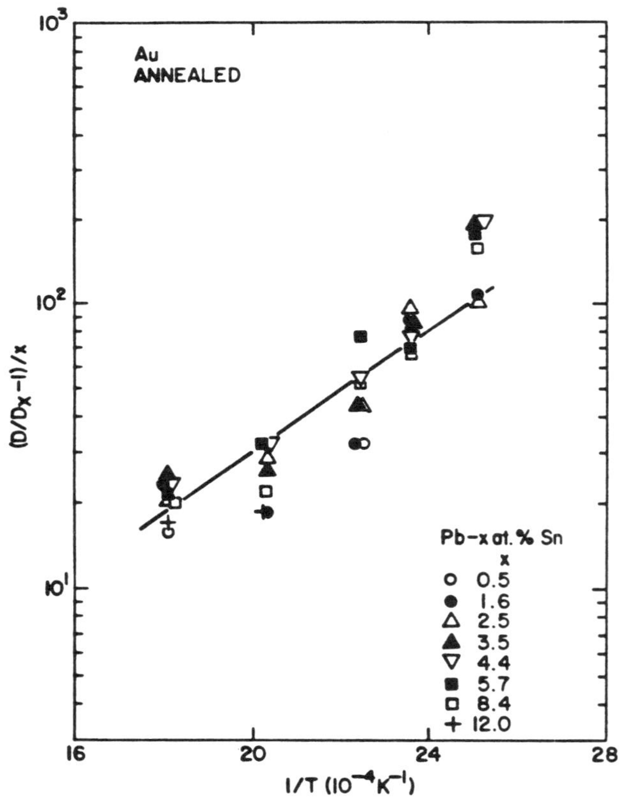

Fig. 14. $(D/D_x-1)/x$ vs. $1/T$ for gold in annealed Pb-Sn specimens.

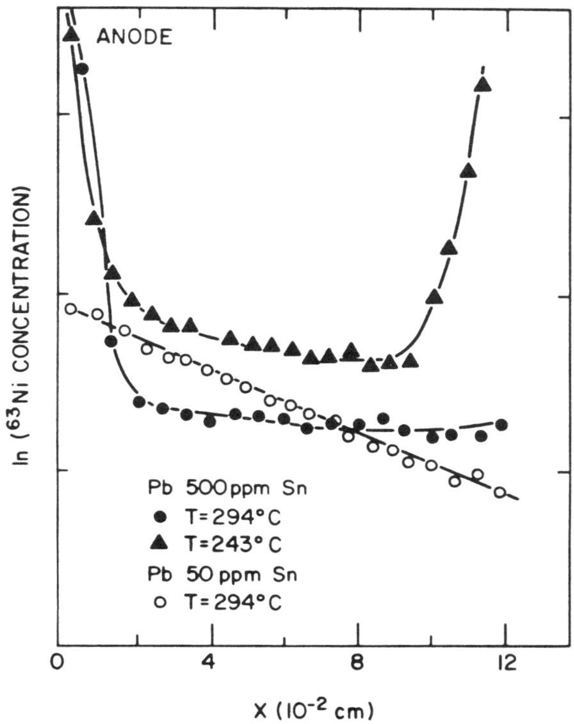

Fig. 15. Electromigration profiles of nickel in lead-tin alloys. Surface effect is large for high concentration of tin.

TABLE II

Fitted parameters Δ/D_f and $\tau(T)$ for the diffusivity as a function of time.

Temperature (°C)	Δ/D_f	$\tau(T)$ (sec)
160	0.41	10^6
170	0.56	7×10^5
179	0.63	2×10^3
190	0	–

Some difficulties were encountered in measuring Z_a* for these three metallic impurities in the lead-tin alloys. Particularly in the case of nickel the addition of tin greatly complicated the measurement. This was because nickel has a strong tendency to move to the surface of specimens heated in an oxygen atmosphere. This does not present much of a problem for pure lead specimens held in a carefully controlled atmosphere, since the diffusion of oxygen in lead is small, but the addition of tin appeared to greatly increase the penetrability of the metal to oxygen, particularly at low temperatures and long times. Several runs under these conditions with 500 ppm tin showed very marked pileups of nickel isotope near the ends of the specimens. Fig. 15 shows the electromigration concentration profiles of these attempts. One run, however, with only 50 ppm tin was quite free of this artifact and gave a value of $Z_a^* = -4.8 \pm 0.2$ in satisfactory agreement with values obtained earlier for Z_a*, for nickel in pure lead.

The situation was somewhat less severe for the electromigration of copper in lead-tin alloys. While a run at 220°C gave an erratic profile with marked segregation toward the ends, runs near 260° were of good

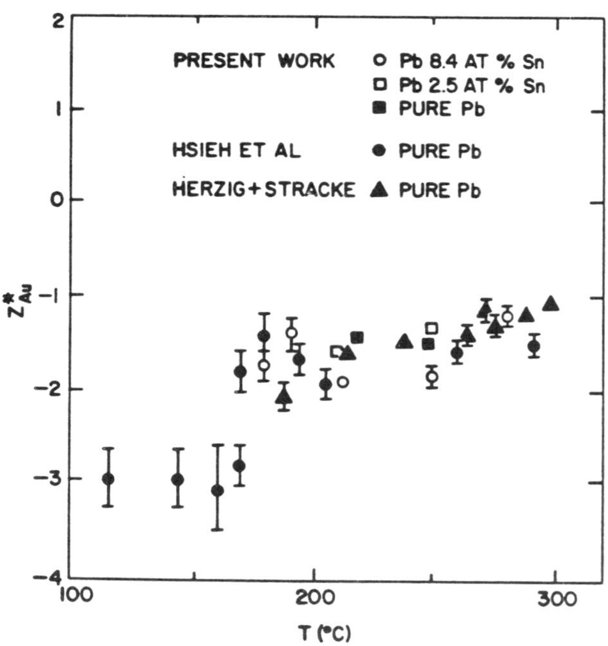

Fig. 16. Comparison of Z_a* values for the solders to earlier values obtained in pure lead.

quality. One specimen with a 50 at. ppm tin gave a Z_a* of 1.1±0.2 and another at the appreciable concentration of 4.4 at.% Sn gave Z_a* = 0.8±0.2, both values being in essential agreement with the literature value (30,31) for Z_a* for copper in pure lead.

With gold as the metallic impurity there was some of the same difficulty in getting reliable runs at low temperatures for long times but the difficulty could be largely avoided by shortening the specimen and thereby reducing the time needed to attain steady state conditions. The Z_a* obtained in this way are shown in Fig. 16 where the Z_a* are plotted against T along with values obtained earlier by other investigators for measurements on gold in pure lead. This figure shows quite clearly that the addition of tin to lead has made no significant change in Z_a* for the gold.

Indeed it appears it can be stated quite generally that, insofar as reliable measurements are available, there appears to be no indication that the drastic trapping attributed to the introduction of the tin on the basis of the reduction in the diffusivity has had any effect on the driving force of the electromigration. Apparently the electromigration flux is carried almost completely by the *untrapped* impurities whose response to the field is independent of the presence of the tin. This hypothesis requires an additional assumption, namely that local equilibrium between trapped and untrapped must be established quickly when electromigration alters their relative concentrations. Of course the presence of the tin delays the attainment of the steady condition but not its eventual nature.

3.2. An Impurity Not Subject to Trapping – Silver

It has been observed that in contrast to most fast diffusers in pure lead the mobility of silver is quite independent of its concentration or indeed of the presence of other impurities (31,27,40). Why the dilute silver impurity should behave so differently from other noble metals is a question that to our knowledge has not been addressed yet theoretically. The studies with lead-tin solders have shown that the susceptibility for trapping exhibited by some metallic impurities and not others is demonstrated in much the same way when the trapping agent is a substitutional impurity as when it is another mobile impurity (28).

A. Diffusion

Measurements of the diffusivity of silver in the solders actually showed an increase with the addition of tin. In Fig. 17 are shown Arrhenius plots for the diffusivity of silver in pure lead, in Pb-8.4 at.% Sn and Pb-12 at.% Sn. At the lowest temperature of measurement the diffusion constant for the twelve percent alloy was double that in pure lead. In Table III are given the corresponding Q and D_o values obtained from these lines. Clearly it is the Q-values that are

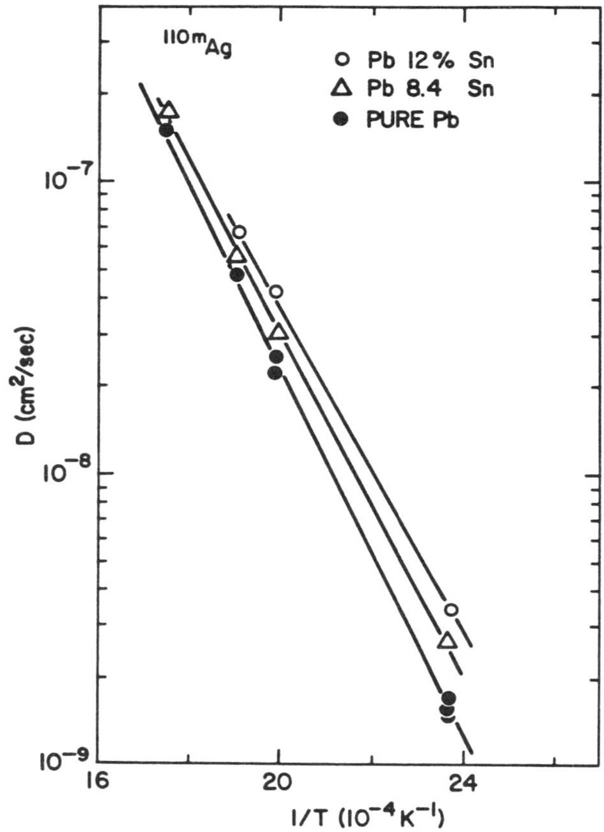

Fig. 17. Arrhenius plots of D_{Ag} at three compositions.

affected and reduced by the addition of the tin. To within the limits of experimental accuracy the D_0's remain unaffected.

There appears to be no unequivocal way to explain this marked increase in diffusivity although there are hypotheses that can be suggested. The size of the tin ion is considerably smaller than the lead and one could argue that this could lead to easier silver motion in the neighborhood of the tin. Also it is known that the addition of tin increases the diffusivity of lead itself (11). Perhaps the tin generally loosens the cohesion of the lattice of the alloy, which might thereby increase the diffusivity of the mobile impurity. We shall have occasion to return to these considerations in a later part of this paper.

B. Electromigration

No problems with attraction to the surface were encountered in the electromigration of silver in the lead-tin alloys. The addition of the tin did, however, cause a marked effect. In pure lead the $Z_a^*(Ag)$ has been shown (30) to be approximately +0.9 and to evince little variation with temperature. With the introduction of tin the Z_a^* was substantially decreased until it appeared to change sign at 12 at.% added tin. The data depicted in Fig. 18 show the observed Z_a^* for several compositions and temperatures. There is some indication, particularly at the higher tin compositions, of a Z_a^* that becomes negative as the temperature decreases. The data shown for silver in the lead-tin alloys seem to indicate the wind force is small for pure lead but the addition of tin makes it appreciable. On the other hand the Z_d is relatively unchanged by the added tin. There is no easy way to determine whether the influence on the wind effect comes about locally or is a band structure effect.

3.3. Experiments With 1000 ppm Nickel

In the work described so far the concentrations of tin have far exceeded the amounts of mobile impurity which have been present

TABLE III

Activation energies Q and preexponential factors D_o for silver in pure lead and lead solders.

	Q(eV)	$D_o(10^{-2}cm^2/sec)$
Pure Pb	0.63±0.02	(4.6±2.2)
Pb-5 at.% Sn	0.62±0.02	(4.8±2.2)
Pb-7 at.% In	0.56±0.02	3.5
Pb-8.4 at.% Sn	0.58±0.02	(2.0±1.1)
Pb-12 at.% In	0.54±0.02	3.7
Pb-12 at.% Sn	0.57±0.02	(2.2±1.1)

only in tracer amounts. To determine what occurs when this situation is reversed a series of analogous experiments were carried out with a stock of lead into which an amount of nickel had been mixed in the molten state to give a nominal concentration of 1,000 ppm nickel. From this stock all specimens were prepared with varying amounts of added tin.

A. Diffusion

The results for nickel diffusion in specimens containing no added tin are shown by the closed circles in Fig. 19. They lie appreciably below the top curve which is an Arrhenius line for nickel in pure lead. The curve through the closed squares has been drawn on the basis of the simple trapping model with parameters B = 0.17 eV and S = 0.45 k gives a consistent fit. At 250°C, $D_{1000}/D = 0.85$.

Next one looks at the diffusion of nickel in specimens to which tin had been added. The results are also shown in Fig. 19 for two sets of specimens carrying tin impurities at the 500 ppm and 1000 ppm levels respectively. The lines are the predictions of the trapping theory which is this time complicated by the need to consider two reactions (Ni+Ni = 2Ni and Ni+Sn = NiSn) and a higher concentration of mobile atoms. The complete expression arising from this analysis is given below:

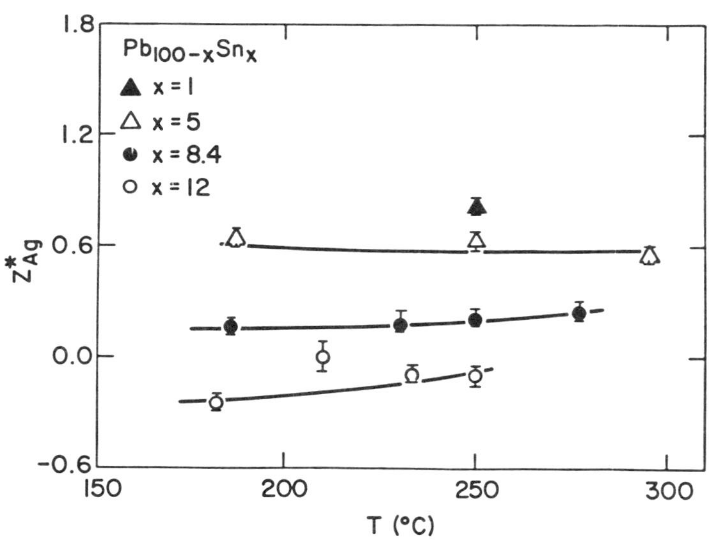

Fig. 18. Z_a for silver as a function of T for several compositions.

$$1 = \frac{D_x}{D} + 12\left(\frac{D_x}{D}\right)^2 C_{Ni} e^{-g_{NN}/kT} + \frac{\frac{D_x}{D} C_{Sn}}{\frac{D_x}{D} C_{Ni} + \frac{1}{6} e^{g_{NS}/kT}} \quad (7)$$

Here D_x and D are the nickel diffusivities in the alloy and in the pure lead respectively. C_{Ni} is the nickel concentration (1000 ppm) and C_{Sn} is the tin concentration. Finally the g_{NN} is the free energy for the nickel-nickel binding and g_{NS} is the same for the nickel-tin binding. The curves were originally drawn for the value of g_{NS} given by Hu and Huntington (see Fig. 5) but for the 1000 ppm tin this gave the dashed line shown in Fig. 19 which lies appreciably above the data. As has been remarked in connection with Fig. 5 there is some curvature to the data. The Hu value for the binding energy was 1.25 eV taken from an average slope over the curve. For the lowest temperatures of Fig. 19 a

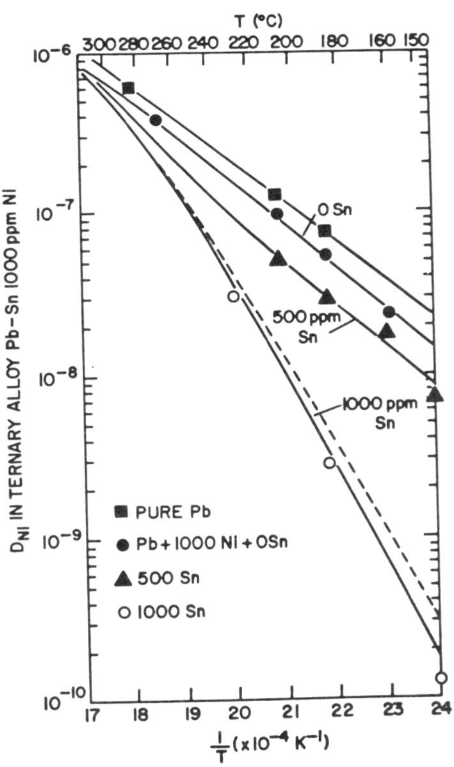

Fig. 19. D_{Ni} in specimens containing 1000 at ppm nickel and various amounts of tin vs. 1/T. Lines near data are the predictions of the simple trapping model.

more appropriate value is B = 1.39 eV and curve predicted by this fitting is the lowest solid line of Fig. 19 which fits the data well. On the whole trapping model fits the results with the 1,000 ppm material satisfactorily. Apparently each tin atom can trap only one nickel since a model postulating multiple trapping on the part of the tin gives widely different predictions.

B. Electromigration

There have been earlier measurements (33) on the electromigration of nickel in lead up to concentrations as high as 1,000 at. ppm nickel. In these cases the Z_a^* increased rapidly as the temperature was lowered and eventually the linear relation between ln c and position

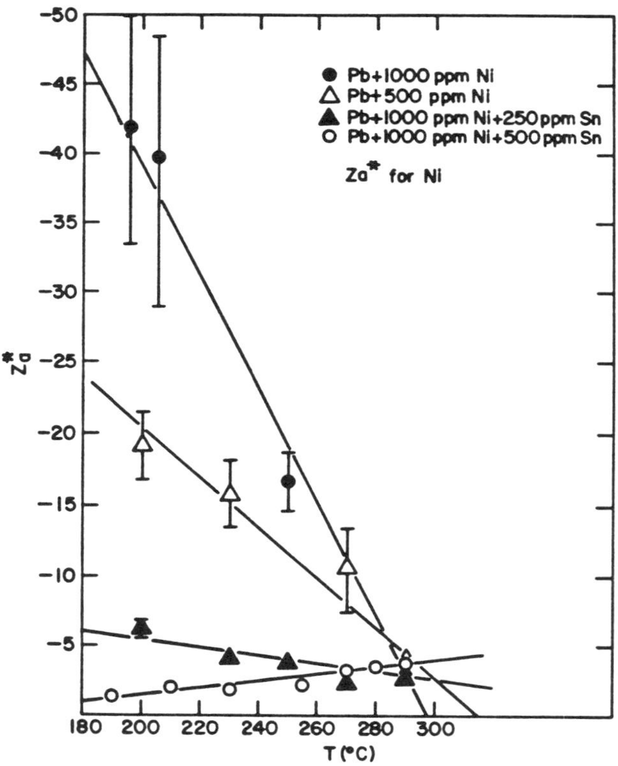

Fig. 20. Z_a^* for nickel in ternary alloys lead-tin-1000 at ppm nickel vs. T^{-1}.

was lost at the high concentration end of the plots, so that meaningful measurements were no longer possible. The quantity of Z_a* can be strongly influenced by the variation of ϕ which appears in the denominator for the expression for Z_a*. A strong concentration dependence of the thermodynnmic factor for nickel in Pb(Ni) system has been observed (41). The Boltzmann-Matano method was employed to determine the nickel chemical diffusivity as a function of nickel concentration in lead for the sample temperature at 226°C. The ϕ for nickel decreased rapidly on adding a small amount of nickel to lead. The value of ϕ drops by about 30 for nickel concentration above 0.02 at.% in lead. Therefore, the rapid increase of Z_a*. values with higher nickel content (33) is attributed to the rapid decrease of ϕ of this system. The result of adding tin to the lead containing 0.1 at.% tin is most interesting as can be seen in Fig. 20. It appears that the addition of the tin removes the nickel from effective solution by trapping, thereby widening the range of solubility. The rapid increase in Z_a* with decrease in temperature is diminished with the addition of tin. Actually the slope of Z_a* vs. temperature is reversed for the highest tin concentration explored (0.05 at.%).

4. DILUTE IMPURITIES IN LEAD-INDIUM SOLDERS

It was decided that a fruitful way to extend these studies would be to make similar measurements with lead-indium solders. By a change in the size and valence of the trapping element one might get some interesting insight into the nature of the mechanism. In addition the lead-indium alloys have considerable technological interest as solders tougher mechanically than the lead-tin alloys (42). The measurements (41,43) have been carried out for same sort of fast diffusers as used in lead-tin solders.

4.1. Diffusion in Lead-Indium Alloys

The diffusion behavior for nickel and gold in lead-indium solders is similar to those diffusion in lead-tin solders. The presence of indium does indeed decrease the diffusion of nickel and gold in lead. The diffusion data seem to be well explained by the simple trapping model. The substitutional impurity indium does not trap the nickel atom as drastically as tin does. The results for 1 and 2 at.% added indium are shown in Fig. 21. At 2 at.% added indium for T = 175°C the diffusivity of Ni is down by a factor of about 15 of its value in pure lead. The application to trapping theory one deduces a binding energy of 1.1 eV and a corresponding entropy of −23k. The simple trapping model is also still a fair approach for diffusion of gold in lead-indium system.

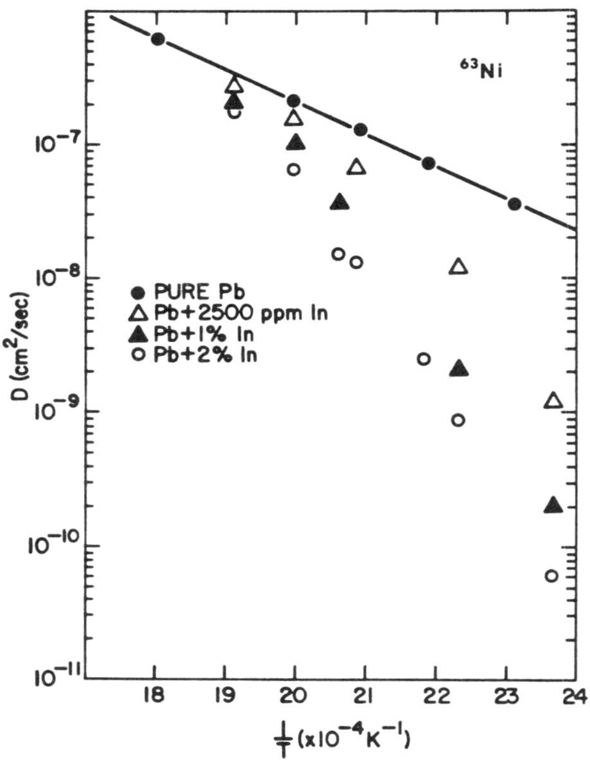

Fig. 21. Diffusion of nickel in three lead-indium compositions.

All work was carried out with well annealed specimens so that no anomaly in curves of D vs. x (the concentration of indium) was manifest. The binding energy between gold and indium is 0.51 eV and the binding entropy is -10k. The comparison of binding energy between nickel, gold and indium, and nickel, gold, copper and tin is shown in Table IV.

Here again the silver shows no evidence of trapping. The diffusivity increases with added indium as can be seen in Fig. 22. In fact the increase is considerably more than that found with tin, being in this case a factor of about 50 at 30 at.% added indium for T = 144°C. Table III shows the activation energies of silver in lead-indium solders are smaller than those in lead-tin alloys. This implies that the enhancement effect of the silver diffusion in lead with addition indium is more pronounced. To some extent the original sugges-

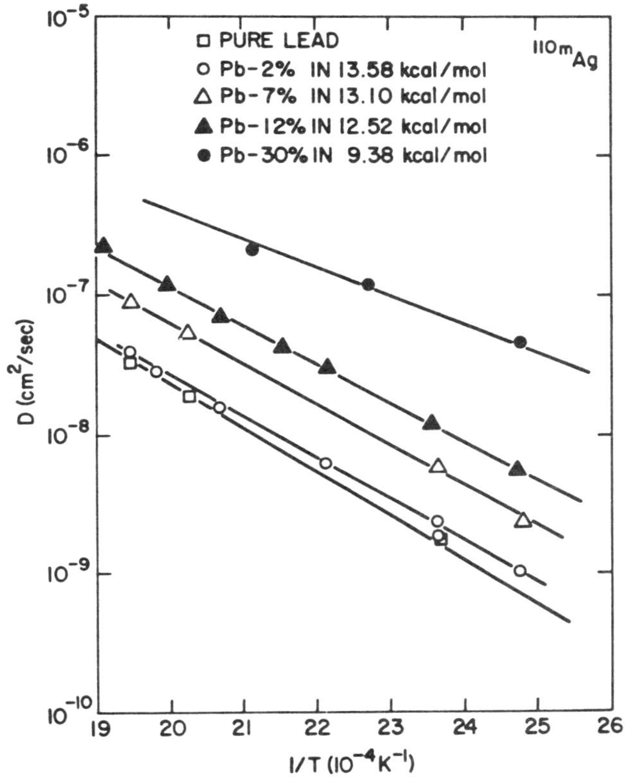

Fig. 22. Diffusion of silver in lead-indium alloys.

TABLE IV Binding Energies

Binding impurities	Binding Energy (eV)
Ni-Sn	1.25
Ni-In	1.10
Cu-Sn	0.32 ± 0.07
Au-Sn	0.21 ± 0.03
Au-In	0.51 ± 0.04

tion that the size of the tin atom (smaller than lead) was responsible for the enhanced diffusion, is hereby refuted, since the indium atom size is larger than the tin. Nor can one attribute the effect to decrease in the melting temperature with alloying as the slope of the solidus in the lead-indium system is less steep than in the lead-tin system. The only remaining simple physical difference is the difference in valence which is nominally three for indium and four for tin. The very marked *increase* of mobility of a tracer impurity, supposedly moving by an interstitial mechanism, with alloying is quite unprecedented and unexpected.

One piece of this puzzle has been uncovered by exploring the increase of silver solubility in lead-indium alloys with increasing indium content as studied at two different temperatures (see Fig. 23). The measurements were made by J. Shi with diffusion runs using sequential sectioning for the penentration of ^{110m}Ag in lead-indium specimens of 0, 12%, 20% and 30% indium. In every case non-radioactive silver was added to the plating bath in the form of $AgNO_3$ to bring the final surface concentration above the solubility limit. For this condition it is

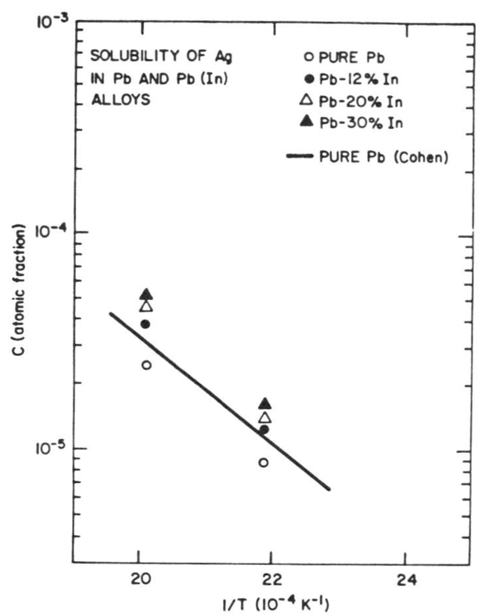

Fig. 23. Solubility of silver in lead-indium alloys.

similar to the case of Cu diffusion in lead (see Sect. 3.1A). The penetration plots were fitted, not by the Gaussian thin-film solution, but by the Complementary Error Function. In this way the saturation composition C was determined as a function of temperature and composition.

To obtain absolute values for C one had to establish an equivalence between the amount of silver and the radioactive counting rate. This was done by calculating the total mass of silver from the product of the plating current and the time and assuming a plating efficiency of 90%. An equivalence was then established between the mass of silver deposited and the sum total activity of the respective sections as counted by a well-type scintillation γ-ray counter. This procedure is certainly subject to appreciable systematic error, so it is not surprising that these solubility values for silver in pure lead lie some 30% below the values obtained by Cohen (27). However the temperature dependence of the solubility as controlled by the heat of solution is in agreement with that of Cohen.

The most interesting aspect of this investigation is the clearly marked increase of solubility with the addition of indium. At about 200°C and 30% indium the solubility is about doubled. Possibly this marked behavior may be a clue to the rapid increase of silver diffusivity with indium content.

Very interesting results are shown from the results of copper diffusion in lead-indium solders. The diffusivity of copper in lead continually decreases as a function of concentration of indium up to 7 at.%. Then the diffusivity rises as more indium is added. One might assume that both de-hancement and enhancement effect take place in this system. If the trapping effect is weak, one would expect the contribution of diffusion from the de-hancement effect should be small. Consequently, the enhancement diffusion will be observed such as in the case of silver and copper.

4.2. Electromigration

From electromigration measurements it is found that Z_a^* decreases for silver in lead-indium and has little effect for gold and copper with the addition of indium. The situation appears to resemble that for the electromigration of silver, gold and copper in the lead-tin alloys. But the electromigration behavior of nickel in lead-indium solders is quite different than that in lead-tin solders. The values of Z_a^* for nickel increases as the concentration of indium is increased. This is a puzzling result for which no reasonable explanation has yet been presented.

5. CONCLUSION

The diffusion and electromigration of Ag, Au, Ni and Cu in pure lead, Pb(Sn) and Pb(In) solders have been studied for sample temperature between 100 and 317°C. The measurements were made by a standard sequential sectioning technique. The concentrations of the impurities were determined by tracing the radioactive isotopes. The addition of the tin and indium brought about several dramatic and unexpected changes in the atom movements of these impurities. For the case of silver diffusivity increased with the tin addition and increased considerably more in the lead-indium alloys. On the other hand the electromigration driving force decreased from nearly +1 to zero, it appears due to an increase in the "electron wind". For Ni, Cu and Au the tin causes strong trapping, particularly in the case of nickel. The similar behavior also shows in the case of nickel and gold in Pb(In) alloys. A straightforward analysis, based on the assumption that the trappers far outnumber the mobile atoms and that the trapped atoms make nil contribution to the diffusivity, yield binding energies for the tin trapping of 1.3 eV for the nickel, 0.3 eV for the copper and 0.2 eV for the gold. Similar to Pb(Sn) alloys, the trapping of nickel and gold in indium was found to have the binding energy of 1.0 eV and 0.5 eV, respectively. For the case of copper in Pb(In), however, a weaker binding between In and copper was observed. Also the enhancement copper diffusion was shown for the indium concentration higher than 7 at.%. The effective charge number, except in the case of nickel in Pb(In) alloys, however, remained unchanged wherever good measurements were effected. Presumably the electromigration flux was carried only by the untrapped atoms and local equilibrium was everywhere maintained between trapped and untrapped atoms. Although the trapping did not affect the effective charge number, it did of course increase the time required to attain the steady state condition.

In the work summarized above the trappers have always substantially outnumbered the mobile impurities. To test the trapping model predictions, a reverse situation of the atom motions of nickel in some lead with 0.1 at.% nickel impurity and various amounts of tin up to 0.05 at.% have been investigated. The diffusion data were well fitted by the trapping model and suggested that only one tin traps only one nickel atom. The Z_a^* in the specimens show a marked increase at low temperature and low tin content which is attributed to the decrease in the thermodynamic factor, but by increasing the tin concentration the temperature dependence tended to disappear.

6. ACKNOWLEDGEMENTS

We are happy to acknowledge the support of the IBM General Technology Division at East Fishkill. We express our gratitude to N. G. Koopman, D. Gupta and P. S. Ho for helpful discussions. We also wish to thank D. Gupta for many useful comments on the manuscript and to Shao-ning Mei and Jinghang Shi for communicating to us their results with the Pb(In) alloys.

REFERENCES

1. P. A. Totta and R. P. Sopher, SLT Device Metallurgy and Its Monolithic Extension, *IBM J. Res. Develop.* **13**, 226 (1969).
2. L. F. Miller, Controlled Collapse Reflow Chip Joining, *IBM J. Res. Develop.* **13**, 239 (1969).
3. J. H. Lau and D. W. Rice, Solder Joint Fatigue in Surface Mount Technology: State of the Art, *Solid State Tech.* **28**, No. 10, 91 (1985).
4. C. A. Steidel, in "VLSI Technology", edited by S. M. Sze (McGraw-Hill, New York, 1983), Chap. 13.
5. L. J. Freed, J. S. Lechaton, J. S. Logan, G. Paal and P. A. Totta, *IBM J. Res. Develop.* **26**, 362 (1982).
6. H. B. Huntington, in "Diffusion in Solids: Recent Developments", edited by A. S. Nowick and J. J. Burton (Academic, New York, 1974), Chap. 6.
7. W. K. Warburton and D. Turnbull, in "Diffusion in Solids: Recent Developments", edited by A. S. Nowick and J. J. Burton (Academic, New York, 1974), Chap. 4.
8. J. W. Miller, Enhancement of Lead Self-Diffusivity by Gold and Silver Addition, *Phys. Rev. B* **2**, 1624 (1970).
9. J. W. Miller, J. N. Mundy, L. C. Robison and R. E. Loess, Unexpected Small Effect of Isotipic Mass on Diffusion of Silver in Lead, *Phys. Rev.* **B8**, 2411 (1973).
10. J. N. Mundy, J. W. Miller and S. J. Rothman, Diffusion of Copper in Lead: Solubility and Isotope Effect, *Phys. Rev. B* **10**, 2275 (1974).
11. D. Gupta and J. Obserschmidt, in "Diffusion in Solids: Recent Development", edited by M. A. Dayanamda and G. E. Murch (A Publication of the Metallurgical Society, Pennsylvania, 1985) p. 121.
12. M. Hansen and K. Anderko, "Constitution of Binary Alloys", (McGraw-Hill 1958).

13. K. C. Norris and A. H. Landzberg, Reliability of Controlled Collapse Interconnections, *IBM J. Res. Develop.* **13**, 266 (1969).
14. W. C. Roberts-Austen, Diffusion of Gold in Solid Lead at Ordinary Temperature, *Proc. Roy. Soc.* **67**, 101 (1900); *Proc. Roy. Soc.* **59**, 281 (1896).
15. W. Seith and H. Etzold, On the Mobility of Gold in Solid Lead, *Z. Electrochem* **40**, 829 (1934).
16. J. W. Miller, Diffusion of Cadmium in Lead, *Phys. Rev.* **18**, 1095 (1969).
17. W. K. Warburton, Modified Model of Diffusion by Interstitial-Vacancy Pairs, *Phys. Rev. B.* **7**, 1330 (1973).
18. A. Ascoli, Diffusion of Au in Pb Single Crystals, *J. Inst. Metals* **89**, 218 (1960).
19. G. V. Kidson, The Diffusion of Gold in Lead Single Crystals, *Phil. Mag.* **13**, 247 (1966).
20. B. F. Dyson, T. Anthony, and D. Turnbull, Interstitial Diffusion of Copper and Silver in Lead, *J. Appl. Phys.* **37**, 2375 (1966).
21. J. A. Weyland, D.L. Decker and H. B. Vanfleet, Effect of Pressure on the Diffusion of Gold in Lead to 50 kbar, *Phys. Rev. B.* **4**, 4225 (1971); C. T. Candland and H. B. Vanfleet, Effect on Pressure on Interstitial Diffusion of Nickel in Lead to 50 kbar, *Phys. Rev. B.* **7**, 575 (1973); H. B. Vanfleet, D. L. Decker and C. T. Candland, Interstitial Diffusion of Pd in Pb for Pressures of up to 50 kbar, *Bull. Amer. Phys. Soc.* **18**, 428 (1973).
22. J. Hudson and R. Hoffman, The Effect of Hydrostatic Pressure on Self-Diffusion in Lead, *Trans. AIME* **221**, 761 (1961).
23. M. Y. Hsieh, H. B. Huntington and R. N. Jeffery, Electromigration of Gold and Silver in Lead, *Cryst. Lat. Def.* **7**, 9 (1977).
24. P. H. Dederichs, C. Lehmann, H. R. Schober, A. Scholtz and R. Zeller, Lattice Theory of Point Defects, *J. Nucl. Matl.* **69**, 176 (1978).
25. S. Ishioka, H. Nakajima and M. Koiwa, Isotope Effect for Anomalous Fast Diffusion, *Phil. Mag A* **55**, 359 (1987).
26. W. K. Warburton, Dehancement Effect of Au on Au Diffusivity in Pb(Au) Alloys, *Phys. Rev. B.* **11**, 4945 (1975).
27. B. M. Cohen and W. K. Warburton, Dehancement Effect of Ag on Ag Diffusivity in Pb(Ag) Alloys, *Phys. Rev. B* **12**, 5682 (1975).
28. D. L. Decker, J. G. Melville and H. B. Vanfleet, De-enhancement of Gold Diffusion in Lead by Impurities in the Lead, *Phys. Rev. B* **20**, 3036 (1979).
29. Richard S. Sorbello, Theory of the Direct Force in Electromigration, *Phys. Rev. B* **31**, 798 (1985).

30. E. Stracke and Chr. Herzig, Electromigration, Thermomigration and Solubility of Copper in Lead, *Status Solidi* (a), **47**, 513 (1978).
Chr. Herzig and E. Stracke, Atomic Transport of Gold and Silver in Lead in a DC Field, *Phys. Status Solidi* (a), **27**, 75 (1975).
31. M. Y. Hsieh and H. B. Huntington, Electromigration of Copper in Lead, *J. Phys. Chem. Solids* **39**, 867 (1978).
32. H. B. Huntington and M. Y. Hsieh, Electromigration of Fast Diffusers in Lead, *J. Nucl. Mat.* **69**, 561 (1978).
33. D. A. Golopentia and H. B. Huntington, A Study of Electromigration of Nickel in Lead, *J. Phys. Chem. Solids* **39**, 975 (1978).
34. C.-K. Hu and H. B. Huntington and, Diffusion and Electromigration of Silver and Nickel in Lead-Tin Alloys, *Phys. Rev. B* **26**, 2782 (1982).
35. C.-K. Hu, H. B. Huntington and and G. R. Gruzalski, Atomic Motions of Copper Dissolved in Lead-Tin Alloys, *Phys. Rev. B* **28**, 579 (1983).
36. W. K. Warburton, The Dehancement of Gold Diffusivity in Dilute Pb(Au) Alloys, *Scripta Met.* **7**, 105 (1973).
37. W. K. Warburton and S. C. Moss, Lattice-Parameter Changes due to Au Dissolution in Pb, *Phys. Rev. Lett.* **42**, 1694 (1979).
38. T. Bolze, H. Metzger, J. Peisl and S. C. Moss, X-ray and Neutron Diffraction Studies of the Structure of Dilute Solutions of Au in Pb: I. The Quenched State, *J. Phys. F: Met. Phys.* **14**, 1073 (1984).
39. C. K. Hu and H. B. Huntington, Atom Movements of Gold in Lead-Tin Solders, *J. Appl. Phys.* **58**, 2564 (1985).
40. B. M. Cohen and D. Turnbull, Kinetics of Precipitation of Ag From Pb(Ag) Alloys, *Acta. Met.* **26**, 113 (1978).
41. S.-N. Mei, J. Shi, and H. B. Huntington, Diffusion and Electromigration in Lead Alloys: I. Nickel as a Mobile Element, *J. Appl. Phys.* **62**, 44451 (1987).
42. R. T. Howard, Optimization of Indium-Lead Alloys for Controlled Collapse Chip Connection Application, *IBM J. Res. Develop.* **26**, 372 (1982).
43. J. Shi, S.-N. Mei, and H. B. Huntington, Diffusion and Electromigration in Lead Alloys: II. Ag, Au and Cu in Lead-Indium, *J. Appl. Phys.* **62**, 451 (1987).

Index